Machine Learning
A Constraint-Based Approach

T0198254

Machine Learning
A Constraint-Based Approach

Second Edition

Marco Gori
Department of Information Engineering and Mathematics
University of Siena
Siena, Italy

Alessandro Betti
Université Côte d'Azur
Inria
CNRS
I3S Laboratory
Maasai team
Nice, France

Stefano Melacci
Department of Information Engineering and Mathematics
University of Siena
Siena, Italy

ELSEVIER

MK
MORGAN KAUFMANN PUBLISHERS
AN IMPRINT OF ELSEVIER

Morgan Kaufmann is an imprint of Elsevier
50 Hampshire Street, 5th Floor, Cambridge, MA 02139, United States

ISBN: 978-0-323-89859-1

For information on all Morgan Kaufmann publications
visit our website at https://www.elsevier.com/books-and-journals

Publisher: Katey Birtcher
Acquisitions Editor: Stephen R. Merken
Editorial Project Manager: Rafael Guilherme Trombaco
Production Project Manager: Shereen Jameel
Cover Designer: Bridget Hoette

Typeset by VTeX

Printed in Bell & Bain in UK

Last digit is the print number:
9 8 7 6 5 4 3 2 1

Working together
to grow libraries in
developing countries

www.elsevier.com • www.bookaid.org

Contents

Contents   **ix**

Preface

MACHINE LEARNING projects our ultimate desire to understand the essence of human intelligence onto the space of technology. As such, while it cannot be fully understood in the restricted field of computer science, it is not necessarily the search of clever emulations of human cognition. While digging into the secrets of neuroscience might stimulate refreshing ideas on computational processes behind intelligence, most of nowadays advances in machine learning rely on models mostly rooted in mathematics and on corresponding computer implementation. Notwithstanding that brain science will likely continue the path toward the intriguing connections with artificial computational schemes, one might reasonably conjecture that the basis for the emergence of cognition should not necessarily be searched in the astonishing complexity of biological solutions, but mostly in higher-level computational laws. The biological solutions for supporting different forms of cognition are in fact cryptically interwound with the parallel need of supporting other fundamental life functions such as metabolism, growth, body weight regulation, and stress response. However, most human-like intelligent processes might emerge regardless of this complex environment. One might reasonably suspect that those processes are the outcome of information-based laws of cognition that hold regardless of biology. There is clear evidence of such an invariance in specific cognitive tasks, but the challenge of artificial intelligence is daily enriching the range of those tasks. While no one is surprised anymore to see the computer power in math and logic operations, the layman is not very well aware of the outcome of challenges on games, yet. They are in fact commonly regarded as a distinctive sign of intelligence, and it is striking to realize that games are already mostly dominated by computer programs! Sam Loyd's 15 puzzle and the Rubik's cube are nice examples of successes of computer programs in classic puzzles. Chess and, more recently, Go clearly indicate that machines undermine the long-lasting reign of human intelligence. However, many cognitive skills in language, vision, and motor control, which likely rely strongly on learning, are still very hard to achieve.

This book drives the reader into the fascinating field of machine learning by offering a unified view of the discipline that relies on modeling the environment as an appropriate collection of constraints that the agent is expected to satisfy. Nearly every task that has been faced in machine learning can be modeled under this mathematical framework. Linear and threshold linear machines, neural networks, and kernel machines are mostly regarded as adaptive models that need to softly satisfy a set of point-wise constraints corresponding to the training set. The classic risk, in both the functional and empirical forms, can be regarded as a penalty function to be minimized in a soft-constrained system. Unsupervised learning can be given a similar formulation, where the penalty function somewhat offers an interpretation of the data probability distribution. Information-based indexes can be used to extract unsupervised features, and they can clearly be thought of as a way of enforcing soft-constraints. An intelligent agent, however, can strongly benefit also from the acquisition of abstract granules of knowledge given in some logic formalism. While artificial intelligence has achieved a remarkable degree of maturity in the topic of knowledge representation and automated reasoning, the foundational theories that are mostly rooted in logic lead to models that cannot be tightly integrated with machine learning. While regarding symbolic knowledge bases as a collection of constraints, this book draws a path toward a deep integration with machine learning that relies on the idea of adopting multivalued logic formalisms, like in fuzzy systems. A special attention is reserved to deep learning, which nicely fits the constrained-based approach followed in this book. Some recent foundational achievements on representational issues and learning, joined with appropriate exploitation of parallel computation, have been creating a fantastic catalyst for the growth of high-tech companies in related fields all around the world. In the book we do our best to jointly disclose the power of deep learning and its interpretation in the framework of constrained environments, while warning from uncritical blessing. In so doing, we hope to stimulate the reader to conquer the appropriate background to be ready to quickly grasp also future innovations.

Throughout the book, we expect the reader to become fully involved in the discipline, so as to mature his own view, more than to settle up into frameworks served by others. The book gives a refreshing approach to basic models and algorithms of machine learning, where the focus on constraints nicely leads to dismiss the classic difference between supervised, unsupervised, and semi-supervised learning. Here are some book features:

a) It is an introductory book for all readers who love in-depth explanations of fundamental concepts.
b) It is intended to stimulate questions and help a gradual conquering of basic methods, more than offering "recipes for cooking."
c) It proposes the adoption of the notion of constraint as a truly unified treatment of nowadays most common machine learning approaches, while combining the strength of logic formalisms dominating in the AI community.
d) It contains a lot of exercises along with the answers, according to a slight modification of Donald Knuth's difficulty ranking.
e) It comes with a companion website to assist more on practical issues.

The book has been conceived for readers with basic background in mathematics and computer science. More advanced topics are assisted by proper appendixes. The reader is strongly invited to act critically, and complement the acquisition of the concepts by the proposed exercises. He is invited to anticipate solutions and check them later in the part "Answers to the Exercises." Our major target while writing this book has been that of presenting concepts and results in such a way that the reader feels the same excitement as the one who discovered them. More than a passive reading, he is expected to be fully involved in the discipline and play a truly active role. Nowadays, one can quickly access the basic ideas and begin working with most common machine learning topics thanks to great web resources that are based on nice illustrations and wonderful simulations. They offer a prompt yet effective support for everybody who wants to access the field. A book on machine learning can hardly compete with the explosive growth of similar web resources on the presentation of good recipes for fast development of applications. However, if you look for an indepth understanding of the discipline, then you must shift the focus on foundations, and spend more time on basic principles that are likely to hold for many algorithms and technical solutions used in real-world applications. The most important target in writing this book was that of presenting foundational ideas and providing a unified view centered around information-based laws of learning. It grew up from material collected during courses at Master's and PhD levels mostly given at the University of Siena, and it was gradually enriched by our own view point of interpreting learning under the unifying notion of environmental constraint. When considering the important role of web resources, this is an appropriate text book for Master's courses in machine learning, and it can also be adequate for complementing courses on pattern recognition, data mining, and related disciplines. Some parts of the book are more appropriate for courses at the PhD level. In addition, some of the proposed exercises, which are properly identified, are in fact a careful selection of research problems that represent a challenge for PhD students. While the book has been primarily conceived for students in computer science, its overall organization and the way the topics are covered will likely stimulate the interest of students also in physics and mathematics.

While writing the book, we were constantly stimulated by the need of quenching our own thirst of knowledge in the field, and by the challenge of passing through the main principles in a unified way. We got in touch with the immense literature in the field and discovered our ignorance of remarkable ideas and technical developments. We learned a lot and did enjoy the act of retracing results and their discovery. It was really a pleasure, and we wish the reader experienced the same feeling while reading this book.

Preface to the second edition

The second edition of this book offers a large number of novel features. Each chapter has been improved by expanding several descriptions, fixing the found errors, and better organizing exercises, thus offering a more accessible presentation of the main concepts. Modern approaches in the context of deep neural networks are now discussed, introducing the most recent trends in the scientific literature. It is important to remark that the most important feature that was added to this new edition is a companion book that can be freely downloaded and that is rooted on a practical approach to machine learning, helping the reader to learn how to solve real-world problems using machine learning software. The companion book is paired with Python code that is fully described and discussed in detail, covering problems in vision, language, and others. It can be downloaded from the following link

https://sailab.diism.unisi.it/mlbook/

We believe that the companion book represents a precious resource that can allow beginners to gain a quick practical feeling with machine learning which, in turn, can be expanded with a full low-level understanding of the theoretical concepts described in the main book.

Siena   Marco Gori, Alessandro Betti, and Stefano Melacci
May 2022

Acknowledgments

As it usually happens, it's hard not to forget people who have played a role in this book. An overall thanks is for all who taught us, in different ways, how to find the reasons and the logic within the scheme of things. It's hard to make a list, but they definitely contributed to the growth of our desire to understand human intelligence and to study and design intelligent machines; that desire is likely to be the seed of this book. Most of what we've written comes from lecturing Master's and PhD courses on Machine Learning, and from re-elaborating ideas and discussions with colleagues and students at the AI lab of the University of Siena (SAILab) in the last decade. Many insightful discussions of Marco Gori with C. Lee Giles, Ah Chung Tsoi, Paolo Frasconi, and Alessandro Sperduti contributed to conquering the view on recurrent neural networks as diffusion machines presented in this book. Marco Gori's viewpoint of learning from constraints has been gradually given the picture that you can find in this book also thanks to the interaction with Marcello Sanguineti, Giorgio Gnecco, and Luciano Serafini. The criticisms on benchmarks, along with the proposal of crowdsourcing evaluation schemes, have emerged thanks to the contribution of Marcello Pelillo and Fabio Roli, who collaborated with Marco Gori in the organization of a few events on the topic. Marco Gori is indebted with Patrick Gallinari, who invited him to spend part of the 2016 summer at LIP6, Université Pierre et Marie Curie, Paris, a very stimulating environment for writing this book. The follow-up of several seminars gave risc to insightful discussions with colleagues and students in the lab. The collaboration with Stefan Knerr contaminated significantly our view on the role of learning in natural language processing. Most of the advanced topics covered in this book benefited from his long-term vision on the role of machine learning in conversational agents. The first edition of the book benefited of the accurate check and suggestions by Beatrice Lazzerini and Francesco Giannini on some parts of the book.

We thank Cristina Melacci and Agnese Gori for their artwork contribution in the cover and in the opening-chapter pictures, respectively. Finally, Marco Gori wants to thank Cecilia, Irene, and Agnese for having tolerated his elsewhere-mind during the weekends of work on the book, and for their continuous support to a Cyborg, who was hovering from one room to another with his inseparable laptop.

Reading guidelines

Most of the book chapters are self-contained, so as one can profitably start reading Chapter 4 on kernel machines or Chapter 5 on deep architecture without having read the first three chapters. Even though Chapter 6 is on more advanced topics, it can be read independently of the rest of the book. The big picture given in Chapter 1 offers the reader a quick discussion on the main topics of the book, while Chapter 2, which could also be omitted at a first reading, provides a general framework of learning principles that surely facilitates an in-depth analysis of the subsequent topics. Finally, Chapter 3 on linear and linear-threshold machines is perhaps the simplest way to start the acquisition of machine learning foundations. It is not only of historical interest; it is extremely important to appreciate the meaning of deep understanding of architectural and learning issues, which is very hard to achieve for other more complex models. Advanced topics in the book are indicated by the "dangerous-bend" and "double dangerous bend" symbols:

research topics will be denoted by the "work-in-progress" symbol:

Notes on the exercises

WHILE READING THE BOOK, the reader is stimulated to retrace and rediscover main principles and results. The acquisition of new topics challenges the reader to complement some missing pieces to compose the final puzzle. This is proposed by exercises at the end of each section that are designed for self-study as well as for classroom study. Following Donald Knuth's books organization, this way of presenting the material relies on the belief that "we all learn best the things that we have discovered for ourselves." The exercises are properly classified and also rated with the purpose of explaining the expected degree of difficulty. A major difference concerns exercises and research problems. Throughout the book, the reader will find exercises that have been mostly conceived for deep acquisition of main concepts and for completing the view proposed in the book. However, there are also a number of research problems that I think can be interesting especially for PhD students. Those problems are properly framed in the book discussion, they are precisely formulated and are selected because of their scientific relevance; in principle, solving one of them is the objective of a research paper.

Exercises and research problems are assessed by following the scheme below which is mostly based on Donald Knuth's rating scheme[1]:

Rating   Interpretation

00   An extremely easy exercise that can be answered immediately if the material of the text has been understood; such an exercise can almost always be worked "in your head."

10   A simple problem that makes you think over the material just read, but is by no means difficult. You should be able to do this in 1 minute at most; pencil and paper may be useful in obtaining the solution.

20   An average problem that tests basic understanding of the text material, but you may need about 15 or 20 minutes to answer it completely.

30   A problem of moderate difficulty and/or complexity; this one may involve more than 2 hours' work to solve satisfactorily, or even more if the TV is on.

40   Quite a difficult or lengthy problem that would be suitable for a term project in classroom situations. A student should be able to solve the problem in a reasonable amount of time, but the solution is not trivial.

50   A research problem that has not yet been solved satisfactorily, as far as the author knew at the time of writing, although many people have tried. If you

---

[1] The rating interpretation is verbatim from D.E. Knuth, *The Art of Computer Programming, Fundamental Algorithms*, vol. 1, Addison-Wesley, 1997.

have found an answer to such a problem, you ought to write it up for publication; furthermore, the author of this book would appreciate hearing about the solution as soon as possible (provided that it is correct).

Roughly speaking, this is a sort of "logarithmic" scale, so as the increment of the score reflects an exponential increment of difficulty. We also adhere to an interesting Knuth's rule on the balance between amount of work required and the degree of creativity needed to solve an exercise. The idea is that the remainder of the rating number divided by 5 gives an indication of the amount of work required. "Thus, an exercise rated 24 may take longer to solve than an exercise that is rated 25, but the latter will require more creativity." As already pointed out, research problems are clearly identified by the rate 50. It's quite obvious that regardless of my efforts to provide an appropriate ranking of the exercises, the reader might argue on the attached rate, but I hope that the numbers will offer at least a good preliminary idea on the difficulty of the exercises. The reader of this book might have a remarkably different degree of mathematical and computer science training. The rating preceded by an M indicates whether the exercises are oriented more to students with good background in math and, especially, to PhD students. The rating preceded by a C indicates whether the exercises require computer developments. Most of these exercises can be term projects in Master's and PhD courses on machine learning (see also the problems described in the Companion Book). Some exercises marked by ▶ are expected to be especially instructive and especially recommended.

Solutions to most of the exercises appear in the answer chapter. In order to meet the challenge, the reader should refrain from using this chapter or, at least, he/she is expected to use the answers only in case he/she cannot figure out what the solution is. One reason for this recommendation is that it might be the case that he/she comes up with a different solution, so as he/she can check the answer later and appreciate the difference.

Summary of codes:		*00*	Immediate
		10	Simple (one minute)
▶	Recommended	*20*	Medium (quarter hour)
C	Computer development	*30*	Moderately hard
M	Mathematically oriented	*40*	Term project
HM	Requiring "higher math"	*50*	Research problem

The big picture

1

Wanna be your main man in the Big Picture
ELTON JOHN, in ***The Big Picture*** (1997)

I am big.
It's the pictures that got small.
NORMA DESMOND, in ***Sunset Boulevard*** (1950)

We can only see a short distance ahead,
but we can see plenty there that needs to be done.
A.M. TURING, ***Computing Machinery and Intelligence*** (1950)

Machine Learning. https://doi.org/10.1016/B978-0-32-389859-1.00008-8

THIS CHAPTER gives a big picture of book. It provides motivation for the study of the discipline and introduces the intriguing topic of induction, by showing its puzzling nature, as well as its necessity in any task, which involves perceptual information. In order to stimulate a parallel learning of lab topics, the chapter contains a section on hands-on experience that focuses on the usage of advanced development tools for dealing with the MNIST classic benchmark of handwritten character recognition. Finally, a few machine learning challenges are discussed with the purpose of promoting the idea of constructing intelligent agents that live in their own environment.

1.1 **Why do machines need to learn?**

Why do machines need to learn? Don't they just run the program, which simply solves a given problem? Aren't programs only the fruit of human creativity, so as machines simply execute them efficiently? No one should start reading a machine learning book without having answered these questions. Interestingly, we can easily see that the classic way of thinking about computer programming as algorithms to express, by linguistic statements, our own solutions isn't adequate to face many challenging real-world problems. We do need to introduce a *metalevel*, where, more than formalizing our own solutions by programs, we conceive algorithms whose purpose becomes that of describing how machines learn to execute the task.

As an example let us consider the case of handwritten character recognition. To make things easy, we assume that an intelligent agent is expected to recognize chars that are generated using black and white pixels only — as it is shown in the figure. We will show that also this dramatic simplification doesn't reduce significantly the difficulty of facing this problem by algorithms based on our own understanding of regularities. One early realizes that human-based decision processes are very difficult to encode into precise algorithmic formulations. How can we provide a formal description of character "2?" The instance of the above picture suggests how tentative algorithmic descriptions of the class can become brittle. A possible way of getting rid of this difficulty is to try a brute force approach, where all possible pictures on the retina with the chosen resolution are stored in a table, along with the corresponding class code. The above 8×8 resolution char is converted into a Boolean string of 64 bits by scanning the picture by rows:

 $\rightarrow$ 0001100000100100000000010000000100000001010000100011111110000000011.

$$(1)$$

Of course, we can construct tables with similar strings, along with the associated class code. In so doing, handwritten char recognition would simply be reduced to the problem of searching a table. Unfortunately, we are in front of a table with $2^{64} = 18446744073709551616$ items, and each of them will occupy 8 bytes, for a total of approximately 147 quintillion (10^{18}) bytes, which makes totally unreasonable the adoption of such a plain solution. Even a resolution as small as 5×6 requires storing 1 billion records, but just the increment to 6×7 would require storing about 4 trillion records! For all of them, the programmer would be expected to be patient enough to complete the table with the associated class code. This simple example is a sort of 2^d *warning message*: As d grows towards values that are ordinarily used for the retina resolution, the space of the table becomes prohibitive. There is more — we have made the tacit assumption that the characters are provided by a reliable segmentation program, which extracts them properly from a given form. While this might be reasonable in simple contexts, in others segmenting the characters might be as difficult as recognizing them. In vision and speech perception, nature seems to have fun in making segmentation hard. For example, the word segmentation of speech utterances cannot rely on thresholding analyses to identify low levels of the signal. Unfortunately, those analyses are doomed to fail. The sentence `computers are attacking the secret of intelligence`, quickly pronounced, would likely be segmented as

`com/pu/tersarea/tta/ckingthesecre/tofin/telligence`.

The signal is nearly null before the explosion of voiceless plosives `p`, `t`, `k`, whereas, because of phoneme coarticulation, no level-based separation between contiguous words is reliable. Something similar happens in vision. Overall, it looks like segmentation is a truly cognitive process that in most interesting tasks does require understanding the information source.

1.1.1 Learning tasks

Intelligent agents interact with the environment, from which they are expected to learn, with the purpose of solving assigned tasks. In many interesting real-world problems we can make the reasonable assumption that the intelligent agent interacts with the environment by distinct segmented elements $e \in E$ of the learning environment, on which it is expected to take a decision. Basically, we assume somebody else has already faced and solved the segmentation problem, and that the agent only processes single elements from the environment. Hence, the agent can be regarded as a function $\chi : E \to O$, where the decision result is an element of O. For example, when performing optical character recognition in plain text, the character segmentation can take place by algorithms that must locate the row/column transition from the text to background. This is quite simple, unless the level of noise in the image document is pretty high.

In general, the agent requires an opportune internal representation of elements in E and O, so that we can think of χ as the composition $\chi = h \circ f \circ \pi$. Here $\pi : E \to X$ is a preprocessing map that associates every element of the environment e with a point $x = \pi(e)$ in the input space X, $f : X \to Y$ is the function that takes the decision $y = f(x)$ on x, while $h : Y \to O$ maps y onto the output $o = h(y)$. In the above handwritten character recognition task we assume that we are given a low resolution camera so that the picture can be regarded as a point in the environment space E. This element can be represented — as suggested by Eq. 1.1–(1) — as elements of a 64-dimensional Boolean hypercube (i.e., $X \subset \mathbb{R}^{64}$). Basically, in this case π is simply mapping the Boolean matrix to a Boolean vector by row scanning in such a way that there is no information loss when passing from e to x. As it will be shown later, on the other hand, the preprocessing function π typically returns a pattern representation with information loss with respect to the original environmental representation $e \in E$. Function f maps this representation onto the *one-hot encoding* of number 2 and, finally, h transforms this code onto a representation of the same number that is more suitable for the task at hand:

$$\boxed{\blacksquare} \overset{\pi}{\to} (0, 0, 0, 1, 1, 0, \ldots, 0, 0, 0, 0, 1, 1)' \overset{f}{\to} (.1, 0., .8, 0., 0., 0., 0.1, 0., 0., 0.)' \overset{h}{\to} 2.$$

Overall the action of χ can be nicely written as $\chi(\boxed{\blacksquare}) = 2$. In many learning machines, the output encoding function h plays a more important role, which consists of converting real-valued representations $y = f(x) \in \mathbb{R}^{10}$ onto the corresponding one-hot representation. For example, in this case, one could simply choose $h\colon \mathbb{R}^{10} \to \{0, 1\}^{10}$ such that $h_i(y) = [i = \arg\max_k y_k]$, where $[A]$ is the Iverson's notation, i.e. $[A] = 1$ if and only if A is true. In doing so, the hot bit is located at the same position as the maximum of y. While this apparently makes sense, a more careful analysis suggests that such an encoding suffers from a problem that is pointed out in Exercise 1.

Functions π and h adapt the environmental information and the decision to the internal representation of the agent. As it will be seen throughout the book, depending on the task, E and O can be highly structured, and their internal representation plays a crucial role in the learning process. The specific role of π is to encode the environmental information into an appropriate internal representation. Likewise, function h is expected to return the decision on the environment on the basis of the internal state of the machine. The core of learning is the appropriate discovering of f, so as to obey the constraints dictated by the environment.

What are the environmental conditions that are dictated by the environment? Since the dawn of machine learning, scientists have mostly been following the principle of *learning from examples*. Under this framework, an intelligent agent is expected to acquire concepts by induction on the basis of collections $L = \{(e_\kappa, o_\kappa) \in E \times O : \kappa = 1, \ldots, \ell\}$, where an oracle, typically referred to as the *supervisor*, pairs inputs $e_\kappa \in E$ with decision values $o_\kappa \in O$. A first important distinction concerns *classification* and *regression* tasks. In the first case, the decision requires the finiteness of O, while in the second case O can be thought of as a continuous set.

First, let us focus on classification. In simplest cases, $O \subset \mathbb{N}$ is a collection of integers that identify the class of e. For example, in the handwritten character recognition problem, restricted to digits, we might have $|O| = 10$. In this case, we can promptly see the importance of distinguishing the physical, the environmental, and the decision information with respect to their corresponding internal representation of the machine. At the pure physical level, handwritten chars are the outcome of the physical process of light reflection. It can be captured as soon as we define the retina R as a rectangle in $\mathbb{R}^2$, and interpret the reflected light by the image function $v\colon R \subset \mathbb{R}^2 \to \mathbb{R}^3$, where the three dimensions express the (R,G,B) components of the color. In doing so, any pixel $z \in R$ is associated with the brightness value $v(z)$. As we sample the retina, we get a set $R^{\sharp} \subset \mathbb{N}^2$ — this is a grid over the retina. The corresponding resolution characterizes the environmental information, namely what is stored in the camera which took the picture. Interestingly, this isn't necessarily the internal information which is used by the machine to draw the decision. The typical resolution of pictures stored in a camera is very high for the purpose of character classification. As it will be pointed out in Section 1.3.2, a significant de-sampling of $R^{\sharp}$ still retains the relevant cues needed for classification.

The output encoding of the decision can be done in different ways. One possibility is to choose function $h = \mathrm{id}$ (identity function), which forces the development of f with codomain O. Alternatively, as already seen, one can use the one-hot encoding. More efficient encodings can obviously be used: In the case of $O = \{0, 1, 2, 3, 4, 5, 6, 7, 8, 9\}$, four bits suffice to represent the ten classes. While this is definitely preferable in terms of saving space to represent the decision, it might be the case that codes which gain compactness with respect to one-hot aren't necessarily a good choice. More compact codes might result in a cryptic coding description of the class that could be remarkably more difficult to learn than one-hot encoding. Basically, functions π and h offer a specific view of the learning task χ, and contribute to constructing the internal representation to be learned. As a consequence, depending on the choice of π and h, the complexity of learning f can change significantly.

In regression tasks, O is a continuous set. The substantial difference with respect to classification is that the decision doesn't typically require any decoding, so that $h = \mathrm{id}$. Hence, regression is characterized by $Y \subset \mathbb{R}^n$. Examples of regression tasks might involve values on the stock market, electric energy consumption, temperature and humidity prediction, and expected company income.

The information that a machine is expected to process may have different attribute types. Data can be inherently continuous. This is the case of classic fields like computer vision and speech processing. In other cases, the input belongs to a finite alphabet, that is, it has a truly discrete nature. An interesting example is the *car evaluation* artificial learning task proposed in the UCI Machine Learning repository https://archive.ics.uci.edu/ml/datasets/Car+Evaluation.

The evaluation that is sketched below in Fig. 1.1 is based on a number of features ranging from the buying price to the technical features.

Here, CAR refers to car acceptability and can be regarded as the higher order category that characterizes the car. The other high-order category uppercase nodes PRICE, TECH, and COMFORT refer to the overall notion of price, technical and comfort features. Node PRICE collects the buying price and the maintenance price, COMFORT groups together the number of doors (doors), the capacity in terms of person to carry (person), and the size of luggage boot (lug-boot). Finally, TECH, in addition to COMFORT, takes into account the estimated safety of the car (safety). As we can see, there is remarkable difference with respect to learning tasks involving continuous feature, since in this case, because of the nature of the problem, the leaves take on *discrete values*. When looking at this learning task carefully, the conjecture arises that the decision might benefit from considering the hierarchical aggregation of the features that is

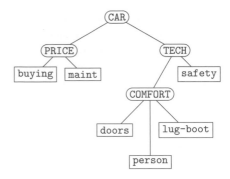

FIGURE 1.1

This learning task is presented in the UCI Machine Learning repository. The values for the attributes `buying` and `maint` are `vhigh`, `high`, `med`, `low`, those for `safety` are `low`, `med`, `high`, `doors` can be 2, 3, 4 or more, `persons` ca assume the values 2, 4 or more while `lug-boot` is either `small`, `med` or `big`.

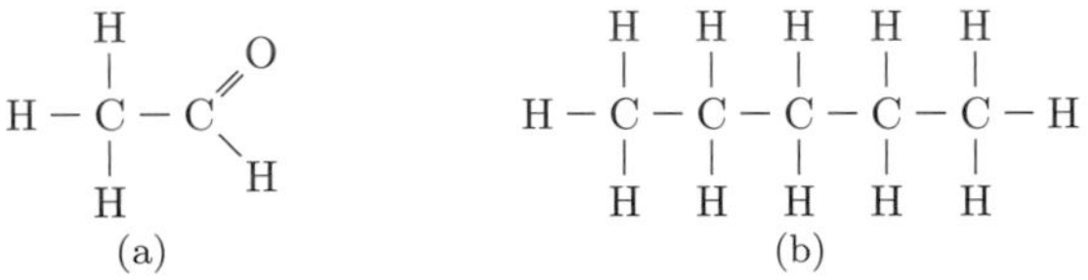

$$
\begin{array}{cc}
\text{(a)} & \text{(b)}
\end{array}
$$

FIGURE 1.2

Two chemical formulas: (a) acetaldehyde with formula CH_3CHO, (b) N-heptane with the chemical formula $H_3C(CH_2)_5CH_3$.

sketched by the tree. On the other hand, this might also be arguable, since all the leaves of the tree could be regarded as equally important for the decision.

However, there are learning tasks where the decision is strongly dependent on truly structured objects, like trees and graphs. For example, Quantitative Structure-Activity Relationship (QSAR) explores the mathematical relationship between the chemical structure and the pharmacological activity in a quantitative manner. Similarly, Quantitative Structure-Property Relationship (QSPR) is aimed at extracting general physical — chemical properties from the structure of the molecule. In these cases we need to take a decision from an input which presents a relevant degree of structure that, in addition to the atoms, strongly contributes to the decision process. Formulas in Fig. 1.2 are expressed by graphs, but chemical conventions in the representation of formulas, like ⬡ for benzene, do require careful investigation of the way $e \in E$ is given an internal representation $x \in X$ by means of function π.

Most challenging learning tasks cannot be reduced to the assumption that the agent processes single entities $e \in E$. For example, the major problem that typically arises in problems like speech and image processing is that we cannot rely on robust segmentations of entities. Spatiotemporal environments that typically characterize human life offer information that is spread in space and time, without offering reliable markers to perform segmentation of meaningful cognitive patterns. Decisions arise because of a complex process of spatiotemporal immersion. For example, human vision can also involve decisions at pixels level, in contexts which involve the spatial regularities, as well as the temporal structure connected to sequential frames. This seems to be mostly ignored in most research on object recognition. On the other hand, the extraction of symbolic information from images that are not frames of a temporally coherent visual stream would have been extremely harder than in our visual experience. Clearly, this comes from the information-based principle that in any world of shuffled frames, a video requires an order of magnitude more information for its storing than the corresponding temporally coherent visual stream. As a consequence, any recognition process is remarkably more difficult when shuffling frames, which clearly indicates the importance of keeping the spatiotemporal structure that is naturally associated with the learning task. Of course, this makes it more difficult to formulate sound theories of learning. In particular, if we really want to fully capture spatiotemporal structures, we must abandon the safe model of processing single elements $e \in E$ of the environment. While an extreme interpretation of structured objects might offer the possibility of representing visual environments, as it will be mostly shown in Chapter 6, this is a controversial issue that has to be analyzed very carefully.

The spatiotemporal nature of some learning tasks doesn't only affect recognition processes, but strongly conditions action planning. This happens for motion control in robotics, but also in conversational agents, where the decision to talk needs appropriate planning. In both cases, the learning environment offers either rewards or punishment depending on the taken action.

Overall, the interaction of the agent with the environment can be regarded as a process of constraint satisfaction. A remarkable part of this book is dedicated to an in-depth description of different types of constraints that are expressed by different linguistic formalisms. Interestingly, we can conquer a unified mathematical view of environmental constraints that enable the coherent development of a general theory of machine learning.

EXERCISES

1. *[16]* Let us consider the output encoding $h_i(y) = [i = \arg\max_\kappa y_\kappa]$. A limit of this encoding function is that, in case there are two or more outputs with similar value, the decision takes the maximum, while disregarding any robustness issue. How can you modify this encoding in such a way to refrain from returning nonrobust decisions?

2. *[M21]* Suppose you are given a supervised learning task χ that is faced by two different input/output representations, so as $\chi - h_2 \circ f_2 \circ \pi_2 = h_1 \circ f_1 \circ \pi_1$. Let π_1 and h_1 be bijective functions. Discuss the asymmetry of learning χ using the two representations.

3. *[18]* Let us consider the car evaluation learning task that is based on features represented in Fig. 1.1. Discuss the construction of function π.

4. *[15]* In the running example on handwritten character recognition, the input is desampled by pooling operations. Since the patterns that come from camera are already of finite dimensions (resolution), why do we bother with pooling operations? Provide a qualitative answer to motivate pooling.

5. [*15*]   Let us consider the output coding function $h \colon Y \to O$. In pattern classification, a reasonable solution seems to be that of choosing the classic binary encoding to express the n classes in which a certain object $e \in E$ must be classified. However, Fig. 1.3 shows a different solution based on one-hot encoding that is clearly more demanding in terms of the number of outputs. Why should one refrain from using only four instead of ten outputs in handwritten digit recognition? Provide qualitative motivations independently of this learning task.

6. [*M45*]   Let us consider the claim of Exercise 5. Can you provide a formal statement of the claim along with a proof?

1.1.2 Symbolic and subsymbolic representations of the environment

The previous discussion on learning tasks suggests that they are mostly connected with environments in which the information has a truly *subsymbolic* nature. Unlike many problems in computer science, the intelligent agent that is the subject of investigation in this book cannot rely on environmental inputs with attached semantics.

Regardless of the difficulty of conceiving a smart program to play chess, it is an interesting case in which intelligent mechanisms can be constructed over a *symbolic* interpretation of the input. The chessboard can be uniquely expressed by symbols that report both the position and the type of piece. We can devise strategies by interpreting chess configurations, which indicates that we are in front of a task, where there is a semantics attached with environmental inputs. This isn't the case for the handwritten character recognition task. In the simple preprocessing that yields the representation shown in the previous section, any pattern turns out to be a string of 64 Boolean variables. Interestingly, this looks similar to the chessboard; we are still representing the input by a matrix. However, unlike chessboard positions, the single pixels in the retina cannot be given any symbolic information with attached semantics. Each pixel is either `true` or `false` and tells us nothing about the category of the pattern it belongs to. This isn't the consequence of the strong assumption of using black and white representations; gray-level images don't help in this respect! For example, gray-level images with a given quantization of the brightness turn out to be integers instead of Booleans, but we are still missing any semantic information. One can promptly realize that any cognitive process aimed at interpreting handwritten chars must necessarily consider groups of pixels by performing a sort of holistic processing. Unlike chess, this time any attempt to provide an unambiguous description of these Boolean representations is doomed to fail. We are in front of subsymbolic information with inherent ambiguous interpretation. Chessboards and retina can both be expressed by matrices. However, while in the first case an intelligent agent can devise strategies that rely on the semantics of the single positions of the chessboard, in the second case, single pixels don't help! The purpose of machine learning is mostly to face tasks which exhibit a subsymbolic nature. In these cases, the traditional construction of human-based algorithmic strategies don't help.

Interestingly, learning-based approaches can also be helpful when the environmental information is given in symbolic form. The *car evaluation* task discussed in the previous section suggests that there is room for constructing automatic induction processes also in these cases, where the inputs can be given a semantic interpretation.

The reason for the need of induction is that the concept associated with the task might be hard to describe in a formal way. Car evaluation is hard to formalize, since it is even the outcome of social interactions. While one cannot exclude the construction of an algorithm for returning a unique evaluation, any such attempt is interwound with the arbitrariness of the design choices that characterize that algorithm. To sum up, subsymbolic descriptions can characterize the input representation as well as the task itself.

The lack of formal linguistic description is what opens the doors for a fundamental rethinking of algorithmic approaches to many problems that are naturally solved by humans. A learning machine can still be given an algorithmic description, but in this case, the algorithm prescribes how the machine is expected to learn to solve the task instead of the way it is solved! Hence, machine learning algorithms don't translate clever human intuitions for solving the given task, but operate at a *metalevel* with the broader view of performing induction from examples.

The symbolic/subsymbolic dichotomy assumes an intriguing form in case of spatiotemporal environments. Clearly, in vision both the input and the task cannot be given a sound symbolic description. The same holds true for speech understanding. If we consider textual interpretation, the situation is apparently different. Any string of written text has clearly a symbolic nature. However, natural language tasks are typically hard to describe, which opens the doors to machine learning approaches. There's more. The richness of the symbolic description of the input suggests that the agent, which performs induction, might benefit from an opportune subsymbolic description of reduced dimension. In text classification tasks, one can use an extreme interpretation of this idea by using *bag of words* representations. The basic idea is that text representation is centered around keywords that are preliminarily defined by an opportune dictionary. Then any document is properly represented in the *document vector space* by coordinates associated with the frequency of the terms, as well as by their rarity in the collection (TF-IDF Text Frequency, Inverse Document Frequency) — see Scholia, Section 1.5.

Machine learning is at the crossroad of different disciplines which share the ambition of disclosing the secret of intelligence. Of course, this has attracted the curiosity of philosophers for centuries, and it is still nowadays of central interest in cognitive science and neuroscience. By and large, machine learning is regarded as an attempt to construct intelligent agents for a given learning task on the basis of artificial models that are mostly rooted on computational models (see Section 1.5 for additional discussions). Yet, the methodologies at the basis of those models are remarkably different. Scientists in the field often come from two different schools of thought, which lead them to look at learning tasks from different corners. An influential school is the one which originates with symbolic AI. As it will be sketched in Section 1.5, one can regard computational models of learning as search in the space of hypotheses. A somewhat opposite direction is the one with focuses on continuous-based representations of the environment, which favors the construction of learning models conceived as optimization problems. This is what is mostly covered in this book. Many intriguing open problems are nowadays at the border of these two approaches.

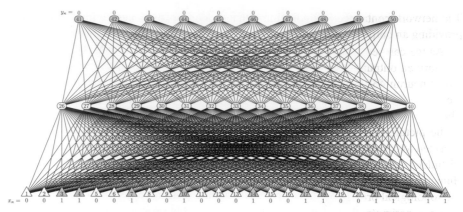

FIGURE 1.3

Recognition of handwritten chars. The incoming pattern $x = \pi($▦$)$ is processed by the feedforward neural network, whose target consists of firing only neuron 43. This corresponds with the *one-hot* encoding of class "2."

1.1.3 Biological and artificial neural networks

One of the simplest models of artificial neurons consists of mapping the inputs by a weighted sum that is subsequently transformed by a sigmoidal nonlinearity according to

$$a_i = b_i + \sum_{j=i}^{d} w_{ij} x_j, \quad y_i = \sigma(a_i). \tag{1}$$

The squash function $\sigma(a) := 1/(1 + \exp(-a))$ favors a decision-like behavior, since $y_i \to 1$ or $y_i \to 0$ as the activation diverges ($a_i \to \pm\infty$). Here, i denotes a generic neuron, while w_{ij} is the weight associated with the connection from input j to neuron i. We can use this building block for constructing neural networks that turn out to compute complex functions by composing many neurons. Amongst the possible ways of combining neurons, the *multilayered neural network* (MLN) architecture is one of the most popular, since it has been used in an impressive number of different applications. The idea is that the neurons are grouped into layers, and that we connect only neurons from lower to upper layers. Hence, we connect j to i if and only if $l(j) < l(i)$, where $l(k)$ denotes the layer where the generic unit k belongs to. We distinguish the input layer I, the hidden layers H, and the output O layer. Multilayer networks can have more than one hidden layer. The computation takes place according to a piping scheme where data available at the input drives the computational flow. We consider the MLP of Fig. 1.3 that is conceived for performing handwritten character recognition. First, the processing of the hidden units takes place and then, as their activations are available, they are propagated to the output. The neural network receives, as input, the output of the preprocessing module $x = \pi(e) \in \mathbb{R}^{25}$.

The network contains 15 hidden neurons and 10 output neurons. Its task is that of providing an output which is as close as possible to the target indicated in the figure.

As we can see, when the network receives the input pattern , it is expected to return an output which is as close as possible to the target, indicated by the 1 on top of neuron 43 in Fig. 1.3. Here one-hot encoding has been assumed, so as only the target corresponding to class "2" is high, whereas all the others are set to zero. The input character is presented to the input layer by an appropriate row scanning of the picture and then it is forwarded to the hidden layer, where the network is expected to construct features that are useful for the recognition process. The chosen 15 hidden units are expected to capture distinctive properties of different classes. Human interpretation of similar features leads to the characterization of geometrical and topological properties.

Area, perimeter, as well as features from mass geometry, help discriminating the chars. The barycenter and the momentum of different order detect nice properties of the patterns but, as already pointed out, the features might also be related to our own intuitions on what is characterizing the different classes, like the roundness and number of holes provide additional cues. There are also some *key points*, like corners and crosses, that strongly characterize some class (e.g., the cross of class "8"). Interestingly, one can think of the neurons of the hidden layer as feature detectors, but it is clear that we are in front of two fundamental questions: Which features can we expect the MLP will detect? Provided that the MLP can extract a certain set of features, how can we determine the weights of the hidden neurons that perform such a feature detection? Most of these questions will be addressed in Chapter 5, but we can start making some interesting remarks. First, since the purpose of this neural network is that of classifying the chars, it might not be the case that the above mentioned features are the best ones to be extracted under the MLP. Second, we can immediately grasp the idea of learning, which consists of properly selecting the weights, in such a way to minimize the error with respect to the targets. Hence, a collection of handwritten chars, along with the corresponding targets, can be used for data fitting. As it will be shown in the following section, this is in fact a particular kind of *learning from example scheme* that is based on the information provided by an oracle referred to as the supervisor. While the overall purpose of learning is that of providing the appropriate classification, the discovery of the weights of the neurons leads to constructing intermediate hidden representations that represent the pattern features. In Fig. 1.3, the pattern which is presented is properly recognized. This is a case in which the agent is expected to get a reward, whereas in case there is an error, the agent is expected to be punished. This *carrot and stick* metaphor is the basis of most machine learning algorithms.

1.1.4 Protocols of learning

Now the time has come to clarify the interactions of the intelligent agent with the environment, which in fact defines the constraints under which learning processes take place. A natural scientific ambition is that of placing machines in the same context as humans', and to expect from them the acquisition of our own cognitive skills. At the sensorial level, machines can perceive vision, voice, touch, smell, and taste with different degree of resolution with respect to humans. Sensor technology has been succeeding in the replication of human senses, with special emphasis on camera and microphones. Clearly, the speech and visual tasks have a truly cognitive nature, which make them very challenging not because of the difficulty of acquiring the sources, but because of the correspondent signal interpretation. Hence, while in principle we can create a truly human-like environment for machines to learn, unfortunately, the complex spatiotemporal structure of perception results in very difficult problems of signal interpretation.

Since the dawn of learning machines, scientists isolated a simple interaction protocol with the environment that is based on the acquisition of *supervised pairs*. For example, in the handwritten char task, the environment exchanges the training set $L = \{(e_1, o_1), \dots, (e_\ell, o_\ell)\}$ with the agent. We rely on the principle that if the agent sees enough labeled examples then it will be able to induce the class of any new handwritten char. A possible computational scheme, referred to as *batch mode* supervised learning, assumes that the training set L is processed as a whole for updating the weights of the neural network. Within this framework, supervised data involving a certain concept to be learned are downloaded all at once. The parallel with human cognition helps understanding the essence of this awkward protocol: Newborns don't receive all their life bits as they come to life! They live in their own environment and gradually process data as time goes by. Batch mode, however, is pretty simple and clear, since it defines the objectives of the learning agent, who must minimize the error that is accumulated over all the training set.

Instead of working with a training set, we can think of learning as an adaptive mechanism aimed at optimizing the behavior over a sequential data stream. In this context, the training set L of a given dimension $\ell = |L|$ is replaced with the sequence $(L_t)_{t \in \mathbb{N}}$, $L_t := \{(e_1, o_1), \dots, (e_t, o_t)\}$, where the index t isn't necessarily upper bounded, that is, $t < \infty$. The process of adapting the weights with L_t is referred to as *online learning*.

We can promptly see that, in general, it is remarkably different with respect to batch-mode learning. While in the first case the given collection L of ℓ samples is everything what is available for acquiring the concept, in case of online learning, the flux of incoming supervised pairs data never stops. This protocol is dramatically different from batch-mode, and any attempt to establish the quality of the concept acquisition also requires a different assessment. First, let us consider the case of batch-mode. We can measure the classification quality in the training set by computing

$$E(w) := \sum_{\kappa=1}^{\ell} \sum_{j=1}^{n} (1 - y_{\kappa j} F_j(w, x_\kappa))_+, \tag{1}$$

where n is the number of classes and $F_j(w, x_\kappa)$ is the jth output of the network fed with pattern x_κ. Throughout the book we use the notation $(\cdot)_+$ to represent the *hinge function*, which is defined as $(z)_+ = z \cdot [z > 0]$, where $[z > 0] = 1$ if $z > 0$ and $[z > 0] = 0$, otherwise. Hence $(1 - y_{\kappa j} F_j(w, x_\kappa))_+ = 0$ holds whenever there is a strong sign agreement between the output $F_j(w, x_\kappa)$ and the target $y_{\kappa j}$. Notice that because of the one-hot encoding, the index $j = 1, \ldots, n$ corresponds with the class index. Now, $F_j(w, x_\kappa)$ and $y_{\kappa j}$ must agree on the sign, but robustness requirements in the classification lead us to conclude that the outputs cannot be too small.

The pair $(\mathcal{N}, L)$, which is composed of the neural network $\mathcal{N}$ and of the training set L, offers the ingredients for the construction of the error function in Eq. (1) that, once minimized, returns a weight configuration of the neural network that well fits the training set. As it will become clear in the following, unfortunately, fitting the training set isn't necessarily a guarantee of learning the underlining concept. The reason is simple: The training set is only sampling the probability distribution of the concept and, therefore, its approximation very much depends on the relevance of the samples, which is strongly related to the cardinality of the training set and to the difficulty of the concept to the learned.

As we move to online mode, the formulation of learning needs some additional thoughts. A straightforward extension from batch-mode suggests adopting the error function E_t computed over L_t. However, this is tricky, since E_t changes as new supervised examples come. While in batch-mode, learning consists of finding $w^* \in \arg\min_w E(w)$, in this case we must carefully consider the meaning of optimization at step t. The intuition suggests that we need not reoptimize E_t at any step. Numerical algorithms, like gradient descent, can in fact optimize the error on single patterns as they come. While this seems to be reasonable, it is quite clear that such a strategy might lead to solutions in which the neural network is inclined to "forget" old patterns. Hence, the question arises on what we are really doing when performing online learning by gradient weight updating.

The link with human cognition early leads us to explore learning processes that take place regardless of supervision. Human cognition is in fact characterized by processes of concept acquisition that are not primarily paired with supervision. This suggests that while supervised learning allows us to be in touch with the external symbolic interpretations of environmental concepts, in most cases, humans carry out learning schemes to properly aggregate data that exhibit similar features. How many times do we need to supervise a child on the concept of glass? It's hard to say, but surely a few examples suffice to learn. Interestingly, the acquisition of the glass concept isn't restricted to the explicit association of instances along with their correspondent symbolic description. Humans manipulate objects and look at them during their life span, which means that concepts are likely to be mostly acquired in a sort of unsupervised modality. Glasses are recognized because of their shape, but also because of their function. A glass is a liquid container, a property that can be gained by looking at the action of its filling up. Hence, a substantial support to the process of human object recognition is likely to come from their *affordances* — what can be done with them. Clearly, the association of the object affordance with the corresponding actions doesn't necessarily require one to be able to express a symbolic description of the action. The attachment of a linguistic label seems to involve a cognitive task that does require the creation of opportune object internal representations. This isn't restricted to vision. Similar remarks hold for language acquisition and for any other learning task. This is indicating that important learning processes, driven by aggregation and clustering mechanisms, take place at an unsupervised level. No matter which label we attach to patterns, data can be clustered as soon as we introduce opportune similarity measures.

The notion of similarity isn't easy to grasp by formal descriptions. One can think of paralleling similarity with the Euclidean distance in a metric space $X \subset \mathbb{R}^d$. However, that metrics doesn't necessarily correspond with similarity. What is the meaning of similarity in the handwritten char task? Let us make things easy and restrict ourselves to the case of black and white images. We can soon realize that Euclidean metrics badly reflects our cognitive notion of similarity. This becomes evident when the pattern dimension increases. To grasp the essence of the problem, suppose that a given character is present in two instances, one of which is simply a right-shifted instance of the other. Clearly, because high dimension is connected with high resolution, the right-shifting creates a vector where many coordinates corresponding to the black pixels are different! Yet, the class is the same. When working at high resolution, this difference in the coordinates leads to large pattern distances within the same class. Basically, only a negligible number of pixels are likely to occupy the same position in the two pattern instances. There is more: This property holds regardless of the class. This raises fundamental issues on the deep meaning of pattern similarity and on the cognitive meaning of distance in metric spaces. The Euclidean space exhibits a somewhat surprising feature at high dimension, which leads to unreliable thresholding criteria. Let us focus on the mathematical meaning of using Euclidian metrics as similarities. Suppose we want to see which patterns $x \in X \subset \mathbb{R}^d$ are close to $\overline{x} \in X$ according to the *thresholding criterion* $\|x - \overline{x}\| < \rho$, where $\rho \in \mathbb{R}_+$ expresses the degree of neighborhood to $\overline{x}$. Of course, small values of ρ define very close neighbors, which perfectly matches our intuition that x and $\overline{x}$ are similar because of their small distance. This makes sense in the three-dimensional Euclidean space that we perceive in real life.

However, the neighbor $B_\rho(\overline{x}) = \{x \in X : \|x - \overline{x}\| < \rho\}$ possesses a curious property at high dimension. Its volume is

$$\mathcal{L}^d(B_\rho(\overline{x})) = \frac{\pi^{d/2}}{\Gamma(1+d/2)}\rho^d, \tag{2}$$

where Γ is the *Gamma function* and $\mathcal{L}^d$ is the d-dimensional *Lebesgue measure* in $\mathbb{R}^d$. Suppose we fix the threshold value ρ. We can prove that the volume approaches zero as $d \to \infty$ (see Exercise 2). There is more: The sphere, regarded as an orange, doesn't differ from its peel! This comes directly from the previous equation. Suppose we consider a ball $B_{\rho-\varepsilon}(\overline{x})$ with radius $\rho - \varepsilon > 0$. When $\varepsilon \ll \rho$, the set $P_\varepsilon = \{x \in X : \rho - \epsilon < \|x - \overline{x}\| < \rho\}$ is the *peel* of the ball. Indeed, for all $\varepsilon > 0$, $\varepsilon < \rho$ as $d \to \infty$, we have

$$\lim_{d\to+\infty}\frac{\mathcal{L}^d(P_\varepsilon)}{\mathcal{L}^d(B_\rho(\overline{x}))} = \lim_{d\to+\infty} 1 - \frac{\mathcal{L}^d(B_{\rho-\varepsilon}(\overline{x}))}{\mathcal{L}^d(B_\rho(\overline{x}))} = \lim_{d\to+\infty} 1 - \left(\frac{\rho-\varepsilon}{\rho}\right)^d = 1. \quad (3)$$

As a consequence, the thresholding criterion for identifying P_ε corresponds with checking the condition $x \in B_\rho(\overline{x})$. However, the above geometrical property, which reduces the ball to its frontier, means that, apart from a set of null measure, we have $x \in P_\varepsilon$. It is instructive to see how $\mathcal{L}^d(B_\rho(\overline{x}))$ scales up with respect to the volume of the sphere which contains all the examples of the training set. Suppose that X is bounded in $\mathbb{R}^d$, if we denote by $x_M \in \mathbb{R}^d$ the point such that $\forall x \in X$ we have $\|x\| \leq \|x_M\|$ and we let $B_{\|x_M\|} := \{x \in \mathbb{R}^d : \|x\| < \|x_M\|\}$, then

$$\lim_{d\to+\infty}\frac{\mathcal{L}^d(B_\rho(\overline{x}))}{\mathcal{L}^d(B_{\|x_M\|})} = \lim_{d\to+\infty}\left(\frac{\rho}{\|x_M\|}\right)^d = 0. \quad (4)$$

To sum up, at high dimension, the probability of satisfying the neighboring condition vanishes, thus making the criterion unusable. While the similarity in the ordinary three-dimensional Euclidean space is somewhat connected with the metrics, as the dimension increases this connection vanishes. It is worth mentioning that this mathematical discussion on oddities associated with high-dimensional spaces relies on the assumption that the points are uniformly distributed in the space, which doesn't really hold in practice. However, as shown in the analysis on right-shifting of handwritten chars, also in cases of the biased distribution induced by a real-world problem, high-dimensional oddities are still a serious issue. This has a direct impact when interpreting the notion of pattern similarity as an Euclidean metrics.

This discussion suggests that the unsupervised aggregation of data must take into account the related cognitive meaning, since metric assumptions on the pattern space might lead to a wrong interpretation of human concepts. As the discussion on the glass example suggests, unsupervised learning is of central importance in human life, and it is likely to be of crucial importance regardless of biology. Hence, just like humans, learning machines must somehow be able to capture invariant features that significantly help clustering processes. The experience with language acquisition in children suggests that parroting is a fundamental developmental step. Could not it be an important skill to acquire well beyond language acquisition? Fig. 1.4 shows a possible way of constructing parroting mechanisms in any pattern space.

Given an unsupervised pattern collection $D = \{x_1, \ldots, x_\ell\} \subset X$, we define a learning protocol based on data auto-encoding. The idea is that we minimize a cost function that expresses the auto-encoding principle: Each example $x_\kappa \in X$ is forced to be reproduced to the output. Hence, we construct an unsupervised learning process, which consists of determining

$$w^* \in \arg\min_w \sum_{\kappa=1}^{\ell} \|F(w, x_\kappa) - x_\kappa\|^2. \quad (5)$$

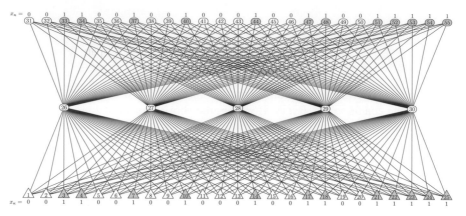

FIGURE 1.4

Pattern auto-encoding by an MLP. The neural net is supervised in such a way to reproduce the input to the output. The hidden layer yields a compressed pattern representation.

In so doing, the neural network parrots x_κ to $F(w, x_\kappa) \simeq x_\kappa$. Notice that the training that takes place on D leads to the development of internal representations of all elements $x_\kappa \in X$ in the hidden layer. In Fig. 1.4, patterns of dimension 25 are mapped to vectors of dimension 5. In doing so, we expect to develop an internal representation at low dimension, where some distinguishing features are properly detected. The internal discovery of those features is the outcome of learning, which leads us to introduce a measure for establishing whether a given pattern $x \in X$ belongs to the probability distribution associated to the samples D and characterized by the learned auto-encoder. When considering the correspondent learned weights w^*, we can introduce the similarity function $s_D \colon X \to \mathbb{R}_+$

$$s_D(x) := \|x - F(w^*, x)\|.$$

This makes it possible to introduce the thresholding criterion $s_D(x) < \rho$. Unlike the Euclidean metrics, the set $B_\rho^D := \{x \in X : s_D(x) < \rho\}$ is learned from examples by discovering w^* according to Eq. (5). In doing so, we give up looking for magic metrics that resemble similarities, and focus attention, instead, on the importance of learning; the training set D turns out to define the class of auto-associated points $B_\rho^D \subset X$. Coming back to the cognitive interpretation, this set corresponds to identifying those patterns that the machine is capable of repeating. If the approximation capabilities of the model defined by the function $F(w^*, \cdot)$ is strong enough, the machine can perform parroting with high degree of precision, while rejecting patterns of different classes. This will be better covered in Section 5.1.

Now, suppose that we can rely on a measure of the degree of similarity of objects; if we know that a given picture A has been labeled as a glass, and that this picture is "similar" to picture B, then we can reasonably conclude that B is a glass. This inferential mechanism somehow bridges supervised and unsupervised learning. If our learning agent has discovered reliable similarity measures in the space of objects, it doesn't need many instances of glasses! As soon as a few of them are correctly classified, they are likely to be identified as similar to incoming patters, which can acquire the same class label. This computational scheme is referred to as *semisupervised learning*. This isn't only cognitively plausible, it also establishes a natural protocol for making learning efficient.

The idea behind semisupervised learning can be used also in contexts in which the agent isn't expected to perform induction, but merely to decide within a close environment, where all data of the environment are accessible during the learning process. This is quite relevant in some real-word problems, especially in the field of information retrieval. In those cases, the intelligent agent is given the access to huge, yet close, collections of documents, and it is asked to carry out a decision on the basis of the overall document database. Whenever this happens and the agent does exploit all the given data, the learning process is referred to as *transductive learning*.

Cognitive plausibility and complexity issues meet again in case of *active learning*. Children actively ask questions to facilitate their concept acquisition. Their curiosity and ability to ask questions dramatically simplifies the search in the space of concept hypotheses. Active learning can take different forms. For example, in classification tasks taking place in an environment where the examples are provided as single entities $e \in E$, an agent could be allowed to pose questions to the supervisor on specific patterns. This goes beyond the plain scheme based on a given training set. The underlying assumption is that there is a stream of data, some of them with a corresponding label, and that the agent can pose questions on specific examples. Of course, there is a budget: The agent cannot be too invasive and should refrain from posing too many questions. The potential advantage for the agent is that it can choose on which examples to ask assistance. It makes sense for it to focus on cases where it has not gained enough confidence. Hence, active learning does require the agent to be able to develop a critical judgment on its decisions. Doubtful cases are those on which to ask assistance. Active learning can become significantly more sophisticated in spatiotemporal environments associated with challenging tasks like speech understanding and computer vision. Specific questions, which correspond to actions, in these cases arise as the outcome of complex focus of attention mechanisms.

Human learning skills strongly exploit the temporal dimension and assume relevant sequential structure. In particular, humans take actions in the environment so as to maximize some notion of *cumulative reward*. These ideas have been exploited also in machine learning, where this approach to concept acquisition is referred to as *reinforcement learning*. The agent is given a certain objective, like interacting with customers for flight tickets or for reaching the exit of a maze, and it receives some information that can be used to drive its actions. Learning is often formulated as a Markov decision process (MDP), and the algorithms typically utilize dynamic programming techniques. In reinforcement learning, neither input/output pairs are presented, nor suboptimal actions are explicitly corrected.

EXERCISES

1. [22]   Consider the error function in Eq. (1). It is based on the principle of imposing strong sign agreement between the agent and the supervisor in classification tasks. Suppose that the agent only needs to decide whether a given pattern x belongs to a certain class. In this case the function reduces to $(1 - yf(x))_+$. Could we replace it with $(s - yf(x))_+$, where s is any positive real number acting as a threshold? Is it the case that large values of s guarantee more robustness? Come back to this exercise after having read Chapter 4.

2. [M20]   Prove that the volume of a d-dimensional ball is given by Eq. (2). What happens as d increases? Discuss and plot the behavior of the volume, with special attention to the asymptotic behavior.

3. [19]   Let us consider a dataset of handwritten chars represented by Boolean vectors as shown in Section 1.1. A possible algorithm for char recognition can be based on the similarity function $s: \{0, 1\}^d \times \{0, 1\}^d \to [0, +\infty)$ defined by $s(x, z) = \sum_{i=1}^d x_i \equiv z_i$, where $\equiv$ is the equivalence of Boolean variables. Write an algorithm for char recognition based on this idea, and discuss its behavior as the char resolution increases ($d \to \infty$). Discuss the relations of this idea with the results discovered in Exercise 2.

4. [C18]   Write a program to recognize MNIST handwritten chars[1] by using the similarity defined in Exercise 3. Then modify the similarity to the classic Euclidean distance $d(x, z) := \|x - z\|$. Produce a technical report on the experimental results corresponding to preprocessing of the chars for growing dimension.

5. [C16]   Using the MNIST handwritten char dataset, write a program to compute the Euclidean distance between the patterns and discuss the outcome of the concentration of the volume in the peel of the ball in this specific case.

6. [15]   Let us consider the auto-encoder of Fig. 1.4 and assume that it is learning over a training set composed of letters of the English alphabet along with the digits. Prove that it is impossible for any learning algorithm to fully learn the task if the returned solution consists of hidden units saturated to asymptotic values, so as they can be regarded as bits.

1.1.5 Protocols of learning

While most classic methods of machine learning are covered in this book, the emphasis is on stimulating the reader towards a broader view of learning that relies on the unifying concept of *constraint*. Just like humans, intelligent agents are expected to live in an environment that imposes the fulfillment of constraints. Of course, there are very sophisticated social translations of this concept that are hard to express into the world of intelligent agents. However, the metaphor is simple and clear: We want our machines to be able to satisfy the constraints that characterize their interactions with the environment. Perceptual data along with their relations, as well as abstract knowledge granules on the task, are what we intend to express by the *mathematical* notion of constraint. Unlike the generic concept used for describing human interactions, we need a unambiguous linguistic description of constraints.

---

[1] The dataset can be downloaded from http://yann.lecun.com/exdb/mnist/.

Let us begin with supervised learning. It can be regarded just as a way of imposing special pointwise constraints. Once we have properly defined the encoding π and the decoding h functions, the agent is expected to satisfy the conditions expressed by the training set $L = \{(x_1, y_1), \ldots, (x_\ell, y_\ell)\}$. The perfect satisfaction of the given pointwise constraints requires us to look for a function f such that $\forall \kappa = 1, \ldots, \ell$ gives $f(x_\kappa) = y_\kappa$. However, neither classification, nor regression can strictly be regarded as pointwise constraints in that sense. The fundamental feature of any learning from examples scheme is that of tolerating errors in the training set, so as the soft-fulfillment of constraints is definitely more appropriate than hard-fulfillment. Error function 1.1.4–(1) is a possible way of expressing soft-constraints, that is based on the minimization of an appropriate penalty function. For example, while error function 1.1.4–(1) is adequate to naturally express classification tasks, in case of regression, one isn't happy with functions that satisfy the strong sign agreement. We want to discover f that transforms the points x_κ of the training set onto a value $f(x_\kappa) = F(w, x_\kappa) \simeq y_\kappa$. For example, this can be done by choosing a *quadratic loss*, so as we want to achieve small values for

$$E(w) = \sum_{\kappa=1}^{\ell} (y_\kappa - F(w, x_\kappa))^2. \tag{1}$$

This pointwise term emphasizes the very nature of this kind of constraint. A central issue in machine learning is that of devising methods for guaranteeing that the fulfillment of the constraints on a sample $X^\sharp$ of the domain X is sufficient for claims on X, namely on new examples drawn from the same probability distribution as the one from which $X^\sharp$ are drawn.

In some learning tasks, instead of presenting labeled examples to an intelligent agent, we can rely on abstract knowledge expressed in terms of properties on intervals in the input space X. Many interesting tasks of this kind come from medicine. A first approach to diagnosis of diabetes is that of considering the *body mass index* and the measure of the *blood glucose*. The *Pima Indian Diabetes Dataset* contains a supervised training set of positive and negative diabetes examples, where each patient is represented by a vector with eight inputs. Instead of relying only on the evidence of single patients, a physician might suggest to follow the logic rules:

$$(\mathtt{mass} \geq 30) \wedge (\mathtt{plasma} \geq 126) \Rightarrow \mathtt{diabetes};$$
$$(\mathtt{mass} \leq 25) \wedge (\mathtt{plasma} \leq 100) \Rightarrow \neg\mathtt{diabetes},$$

These rules represent a typical example of what draws expert inference in medical diagnostic processes. In a sense, these two rules define somehow a training set of the class `diabetes`, since the two formulas express positive and negative examples of the concept. Interestingly, there is a difference with respect to supervised learning: These two logic statements hold for infinite pairs of virtual patients! One could think of reducing this interval-based learning to supervised learning by simply griding the domain defined by the variables `mass` and `plasma`. However, as rules begin

FIGURE 1.5

Mask-based class description. Each class is characterized by a corresponding mask, which indicates the portion of the retina which is likely not to be occupied by corresponding patterns.

involving more and more variables, we experience the *curse of dimensionality*. As a matter of fact, griding the space isn't feasible anymore and, at most, we can think of its sampling. A full space griding fails converting exactly interval-based rules into supervised pairs. Basically, knowledge granules which rely on similar logic expressions identify a non-null measure subset $M \subset X$ of the input space X that, because of the curse of dimensionality, cannot be reasonable by griding. Interval-based rules are common also in prognosis. The following rules:

$$(\texttt{size} \geq 4) \wedge (\texttt{nodes} \geq 5) \Rightarrow \texttt{recurrent};$$
$$(\texttt{size} \leq 1.9) \wedge (\texttt{nodes} = 0) \Rightarrow \neg\texttt{recurrent},$$

draw the tumor recurrence on the basis of the diameter of the tumor (`size`) and of the number of metastasized lymph nodes (`nodes`).

Interval-based rules can be used in very different domains. Let us consider again the handwritten character recognition task. In Section 1.1.1, we had already carried out a qualitative description of the classes, but we ended up with the conclusion that they are hard to express from a formal point of view. However, it was clear that, based on qualitative remarks on the structure of the patterns, we can create very informative features to be used for classification. In this perspective, it is up to the preprocessing function π to extract the distinguishing features. Given any pattern e, the processing module returns $x = \pi(e)$, which is the pattern representation used for standard supervised learning.

However, we can provide another qualitative description of the patterns that very much resembles the structure of the two learning tasks on diagnosis and prognosis of diabetes. For example, when looking at Fig. 1.5, we realize that the "0" class is characterized by drawings that don't typically occupy $M_1 \cup M_2$.

Likewise, any other class is described by expressing the portions of the retina that any pattern of a given class is expected not to occupy. Those forbidden portions are expressed by appropriate masks that represent the prior knowledge on the pattern structure. Now, let $m^c \in \mathbb{R}^d$ be a vector coming from row scanning of the retina such that $m_i^c = 1$ if the pixel corresponding to index i isn't expected to be occupied by character c and $m_i^c = 0$, otherwise. Let

$\delta > 0$ be an appropriate threshold. We have

$$\forall x : \frac{x \cdot m^c}{\|x\| \|m^c\|} > \delta \Rightarrow (h(f(x)) \equiv \neg c(x)), \tag{2}$$

where $c(x) = 1$ if x represent character c and 0 otherwise. This constraint on f states that whenever the correlation of pattern x with the class mask m^c exceeds the threshold δ, the agent is expected to return a coherent decision with the mask, that is, $h(f(x)) \equiv \neg c$. Notice that an extended interpretation of Boolean masks, like those adopted in Fig. 1.5, can lead to the replacement of the correlation condition expressed by Eq. (2) with the statements on the single coordinates of x. For instance, we could state constraints on the generic pixel of the retina of the form $x_{m,i} \leq x_i \leq x_{M,i}$. In doing so, we are again in front of a *multiinterval constraint*. The values of $x_{m,i}$ and $x_{M,i}$ represent the minimum and the maximum value of the brightness that one has discovered in the training set.

Both the learning tasks on medicine and on handwritten character recognition indicate that need for a generalization of supervised learning that involves sets instead of single points. Within this new framework, the supervised pairs (x_κ, y_κ) are replaced with (X_κ, y_κ), where X_κ is the set with supervision y_κ. Throughout this book, these are referred to as *set-based constraints*.

The need for robust and reliable decisions can also nicely be expressed by appropriate constraints. Suppose we want to preprocess the handwritten chars by two different schemes carried out by functions π_1 and π_2. They return two different patterns $x_1, x_2 \in X$ that are used by functions $f_1 \colon X \to \{-1, +1\}$ and $f_2 \colon X \to \{-1, +1\}$ for the classification. Clearly, we can enforce robustness in the classification by imposing that for every $e \in E$ such that $x_1 = \pi_1(e)$ and $x_2 = \pi_2(e)$ we have

$$(1 - f_1(x_1) f_2(x_2))_+ = 0. \tag{3}$$

Notice that this constraint doesn't involve supervisions, since it only checks the decision coherence of the two functions, no matter how they are constructed. How can we impose such a satisfaction imposed by the universal quantification over all the pattern domain E? The simplest solution is that of imposing such a quantification over a sampling $E^\sharp$ of E, and relaxing the constraint by imposing the minimization of the corresponding penalty function

$$C(f_1, f_2) = \sum_{e \in E^\sharp} (1 - f_1(\pi_1(e)) f_2(\pi_2(e)))_+. \tag{4}$$

Exercise 1 discusses the form of the penalty term when $f_1 \colon X \to [0, +1]$ and $f_2 \colon X \to [0, +1]$, while Exercise 2 deals with the new face of joint decision in case of regression. In both classification and regression, whenever the decision is jointly made by means of two or more different functions we refer to *committee machines*. The underlying idea is that more machines jointly involved in prediction can be more effective than a single one, since each of them might be able to better capture specific cues. However, in the end, the decision requires a mechanism for balancing their specific qualities. If the machines can provide a reliable score of their degree of uncertainty in the decision, we can use those scores for combining the decisions. In classification, we can also use social inspired solutions, like those based on majority voting decisions. The approach behind Eq. (4) and, generally, behind coherent decision constraints is that of imposing the single machines to come out with a common decision. As it will be more clear in Chapter 6, the resulting learning algorithms let the machines to freely express their decision at the beginning, while enforcing a progressive coherence.

While the discussed approach to committee machines deals with coherent decision, in other cases, the availability of multiple decisions may require satisfaction of additional constraints that have an important impact on the overall decision. As an example, consider an asset allocation problem connected with portfolio management. Suppose we are given a certain amount of money T that we want to invest. We assume that the investment will be in Euro and USD by an appropriate balance of cash, bond, and stock. The agent perceives the environment by means of segmented objects e that are preprocessed by π so as to return $x = \pi(e)$. Here, x encodes the overall financial information useful for the decision, like, for instance, the recent earning results for stock investments and stock series. The overall decision is based on $x \in X$, and involves functions f_c^d, f_b^d, f_s^d that return the money allocated in cash, bond, and stock, respectively. The allocation of this functions involved USD, while the functions f_c^e, f_b^e, f_s^e carried out the same decision by using Euro. Another two functions, t_d and t_e, decide on the overall allocation in USD and Euro, respectively. Since this is a regression problem, we assume that for any task $h = \text{id}$. The given formulation of this asset allocation problem requires satisfaction of the constraints:

$$f_c^d + f_b^d + f_s^d = t_d / T, \qquad f_c^e + f_b^e + f_s^e = t_e / T, \qquad t_d + t_e = T. \qquad (5)$$

Clearly, the single decisions of the functions are also assumed to be drawn from other environmental information. For example, from each of these functions we can have a training set, which results in an additional collection of pointwise constraints. Notice that the decision based on supervised learning cannot guarantee the above stated consistency conditions. The fulfillment of Eq. (5) contributes to shaping the decision, since the satisfaction of the above constraints results in feedback information onto all decision functions. Roughly speaking, the prediction on the amount of money in dollars to allocate for stocks by function f_s^d clearly depends on financial data $x \in X$, but the decision is also conditioned by concurrent investment decisions on bonds and cash, as well as by an appropriate balancing of euro/dollar investment. Hence, the machine could be motivated not to invest in stocks in dollars in case there are strong decisions on other investments. While one could always argue that each function can in principle operate on an input x sufficiently rich to contain the information on all the assets, it is of crucial importance to notice that the corresponding learning process can be considerably different. To stress this issue, it is worth mentioning that the decision on asset allocation typically depends on different information, so as the argument of the corresponding functions could conveniently be different and properly carved to the specific purpose, which likely makes the overall learning process more effective.

The satisfaction of constraints in environments of multiple tasks is also of crucial importance in classification tasks. Let us consider a simple text classification problem that involves articles in computer science. As usual, we simply assume that an opportune preprocessing of the documents $e \in E$ results in the corresponding vector $x = \pi(e) \in X$. In information retrieval, this vector-based representation is at the basis of most classic approaches. In particular, as already noticed, the TF-IDF representation is very popular. Suppose that the articles only belong to the four classes, numerical analysis, neural networks, machine learning, and artificial intelligence, that are represented by functions $f_{na}: X \to \{0, 1\}$, $f_{nn}: X \to \{0, 1\}$, $f_{ml}: X \to \{0, 1\}$, and $f_{ai}: X \to \{0, 1\}$, respectively. These functions are expected to express the classification of any document $x \in X$. The knowledge on the subjects involved in the classification can naturally be translated into the following constraints:

$$\forall x \in X \quad f_{na}(x) \wedge f_{nn}(x) \Rightarrow f_{ml}(x), \qquad f_{ml}(x) \Rightarrow f_{ai}(x). \tag{6}$$

Notice that the difference with respect to the previous constraints is that these expressions involve Boolean variables and propositional logic statements. Interestingly, as it will be shown in Chapter 6, we can systematically express logic constraints in the continuous setting. Now, it suffices to notice that any of the above functions can be associated with a corresponding real-valued functions. For instance, $f_{na} \leftrightarrow f_{na}$, where $f_{na}: X \to [0, 1] \subset \mathbb{R}$. Intuitively, $f_{na}(x) \simeq 1$ indicates the truth state, whereas $f_{na}(x) \simeq 0$ corresponds to the false state. We can promptly see that the translation

$$f_{na}(x) \wedge f_{nn}(x) \Rightarrow f_{ml}(x) \longleftrightarrow f_{na}(x)f_{nn}(x)(1 - f_{ml}(x)) = 0, \tag{7}$$

serves the purpose. The good news is that this allows us to use a unified computational setting, since while this is an inherently logic constraint it can be paired with perceptual constraints, like those of supervised learning.

We can go one step further by starting the exploration of learning environments that possess an inherently temporal nature. In the discussion carried out so far, the learning tasks are defined over feature-space domains $X \in \mathbb{R}^d$, where there is no presence of time. However, challenging fields like speech and video understanding are interwound with time, which suggests a further development of the idea of constraint. In order to give an insight on the power of this unified way of thinking, let us consider problem of estimating the optical flow in computer vision. It can be stated as follows: Given a video signal, $v: T \times R \to \mathbb{R}$ that maps $(t, z) \mapsto v(t, z)$ over a certain temporal basis $T \subset \mathbb{R}$ and retina $R \subset \mathbb{R}^2$ determines the optical flow $f(t, z) \in \mathbb{R}^2$ for any pixel $z \in R$. A classic solution is based on imposing the *brightness invariance* condition. It states that any pixel $z \in R$ doesn't change its brightness while the object it belongs to is moving. Hence, for any frame, defined by $t \in T$ and for a given video signal, f must satisfy the above constraint:

$$v_t + f \cdot \nabla v = 0. \tag{8}$$

While in computer vision it isn't customary to impose additional pointwise constraints, in principle, a training set of velocity values $f(t, x)$ could also be given to contribute to the computation of the optical flow. The classic solution is based on the idea of regarding $f(t, x)$ as a value that only respects brightness invariance, regardless of any hypothesized model. This is due to the fact that it is hard to collect a training set of values $f(t, x)$, though it could always be done. In this framework, the agent is asked to learn a function f that satisfies the constraints. However, the process of learning cannot be simply regarded as constraint satisfaction. This topic is fully covered in Chapter 6, but basic ideas on the importance of parsimonious solutions are also given in this chapter.

EXERCISES

1. [*17*]   Suppose a committee of machines is based on $f_1 : X \rightarrow [0, 1]$ and $f_2 : X \rightarrow [0, 1]$. Modify the penalty function (4) by using the same principle of coherent decision.

2. [*15*]   Consider a committee of machines for regression. Choose a penalty to keep the decisions of two machines close in the framework of regression.

3. [*16*]   Let us consider the constraint expressed by Eq. (2). One might find it more opportune to change it to

$$\forall x : \frac{x \cdot m^c}{\|x\| \|m^c\|} \leq \delta \Rightarrow (h(f(x)) \equiv c(x)).$$

Is this equivalent to Eq. (2)? If not, then discuss qualitatively the difference and which one is likely to be more adequate to model the idea of retina pattern occupancy that the constraint is expected to express.

4. [*13*]   Propose output functions h for the discussed computer science document classification problem.

1.2 Principles and practice

Machines' life in the environment is based on the information process that arises when following the protocols sketched in Section 1.1.4. But what drives their learning? How can they satisfy the environmental constraints? Even though they can, who can ensure that the current constraint satisfaction is supported by a deep understanding that allows them to succeed also with new experiences in the environment? Couldn't they simply learn by heart the concepts that are instead expected to be generalized to new examples? How are principles translated into practice? While the whole book is about addressing these questions, in this section, we provide insights that help driving the understanding of subsequent more technical issues.

1.2.1 The puzzling nature of induction

Needless to say, the development of both human and artificial learning skills is tightly interwound with principles of induction, which have been the subject of investigation for centuries. Regardless of its multifaceted manifestation, the process of induction presents a puzzling face. The sequence

$$0, 1, 1, 2, 3, 5, 8, 13, \ldots, \tag{1}$$

can be used as a nice example of induction. Given the above numbers, what comes next? The sequence is known to be recursively generated by the Fibonacci recurrence rule

$$\begin{cases} F_0 = 0, & F_1 = 1; \\ F_{n+2} = F_{n+1} + F_n & n \geq 0. \end{cases} \tag{2}$$

Apparently, this argument isn't the subject of discussion. However, a more accurate analysis reveals that the above induction is indeed arguable! How can we really trust the above recurrent model if we can only see the eight numbers of the partial sequence (1)? One could have in fact come into the conclusion that the underlying generative rule of the sequence is instead

$$\begin{cases} F'_0 = 0, & F'_1 = 1, & F'_2 = 2; \\ F'_{n+3} = F'_{n+1} + \lfloor (F'_{n+2} + F'_{n+1} + F'_n)/3 \rceil & n \geq 0, \end{cases} \tag{3}$$

where the $\lfloor x \rceil$ operator denotes the round-off of $x \in \mathbb{R}$. A prompt check allows us to see that also this rule offers a correct interpretation of the partial sequence (1). So, which one is right? Of course, the question is ill-posed! Both rules are fine, and the one which better interprets the sequence can only be discovered when we replace the dots with the actual numbers! If an oracle tells us that the 8th value of the sequence is 22, then only (3) is still perfectly coherent, whereas (2) fails, since it predicts $F_8 = 21$. Here are the values for $n \leq 20$ of the two sequences; as we can see, for $n > 7$ the two sequences keep diverging.

$$
\begin{aligned}
n &= 0\ 1\ 2\ 3\ 4\ 5\ 6\ 7\ \ 8\ \ 9\ \ 10\ 11\ \ 12\ \ 13\ \ 14\ \ 15\ \ 16\ \ \ 17\ \ \ 18\ \ \ 19\ \ \ 20 \\
F_n &= 0\ 1\ 1\ 2\ 3\ 5\ 8\ 13\ 21\ 34\ 55\ 89\ 144\ 233\ 377\ 610\ 987\ 1597\ 2584\ 4181\ 6765 \\
F_n' &= 0\ 1\ 1\ 2\ 3\ 5\ 8\ 13\ 22\ 36\ 60\ 99\ 164\ 272\ 450\ 745\ 1234\ 2044\ 3385\ 5606\ 9284
\end{aligned}
$$

Whenever we access a finite portion of the sequence, and we possess no additional information, we don't really know which numbers will come next. It is just like tossing a fair coin or a fair die; it seems that the prediction is doomed to be a true guess! However, this extreme view on randomness might also be just the outcome of the lack of additional underlying knowledge on the generating process. For example, there is nothing strange with the motion of coins and dies, since they obey mechanical laws. Once the initial conditions are given, the corresponding flight is doomed to follow a precise trajectory that can be predicted very well also by neglecting the air friction. As coins or dies reach the floor, their bouncing obeys elastic or inelastic models of materials, which allow us to predict how the trajectory evolves. Hence, in principle we can definitely bias coins and dies by exploiting physical laws! A motion that is apparently truly random can, on the other hand, be given an explanation rooted in mechanical models of tossing. While in general, we don't achieve a perfect prediction, we can definitely bias the probability of these events. Since even the supposed randomness of tossing coins and dies is arguable, the same principle can be applied to any induction problem, including that of predicting the numbers which complete the partial sequence (1). Hence, the arguments against the Fibonacci recurrent model of partial sequence (1) make sense only provided that we are doomed to live in the dark, without any additional knowledge on the underlying generative model. However, we might think of methods for ranking different models of the given data, since not all of them are equally likely.

A classic way of dealing with this concurrent explanations of the given data is that of using the parsimony principle (lex parsimoniae), according to which *among competing hypotheses, the "simplest one" is the best*. We can promptly see that Eq. (2) is simpler than Eq. (3): There are less initializations, less variables involved in the recurrent step, and there is nether the need of performing division nor of rounding off. Hence, the parsimony principle suggests that the classic interpretation of Fibonacci numbers is preferable in terms of simplicity. As shown in Exercise 2, we can also easily see that

$$
F_n = \frac{1}{\sqrt{5}} \left(\phi^n - \widehat{\phi}^n \right), \tag{4}
$$

where $\phi = (1 + \sqrt{5})/2$ and $\widehat{\phi} := -1/\phi = (1 - \sqrt{5})/2$. While the previous rule was based on a recursive interpretation, the above equation offers a direct interpretation on the basis of a global computational scheme that, like in the previous case, is extremely compact. Interestingly, we can see that the same (Fibonacci) generative model can be given different linguistic description by either local [see Eq. (2)] or global [see Eq. (4)] equations. Which one is simpler? The deep difference between local and global models results in a remarkably different structure of the corresponding equations, so that comparing their complexity is not as straightforward as in the previous case. This remark indicates that we need some good formalisms for a precise definition of model complexity.

But induction is really puzzling! If we replace the original sequence (1) with the partial sequence $0, 1, 1, 2, 3, 5, \ldots$ then yet another rule that correctly interprets these data is:

$$\begin{cases} F_0'' = 0, \quad F_1'' = 1; \\ F_{n+2}'' = F_{n+1}'' + \lfloor F_{n+1}'' F_n' \rfloor \quad n \geq 0, \end{cases} \tag{5}$$

However, for $n = 6$, we have $F_6'' = 5 + \lfloor \sqrt{5 \times 3} \rfloor = 9$, which indicates that the model defined by Eq. (5) agrees with Eqs. (2) and (3) only provided that $n < 6$. The application of the parsimony principle in this case collocates the complexity of this model[2] in between the models defined by Eqs. (2) and (3).

Now we jump to an apparently unrelated topic that, however, fully shares the puzzling nature of induction discussed for the sequence (1). We explore an induction problem that very much resembles the spirit of many IQ tests. Consider the following sequence of stylized clocks:

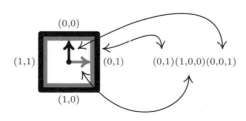

As in many related IQ tests, one is asked to guess the next clock of the sequence — something that is clearly similar to predicting which number comes next in the sequence (1). This time, however, things are more involved. Clock drawings are not numbers! Making inferences in this case requires us to capture some distinctive compositional elements so as to provide a simple, yet effective symbolic description. This requires us to gain a considerable level of abstraction, since we need to go beyond pixel-based representations. The inspection of Eq. (6) leads us to regard stylized clocks as entities composed of the frame and of the hands. Furthermore, the frame is simply composed of two "annuluses" that can be recognized from their black/gray color. Notice that conquering this abstraction capability is one of the most challenging problems in computer vision, where the perception of structured objects, just like for stylized clocks, requires converting the pixel-based information on the retina to a compact symbolic description.

Formally, stylized clocks can be elegantly described by a simple Boolean string which defines the frame and the hands of the clock. The frame is defined by the pair of bits (f_1, f_2) and can be $\{(0, 0), (0, 1), (1, 0), (1, 1)\}$. For instance, $(0, 1)$ defines the frame of the second clock in Eq. (6), since it indicates that the external "annulus" of the frame is gray, while the internal one is black.

---

2   Other interesting induction rules for the sequence (1) are proposed in Exercise 1.

In this stylized clock world, both the gray and black hands can only be in the four cardinal positions, so that any hand configuration can be represented by the triple (c, p_1, p_2), where the first bit is used for coding the color, while the other two bits are used for coding the four possible positions. Overall, any possible clock picture is defined by a byte, since two bits are for the frame and six for the position of the two hands. The coding construction is shown in the figure for another clock of the collection that isn't in the sequence. It is tacitly assumed that one can distinguish the first from the second hand. This can univocally[3] be done whenever the two hands are not in opposite direction. In that case, the same clock configuration may correspond to different codes that correspond with swapping the first with the second triple in the hand position. Like for the partial sequence (1), we want to predict which clock comes next. However, unlike in Eq. (1), in which any integer can be expected to come, here we know in advance the finite universe of possible clock configurations. Now let us get into Bob's shoes, the expert who prepared such a test, so as we can understand the rationale behind the partial sequence (1). The construction was driven by a distinctive property to be exposed for testing induction capabilities. To make things interesting, such a property was conveniently hidden by confusing features that, at least for an intentional point of view, play no role in the induction.

The key idea is that of exposing an induction rule that is connected with the hand movement: The hands are rigidly connected and perform four clockwise movements until they restore the first configuration. Then the same sequence continues counterclockwise, and so on and so forth. Interestingly, this happens regardless of the frame, which is just for confusing inductions. While the hands reflect a driving principle, the annulus of the frame is just drawn randomly. Hence Bob knows that there are four different answers composed of the different annuli in the frame, each one equally acceptable, where only the position of the hands is important. Hence a correct answer is

 . . . . (7)

Now, the IQ test was proposed to Jack, who is expected to capture the inductive rule that Bob injected into the test. Interestingly, Jack didn't focus attention on the clock hands, since he couldn't find anything regular in the given six clocks. On the other hand, he looked at the clock frames trying to discover what they have in common. He found that they deliver less information than the hands and, therefore, he conjectured that there is more chance to discover regularities. He immediately noticed that the frames can be represented by the Boolean code (f_1, f_2). He is familiar with Boolean functions, and while looking at the sequence $(1, 1), (0, 1), (1, 0), (1, 0), (1, 0), (1, 0)$ he realized that there is in fact a simple rule that generates the above six Boolean pairs $((f_1^k, f_2^k))_{k \in \{1, \dots 6\}}$ of the sequence. While Bob injected a rule on the hand positions

---

[3] For example, we can decide the first hand on the basis of the counterclockwise assumption.

based on the previous watch, Jack notices that

$$(f_1^{k+1}, f_2^{k+1}) = (f_1^k \oplus f_2^k, f_1^k \wedge f_2^k), \quad k = 1, \ldots, 5,$$

where $\oplus$ is the XOR function. Since he likes the stylized clocks, he went beyond the above formal statement of the induction rule. He though of a story where these stylized clocks are in fact depicting real clocks on the market. He likes to regard the step k as a temporal indicator synchronized with the acquisition of the liking of the customers. At the beginning, $k = 1$, the frame was pretty big — basically a double black "annulus." The customers didn't like it very much, so the designers were forced to change it. They also felt that the customers would have appreciated more than a new frame with the final appropriate pair of colors, a gradual process based on more updating steps according to the following rule: *The external annulus is black only in case the two annuli were different at the previous step, while the internal annulus is black only in case both annuli were black.* Finally, Jack made his prediction, which is independent of the hands position; in his own view they are randomly drawn, and there is nothing to induce. Hence, he ended up with the decision:

(8)

Bob, the expert, corrected the test, and rejected Jack's answer! Needless to say, Jack was very disappointed, since he couldn't understand what went wrong. As he presents his own interpretation, Bob simply confirmed his decision. In the end, Jack didn't really discover what Bob had in mind, and the negative result impacts on Jack's IQ. Moral of the story: *According to this IQ test, somebody can claim that Jack is unintelligent and/or sluggish!* This story clearly shows some fallacies and pitfalls of induction, and it also sheds light on the ill-position of some IQ tests. Bob and Jack both possess a coherent interpretation of the sequence of stylized clocks. Like in the sequence (1), multiple interpretations are indeed possible! An induction scheme, to be well-posed and technically sound, doesn't only need to discover coherent interpretations, but must also be paired with a ranking method to select which one is preferable. Bob and Jack interpreted different regularities while regarding others as random. Hence, one might wonder which regularity is preferable — if any — and the extent to which what was neglected is truly random. If we follow the parsimony principle, both regularities appear to be quite simple. In particular, Bob's induction ranges in a larger universe of clock configurations, but it is still quite simple.[4] In a sense, since the chosen universe of induction is larger than in Jack's induction, it is clear that Bob regularity is more difficult to capture. Jack decided to focus on the simplest one, which isn't necessarily the sign of lacking intelligence. As far as randomness is concerned, the issue is more tricky. To Bob's eyes, the frames appear random, whereas to Jack's eyes the hands of the clock are not driven by any regularity. Their own different interpretations of randomness might be regarded just as

---

[4] See Exercise 6 for a formal description of Bob's underlying rule.

the outcome of lack of knowledge on the generation process. Of course, this example underlines a very general critical issue in induction, which becomes more and more important as the cognitive task grows in complexity.

These examples on the puzzling nature of induction pose questions on the true nature of randomness. We noticed that tossing coins and dies, in the end, isn't a random process as we might have originally believed. Now, let us stress this issue and consider the following sequence composed of integer numbers in $[0..7]$

$$(X_k)_{k \in \{0,...7\}} = 1, 4, 3, 6, 5, 0, 7, 2.$$

Is there any regularity to capture? Unlike previous examples, here we provide all the elements of the finite sequence, yet there is no apparent generative rule behind. To what extent can this be supported? The fallacies and pitfalls in the IQ test suggest that we must be very careful about statement on the lack of regularities, since it might be the case that we are just ignoring them. The above sequence can in fact be generated by

$$X_{k+1} = (aX_k + c) \mod m \tag{9}$$

where $X_0 = 1$, $m = 8$, $c = 7$ and $a = 5$. Similar sequences can be generated with any $m \in \mathbb{N}$ (see Section 1.5 for additional details). This is in fact the basis of classic algorithms that generate pseudorandom numbers. It's better not to drop the attribute "pseudo!" What is more regular than a computer program that returns numbers? This is a nice example to underline the controversial notion of randomness. When making this point extreme, one could be led to believe that randomness is just a simple manifestation of ignorance. While this is intriguing, it's hard to fully subscribe to this point, since it is likely to quickly draw us outside the perimeter of science and, surely, of machine learning.

EXERCISES

1. [*M15*] Given sequence (1), provide another interpretation in addition to those given in the text.

2. [*M16*] Compute explicitly F_n on the basis of the recurrent definition given by Eq. (2).

3. [*M15*] For any $n \geq 1$ prove that

$$\begin{pmatrix} 1 & 1 \\ 1 & 0 \end{pmatrix}^n = \begin{pmatrix} F_{n+1} & F_n \\ F_n & F_{n-1} \end{pmatrix}$$

4. [*M16*] (*Cassini's identity.*) Prove that, if we set $F_0 = 0$ and $F_1 = 1$ for $n > 1$ sequence (1), is interpreted by the generative scheme given by $F_{n+1} = (F_n^2 + (-1)^n)/F_{n-1}$. Compare the sequence generated by this recurrent equation with Fibonacci numbers.

5. [*30*] Consider the clock induction test. Can you find another induction rule for the sequence in Eq. (6)?

6. [*13*] Provide a formal description of Bob's stylized clock IQ intelligent test by resembling what is proposed to describe Jack's interpretation.

7. [*M30*] Consider the congruential random generation scheme defined by recurrent Eq. (9). Can you suggest an initialization which makes the numbers more predictable?

1.2.2 **Learning principles**

So far we have mostly discussed the learning environment that characterizes the agent's life and the nature of induction. While the statement of the corresponding constraints suggests schemes for their satisfaction, we haven't discussed the mechanisms that drive the learning process. Because of the puzzling nature of induction, it has already been pointed out that those processes cannot simply be the outcome of constraint satisfaction. The symbolic/subsymbolic dichotomy also suggests the adoption of different learning principles. In some learning tasks, we can play with symbols (see, e.g., *car evaluation task*), whereas in others the processing has mostly a subsymbolic nature. When working with symbols, one can construct learning schemes that are based on searching in the hypothesis space, whereas in case of analog data, learning is typically regarded as an optimization problem.

A classic approach to the multiple satisfaction of environmental constraints is to pose learning processes into the framework of statistics. The collection of samples (training data) opens the door to the construction of inferential processes that take place at both supervised and unsupervised level. As it will be seen in Section 2.1, statistical machine learning mostly relies on the Bayesian approach that is based on the tacit assumption of believing in a true state of nature.

Though rooted on a statistical concept, the *maximum entropy principle* nicely responds to the general view of modeling the environment as a collection of constraints to be satisfied. Amongst all configurations which meet the constraints the one with maximum entropy is selected. In this framework, unsupervised data can nicely be modeled by the maximization of the mutual information between the input and the encoded representation that we want to extract.

Unlike most statistical approaches to learning, the adoption the *parsimony principle* doesn't require making specific assumptions on the probability data distributions. It is inspired by the *Occam razor principle* and relies on the idea that whenever the given set of environmental constraints are satisfied by a set of different hypotheses, the *simplest* one is selected. As it will be shown in Section 2.4, one needs to make clear the meaning of simplicity, which might take different forms in different contexts.

1.2.3 **The role of time in learning processes**

Most approaches to machine learning consist of devising methods for performing inference on the basis of a finite collection of samples of the concept to be learned. No matter what solution is adopted, they share the idea that an intelligent agent is expected to acquire the concept based on a batch of data that is given all at once. This is a very natural context for statistical interpretations, which offers plenty of methods for formulating appropriate predictions on the basis of a sample from the data probability distribution. However, in most challenging learning tasks, this idea, which is supported by well-established statistical theories, is likely to fail. In order to put things in context, it might help to parallel our learning tasks with those that humans regularly face. Scientists in machine learning have the ambition of providing the basic instruments to replicate visual and language human skills in machines. The rising interest in these challenges suggests us to parallel the batch mode scheme in humans. Could we expect a child to learn to see in a batch mode from a visual data collection that contains all his life? Most approaches to computer vision that are based on machine learning are in fact based on this strong assumption, where, on top of that, since the processing takes place at image level, we essentially pretend that the child learns to see from a collection of shuffled video frames! Human acquisition of language seems to be trapped in similar problems, which suggest that batch mode might not be the right direction for facing challenging human-like cognitive skills.

One might erroneously conjecture that the formulation of online learning mode schemes represents an answer to the inherent limitation of batch mode. Unfortunately, as it will be seen in the remainder of the book, most classic formulations of online learning are approximations of batch mode, so as they don't necessarily capture the natural temporal structure of the learning task. Time plays a crucial role in cognition. One might believe that this is restricted to human life, but more careful analyses lead us to conclude that the temporal dimension plays a crucial role in the well-positioning of most challenging cognitive tasks, regardless of whether they are faced by humans or machines. It looks like that, while ignoring the crucial role of time, the positioning of some of current computer vision tasks is more difficult that what nature has prepared for humans! Time provides an ordering to the visual frames, which dramatically reduces the complexity of any inferential process. As a consequence, ignoring time makes sense only for restricted computer vision tasks.

1.2.4 Focus of attention

The crucial role of time in learning isn't restricted to the importance of the structure that is imposed to perceptual data. The life of an intelligent agent in human-like environments does require developing clever mechanisms for optimizing the complexity of the learning algorithms. In order to grasp the importance of saving time in the agent's life, let us consider the typical situation that happens when a teenager nerd enters a research lab where a computer science researcher is describing his last results to be included in a paper that he's going to present to the next top level conference. At the beginning, the teenager is strongly attracted by a few computer science keywords, and he does his best to understand the subject of the discussion, which he finds to be very attractive. However, it takes only a few minutes for him to realize that he cannot really get the essence of the talk, and then he leaves the lab and goes back to his ordinary social interactions. Children ordinarily act in a similar way when posed with advanced topics that they cannot grasp: They simply escape complex information sources and focus attention on what they can concretely understand at a certain stage of cognitive development.

When children are not exposed to teaching activities, they tend to focus attention on pleasant tasks, which are the easiest for them to face. While they can choose to be involved in more complex tasks, they abandon them in case the results are very hard to obtain. In a sense, they follow an *easy-first* strategy that doesn't only make them happy; it also leads to a smooth acquisition of learning skills and, consequently, to a very efficient use of time. Machines can just do the same when acting in an environment without any human supervision.

The systematic focus on easy tasks leads to acquiring concepts to be used later on to grasp more advanced concepts by a sort of compositional mechanism. Instead of posing all concepts at the same level, an agent which uses the easy-first strategy ignores complexity at the beginning, thus saving precious time for focusing on what it can concretely grasp! This clever filtering on what is complex at a certain stage of the agent evolution is one of the most important secrets of learning, something which prevents from getting stuck into inescapable *cul de sac*. This seems to reflect studies in developmental psychology, which clearly indicate the importance of a gradual acquisition of learning skills. Interestingly, the structural organization in developmental stages is likely to be interwound with the pressure for learning that also comes from social interactions (see Section 1.5 for additional comments). It's quite easy to realize that most advanced human skills are not conquered by following an easy-first strategy only. While the teenager nerd escapes a top level scientific talk, he strongly benefits from other social interactions. Unlike other animals, humans crucially experience the benefit of education. Interestingly, easy-first driven curiosity is nicely integrated by appropriate teaching mechanisms that mostly stimulate novelty and allow humans the acquisition of highly structured concepts. Clearly, *teaching policies* play a fundamental role in the success of learning. In a sense, the focus of attention and teaching policies shouldn't be regarded as independent processes. For example, in most machine learning approaches, the act of teaching is reduced to providing supervised pairs of the concept to be learned. On the opposite, the pairing of learning and teaching, along with their mutual relationships, does require the formulation of joint strategies to meet a common goal.

EXERCISES

1. [*12*] Write one sentence to distinguish between statistical-based and parsimony-based learning algorithms.

2. [*15*] Write one sentence to state the reason why online learning algorithms don't fully reflect the temporal dimension of learning.

3. [*20*] Consider the easy-first policy in focus of attention. Can you see any drawbacks of this approach?

4. [*33*] Can you provide arguments to explain the emerge of developmental learning? Why shouldn't an agent just learn gradually instead of passing through different stages with remarkable qualitative difference?

5. [*29*] Discuss the relations between the easy-first policy and different teaching schemes. The active learning with teacher support improves learning dramatically. What can stimulate active learning?

6. [*20*] (*Supervisor paradox.*) Discuss the following claim: *Since an agent, which performs supervised learning in a given task, at most achieves an approximation of human supervisor decisions, the agent will never be able to overcome human skills.* This can be concisely stated by *the student cannot overcome the teacher*. Does this hold in the machine-supervisor relationship?

1.3 Machine learning experiments

While the discussions on computational aspects of learning carried out so far have sketched some foundational topics, now we want to go straight to practical issues. This is in fact what makes machine learning so attractive in different applicative contexts. While some foundational topics gave rise to the first wave of interest at the end of the 1980s, nowadays resurgence of interest in machine learning seems to be mostly driven by the systematic development of fantastic software tools that allow people to quickly develop apps based on the exploitation of machine learning algorithms. They allow also people with scarce background in the discipline to concretely attack the problem at hand, without bothering with sophisticated algorithmic issues. When dressing software engineering cloths, one would very much like to write software while seated on a solid machine learning layer. This is in fact what has been happening! A growing number of software developers can use developmental frameworks that dramatically facilitate their task. It looks like an "instant déjà vu": Low level assembly languages were replaced with high level languages that have progressively been framed into object oriented programming, so that we needn't care about complex computer resources anymore. Couldn't the same replacement process take place for machine learning? Nowadays, explosion of interest in software development environments suggests that a somewhat similar trend of replacement has already been launched! Why should one continue to care about technical details of learning algorithms when they are already nicely incorporated into software developmental systems? Authors of this book believe that while similar environments have been playing a crucial role in the massive development of specific apps, the overall progress in the field cannot rely on those environments. On the opposite, we still require gaining an in-depth understanding of foundational topics that might enable a true breakthrough also in fields like computer vision and natural language understanding that are so deeply connected with machine learning. Hence, while you can significantly benefit from those high level software environments, keep working with "assembly languages!"

Needless to say, hands-on experience based on nowadays machine learning developmental environments is very important. In addition to enabling us to quickly develop applications, the easy setup of experiments can likely inspire new ideas to push foundational topics. The purpose of the practical guidelines that will be described in the next sections is that of stimulating the interest in the experimental phase, and to emphasize its importance in the acquisition of the discipline. However, a warning is in order about the details that you can expect to find to set up your experiments and construct your applications: The following brief presentation is not expected to fully cover your technical questions. It was done purposely, while thinking about this book. It's in fact our opinion that there's nothing better than web resources when trying to serve specific application purposes. It's a different story if you are looking for an in-depth understanding of the field, which is likely to be the only road to follow to quickly grasp the novelties and face competitive applications.

1.3.1 **Measuring the success of experiments**

Before handling experiments, one should be very well aware of evaluation schemes that drive the actual development and allow us to end up with the conclusion that things work properly. This is a topic whose importance is often neglected, since the scientific community typically agrees on performing the experimental evaluation on appropriate benchmarks, which have received a sort of "blessing." Because of their quality and diffusion, some of them become a "de facto standard" and strongly motivate scientists to exhibit their results. While the assessment driven by benchmarks has solid statistical roots, one shouldn't give up exploring other evaluation schemes, like crowdsourcing, that better reflect some human skills, like vision and speech understanding (see Chapter 7). In this section, we focus on benchmarks, which typically collect and organize data to test supervised learning algorithms. The success of the recognition process carried out by the classifier needs to be properly evaluated on the test set. As we think of such an assessment, we promptly realize that the patterns can be partitioned in a natural way. When focusing on a certain class, we can distinguish the *true positive* P_t, the *false positive* P_f, the *true negative* N_t, and the *false negative* N_f. True positive patterns are those which are correctly classified as members of the certain class, whereas false positives are those patterns of the same class that were not correctly recognized. In the notations P_t, P_f, N_t, N_f, the capital letters P and N refer to the predictions of the classifier, while the indices t and f indicate correct and wrong decisions, respectively." The *accuracy*, defined as

$$a = \frac{|P_t| + |N_t|}{|P_t| + |N_t| + |P_f| + |N_f|} \tag{1}$$

turns out to be a natural way of assessing the performance of the classifier. It is in fact the ratio of the number of the successfully recognized patterns to the number of decisions. Clearly, $0 \leq a \leq 1$ and $a = 1$ is achieved only in case in which there are no errors, that is, there are neither false positives ($|P_f| = 0$) nor false negatives ($|N_f| = 0$). Of course, the above definition holds also in case of multiclassification (see Exercise 8 for an interesting discussion on the relations with the accuracy of the single classes). The accuracy is related to the value of the loss function (2), but the connection is not trivial (see Exercise 9).

The accuracy offers an overall ranking of the classifier. However, one might be interested in the precision on positive examples. Hence suppose we want to recognize patterns of a certain class. We might disregard the behavior of the classifier with respect to the rest of the classes and focus on the single class. The ratio

$$p = \frac{|P_t|}{|P_t| + |P_f|} \tag{2}$$

is referred to as the *precision* of the classifier. We have $0 \leq p \leq 1$, and $p = 1$ only in case there is no error on positive examples. When comparing with the accuracy, we can see that the precision can reach 100% more easily than accuracy, since the precision measure does not care about the outcome on negative examples. Hence, a

classifier can easily achieve high precision by following very strictly decision policies, which state the membership only in cases of very low degree of uncertainty. In
so doing, there might be a remarkable number of false negatives N_f, which might be
acceptable in some applications. A natural way to recover the limit inherent with the
given definition of precision is to introduce the *recall* measure

$$r = \frac{|P_t|}{|P_t| + |N_f|}.$$

(3)

Again $0 \leq r \leq 1$, but this time $r = 1$ arises only in case $|N_f| = 0$. Hence, a classifier
achieves high recall only if it collects few false negatives. The value of recall becomes
very important in case of many classes and in case of patterns that do not belong to
those under attention.

Since precision and recall express two different qualities of classifiers, we are often interested in expressing an overall measure to summarize their value. A common
choice is to consider

$$F_1 = 2\frac{pr}{p+r}.$$

(4)

Notice that the F_1 measure can be thought as twice the equivalent value of a pair of
parallel resistors with values p and r. Alternatively, one can think of F_1 as the ratio
between the geometric and the arithmetic means. We can easy see that $\min\{p,r\} \leq
F_1 \leq \max\{p,r\}$ (see Exercise 3).

1.3.2 The MNIST benchmark

Once the chars have been segmented, we can start thinking about their recognition. In
the subsequent discussion we consider the popular MNIST handwritten recognition
dataset.[5] It is composed of gray-level images sampled with 28×28 resolution. The
corresponding patterns are real vectors of dimension 784 obtained by scanning the
rows of the images. Of course, this is not necessarily the ideal resolution to be used for
performing experiments. As it will be clear in the rest of the book, there are a number
of variables that affect this design choice. It might be the case that a de-sampling of
the picture doesn't compromise the class discrimination, while the reduction of the
dimension of the input space improves the efficiency of learning and inference. Put
it simply, it means that the chars, as they appear at the given resolution, are likely
to contain a lot of information that might not be necessary for their discrimination.
For example, in its simplicity, handwritten chars somehow convey also the writer's
identity, which is not the purpose of their recognition. This typically happens also in
speech recognition, where the voice signal can also allow us to identify the speaker.

Formally, we are given a segmented pattern $e \in E$ and we want to find an appropriate internal representation $x = \pi(e)$ that is effective for recognition purposes.

---

[5] See also Exercise 1.1.4–4.

Because of the previous remarks, the preprocessing carried out by function π should lead to representations for which $|E| > |X|$. In order to reduce the input dimensionality, we can perform a de-sampling of the pictures. We can use *max-pooling*, which partitions the input image into a set of nonoverlapping boxes and, for each such box, outputs the maximum value. Alternatively, the de-sampling can be obtained by *average pooling* where, for each box, the average value is returned. The de-sampling consists of setting the gray level of the generated boxes to the value returned by the pooling process. Here's an example of the maximum and average pooling that can be regarded as realizations of function π when dealing with an instance of class "2" char:

The original MNIST char with resolution 28×28 is transformed by π with average and maximum pooling at resolutions[6] $14 \times 14, 7 \times 7$, and 5×5. The human perceptual effect of the two different schemes of pooling is very clear: At low resolutions, the averaging results in a dramatic blurring, while filling with the maximum yields a sort of darkness effect. We can see the effect of the resolution with both pooling schemes; in some cases, the discrimination with other classes can become critical. For example, in the above instances, the human perception of class "2" and class "3" after having performed max pooling with resolution 5×5 yields patterns that look pretty similar:

$$\pi_{mp}^{5\times5}\left(\mathbf{2} \right) = \blacksquare \simeq \blacksquare = \pi_{mp}^{5\times5}\left(\mathbf{3} \right).$$

Similar qualitative remarks can provide insights on appropriate choice of the preprocessing. The adoption of average pooling yields

$$\pi_{ap}^{5\times5}\left(\mathbf{2} \right) = \blacksquare \ ; \qquad \pi_{ap}^{5\times5}\left(\mathbf{3} \right) = \blacksquare \ .$$

which seems to be slightly better than max pooling in terms of discrimination. However, decisions based on single instances can bias significantly the performance, and therefore the appropriate design of pattern classifiers requires more solid and systematic experimental analysis. This kind of design choice needs foundations on inductive inference that are covered throughout the book.

---

[6] Notice that, while the resolutions 14×14 and 7×7 are obtained by creating perfect boxes, the resolution 5×5 necessarily requires us to use an approximation in the pooling process.

1.3.3 Design of a machine learning experiment

Now, let's see how to design a machine learning experiment for handwritten char recognition. As already pointed out, the diffusion of software development environments for machine learning has dramatically facilitated the setup of standard machine learning experiments. In what follows we sketch the main step that one needs to take in order to implement a two layer neural network classifier, similar to the one in Fig. 1.3, with 100 hidden units. In this example we don't use any kind of additional preprocessing, so that the input layer consists of $28 \times 28 = 784$ units.

First of all we need to download and read the handwritten chars from Yann Le-Cun's website (http://yann.lecun.com/exdb/mnist/). Then we can, using our preferred software, store the training set digits in a tensor and convert the corresponding labels into one-hot encoding. We then choose the nonlinearity of the single neuron; in this case, we do not choose sigmoidal function as indicated in Eq. 1.1.3–(1), but on the rectifier or "relu" function defined by $r(a) := a[a > 0]$. For the output layer, we choose as the smx activation function:

$$(\text{smx}(a_1, \ldots, a_n))_i = \frac{e^{a_i}}{\sum_{j=1}^{n} \exp(a_j)} \tag{1}$$

This way of processing the activations of the output layer forces an almost one-hot encoding since, as it will be discussed in more details in Section 3.2, the smx function is nearly 1 on the outputs that have got the highest activation. This behavior is particularly suitable when we want to train our network for classification through one-hot encoded targets.

As briefly outlined in Section 1.1.4, the last missing ingredient to define the learning process is the loss function that measures the quality of the classification. Here we choose the cross-entropy function, which for any pattern x_κ, $\kappa = 1, \ldots, \ell$ of the training set enforces the matching of the output $F(w, x_\kappa)$ (here w are the parameters of the network) and the corresponding target y_κ by

$$V(x_\kappa, y_\kappa, w) = - \sum_{j=1}^{n} y_{\kappa j} \log F_j(w, x_\kappa). \tag{2}$$

Notice that this loss makes sense if we can ensure that whenever an output is found that is close to 1 then the others must be nearly close to 0, otherwise if there exist more values of $f(x_\kappa)$ close to 1, but disagreeing with the corresponding supervision value y_κ (close to 0), this function wouldn't detect the discrepancy! In our experiment the previous property required for f is in fact guaranteed by the smx function, which imposes the probabilistic normalization. In case there is no output normalization, related definitions of the loss functions can be used (see Section 2.1.1).

Lastly we need to choose an optimization method to minimize the accumulated error on the training set as a function of the parameters w of the network

$$E(w) := \sum_{\kappa=1}^{\ell} V(x_\kappa, y_\kappa, w). \tag{3}$$

In the next section, we give a preliminary discussion on experimental issues and their connection with design choices.

1.3.4 Test and experimental remarks

One can easily experiment with different neural network architectures, thus following a sort of *trial and error* approach to discover an ideal solution. There're a number of design choices that can remarkably affect the performance. Here we discuss some of those choices, but we don't pretend to provide an exhaustive treatment of this critical issue. In order to set up conscious design choices, one can surely benefit from an in-depth understanding of these topics that can be gained in the next chapters of the book. The neural network defined in the previous section is the outcome of well-known related experiments, but one can challenge this choice and look for different numbers of hidden layers and different numbers of neurons per hidden layer. It's worth mentioning that, just like in Fig. 1.3, the pyramidal networks (number of units decreasing while going towards the output) are typical chosen, since they naturally carry out the compression of information that one expects to see when going from the input to the output, throughout the hidden layers.

If we stay with the choice of the "relu" neural activation function, then the important design choices made in the previous section include the optimization algorithm, the learning rate (degree of updating in gradient descent), and the batch size. The formal optimization should take place on the overall error function 1.1.3–(3). The classic batch-mode algorithm minimizes E, which accumulates the loss the batch of the overall training set. However, one can use online learning, or something in between online and batch mode. In so doing, for each iteration, a certain number of patterns (100 patterns in the code) of the training set are drawn at random and are used for composing a mean-error function, which restricts the assessment of the error to a *mini-batch*. Like with the online mode, we don't implement the correct gradient descent, since the function that we optimize does change at every iteration.

In the experimental results that are reported below, we use a more sophisticated numerical algorithm, referred to as "Adam," which adapts the learning rate automatically with parameter 0.003 [for further details on the meaning of this parameter, we refer to D.P. Kingma and J.L. Ba, "Adam: A Method for Stochastic Optimization," https://arxiv.org/pdf/1412.6980.pdf]. We run experiments using both mini-batch (as previously described) and batch modes. An important experimental issue concerns the way we select the training data from the overall MNIST collection.

From the database of 55,000 labeled examples, we need to isolate those to be used for carrying out the learning process as previously discussed. The meaning of induction gained in Section 1.2.1 clearly indicates that once we have learned the appropriate weights of the neural network, we need to see how it works on different handwritten digits. Hence, along with the training set we need to define a *test set* where to evaluate the performance.

Fig. 1.6 shows experimental results connected with the choice of a training set with 5,000 examples, once we adopt batch mode. There are two plots which report

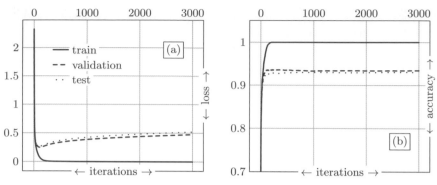

FIGURE 1.6

Loss (a) and accuracy (b) versus iterations for the same experiment. In order to put into evidence the overfitting, training has been done in batch mode on a set of only 5000 examples. We can see how overfitting is noticeable in the loss curves where the error on the test set almost doubles with respect to the point of minimum, whereas this phenomenon is much less evident on the accuracy curves.

the results on the overall error E [plot (a)] and on the accuracy [plot (b)]. Let's focus on the curves identified with the label "train." We experience a nearly monotonic decrease of the error, with the correspondent increase of accuracy. This is exactly what we expect from the learning agent, which stresses the optimization process while burning CPU time. Now the error plot on the test set presents a somewhat surprising result, which suggests refraining from continuing the optimization beyond a certain limit. Curiously, the error on the test set presents an abrupt decrease until it reaches a minimum, from which it increases while continuing the optimization on the training set. Put in simple words, "the more we fit the training set the worst it gets!" It's one of the many intriguing faces of induction, that is referred to as *overfitting* — a phenomenon that one should carefully avoid. What is the reason of overfitting? We can promptly come up with the answer when thinking of the optimization process, that is tuned on the training set. If it is not statistically representative of the overall probability distribution, it's quite obvious that the perfect learning of the training set doesn't necessarily result in the best solution that we can get on real-world data — performance on the test set. Basically, if we "learn by heart" a few examples of the training set, we will likely discover a very biased solution, and therefore it might be the case that it's not a good strategy to stress learning. On the other hand, as indicated in Fig. 1.6, an *early stopping* allows us to achieve better performance with respect to using an asymptotic solution gained with waste of computational resources. When should we stop learning? Ideally, one would like to stop when the performance on the test set starts increasing. Hence, one could inspect the first plot of Fig. 1.6 and stop the algorithm at the minimum value of the error in test set, which is about 0.25.

However, there're at least a couple of problems with this solution. First, if we also want to avoid wasting computational resources, detecting the minimum of the

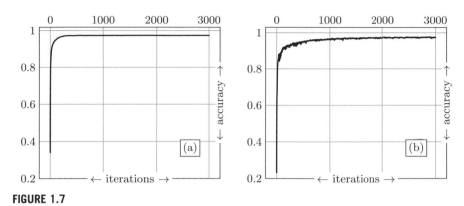

FIGURE 1.7

Diagram of the accuracy on the test set gained by the neural network over iterations. In (a) we have used batch mode on the whole training set dataset with a final accuracy (after 3000 iteration) of 0.9737. In (b) we used mini-batches of 100 examples; the final accuracy reached is 0.9736.

error is not a trivial issue. If we check the minimum just by controlling whether the error on the test increases at the next step, it's quite obvious that we might simply end up with strongly suboptimal solutions — especially if the error curve is pretty noisy, an issue that will be better understood later. Second, and more important, in no way we can use the test set during in the learning algorithm; the implementation of early stopping is in fact using information that is not available. This naturally explains the reason why benchmark data are typically split into training, validation, and test sets. The *validation set* in this case is an accessible data collection, used for training, with the specific purpose of implementing the early stopping criterion. Basically, the validation set is just an accessible test set that hasn't been used for updating the weights, while it serves the purpose of detecting the overfitting process. As we can see from Fig. 1.6, the plot of the error on the validation set very much resembles that of the test set. Now we are back to discussing the solid procedures to perform the stop. In case we don't care about computational resources, we can easily determine the minimum of the error on the validation by inspecting all the plot. However, if computational cost is an issue, then the solution to be adopted for stopping the algorithm can hardly be regarded as general. One surely fails detecting the minimum if the plot is multimodal. But even though we want a sort of unimodal behavior, in some cases the curve is pretty noisy [see Fig. 1.7–(b) concerning the accuracy, which clearly yields something similar for the error]. The noise might be enough to lead to a brittle behavior in the detection of the minimum for stopping the algorithm.

Unlike the other plots, the plot of the accuracy in Fig. 1.7–(b) is pretty noisy. Why? It refers to a learning algorithm based on mini-batches, whereas the others use batch-mode. The noise is in fact due to the change of the mini-batch in different iterations, whereas, when using batch mode, the error function used for the optimization is clearly the same, and the change of the error follows the smoothness of the function. It's worth mentioning that such a smoothness depends on the appropriate dynamics of the gradient descent algorithm, whereas it is quite common to experience the presence of a different kind of noise which depends on the too large gradient steps.

The analysis of the plots of Fig. 1.6 indicates that the overfitting of the training set has a different effect on the error and on the accuracy. For a first analysis, this sounds odd, since, as we expect that when the error increases the accuracy decreases. Moreover, the claims on the emergence of overfitting hold concerning the performance of the classifier, which are clearly expressed by the accuracy. However, after a closer look, we can understand that the reason of the different behavior is due to the robustness of the thresholding scheme used to quantize the output of the network for performing the decision. Even though the error on the test set begins its departure from the minimum value, the accuracy doesn't change until we end up with a quantization error. As a matter of fact, in the reported experiments, the overfitting of the training set has an effect on the loss, but it doesn't affect the accuracy significantly. Of course, while this behavior might not be observed in other experiments, the different effect of overfitting on the loss and on the accuracy holds in general.

EXERCISES

1. [*11*] Suppose you are given the precision p and the recall r for a certain experiment. Can we determine the accuracy from the values of p and r?

2. [*17*] Let us introduce the notions of precision and recall also for the negative class. Then prove that if we know p^+, r^+ (precision and recall related to the positive class) and p^-, r^- (precision and recall related to the negative class) then this suffices to compute the accuracy.

3. [*16*] Prove that $\min\{p, r\} \leq F_1 \leq \max\{p, r\}$.

4. [*13*] Discuss the reasons for choosing smx instead of the simpler

$$(a_1, \ldots, a_n) \mapsto \frac{a_i}{\sum_{j=1}^{n} a_j}.$$

5. [*15*] Provide a qualitative description of the differences between average and max-pooling.

6. [*C18*] Write two algorithms for max and average pooling and write down the corresponding code. Discuss your choice when integer division is not possible.

7. [*15*] Comment the following statements: "The higher the resolution, the better performance we get in handwritten character recognition."

8. [*17*] Propose an extension of the definition of accuracy to n classes and relate it to the case of a single class.

9. [*22*] Discuss the relation between the cross-entropy loss function and accuracy.

10. [*C23*] Write an algorithm for early stopping and write down the corresponding code.

1.4 **Challenges in machine learning**

In addition to a potential breakthrough in biology and medicine, machine learning has been facing fantastic scientific challenges that aim to grasp most recondite secrets of intelligence. While vision and speech understanding have already significantly benefited from the progress of machine learning, there are still fundamental open issues that are likely to be attacked when opening the discipline to a truly spatiotemporal context that characterizes human perception. Hence it might not be enough to rely on the explosive growth of computational resources, as we will likely need an in-depth rethinking on foundational topics. It is quite a recurrent situation that has been experienced a number of times in the broad field of artificial intelligence. It looks like an "instant déjà vu" which nicely emerges in Marvin Minsky's quote:

> *Despite that, the theory of machines does not excite the public imagination the way that sport contexts do, or games shows, rock concerts, or the scandals of the private lives of celebrities. No matter. When we're doing mathematics, we don't have to consider the weights of the opinion of the majorities.*

1.4.1 **Learning to see**

The acquisition of visual skills can mostly be regarded as a learning process. The supervised metaphor, which requires us to accumulate an enormous amount of image labels, enables the application of sophisticated learning and reasoning models that have been proving their effectiveness in related applicative fields of AI. However, so far the semantic labeling of pixels of a given video stream has been mostly carried out at the frame level. This seems to be the natural outcome of well-established methods working on images, which grew up when computational models on video were not regarded as a viable direction because of complexity issues. While algorithms operating at the frame level are still driving the state-of-the-art in object recognition, there are strong arguments to start exploring the more natural visual interaction that humans experience in their own environment. As previously pointed out, one might figure out what human life could have been in a world of visual information with shuffled frames. Any cognitive process aimed at extracting symbolic information from images that are not frames of a temporally coherent visual stream would have been extremely harder than in our visual experience. Clearly, this comes from the information-based principle that in any world of shuffled frames, a video requires an order of magnitude more information for its storing than the corresponding temporally coherent visual stream. As a consequence, any recognition process is remarkably more difficult when shuffling frames. It seems that most of current state-of-the-art approaches have been attacking a problem which is harder than the one faced by humans. This leads us to believe that the time has come for an in-depth rethinking of machine learning for semantic labeling of images. We likely need to start facing the challenge of disclosing the computational basis of vision by regarding it as a true learning field that needs to be attacked by an appropriate vision learning theory.

When the target is moved to unrestricted visual environments and the emphasis is shifted from huge labeled databases to a human-like protocol of interaction, we need to go beyond the current peaceful interlude that we are experimenting in vision and machine learning. A fundamental question a good theory is expected to answer is why children can learn to recognize objects and actions from a few supervised examples, whereas nowadays machine learning approaches strive to achieve this task. In particular, why are they so thirsty for supervised examples? Interestingly, this fundamental difference seems to be deeply rooted in the different communication protocol at the basis of the acquisition of visual skills in children and machines. A video stream provides a huge amount of information which comes from imposing coherent labeling, which is likely to be the essential information associated with visual perception experienced by any animal. Roughly speaking, once a pixel has been labeled, the constraint of coherent labeling virtually generates tons of other supervisions that are essentially ignored in most machine learning approaches on big databases of labeled images. It turns out that most of the visual information to perform semantic labeling comes from the motion coherence constraint, which explains the reason why children learn to recognize objects from a few supervised examples. The linguistic process of attaching symbols to objects takes place at a later stage of children development, when they have already developed strong pattern regularities. The enforcement of motion coherence constraints is a high level computational principle that, regardless of biology, plays a fundamental role in the discovering of pattern regularities. Once a representation gained by motion coherence is obtained, the mapping to linguistic descriptions is dramatically simplified with respect to current brute force machine learning approaches, where there are no such built-in constraints. Interestingly, the enormous literature on tracking might be a mine of precious results for devising successful methods for semantic labeling.

1.4.2 Speech understanding

One of the most striking human capabilities is the ability of constructing linguistic structures to express the semantics of the world. Interestingly, this takes place in different languages, which suggests us that the representation of the world, along with the attached semantics, is independent of language. While there are plenty of experiments to support the principle that the overall semantics is the outcome of different perceptual experiences, the distinctive role of language can hardly be neglected, since it is in fact the one which largely sets humans apart from all other life on earth. Semantics can be extracted from linguistic sentences by following the underlying principle of invariance of the constructed concepts. We start noticing that the listening mechanism leads to focusing attention on the part of speech which is being listened. Once we conquer the ability of word spotting, this mechanism migrates at word level, along with its corresponding meaning within the sentence.

Hence, to a large extent, the construction of semantics corresponds with the creation of *word concepts*. Interestingly, this is not the only abstraction which takes place. The need to express relations among the word concepts leads to the emergence of sentence structures. This might depend somewhat on the asymmetric structure of the pulmonic airstream at the basis of the production of speech [see *An Introduction to the Science of Phonetics* by Hewlett and Beck (Routledge, 2006)]. In their own words:

> It is possible to speak on a pulmonic ingressive airstream, *that is, talking while breathing in, but this is generally much less efficient. This is largely due to the shape of the vocal folds, which are not adapted to vibrate efficiently on an ingressive airstream, but it is also due to habitual patterns of controlling air flow from the lungs.*

This indicates that ordinary speech is ruled out by the principle of *breathe or talk*, according to which we regularly pause while talking to breathe. Since infants start experimenting language production by simple separated parts of speech, this favors the development of asymmetric vocal folds. Basically, the symmetry would only be required in case of endless utterances and, most importantly, symmetry breaking likely favors the learning processes.

The outcome of this asymmetry is daily spoken experience of pausing the phonetic emission to breath before continuing. The decision on where to stop for breathing is somewhat associated with the linguistic notion of sentence. While sentences are not necessarily separated by pauses, the duration of ordinary pulmonic airstream suggests that parts of speech are composed in "sentences" so as to express higher-level concept. When linguistic sentences are properly spotted, then, in addition to word concepts, we can access a new type of concept that somewhat summarizes the words accumulated during a breath-suspended utterance. A sentence concept expresses the semantics connected with the corresponding collection of words. Clearly, while word concepts have a local nature, which is reflected by the current focus of attention, sentence concepts express more abstract forms of meaning that arises globally when considering all the words of the sentence together. Word and sentence concepts turn out to be the outcome of linguistic regularities that can likely be captured in the general setting of learning from/of constraints.

1.4.3 Agents living in their own environment

By and large, the multifacet solutions that have been devised in nowadays different applications of machine learning are mostly based on a lab training phase, which is required to define the parameters for the operational phase. However, in many challenging tasks, things are likely to change quickly. Intelligent agents are expected to live in their own environment and react to the stimuli just like humans do. Visual services in unrestricted environments would be very welcome, but, as of now, the technology doesn't allow us to come up with very satisfactory results.

Conversational agents offer a perfect example of the importance of computational models capable of handling the agent life. The social interactions of the agent and the need for active learning processes are better sustained in a fully dynamic environment that is not simply defined as a huge collection of data. The learning agent is likely to benefit from appropriate focus of attention mechanisms and from opportune developmental stages. The teacher cannot simply be regarded as a mere supervisor on the label to be attached to a collection of examples. We need a substantial pressure for learning new concepts that only appropriate teaching policies can implement.

EXERCISES

1. [25] Segmentation techniques based on spatiotemporal locality are doomed to fail in most complex tasks of speech and vision understanding. However, they can be useful in restricted application tasks — a good reason for which they still deserve attention. The segmentation of speech bursts can be based on checking the presence of "silence" in a given utterance, that is, checking when the audio signal $t \mapsto s(t)$ is such that $s(t) \simeq 0$. For a video signal v, a simple segmentation, which aggregates regions with uniform gray level, can be driven by checking $|\nabla_x v(t, x)|^2 \simeq 0$. For a speech signal, one might think of the dual condition $(s'(t))^2 \simeq 0$.

2. [46] Future challenges in machine learning are to devise methods for pattern segmentation that make it possible to extract objects $e \in E$. Discuss qualitatively the main problems that we need to face in order to regard pattern segmentation and recognition as a unified process.

3. [22] Future challenges in machine learning are to devise methods capable of learning directly the $\chi = h \circ f \circ \pi$ without defining π and h in advance. Can you address the main problems that we need to face?

4. [25] Provide a qualitative description of the difficulties connected with the problem of extracting information from optical documents (e.g., an invoice) and ordinary pictures, like those that you take with a camera.

5. [26] After having played with commercial conversational agents, which you can access, make a list of relevant cognitive features that they are missing with respect to humans performing the same task.

6. [40] After having played with commercial conversational agents, which you can access, provide a qualitative description of their adaptation qualities to different tasks. Why are they mostly failing? There are conversational agents involved in question & answering and conversational agents assisting during the process of buying a flight ticket. How can they exchange their work? What are we still missing to achieve such a cognitive capability?

7. [20] Write one sentence to explain why spatiotemporal learning tasks are essentially outside the common pattern recognition model, where patterns are represented by $x \in \mathbb{R}^d$.

8. [19] To what extent do you think artificial neural networks, and particularly the multilayer networks presented in this chapter, represent a scientific model of human brain?

9. [25] Nowadays dominant approaches to machine learning and related disciplines like computer vision and natural language processing are based on benchmarks. However, as discussed in Section 1.3.1, a possible new direction could involve a crowdsourcing based evaluations. Discuss the pros/cons of this method according to your own view when considering its possible impact on the process of facing machine learning challenges.

1.5 Scholia

Section 1.1 Machine learning is at the crossroads of many disciplines that involve understanding of intelligent as well as neurobiological processes. As such, its roots and historical evolution have been significantly marked by progress also in these related disciplines.

Scientists early became to suspect that also machines need to learn if they want to face some challenging human-like tasks. They also early realized that learning a concept is not only a metric issue. The straightforward idea of modeling patterns by appropriate centroids and determining their membership on the basis of distance-based algorithms is doomed to fail. This is due to bizarre behavior of Euclidean and other metrics at high dimensions [see J. Barhen, J.W. Burdick, B.C. Cetin, *IEEE Int. Conf. on Neural Netw.* **2** (1993), 836–842 for a gentle introduction]. For example, when choosing the Euclidean distance, there is little difference in the distances between different pairs of samples. This property on the concentration of points in the peel of a multidimensional Euclidean sphere is clearly expressed by Eq. 1.1.4–(3). A straightforward way of describing the impact of this behavior in pattern recognition is to consider the distance of a collection of ℓ points from a given reference point x_r. Suppose they are randomly drawn and denote by X_m and X_M the random variables that represent the positions of the points whose distance from x_r are minimum and maximum, respectively. It can be proven that [see K.S. Beyer, J. Goldstein, R. Ramakrishnan, U. Shaft, *Proc. of the 7th Int. Conf. on Database Theory,* (1999), 217–235]

$$\lim_{d \to +\infty} \mathrm{E}\, \frac{\|X_M - x_r\| - \|X_m - x_r\|}{\|X_m - x_r\|} = 0,$$

which is yet another way of stating that the distance of any point to the reference x_r becomes indiscernibles compared to the minimum distance. Hence, at high dimensions, distance functions lose their usefulness in any nearest-neighbor criterion in feature-comparison algorithms. These negative results, which fall under the umbrella of "curse of dimensionality," have been recently mitigated, since it has been argued that when attributes are correlated then data provide higher distance contrast [A. Zimek, E. Schubert, H. Kriegel, *Stat. Anal. Data Min.* **5** (2012), 363–387]. This brings us back to the discussion on the Euclidean distance of handwritten chars at high dimensions; while the specific probability distribution associated with the problem at hand yields attribute correlations, the distance computed at high dimension still yields a classification method that is scarcely accurate. The presence of attribute correlation suggests providing lower-dimensional representations capable of capturing the truly free variables that represent the patterns. This is in fact what is done when using auto-encoders. The bizarre geometrical behavior that arises at high dimensions and dramatically complicate nearest neighbor search is discussed in R.B. Marimont, M.B. Shapiro, *IMA J. Appl. Math.* **24** (1979), 59–70. Clearly, the effect of high dimensionality on distance functions is even more serious in the construction of k-nearest neighbor (k-NN) graphs. At high dimension, the in-degree distribution of the digraph is characterized by a huge number of hubs, namely points that appear in many more k-NN lists. Because of the massive application of the k-NN classifier, there is a remarkable impact of this phenomenon in the overall field of machine learning and of all related disciplines [E. Chávez, G. Navarro, R. Baeza-Yates, J.L. Marroquín, *ACM Comput. Surv.* **33** (2001), 273–321].

The development of machine learning is interwound with an in-depth understanding of the relationships between symbols and subsymbols, and with the meaning of symbol-grounding. According to Stevan Harnad [*Physica D: Nonlinear Phenomena 42* **42** (1990), 335–346], symbols are

> *a set of arbitrary "physical tokens" scratches on paper, holes on a tape, events in a digital computer, etc., that are manipulated on the basis of "explicit rules" that are likewise physical tokens and strings of tokens. The rule-governed symbol-token manipulation is based purely on the shape of the symbol tokens (not their "meaning"), i.e., it is purely syntactic, and consists of "rulefully combining" and recombining symbol tokens. There are primitive atomic symbol tokens and composite symbol–token strings. The entire system and all its parts — the atomic tokens, the composite tokens, the syntactic manipulations both actual and possible and the rules — are all "semantically interpretable:" The syntax can be systematically assigned a meaning, e.g., as standing for objects, as describing states of affairs).*

The discussion on the symbolic vs subsymbolic representations of the environments has immediately emerged at the dawn of machine learning, but it became a well-established research topic by the workshop series on Neural-Symbolic Learning and Reasoning http://www.neural-symbolic.org/.

The connections between biological and artificial neural networks have been explored in different contexts, where scientists have outlined intriguing analogies that arise between natural and artificial cognitive processes. A very good collection of seminal papers on this topic, where the reader can also get a consistent historical view, can be found in the book *Neurocomputing: Foundations of Research* edited by Anderson and Rosenfeld (MIT Press, Cambridge, MA, 1988) and in *Neurocomputing 2, Volume 2 Directions for Research* edited by Anderson, Rosenfeld and Pellionisz (MIT Press, Cambridge, MA, 1988).

Machine learning finds its fundamental roots in the protocol of learning that we establish with the intelligent agent. Although supervised learning, which is the most straightforward protocol, has been advocated in many different contexts, Frank Rosenblatt's contribution is by far the most significant one [*Principles of Neurodynamics: Perceptrons and the Theory of Brain Mechanisms* (Spartan Books, 1962)]. First foundations on reinforcement learning only appeared at the rise of the first wave of neural network popularity during mid-1980s [see the PhD thesis by Richard S. Sutton *Temporal Credit Assignment in Reinforcement Learning* (University of Massachusetts, 1984)]. First steps towards unsupervised learning were moved by means of the popular k-means algorithm that was introduced by James Mac Queen [*Proc. of the Fifth Berkeley Symp. on Math. Statist. and Prob.* **1** (1967), 281–297]. The algorithm is the unsupervised counterpart of nearest neighbor classifiers for supervised learning. Given a certain integer k, this time there is no reference vector of a certain class for determining the distances, whereas we want to construct k clusters of points. The algorithm is pretty easy to describe. We start choosing k different points (centroids) and, then, for each centroid we determine the points that turn out to be the closest neighbors with respect to the other centroids. Basically, any centroid turns out to define a cluster in the feature space. Now, for each cluster, since the initial centroids are not necessarily the "best representative patterns," we update the centroid by averaging over all points of the cluster. Because of the updating of the centroid, it might be the case that when we aggregate the patterns to the closest centroid, the previous cluster structure is modified. Hence we repeat the process of centroid updating and continue cycling until there is no change in the cluster composition. Apart from convergence and related issues, it is clear that k-means inherits all the curse of dimensionality issues discussed for the nearest neighbor algorithm. Among others, pioneering work in unsupervised learning has been carried out by Tuevo Kohonen using his notion of self-organizing map [*Self-Organization and Associative Memory*, third edition (Springer-Verlag, Berlin, 1989)].

Studies on transductive inference were introduced in the mid-1970s, but more solid results emerged years later [see A. Gammerman, V. Vovk, V. Vapnik, *Proc. of the Fourteenth Conf. on Uncertainty in Artif. Intell.* (1998), 183–192]. Interesting contributions in semisupervised learning were introduced in the mid-1990s when also the differences with respect to transduction were spread in the scientific community [see e.g. D. Zhou, B. Schoelkopf, *Proc. of the ICML-2004 Workshop on Statistical Relational Learn.* (2004), 126–131 and M. Belkin, P. Niyogi, V. Sindhwani, *J. Mach. Learn. Res.* **7** (2006), 2399–2434]. The term active learning, as intended in machine learning, has been the subject of a remarkable investigation that is nicely summarized in Burr Settles Univ. of Wisconsin-Madison Computer Sciences Tech. Report 1648 (2009), while Dana Angluin provided a clear picture on the importance of queries in learning [*Mach. Learn.* **2** (1988) 319–342].

The richness and the expressiveness of the communication protocol that governs the interactions of the intelligent agent with the environment strongly affect the conception of learning algorithms and, later on, of real-world applications. The tradition of machine learning is somewhat on the opposite site of knowledge representation and reasoning that characterize symbolic AI. Hence the descriptions of the interaction with the environment has been mostly relegated to the presentation of examples. In addition to what has already been mentioned on the relations between symbolic and subsymbolic representations, the studies on graphical models and causality have been shedding light on the long-term goal of providing a unified view of learning and reasoning processes [see, e.g. *Probabilistic Reasoning in Intelligent Systems: Networks of Plausible Inference* by J. Pearl (Morgan Kauffman, 1988), *Models, Reasoning, and Inference* by J. Pearl (Cambridge Univ. Press, 2000) and *Learning in Graphical Models* edited by M.I. Jordan (MIT Press, 1999)]. While these studies are basically oriented to provide a probabilistic interpretation of logic descriptions, more recent research on the abstract notion of constraint offers a related, yet different perspective [M. Gori, *Lecture Notes in Comp. Sci.* **6911** (2011), 6–6 and G. Gnecco, M. Gori, S. Melacci, M. Sanguineti, *Neural Comput.* **27** (2015), 388–480]

Section 1.2 The puzzling nature of induction is a recurrent challenge of philosophers and mathematicians. It slides like an eel; as soon as it is in your hands, it slips away! Of course, regardless of the approach that we have been following, it's a fundamental issue in machine learning. This was early realized in the formal context of *inductive inference* that was nicely expressed in terms of language identification in the limit [E.M. Gold, *Information and Control* **10** (1967), 447–474]. An intelligent agent (learner), who is provided with a string of a certain language, must discover a generative description of the language. It was early realized that inductive inference of grammars is a remarkably difficult computational problem [D. Angluin, C.H. Smith, *ACM Comput. Surv.* **15** (1983) 237–269]. While the adoption of a truly statistical approach to induction seems to be the unarguable direction to follow, because of typical human induction processes, one might suspect that there are good reasons for thinking of computational processes driven by different principles. Humans come out with induction processes that seem to be hard to be formally represented. Needless to say, they are regarded as a distinctive sign of intelligence. Binet and Simon early realized that inductive processes, along with other qualities of human intelligence, can indeed be measured [*The development of intelligence in children (The Binet-Simon Scale)* translated by E.S. Kite (Williams & Wilkins Co. 1926)]. While their scale has had an enormous impact, as pointed out in Section 1.2.1, we must be careful about the assessment of induction qualities. A striking difference with respect to human learning is that natural human environments possess the spatiotemporal dimension. In addition, focus of attention and teaching strategies play a crucial role in the actual acquisition of concepts.

Section 1.3 Setting up experiments of machine learning is quite easy when using current software development support. The running example on handwritten character recognition presented in this chapter is likely to be the most popular benchmark in the field. The MNIST database, along with useful information on the pattern representation and on the state-of-the-art, can be found at http://yann.lecun.com/exdb/mnist/. More details on the architectural and algorithmic choices are given in Chapter 5.

The assessment of experimental results is based on measures that are used mostly in statistics and related disciplines. Notice that the F_1 measure is a special F_β measure that is defined by

$$F_\beta = (1 + \beta) \frac{pr}{\beta^2 p + r}.$$

These measures were introduced in the field of information retrieval [*Information Retrieval* by C.J. van Rijsbergen (Butterworth, 1979)]. We can easily see that values of $\beta < 1$ yield to scoring precision more than recall, whereas for $\beta > 1$ recall is scored more than precision. Typical measures are $F_{0.5}$ to score mostly precision and F_2 to score mostly recall.

The impressive growth of spectacular web resources for the acquisition of concepts and for quick development of applications has been changing dramatically the way machine learning is thought. This raises questions on the role of books, especially if one wants to develop quickly the expertise required to develop simple applications.

Among others, TensorFlow (https://www.tensorflow.org/), Torch (http://torch.ch/), Caffe (http://caffe.berkeleyvision.org/), Matlab® (https://mathworks.com) offer effective developmental environments, where the support to software coding is paired with wonderful simulation tools and very good tutorials. By and large, there is an agreement that the resurgence of interest in neural networks in the mid-1980s was strongly driven by three seminal books by the Parallel Distributed Research Group (PDP) [*Parallel Distributed Processing* vol. 1, 2, and 3 by J.L. McClelland and D.E. Rumelhart (MIT Press, 1988)]. In addition to the first book on foundations and to the second on cognitive topics, the third book played a crucial role in spreading the idea. It came up with floppy disks with software simulators of most relevant PDP models, which facilitated everyone interested in the subject to quickly run experiments and think of new ideas and applications. The book contained a code description, just like what has been sketched in Section 1.3. Needless to say, a similar book does not make sense anymore. The quality of the tutorials that are available with the above mentioned web resources, however, might dramatically reduce also the importance of introductory descriptions of learning algorithms. What might not be easy to cover without books, however, is the presentation of unifying approaches that lead to acquisition of an in-depth understanding of the field, with the purpose of playing a primary role in the development of applications and quickly intercepting emerging novelties.

Section 1.4 Most of the difficulties faced in challenging problems of computer vision and natural language processing seem to be related to their inherent spatiotemporal nature that has not yet been fully captured in machine learning models. Principles of cognitive development stages that have been the subject of an in-depth analysis in children by Jean Piaget [*The Growth of Logical Thinking from Childhood to Adolescence* by B. Inhelder and J. Piaget (New York: Basic Books, 1958) and *La psychologie de l'intelligence* by J. Piaget (Paris: Armand Colin, 1961)] are likely to inspire important advances in machine learning. He pointed out that we can identify four major stages or periods of development in child learning, where each stage is self-contained and builds upon the preceding stage. In addition, children seem to proceed through these stages in a universal, fixed order. They start developing *sensorimotor* and *preoperational skills*, in which the perceptual interactions with the environment dominate the learning process, and evolve by exhibiting *concrete* and *formal operational skills*, in which they start to think logically and develop abstract thoughts. When observing human and current artificial minds on the same play, one quickly realizes that machines do not take into account most of the rich human communication protocols. In most of the studies of machine learning, the agent is expected to learn from labeled and/or unlabeled examples finalized to a specific task. There are, however, a number of other crucial interactions of the agent that are rarely taken into account. Human learning experiences witness the importance of asking questions and of learning under a teaching plan. While the first interaction has been considered in a number of machine learning models, apart from a few exceptions, teaching plans have not been significantly involved in learning algorithms. What is often neglected in machine learning is that most intriguing human learning skills are due, to a large extent, to the acquisition of relevant semantic attributes and to their relations. This makes learning a process which goes well beyond pure induction; the evidence provided by the induction of a semantic attribute is typically propagated to other attributes by formal rules, thus giving rise to a sort of reinforcement cyclic process.

Though with a different perspective, developmental issues in machine learning have been covered in the studies on curriculum learning [Y. Bengio, J. Louradour, R. Lollobert, J. Weston, *Proc. Twenty-Sixth Int. Conf. on Machine Learning* (2009), 41–48] and in preliminary studies on computational laws at the basis of intelligence [M. Gori, *Neural Netw.* **22** (2009), 1035–1036].

Learning principles

Err
and err
and err again
but less
and less
and less.
PIET HEIN, The Road to Wisdom (1966)

Machine Learning. https://doi.org/10.1016/B978-0-32-389859-1.00009-X

IN THIS CHAPTER, we give an introduction of fundamental principles of machine learning, where intelligent agents are expected to live in an environment that is described in terms of constraint satisfaction. We present different approaches to capture regularities by induction and explore their relations. We begin with an in-depth discussion on the interaction between the agent and the environment, and then we discuss learning methods mostly inspired by statistics and information-based theories.

2.1 Environmental constraints

AS ALREADY pointed out in Section 1.1.5, intelligent agents, just like humans, are expected to live in an environment that imposes the fulfillment of constraints. We want our machines to be able to satisfy an appropriate mathematical notion of environmental constraint. Most of machine learning has focused attention on the special case of supervised learning, but a more general view suggests an extension to deal with constraints that enable the interpretation and exploitation of prior knowledge on the problem at hand.

2.1.1 Loss and risk functions

In this section, we start discussing environmental constraints by focusing attention on *supervised learning*. In the running example on handwritten character recognition shown in the previous chapter, it has been shown that learning algorithms are driven by the specific choice of the error function that clearly depends on the way we express the error with respect to the given targets. In general, we assume that our intelligent agent lives in a learning environment that is composed of the collection of training data $L = \{(x_k, y_k) \in X \times Y : k = 1, \ldots, \ell\}$, where $X \subset \mathbb{R}^d$ and $Y \subset \mathbb{R}^n$. If $Y \subset \{-1, +1\}^n$ then the learning task is referred to as a *classification task*, whereas if $Y \subset \mathbb{R}^n$, we talk about *regression tasks*. It is quite clear that supervised learning for the classifiers and regressors needs to be treated in a different way. Intuitively, in regression one is typically interested in keeping $y - f(x)$ close to zero. The classic quadratic loss associated with the prediction is clearly a good measure to force the agent to track the supervisor as closely as possible. For example, in automatic driving, a machine that is tracking the supervisor, whose action is defined by y under sensorial conditions x, must keep $\|y - f(x)\|^2$ as small as possible. Of course, this quadratic loss works fine for classification, too. However, forcing the agent towards the minimization of this loss is not necessary to perfectly acquire classification skills. This has to do with the discrete nature of the target, so that an agent which returns $+0.3$ or $+1.8$ reports the same decision! The same for negative values. We can promptly see that what does really matter is $yf(x)$: If the agent returns an output $f(x)$ whose sign agrees with the sign of y then the decision is correct. Hence the loss must be returned only in case there is sign disagreement, namely whenever $yf(x) < 0$. There is more. There are learning tasks in which the way we penalize agent behavior with respect to the supervisor might depend on the point x itself. Coming back to the running example on handwritten chars, the value of $\alpha_c(x) = x'm^c/(\|x\|\|m^c\|)$ can conveniently be used to drive the strength of the supervision. Clearly, whenever $\alpha_c(x)$ is close to one, it means that we are dealing with patterns that are mostly in the area of the retina that is not typically occupied by a pattern of class c. As a result one might reasonably argue that for patterns x, for which $\alpha_c(x)$ is close to one, the corresponding loss be small. This is in fact the case of odd patterns in the environment, which one could decide not to consider in the construction of the loss. Notice that these patterns are likely to be rare, but rarity in itself must not be necessarily penalized. It is quite common in practice to work with learning environments in which the classes are not balanced. Of course, in these cases, a good idea is to associate higher loss with rare patterns (see Exercise 1). To sum up, in general, in addition to y and f, the loss depends on the specific point x of the training set.

Now let us introduce the notion of loss in a more formal way. We assume that our intelligent agent is modeled by a function[1] $f \in H^k(X; Y)$ — for some suitable $k \in \mathbb{N}$. Given any $x \in X$, we assess the quality of the learning process by means of an appropriate measure (loss) associated with the guess $f(x_k)$ of y_k. Formally, any function

$$V: X \times Y \times H^k(X; Y) \to \overline{\mathbb{R}}_+, \qquad (x, y, f) \mapsto V(x, y, f), \qquad (1)$$

such that if $f(x) = y$ then $V(x, y, f) = 0$ (consistency) is referred to as a *loss function*. Alternatively, we can regard the loss as the function with image $\bar{V}(x, y, f(x))$ under the same consistency condition. Clearly, the only difference is that the definition given by Eq. (1) emphasizes that the role of learning is to determine $f \in H^k(X; Y)$, namely the functional nature of problem. Notice that the consistency does not imply that whenever $V(x, y, f) = 0$ one necessarily has $y = f(x)$. As discussed on classification losses, one can simply check the sign agreement, so that we can define $V(x, y, f) = 0$ simply whenever $yf(x) > 0$. Moreover, as shown in Exercise 2, the loss function V does not necessarily satisfy the symmetry property $V(x, y, f) = V(x, f, y)$.

We begin considering classification tasks, where we assume that the targets are in $Y = \{-1, +1\}$. A natural loss is

$$V(x, y, f) = [y \neq f(x)], \qquad (2)$$

which returns an error $V(x, y, f) = 1$ whenever the machine is wrong. Notice that sometimes one might want to replace the above equation with

$$V(x, y, f) = \tilde{V}(x)[y \neq f(x)] \qquad (3)$$

---

[1] Following classic notations in functional analysis $H^k(X; Y)$ denotes the Sobolev space of functions that are in $L^2(X; Y)$ and admit weak derivatives up to k.

where $\tilde{V}: X \to \mathbb{R}^+$ returns a loss in case of misclassification that, however, might depend on the pattern being classified. As already pointed out, in classification problems, more than checking $y = f(x)$ one is interested in checking $\text{sign}(f(x))$, where here sign is the component-wise sign function. If we assume that $f(x) \in \mathbb{R}$ (learning machines with one output), then we can reasonably assume that the correct classification can be checked on the basis of the outcome of $p(x) = yf(x)$, which expresses the sign agreement between the supervisor and the machine decision. However, it doesn't take too much to realize that this is not the best we can do; clearly, the pairs $(f(x), y) = (0.01, +1)$ and $(f(x), y) = (0.99, +1)$ are not reporting the same classification quality! The *hinge loss*

$$V(x, y, f) := \sum_{j=1}^{n} \left(\theta - y_j f_j(x) \right)_+ \tag{4}$$

addresses the requirement of a *strong sign agreement*, since it still reports an error in case the decision is not "strong" enough. In this equation $(\cdot)_+$ is the component-wise extension of $(a)_+ = [a > 0]a$. In case of a single output we get $V(x, y, f) := (\theta - yf(x))_+$. Clearly, $\theta > 0$ introduces the required robustness in the loss that, in case of mistake, grows linearly with $yf(x)$. The loss does not return any error in case of strong sign agreement, that is, when $\theta - yf(x) \geq 0$. When using this kind of loss, the correspondent classifier is referred to as a *margin classifier*. A related loss is the one in which $(\theta - yf(x))_+$ is replaced by $((\theta - yf(x))_+)^2$ which is differentiable everywhere. Likewise, logistic loss

$$V(x, y, f) := \log(1 + \exp(\theta - yf(x))) \tag{5}$$

is another smooth approximation of Eq. (4) for the scalar case; the multiclass extension is straightforward. If $Y = \{0, +1\}$ then we need a redefinition of Eqs. (4) and (5), which is covered in Exercise 4.

Another nice way of expressing the classification disagreement is to consider the loss[2] $V(x, y, f) = d(y, f(x))$, where

$$d(y, z) := -\frac{1+y}{2} \log \frac{1+z}{2} - \frac{1-y}{2} \log \frac{1-z}{2} \tag{6}$$

where we assume that $f: X \to [-1, 1]$. For reasons that will be clear in the following, this is referred to as the *relative entropy loss*. We can promptly see that $d(-1, -1) = 0$ while $\lim_{z \to 1^-} d(-1, z) = +\infty$, $d(1, 1) = 0$ and $\lim_{z \to -1^+} d(1, z) = +\infty$. The discussion on the case of multiclassification is proposed in Exercise 5.

To some extent, the case of regression can incorporate classification. The *quadratic loss*, defined as $V(x, y, f) = \frac{1}{2}(y - f(x))^2$, is a natural way to express the error

---

[2] If the targets are chosen in $Y = \{0, +1\}$, we can easily see that $V(x, y, f) := -y \log f(x) - (1 - y) \log(1 - f(x))$ plays the same role.

with respect to the target. It is in fact a special case of

$$V(x, y, f) = \frac{1}{p} \|y - f(x)\|_p^p, \tag{7}$$

where $\|u\|_p := \sqrt[p]{\sum_{i=1}^d u_i^p}$. It satisfies the symmetry $V(x, y, f) = v(x, f, y)$ and is *translation independent* with respect to the pair $(y, f(x))$. Basically, it is a special case of the class of loss defined by $V(x, y, f) = \gamma(y - f)$, where the loss only depends on the difference $y - f$. The value $p = 2$, which restores the quadratic loss, is not the only interesting one; the cases $p = 1$ and $p = \infty$ also give rise to remarkable properties. Notice that loss function (7) can also be used for classification. However, as it will be seen, there are a number of arguments that led to prefer the previously indicated loss functions.

High values of p return a nearly null loss whenever $|y - f(x)| < 1$; in the extreme case, as $p \to \infty$ the loss returns $\max_j |y_j - f_j(x)|$. Loss functions defined according to Eq. (7) implement the principle that, unless $f(x) = y$, an error is reported. Sometimes this is not very good: One might be happy with reporting error only when the prediction is outside a certain margin. This arises naturally from the extension of (4). Let us restrict to single output regressors; we can symmetrize the hinge function by choosing

$$V(x, y, f) = (|y - f(x)| - \epsilon)_+ \tag{8}$$

that is referred to as the *bathtub loss*. The extension to multiclass domains is proposed in Exercise 7.

So far we have discussed how to return a loss when the machine takes a decision or make a prediction that is affected by an error. Since the process of supervised learning takes place on a collection of supervised examples, it is clear that for the agent to perform the learning task satisfactorily, the error over all the training set must be kept small. Suppose the learning environment can be modeled by the probability measure P that can be represented in terms of a density function $p = dP/dxdy$ with respect to the usual Lebesgue measure $dxdy$ on $X \times Y$. Once f is defined, the random variables X and Y generate the random variable V, whose instances are the values of the loss function $V(x, y, f)$. The expected behavior of the agent turns out to be characterized by the minimization of

$$E(f) = \mathrm{E}\, V(X, Y, f) = \int_{X \times Y} V(x, y, f)\, dP(x, y), \tag{9}$$

where $X(x, y) = x$ and $Y(x, y) = y$ are the projection random variables, which is the expected value of V. This is the expected risk, which is also referred to as the functional or structural risk. Unfortunately, the minimization of E is generally hard, since P is not explicitly given. A comment is in order when considering the distribution P that has a significant impact on the way regression and classification problems are modeled. In both cases the patterns exhibit an inherent randomness that is characterized by the random variable V. In classification, one can also think of the supervisor as an oracle that makes no mistakes during the decision process. In this extreme case, randomness is reduced to X since it is possible to express the supervision deterministically. Of course, a more practical view of the supervision process suggests that errors are possible during the labeling, so that Y is a random variable that cannot be generated deterministically from X. In regression tasks, the joint probability distribution of X and Y has typically a different flavor. This can be promptly understood by an example. Suppose we want to construct a machine that supports autonomous navigation by outputting the rotation angle y of the steering wheel on the basis of a vector $x \in \mathbb{R}^d$ properly extracted from the image that is being acquired from a camera. Furthermore, suppose that a human driver acts as the supervisor by reporting a training set of pairs (x, y) acquired during his driving lesson. Because of the nature of the problem, which is typically shared in regression tasks, the supervisor typically reports different actions, defined by angles y, in the same image conditions defined by x. Hence, there is a truly joint probability distribution that is somewhat different to what happens in classification.

In real-world problems, the measure P only arises from the collection of samples of a training set $\mathcal{L}$. This leads to estimate the measure P in terms of a concentrated measure P_ℓ defined by $P_\ell = (1/\ell) \sum_{\kappa=1}^{\ell} \delta_{x_\kappa} \delta_{y_\kappa}$, where δ_x is the delta function (measure) concentrated on the point x. If we replace P with P_ℓ the risk function in Eq. (9) becomes

$$\int_{X \times Y} V(x, y, f)\left(\frac{1}{\ell}\sum_{k=1}^{\ell} d\delta_{x_k}(x) d\delta_{y_k}(y)\right) = \frac{1}{\ell}\sum_{k=1}^{\ell} V(x_k, y_k, f).$$

Unlike the expected risk defined by Eq. (9), its approximation over the training set

$$E_\ell(f) = \frac{1}{\ell}\sum_{\kappa=1}^{\ell} V(x_\kappa, y_\kappa, f), \tag{10}$$

which is referred to as the *empirical risk*, is a functional that can be directly optimized for returning the behavior of the agent, defined by function f. As we can see, the probability distribution induced by the training set makes it possible to define P_ℓ, so that E_ℓ can be computed for any choice of f.

When focusing on this *loading problem* (discovery of f), one should not forget that the empirical risk somewhat departs from the agent's ideal behavior modeled by the minimization of the expected risk, since the training set might not adequately represent the distribution P. This is a central problem in statistics and machine learning, which can be badly ill-posed. We can promptly realize that we are in presence of an ill-position when thinking of the fitting of the training set by function f. Basically, if f is chosen from a rich class of functions, and if it has got a high degree of freedom in mapping the input x, then a training set composed of a few pairs (x, y) only yields a few constraints on the selection of the most appropriate function f. In other words, the information provided by T is not enough for the ambition of discovering a function f with too many degrees of freedom. Hence, if we stress the idea of optimizing the empirical risk, we are essentially looking for the moon in the well! This problem will be addressed in different ways in the book, but the first nice way of fully grasping the ill-conditioning of this problem is presented in Section 3.1.2 concerning the class of linear machines. It is worth mentioning that most of machine learning algorithms rely on the hypothesis of providing a parametric representation of f. Linear machines are just a way of providing a model of the agent which is based on an appropriate selection of parameters. In general, a learning machine that relies on a parametric representation can be regarded as a function $F: W \times X \to Y$ that maps $(w, x) \mapsto F(w, x)$, where $W \subset \mathbb{R}^m$. In so doing, the loading problem in an infinite dimensional space is transformed into one in finite dimension, that is

$$\underset{f \in H^k(X;Y)}{\arg\min} \frac{1}{\ell} \sum_{\kappa=1}^{\ell} V(x_\kappa, y_\kappa, f) \quad \text{becomes} \quad \underset{w \in W}{\arg\min} \frac{1}{\ell} \sum_{\kappa=1}^{\ell} V\big(x_\kappa, y_\kappa, F(w, x_\kappa)\big).$$

Notice that these two optimization problems differ significantly because of the underlying assumptions on the optimization space. In the first case, we are in front of a functional optimization problem, whereas in the second case, we are reduced to dealing with a finite-dimensional optimization.

Now let us give a different interpretation of learning that, as it will be shown, is related to risk minimization. Given any pattern x, we can think of a *generative model* of our data in $X \times Y$, represented by $f: X \to Y$, aimed at maximizing the conditional distribution density $p_2(y|x)$, where $E(Y \mid X = x) = f(x)$ with for instance $f \in H^k(X; Y) =: \mathcal{F}$. This means also that the joint distribution density depends on the choice of the model f; to remember this we will write $p_2(y \mid x; f)$. Now suppose we are given the training set $\mathcal{L} = \{(x_1, y_1), \dots, (x_\ell, y_\ell)\}$ as a sampling of the probability measure P. If we assume that the measurements (x_κ, y_κ), $\kappa = 1, \dots, \ell$, come from observations of random variables (X_i, Y_i), $\kappa = 1, \dots, \ell$, *independently and identically distributed* (i.i.d.) then if we ask to choose a model that maximizes the conditional joint probability density of observing $y_1, \dots, y_\ell$ given $x_1, \dots, x_\ell$ we simply have:

$$f^* \in \underset{f \in \mathcal{F}}{\arg\max} \prod_{\kappa=1}^{\ell} p_2(y_\kappa \mid x_\kappa; f) = \underset{f \in \mathcal{F}}{\arg\min} -\mathcal{L}(f) \tag{11}$$

where

$$\mathcal{L}(f) := \sum_{\kappa=1}^{\ell} \log p(x_\kappa, y_\kappa; f) \tag{12}$$

is the *log-likelihood* associated with the loading problem. Indeed, since the marginal probability distribution density $p_1(x) := \int_Y p(x, y)\, dy$ does not depend on f we have

$$\arg\max_{f \in \mathcal{F}} \prod_{\kappa=1}^{\ell} p_2(y_\kappa \mid x_\kappa; f) = \arg\max_{f \in \mathcal{F}} \left(\prod_{\kappa=1}^{\ell} p_1(x_\kappa) \right) \left(\prod_{\kappa=1}^{\ell} p_2(y_\kappa \mid x_\kappa; f)) \right).$$

Also since the log function is a strictly monotonically increasing function and it does not change the point where the maximum is attained, we immediately get Eq. (11). Notice that the maximization of the log-likelihood can be thought of as a problem of empirical risk minimization, when we define the loss as

$$V(x, y, f) := -\log p_2(y \mid x; f).$$

The corresponding structural risk is

$$E(f) = -\mathrm{E}\log p_2(Y \mid f(X)) =: H(Y \mid X; f), \tag{13}$$

which reduces to the conditional entropy of variable Y conditioned to X. The information-based interpretation of the minimization of the conditional entropy is straightforward: The aim of supervised learning is that of minimizing the uncertainty on random variable Y when X is available by an optimal choice of f. The optimal value corresponds with the lack of information, namely with $H(Y \mid X; f) = 0$, that is, to the ideal case in which the agent is an oracle that always guesses Y.

Like for the general formulation of learning based on the minimization of the expected risk, we cannot afford to attack this problem directly, unless we know the probability distribution, which is unlikely to happen in practice. However, once we make assumptions on the structure of the density distribution of $Y \mid X_\kappa, \kappa = 1, \ldots, \ell$, the optimization problem in Eq. (11) collapses. Let us consider the case in which $Y \mid X_\kappa \sim \mathcal{N}(f(X_\kappa), \Sigma)$, that is,

$$p_2(y_\kappa \mid x_\kappa; f) = \frac{1}{(2\pi)^{n/2}(\det \Sigma)^{1/2}} \exp\left[-\frac{1}{2}\big(y_\kappa - f(x_\kappa)\big)' \Sigma^{-1}\big(y_\kappa - f(x_\kappa)\big) \right],$$

where n is the dimensionality of the output space Y. From Eq. (11), we get the minimum problem

$$\min_{f \in \mathcal{F}} - \prod_{\kappa=1}^{\ell} \frac{1}{(2\pi)^{n/2}(\det \Sigma)^{1/2}} \exp\left[-\frac{1}{2}\big(y_\kappa - f(x_\kappa)\big)' \Sigma^{-1}\big(y_\kappa - f(x_\kappa)\big) \right]$$

$$= \min_{f \in \mathcal{F}} - \frac{1}{(2\pi)^{n\ell/2}(\det \Sigma)^{\ell/2}} \exp\left[-\frac{1}{2} \sum_{\kappa=1}^{\ell} \big(y_\kappa - f(x_\kappa)\big)' \Sigma^{-1}\big(y_\kappa - f(x_\kappa)\big) \right].$$

Hence the related minimization of the log-likelihood leads to the minimization of a quadratic loss problem:

$$\min_{f \in \mathcal{F}} \frac{1}{2} \sum_{\kappa=1}^{\ell} \left(y_\kappa - f(x_\kappa)\right)' \Sigma^{-1} \left(y_\kappa - f(x_\kappa)\right).$$

Now we consider the case of classification, in this case the conditional distribution $p(y \mid x; f)$ is much simpler since the random variable Y only takes values in $\{+1, -1\}$. Hence, a single point x returns the loss

$$V(x, y, f) = -\frac{1+y}{2} \log p(1 \mid x; f) - \frac{1-y}{2} \log p(-1 \mid x; f). \qquad (14)$$

Now if

$$p(1 \mid x; f) = \frac{1 + f(x)}{2}, \qquad p(-1 \mid x; f) = \frac{1 - f(x)}{2}, \qquad (15)$$

then we recover the loss given by Eq. (6). Here, we implicitly assume that $-1 \leq f(x) \leq 1$, which can automatically be obtained when posing $f(x) = \text{Th}(a(x))$, with $a \colon \mathbb{R} \to \mathbb{R}$. We end up with the conclusion that in classification problems, as soon as we choose functions $f \in \mathcal{F}$ which satisfy the probabilistic condition (15); the maximization of the likelihood, which corresponds with the minimization of the conditional entropy, corresponds with the minimization of the cross-entropy given by Eq. (6).

We can extend to *multiclassification* in different ways. A straightforward idea is to update the loss function so as to accumulate the errors over all the categories. The learning environment turns out to be

$$\mathcal{L} = \left\{ (x_\kappa, y_\kappa) \in \mathbb{R}^{d+1} \times Y^n : \kappa = 1, \ldots, \ell \right\}.$$

For two classes, we can model the conditional probabilities as

$$p_2(1 \mid x; f) = \frac{\exp(f(x))}{1 + \exp(f(x))}, \qquad p_2(-1 \mid x; f) = \frac{1}{1 + \exp(f(x))}. \qquad (16)$$

It can be proven that this probabilistic assumption leads to restoring the logistic loss (see Exercise 9). The extension to multiclassification follows the equations

$$P(Y_i = +1 \mid f_i(x)) = \frac{\exp(f_i(x))}{1 + \sum_{j=1}^{n} \exp(f_j(x))},$$

$$P(Y_i = -1 \mid f_i(x)) = \frac{1}{1 + \sum_{j=1}^{n} \exp(f_i(x))}. \qquad (17)$$

Notice that this is similar to softmax, but here we better model class rejection. The extension to n classes yields a loss function that is studied in Exercise 11.

In case of quadratic loss we have

$$E(\hat{w}) = \frac{1}{2} \sum_{i=1}^{n} \sum_{\kappa=1}^{\ell} (y_{\kappa,i} - f_i(x_\kappa))^2, \tag{18}$$

while related extensions arise for different metrics.

EXERCISES

1. [*12*] Suppose you are given a classification problem with three classes, where the number of patterns per class is $n_1 = 10$, $n_2 = 20$, and $n_3 = 30$, respectively. Write down an empirical risk function to balance the different number of examples in the classes.

2. [*15*] Provide a counterexample to the symmetry claim $V(x, y, f) = V(x, f, y)$.

3. [*M22*] Let us consider the cross-entropy loss function given by Eq. (6). It is typically used for classification. Is it possible to express it as a function of $p(x) := yf(x)$?

4. [*17*] Let us consider the hinge and the logistic loss functions that have been defined with $Y = \{-1, +1\}$. How can we extend their definition to the case of $Y = \{0, +1\}$?

5. [*11*] Let us consider the loss function (6) (relative entropy). Extend the definition to multiclassification.

6. [*13*] Let us consider the function

$$V(x, y, f) := 1 - \exp(-\exp(-yf(x))).$$

Prove that it is an adequate loss function in classification tasks. Why does this function exhibit good robustness to noise?

7. [*16*] Extend the loss function defined by Eq. (8) to multiclass domains.

8. [*14*] Discuss the general interpretation of the log-likelihood in terms of conditional entropy prescribed by Eq. (13).

9. [*17*] Prove that the probabilistic assumptions (16) in the maximization of log-likelihood corresponds to the minimization of the empirical risk with the logistic loss (5) in the case $\theta = 0$.

10. [*16*] Discuss the quadratic optimization for the case of multioutput functions.

11. [*17*] Extend the result given in Exercise 9 to multiclassification, where the probabilistic hypothesis is based on softmax. What is the loss functional in this case?

2.1.2 Ill-position of constraint-induced risk functions

The discussion on the risk function carried out in the previous section covers most classic formulations of supervised learning. The risk comes out from averaging loss functions $V(x, y, f)$ that measure the degree of fit of the given target y. It has been shown that such a formulation of learning assumes inherently that the optimization takes place by using the empirical risk. The target typically comes out from human supervision, so that we typically deal with a finite collection of supervised data more than with a continuous target y. As pointed out in Section 1.1.5, however, one can formulate learning processes in a more general framework when considering the crucial role of the interaction amongst the learning tasks. Basically, we need to go beyond the formalization of supervised learning by modeling the task interactions in terms of constraints. A remarkable part of this book is devoted to exploring this extension of the notion of loss function along with the correspondent implications. While the core of this topic will be covered in Chapter 6, we begin with a preliminary discussion on *constraint-induced risk*. We start noticing that some of the cases discussed in Section 1.1.5 have been already paired with a corresponding loss. In particular, supervised learning turns out to be a special form of pointwise constraint. However, in other cases, like for committee-based decision and asset allocation, the underlying model continuously involves the environmental interactions amongst the tasks, without needing human supervision. Constraints can also arise when bridging logic to real-valued representations, which is an important issue that will be covered in Chapter 6.

To sum up, when thinking of a constraint-based environment, the loss function V defined by Eq. 2.1.1–(1) is not always an adequate index for the construction of learning processes. In order to get an insight on this important issue, the classic problem of brightness invariance for optical flow estimation in computer vision (see Eq. 1.1.5–(8)) is a very good example. In this case, the learning task is the velocity field in a given retina, that is, $f = \dot{x}$, with x being the coordinate of a generic pixel. Then we can regard y as a *tracking function* that corresponds with the video signal $y(t, x) := v(t, x)$, that is a gray-level information source. Now suppose that we are given a probability distribution P that measures where we pose attention during learning. In the simplest case, one can simply look at all pixels and frames with the same probability, so that P is uniform. The fulfillment of the constraint over all the retina $X \subset \mathbb{R}^2$ and over a certain time horizon T can be translated into the minimization of

$$E(f) = \mathrm{E}(y_t + f \cdot \nabla y)^2 = \int_{X \times [0,T]} (y_t + f \cdot \nabla y)^2 dP(x, t). \tag{1}$$

The assumption of uniform P, along with the additional simplification in which the integral is computed only over the retina, leads to imposing the fulfillment of the constraints for any $t \in [0, T]$, without any relation between the frames. Regardless of this simplification, this example suggests that loss criteria may need to involve environmental information that, like the video signal y, are not necessarily elements of the domain of f. In this learning task, $f \colon X \times [0, T] \to \mathbb{R}$, which is the optical flow, from a formal point of view, operates on the temporal unfolding of the retina, where the domain does not involve the video signal. The loss is in fact expressing a property which is related to the tracking function y. This more general view of learning poses additional challenges. In supervised learning, the minimization of $E(f)$ has a clear meaning, and it captures the essence of learning very well. The same does not hold in this case, since the condition $E(f) = 0$ is not enough to properly set the optical flow estimation problem. We can promptly see the difference when considering static visual information. In this case, $y_t = 0$, so that the above condition is verified by any optical flow f, which is orthogonal to ∇y. Clearly, such a freedom is not acceptable! Static vision means that there is no pixel movement, that is, $f(x) = 0$ over all the retina. If we immerse the loss into a probabilistic setting, this does not save us from this inherent ill-position of the problem. Unlike ordinary loss functions for supervised learning, the associated expected risk does not offer a proper formalization of learning. The corresponding constraint expressed by the penalty term of Eq. (1) only contributes to determining the optical flow, but it does require something more! This is quite common whenever we involve more abstract and sophisticated constraint-based descriptions of the learning environment. In case of the optical flow problem, an appropriate learning formulation can be given under the framework of the parsimony principle. In doing so, we promote the preference for smooth changes of velocity, so that a classic formulation turns out to be one in which we minimize

$$E(f) = \int_{X \times [0,T]} (y_T + f \cdot \nabla y)^2 \, dx dt + \mu \|\nabla f\|^2,$$

where $\mu > 0$ is an opportune parameter that weighs the importance that we want to attribute to the smoothness in the solution.

The ill-position of optical flow is not a pathological case. Whenever the available constraints weakly characterize the tasks, their satisfaction might take place in huge solution spaces, so that statistical formulations based on extending the minimization of Eq. 2.1.1–(9) are ill-posed. Any deriving learning algorithm, like for the discussed case of optical flow, is doomed to fail. This clearly indicates that in complex learning tasks where the loss function does not clearly guarantee the well-positioning, risk-based minimization algorithms might not be appropriate. The classic framework of risk-based minimization in machine learning relies on the principle that whenever $V(x, y, f) = 0$ the decision (prediction) $f(x)$ is satisfactory for the given learning task. While the discussion in the previous section indicates that this makes sense for supervised learning, when we shift attention to the satisfaction of general types of constraints, we are not covered with the same guarantee.

The parsimony principle, which will be discussed in detail in Section 2.4, offers a solid technical support to address this kind of ill-positioning, thus offering a more general approach to learning in the context of environmental constraints.

2.1.3 Risk minimization

We can interpret learning as the outcome of the minimization of the expected risk. In the previous section, we have analyzed different kinds of loss function, that are used to construct the expected and the empirical risk. In this section, we discuss the minimization of the risk with the main purpose of understanding the role of the chosen loss in the optimal solution. In general, we would like to minimize the expected risk $E(f)$ as given by definition 2.1.1–(9). This can be carried out by an elegant analysis based on variational calculus that is given in Section 2.5. Here, we will concentrate on the minimization of the empirical risk, which is what is used in the real-word. Interestingly, the study on the empirical risk will also indicate the structure of the minimum of the functional risk.

We begin discussing the case of regression. Let us consider the class of loss functions defined by Eq. 2.1.1–(7). We discuss the cases $p = 0, 1, 2$, and $+\infty$, which are more commonly adopted. We will also assume that the marginal probability density p_1 is strictly positive: $p_1 > 0$. The process of learning is converted into the problem of minimizing

$$E = \mathrm{E}\,\|Y - f(X)\|_p^p = \frac{1}{p}\int_{X \times Y} |y - f(x)|^p dP(x, y). \qquad (1)$$

Now the computation of this risk benefits from the replacement of $p(x, y) = p_2(y \mid x)p_1(x)$. In so doing, we get

$$E = \frac{1}{p}\int_X \left(\int_Y |y - f(x)|^p p_2(y \mid x)dy \right) p_1(x)dx.$$

This can be written in a more compact statement as

$$\mathrm{E}(|Y - f(X)|^p) = \mathrm{E}\big(\mathrm{E}(|Y - f(X)|^p \mid X)\big).$$

Now the functional to be minimized has the structure $\int_X (\cdot)p(x)dx$ and, since $p_1(x) > 0$, we have

$$f \in \underset{f \in \mathcal{F}}{\arg\min}\, \mathrm{E}\,|Y - f(X)|^p \iff f(x) \in \underset{a \in \mathbb{R}}{\arg\min}\, \mathrm{E}(|Y - a|^p \mid X = x). \qquad (2)$$

This offers a straightforward way to discover the optimum, which can be restricted to finding $f(x)$ for any given x instead of the infinite-dimensional optimization over f. Once we set $X = x$, we must find $f(x)$ such that it gets as close as possible to $Y = f(x)$ according to the metrics induced by $\|\cdot\|_p$. A remark is in order concerning the assumption on the probability density. Let $N \subset X$ be a region of X for which $p_1(x) = 0$. Clearly, $\int_X (\cdot)p_1(x)\,dx = 0$ is satisfied even in case in which $\mathrm{E}(|Y - f(X)|^p \mid X = x) \neq 0$. Hence, we can promptly see that the solution f which comes out from Eq. (2) can take any value over N. While this is quite reasonable, there are cases in which it is not acceptable (see Exercise 1).

Now let us explore the optimization with different choice of p. We begin with the case $p = 2$. We need to find $\arg\min_a \mathrm{E}((Y - a)^2 \mid X = x))$. Let us restrict the analysis to single output functions (see Exercise 10 for the extension). The idea of the solution can promptly be gained when considering the associated empirical risk. Notice that it turns out to approximate the functional risk only for a specific realization of the random variable $X = x$. Hence, we can look for the critical points by imposing

$$\frac{d}{df(x)}\frac{1}{2}\sum_{\kappa=1}^{\ell}(y_\kappa - f(x))^2 = -\ell f(x) + \sum_{\kappa=1}^{\ell} y_\kappa = 0 \qquad (3)$$

from which $f(x) = (1/\ell)\sum_{\kappa=1}^{\ell} y_\kappa$. For large ℓ (see Exercise 2), we have

$$f^*(x) = \mathrm{E}(Y \mid X = x). \qquad (4)$$

If $p = 1$, from Eq. (2), the problem is reduced to finding

$$f^*(x) \in \arg\min_{f(x)} E(|Y - f(X)||X = x).$$

Again, we restrict to single output functions and analyze the associated empirical risk to gain insights on the solution. Unlike the quadratic case, this minimization is more involved, since the function $(1/\ell)\sum_{\kappa=1}^{\ell}|y_\kappa - f(x)|$ is not regular enough to allow the straightforward search of critical points carried out by Eq. (3). Exercise 4 proposes an example in which we discuss the cases of even and odd cardinality of the collection of targets. However, we can shed light on the general solution by rewriting the loss as

$$\sum_{\kappa=1}^{\ell}|y_\kappa - f(x)| = \sum_{\alpha=1}^{m}(y_\alpha - f(x))[y_\alpha - f(x) > 0]$$

$$+ \sum_{\alpha=m+1}^{\ell}(f(x) - y_\alpha)[y_\alpha - f(x) < 0].$$

Here the case $f(x) = y$ has been omitted since it does not return any loss. When differentiating both sides with respect to $f(x)$, we get

$$\frac{d}{df(x)}\sum_{\kappa=1}^{\ell}|y_\kappa - f(x)| = -\sum_{\kappa=1}^{m}[y_\kappa - f(x) > 0] + \sum_{\kappa=m+1}^{\ell}[y_\kappa - f(x) < 0].$$

Without loss of generality, we can assume that the samples are presented in such a way that the targets y_κ are sorted in descending order. If we choose $f(x)$ such that $y_m > f(x) > y_{m+1}$, then we get

$$\frac{d}{df(x)}\sum_{\kappa=1}^{\ell}|y_\kappa - f(x)| = \ell - 2m = 0,$$

which indicates that the solution $f(x)$ is located in such a way that half of the targets are lower and half are higher. When extending to the functional risk, we can prove that

$$f^*(x) = \text{med}(Y \mid X = x). \tag{5}$$

In order to get an insight on the proof, notice that the above analysis assumes that ℓ is even, so that $m = \ell/2 \in \mathbb{N}$. It takes a while to realize that this hypothesis is not restrictive, since the approximation of the expected risk by the empirical risk can always be obtained by even numbers! The details of the proof are discussed in Exercise 5.

Finally, we consider the case $p = 0$. From Eq. (2), we know that the problem is reduced to finding $\min_{f(x)} E(|Y - f(X)|^0 \mid X = x)$. This time, the associated empirical risk is proportional to

$$\sum_{\kappa=1}^{\ell} |y_\kappa - f(x)|^0 = \sum_{\kappa=1}^{\ell} [y_\kappa \neq f(x)]. \tag{6}$$

Basically, $|y_\kappa - f(x)|^0$ corresponds with the 0–1 loss. If $y_\kappa \neq f(x)$ then $|y_\kappa - f(x)|^0 = 1$, whereas if $y_\kappa = f(x)$ then $|y_\kappa - f(x)|^0 = 0$. This corresponds with the formal adoption of $0^0 = 0$. Why can we make such an arbitrary choice on 0^0, which is clearly an undetermined form? To answer this question, first we need to make clear the meaning of $y_\kappa = f(x)$, when considering that $y_\kappa, f(x) \in \mathbb{R}$. Any check of this condition in nature does require its restating as $|y_\kappa - f(x)| < \delta$, with $\delta > 0$ being an arbitrary threshold chosen small enough to claim that $f(x) \simeq y_\kappa$. Now $\forall \epsilon > 0$ and $p > 0$ arbitrarily close to 0, if we choose $\delta = \exp(\log \epsilon / p)$ then

$$|y_\kappa - f(x)|^p < \delta^p < \epsilon.$$

Hence, the rationale behind Eq. (6) is that no matter how small we choose p, a strong enough interpretation of the equality relation $|y_\kappa - f(x)| < \delta$ by an appropriate "small" δ makes $|y_\kappa - f(x)|^p$ arbitrarily small. This motivates the statement

$$f^*(x) \in \arg\min_{f(x)} \sum_{\kappa=1}^{\ell} |y_\kappa - f(x)|^0 = \text{mode}(Y \mid X)$$

since the most common value (mode) y_κ is the one which minimizes the function defined by Eq. (6). This analysis indicates that $\text{mode}(Y \mid X)$ is in fact an approximate solution for values of $p \to 0$. It is worth mentioning that for $p = 0$ the property cannot be formally stated. We are in fact back to the mentioned undetermined condition 0^0. Exercise 6 suggests a more sophisticated way to escape the trap of 0^0.

Finally, suppose we take a big value of p and, in particular, consider $p \to \infty$. In this case

$$f^*(x) \in \arg\min_{f(x)} \mathbb{E}_{Y|X}(|Y - f(X)|^p \mid X = x) \overset{p=\infty}{\longrightarrow} \arg\min_{f(x)} \max |y - f(x)|.$$

Overall, the analysis with different exponents p indicates that we can control the errors in different ways. While for small p the optimal solution is biased towards the conditional mode, as p increases, the solution takes the form of the conditional median and the conditional average. Interestingly, as $p \to \infty$, the optimal solution is one which disregards the distribution, and only concerns the maximum error. Clearly, the presence of noisy data makes the choice of high values of p very dangerous, since there is a very strong association with a few patterns of the distribution.

The previous analysis holds for regression. In case of classification, it is opportune to introduce a different loss function to penalize the errors. First, we start by considering that the output needs to encode the class of the pattern, which can clearly be done in different ways. Suppose we use a one-hot encoding for n classes, so that given any pattern x, the decision is based on $f(x) \in \mathbb{R}^n$. Basically, we can think of n independent classifiers where class $i = 1, \ldots, n$ is modeled by the corresponding function f_i. We replace the index (1) with

$$E = \mathrm{E}_{XY} \, V(X, Y, f) = \mathrm{E}_X \, \mathrm{E}_{Y|X} \, V(X, Y, f)$$

$$= \mathrm{E}_X \, \mathrm{E}_{Y_i|X} \sum_{i=1}^{n} V(X, Y_i, f_i) \Pr(Y_i \mid X).$$

We can carry out a point-wise minimization with respect to $f(x)$, so that we get

$$f^*(x) \in \arg\min_{f(x)} \mathrm{E}_{Y_i|X} \sum_{i=1}^{n} V(X, Y_i, f_i) \Pr(Y_i \mid X = x).$$

Now let us assume that the loss V is defined as

$$V(x, y_i, f_i) = [f_i(x) \neq y_i].$$

If $X = x$, for the random variable V, we have $V(X, Y, f_i)|_{X=x} = V(x, y_i, f_i)$. This comes from assuming that whenever the random variable X takes on the specific value $x \in \mathbb{R}^d$, the associated classification is unique, that is, we assume that there is no noise in the decision. In other words, the joint distribution $\Pr(X, Y)$ in this case is only driven by the random nature of X, since Y takes on a consequent value associated deterministically. As a consequence, we have $\Pr(f_i(X) \mid X = x) + \Pr(Y_i \mid X = x) = 1$. This comes from using the 0–1 loss, that is, either $f_i(x) \neq y_i$ or $f_i(x) = y_i$ is true. Hence we have

$$f^*(x) \in \arg\min_{f(x)} \mathrm{E}_{Y_i|X} \sum_{i=1}^{n} V(X, Y_i, f_i) \Pr(Y_i \mid X = x)$$

$$= \arg\min_{f(x)} \sum_{i=1}^{n} [f_i(x) \neq y_i] \, (1 - \Pr(f_i(X) \mid X = x)) .$$

The function to be minimized is a sum of n positive terms, which is minimized whenever all single terms are minimized, which happens if $\forall i = 1, \ldots, n$ we choose $f_i^*(x) \in \arg\max_{f(x)} \Pr(f_i(X) \mid X = x)$. Hence, we have

$$f^*(x) \in \arg\max_{f(x)} \Pr(f(X) \mid X = x). \tag{7}$$

This is fundamental result. It states that once we choose the 0–1 loss function, the optimal decision based on the risk minimization corresponds with Bayesian decision.

Finally, it is worth mentioning that the analyses carried out in this section have been mostly based on scalar functions. When the loss involves vectorial functions, its structure also plays an important role in the learning process. In addition, instead of simply thinking of the risk as averaging of the loss contributions, one might use different strategies. As an example, a learning process which cycles over the training set might provide a different weight to the errors; it could focus attention only on large errors, while refraining from updating the parameters for patterns reporting small errors. Some more comments on this issue can be founded in Section 2.5.

EXERCISES

1. [15]   Can you think of cases in which there are patterns such that $p(x) = 0$ but the response $f(x)$ of a machine is important?

2. [17]   Prove that if for any number of examples the optimum of the quadratic empirical risk is the conditional mean then the optimum of the expected risk is still the conditional mean $E(Y \mid X = x)$.

3. [16]   Consider the solution arising from the quadratic risk. We can think of an approximation of $f^*(x) = E(Y \mid X = x)$ by assuming that for any $x \in X$ the machine returns an output which is simply the average of random variable Y, conditioned to $X = x$. It suggests a direct computational scheme based on

$$\hat{f}(x) = \frac{1}{m} \sum_{x_\kappa \in \mathcal{N}_m(x)} y_\kappa \mid x_\kappa.$$

Here $\mathcal{N}_m(x)$ is an m-neighborhood of x that is typically constructed by regular geometrical structures, while $y_\kappa \mid x_\kappa$ denotes the supervision y_κ that is associated with x_κ. Discuss this approximation in real-world problems.

4. [22]   Given a finite set Y of numbers $y_\kappa \in \mathbb{R}$, we call *medians of the set* a value for which half the numbers are larger and half are smaller. Intuitively, the median of a set of data is the middlemost number in the set. The median is also the number that is halfway into the set. Without loss of generality, it is convenient to regard Y as a set sorted in ascending order. Hence med Y satisfies $y_h \le$ med $Y \le y_\kappa$ where $h < \lfloor |Y|/2 \rfloor$ and $\kappa > \lfloor |Y/2| \rfloor$. If $|Y|$ is even, then there are two middlemost numbers and med Y is defined as the mean of the two middlemost numbers. Let $s := f(x)$ and consider the sorted (in ascending order) collection $Y := \{y_1, y_2, y_3\}$. Now let us assume we are given the loss $V(x, y, f) = v(Y, s)$ expressed by $v(s) = |y_1 - s| + |y_2 - s| + |y_3 - s|$. Prove that

$$\text{med } Y = \min_{s \in \mathbb{R}} v(s).$$

What if $Y := \{y_1, y_2, y_3, y_4\}$? Does the solution in both cases depend on the specific choice of y_κ?

5. [M25]   Prove that function (5) represents the minimum of the expected risk (1) for $p = 1$.

6. [20]   Consider the function $\delta(x)^{p(x)}$, where $\lim_{x \to 0} \delta(x) = 0$ and $\lim_{x \to 0} p(x) = 0$. Under which condition $\lim_{x \to 0} \delta(x)^{p(x)} = 0$? Relate this analysis to the property of the mode in case of risk minimization corresponding to $p = 0$.

2.1.4 The bias–variance dilemma

In this section we discuss how different parametric models that are tuned on a certain training set approximate the expected risk. This is clearly what one would like to know in order to fully capture the outcome of the experiments. In general, a learning machine operating in the framework of supervised learning can be regarded as a parametric function

$$F : W \times X \to Y, \quad x \mapsto F(w, x). \tag{1}$$

For example, neural networks are learning machines whose parameters (The learning parameters of neural networks are typically called weights) are learned from a given training set $\mathcal{L}$. A typical learning algorithm selects the appropriate parameter w which better fits the training data, but this takes place once we have chosen the class of models defined by Eq. (1). For example, in a neural network, the chosen architecture corresponds with defining a specific model, $F(w, \cdot)$, that is subsequently adapted by the learning parameters. Hence the layered structure or, more generally, the pattern of connections defines F, while the learning algorithm discovers the value w. Any function $F(w, \cdot)$ presents an inherent *model complexity* that can intuitively be understood when thinking of the number of weights of a neural network. It is quite obvious that, because of their high inherent complexity, architectures with many weights are likely to better fit the training set. As it will be clear in the discussion on deep networks, the issue of model complexity is quite involved, but the intuitive connection with the number of parameters of the model clearly helps. More formally, a possible expression of the model complexity will be given in Chapter 4 in the framework of the parsimony principle.

Here we focus attention on the case of quadratic loss and scalar functions, while other cases are briefly discussed in Section 2.5. We begin to notice that the agent behavior on pattern $x \in \mathbb{R}^d$, which is characterized by function $F(\cdot, x)$, depending on the specific value of $w \in W$, departs to a different extent from the optimal solution $f^*(x) = \mathrm{E}_{Y|X}(Y \mid X = x)$. Any function $F(w, \cdot)$ can be regarded as an element of a family of parametric functions that are characterized by the correspondent parameter w, which is learned from the corresponding training set $\mathcal{L}$. Now, in order to grasp the behavior of the machine, we need to think of its training on a collection of training sets $\mathcal{L}_\alpha$ that can be regarded as instances of the random variable L. Let w_α be the weight vector learned from the generic realization L_α, $\alpha \in \mathbb{N}$ of $\mathcal{L}_\alpha$, and consider the correspondent function $F(w_\alpha, \cdot)$. In order to provide its statistical interpretation, it is convenient to introduce the random variable $\Phi(X, L)$ that can be regarded as a function of random variables (X, L). Random variable Φ expresses the agent decision that has to be compared with the value Y imposed by the supervisor. This yields the quadratic error defined by the random variable $Q^2(F) := \mathrm{E}_{Y|X}(Y - \Phi(X, L))^2$. Notice that because of the correspondence between $\mathcal{L}_\alpha$ and w_α we have $\Phi(X, L = \mathcal{L}_\alpha) = F(w_\alpha, X)$. The different value of $F(w, \cdot)$ and f^* is expressed, over all the training sets, by $E_o := \mathrm{E}_X(\mathrm{E}_{Y|X}(Y \mid X) - \Phi(X, L = \mathcal{L}))^2$.

What makes $\Phi(X, L)$ different from Y? As already noticed, $\Phi(X = x, L = \mathcal{L})$ departs from the conditional mean $f^*(x) = \mathrm{E}_{Y|X}(Y \mid X = x)$, which, in fact, minimizes the expected risk. Any realization of $\Phi(X, L)$ yields different results, since the instance $\mathcal{L}_\alpha$ of the training set affects the parameters w_α which, in turn, leads to the realization $F(w_\alpha, \cdot)$. Hence, there is an error that depends on the specific run of the experiment. It can be nicely interpreted when comparing $\Phi(X, L)$ with the mean

$E_L \Phi(X, L)$ over random variable L. Clearly, $E_L \Phi(X, L)$ is also generally different from $E_{Y|X}(Y \mid X)$. In a sense, while $V^2(F) = (\Phi(X, L) - E_L \Phi(X, L))^2$ expresses the variance of different runs, the error $B^2(F) = (E_L \Phi(X, L) - E_{Y|X}(Y \mid X))^2$ indicates a sort of bias of the predictor with respect to the optimal solution $E_{Y|X}(Y \mid X)$. Moreover, f^* does not necessarily reproduce targets in a perfect way, so that there is yet another source of error. One should not confuse the fact that $F(w, \cdot)$ is optimal with the property of reporting null error with respect to supervision modeled by Y. We can get an insight on the reason why such an error may emerge by thinking about any empirical risk associated with the expected risk. Suppose that a certain example $x_{\overline{\kappa}}$ is contained in the two pairs $(x_{\overline{\kappa}}, y_1)$ and $(x_{\overline{\kappa}}, y_2)$ with $y_1 \neq y_2$. Clearly, no function $F(w, \cdot)$ can perfectly fit these two pairs of supervised examples, which results in an overall error that is different from zero. Hence, in general there might be an inherent source of noise in the training data, which inevitably leads to a non-null error defined by $N^2 = (Y - E_{Y|X}(Y \mid X))^2$. As already seen in the previous section, this naturally happens in regression tasks, but it might be present also in classification tasks. Now let us analyze the problem in a more formally way. Suppose we are training our agent by using the set $\mathcal{L}$. The error corresponding with the experiment training set $\mathcal{L}$ is $E_{XY}(Y - \Phi(X, L = \mathcal{L}))^2$, while $E_L E_{XY}(Y - \Phi(X, L))^2$ returns the error averaged over all possible training sets. We start noticing that (see Exercise 1 for a different proof)

$$
\begin{aligned}
E_{X|Y}(Y - E_{Y|X}(Y \mid X)) &= E_X(E_{Y|X}(Y \mid X) - E_{Y|X} E_{Y|X}(Y \mid X)) \\
&= E_X(E_{Y|X}(Y \mid X) - E_{Y|X}(Y \mid X)) = 0.
\end{aligned}
\tag{2}
$$

This is in fact a strong property deriving from the optimality of f^*. Even though the optimal solution might depart from Y, the averaging over the joint distribution is null. Moreover, we have $E_{Y|X} \Phi(X, L = \mathcal{L}) = \Phi(X, L = \mathcal{L})$. This is due to the fact that function $\Phi(X, L = \mathcal{L})$ encodes all possible dependencies on Y by means of the specific realizations of X. As a consequence, when using Eq. (2) we get

$$
\begin{aligned}
E_{XY}((Y &- E(Y \mid X))\Phi(X, L = \mathcal{L})) \\
&= E_X E_{Y|X}((Y - E_{Y|X}(Y \mid X))\Phi(X, L = \mathcal{L})) \\
&= E_X(\Phi(X, L = \mathcal{L}) E_{Y|X}(Y - E_{Y|X}(Y \mid X))) = 0,
\end{aligned}
\tag{3}
$$

where we can see that the error $Y - E_{Y|X}(Y \mid X)$ and the random variable $\Phi(X, L = \mathcal{L})$ associated with the model are uncorrelated. This is referred to as the *orthogonality principle*. From Eqs. (2) and (3) we immediately conclude that[3]

$$
\begin{aligned}
E_{XY}((Y &E_{Y|X}(Y \mid X))(E_{Y|X}(Y \mid X) - \Phi(X, L = \mathcal{L}))) \\
&= E_{XY}((Y - E_{Y|X}(Y \mid X)) E_{Y|X}(Y \mid X)) \\
&\quad - E_{XY}((Y - E_{Y|X}(Y \mid X))\Phi(X, L = \mathcal{L})) = 0.
\end{aligned}
\tag{4}
$$

---

[3] Here, the orthogonality principle is applied also to conclude that $E_{XY}((Y - E_{Y|X}(Y \mid X)) E_{Y|X}(Y \mid X))$. It is easy to realize that, just like for $\Phi(X, L = \mathcal{L})$, it comes from $E_{Y|X} E_{Y|X}(Y \mid X) = E_{Y|X}(Y \mid X)$.

We are now ready to calculate $E_{XY}(Y - \Phi(X, L = \mathcal{L}))^2$, which allows us to get an idea on the behavior of the specific agent, characterized by function $F(w, \cdot)$ that has been constructed by learning on the specific instance $\mathcal{L}$ of L. When using Eq. (4), which is a direct consequence of the orthogonality principle, the overall error $E_t :=$ $E_{XY}(Y - \Phi(X, L = \mathcal{L}))^2$ can be split as follows:

$$
\begin{aligned}
E_t &= E_{XY}(Y - E_{Y|X}(Y \mid X) + E_{Y|X}(Y \mid X) - \Phi(X, L = \mathcal{L}))^2 \\
&= E_{XY}(Y - E_{Y|X}(Y \mid X))^2 + E_{XY}(E_{Y|X}(Y \mid X)) - \Phi(X, L = \mathcal{L}))^2 \\
&\quad + 2 E_{XY}((Y - E_{Y|X}(Y \mid X))(E_{Y|X}(Y \mid X) - \Phi(X, L = \mathcal{L}))) \\
&= E_{XY}(Y - E_{Y|X}(Y \mid X))^2 + E_X(E_{Y|X}(Y \mid X) - \Phi(X, L = \mathcal{L}))^2.
\end{aligned}
\tag{5}
$$

As we can see, the overall risk E_t is split into two terms: One is the *noise term* N^2, while the other one is the error E_o with respect to the optimal solution f^*. Now, for any L, we can go one step further by splitting E_o as follows:

$$
\begin{aligned}
E_o &= E_X(E_{Y|X}(Y \mid X) - E_L \Phi(X, L) + E_L \Phi(X, L) - \Phi(X, L))^2 \\
&= E_X(E_{Y|X}(Y \mid X) - E_L \Phi(X, L))^2 + E_X(E_L \Phi(X, L) - \Phi(X, L))^2 \\
&\quad + 2 E_X((E_{Y|X}(Y \mid X) - E_L \Phi(X, L))(E_L \Phi(X, L) - \Phi(X, L))).
\end{aligned}
\tag{6}
$$

If we apply the E_L operator to half of the last term, we get

$$
\begin{aligned}
&E_L E_X((E_{Y|X}(Y \mid X) - E_L \Phi(X, L))(E_L \Phi(X, L) - \Phi(X, L))) \\
&= E_X E_L((E_{Y|X}(Y \mid X) - E_L \Phi(X, L))(E_L \Phi(X, L) - \Phi(X, L))) \\
&= E_X((E_{Y|X}(Y \mid X) - E_L \Phi(X, L)) E_L(E_L \Phi(X, L) - \Phi(X, L))) \\
&= E_X((E_{Y|X}(Y \mid X) - E_L \Phi(X, L))(E_L \Phi(X, L) - E_L \Phi(X, L))) = 0.
\end{aligned}
\tag{7}
$$

As a consequence, from Eqs. (6)–(7) we get

$$
\begin{aligned}
&E_L E_X(E_{Y|X}(Y \mid X) - \Phi(X, L))^2 \\
&= E_L E_X(E_{Y|X}(Y \mid X) - E_L \Phi(X, L))^2 + E_L E_X(E_L \Phi(X, L) - \Phi(X, L))^2 \\
&= E_X(E_{Y|X}(Y \mid X) - E_L \Phi(X, L))^2 + E_L E_X(E_L \Phi(X, L) - \Phi(X, L))^2.
\end{aligned}
\tag{8}
$$

If we replace the expression for $E_L E_X(E_{Y|X}(Y \mid X) - \Phi(X, L))^2$ given by Eq. (8) into Eq. (5), after having averaged using E_L, we get

$$
\begin{aligned}
E_L E_{XY}(Y - \Phi(X, L))^2 &= E_L E_{XY}(Y - E_{Y|X}(Y \mid X))^2 \\
&\quad + E_L E_X(E_{Y|X}(Y \mid X) - \Phi(X, L))^2 \\
&= E_L E_X(E_{Y|X}(Y \mid X) - E_L F(X, L))^2 \\
&\quad + E_L E_X(E_L \Phi(X, L) - \Phi(X, L))^2 \\
&\quad + E_L E_{XY}(Y - E_{Y|X}(Y \mid X))^2.
\end{aligned}
$$

Hence we get

$$E_L E_{XY}(Y - \Phi(X, L))^2 = E_X(E_{Y|X}(Y \mid X) - E_L \Phi(X, L))^2$$
$$+ E_L E_X(\Phi(X, L) - E_L \Phi(X, L))^2 \qquad (9)$$
$$+ E_{XY}(Y - E_{Y|X}(Y \mid X))^2.$$

This decomposition of the risk, averaged over all training sets, is coherent with the intuitive picture given at the beginning of this section. The noise term

$$E_{XY}(Y - E_{Y|X}(Y \mid X))^2 = E_L E_X N^2 \qquad (10)$$

turns out to be the variance of the optimal solution $E_{Y|X}(Y \mid X)$. As already pointed out, it is reflecting the inherent random nature of the problem at hand, that is independent of experimental issues. Now the other two terms,

$$E_X(E_{Y|X}(Y \mid X) - E_L \Phi(X, L))^2 = E_L E_X B^2(F), \qquad (11)$$
$$E_L E_X(\Phi(X, L) - E_L \Phi(X, L))^2 = E_L E_X V^2(F), \qquad (12)$$

are, on the other hand, the outcome of specific experimental choices.

The variance term (12) expresses the variability of results gained with the same model whenever function F is learned on different realizations of L. It is exactly a variance, which refers to the mean value $E_L \Phi(X, L)$. Of course, one would very much like to keep this value as small as possible. The above equation can be rewritten in a more compact way as

$$E_L E_X Q^2(F) = E_L E_X (B^2(F) + V^2(F) + N^2). \qquad (13)$$

Now the choice of the function F to be used for training defines the value of $E_L E_X Q^2(F)$. In particular, the outcome of our experiment depends on the specific choice of $F(w, \cdot)$. As a consequence, apart from the noise term, the above equation states that there is the overall error coming from balancing bias and variance, which is typically referred to as a bias-variance tradeoff. A machine with small bias $E_L E_X B^2(F)$ is clearly desirable, but in this case the overall error $E_L E_X Q^2(F)$ is likely to be moved to a high value of variance $E_L E_X V^2$, which is clearly denoting that something is wrong! In that case, different runs of the same model likely lead to remarkably different results, which indicates that we are not providing a good generalization to new examples. On the other hand, a small value the variance can be obtained when learning model with a few parameters (small dimension of w) on large training sets $\mathcal{L}_\alpha$. Unfortunately, small dimension of w likely moves all the error $E_L E_X Q^2(F)$ to the bias term, since the model $F(w, \cdot)$ might not be expressive enough from a computational point of view.

We are in front of a dilemma in the model selection that is typically referred to as the *bias–variance dilemma*. Either the bias or the variance can easily be kept small, but unfortunately, as we keep the bias small we may get high variance and vice versa.

It is worth mentioning that the *bias–variance dilemma* can also be stated when referring to a singe pattern x. This can promptly be seen when looking at the term splitting expressed by Eq. (6). It also holds in case in which we remove the E_X operator, since we get

$$
\begin{aligned}
(E_{Y|X}(&Y \mid X) - \Phi(X, L))^2 \\
&= (E_{Y|X}(Y \mid X) - E_L \, \Phi(X, L) + E_L \, \Phi(X, L) - \Phi(X, L))^2 \\
&= (E_{Y|X}(Y \mid X) - E_L \, \Phi(X, L))^2 + (E_L \, \Phi(X, L) - \Phi(X, L))^2 \\
&\quad + 2(E_{Y|X}(Y \mid X) - E_L \, \Phi(X, L))(E_L \, \Phi(X, L) - \Phi(X, L)),
\end{aligned}
\tag{14}
$$

and the application of E_L yields

$$
E_L(E_{Y|X}(Y \mid X) - \Phi(X, L))^2 = E_L \left(B^2(F) + V^2(F) \right).
\tag{15}
$$

This is in fact the statement of the bias–variance dilemma for single patterns.

EXERCISES

1. [22] Provide a different proof that $E_{XY}(Y - E_{Y|X}(Y \mid X)) = 0$.

2.2 **Statistical learning**

Machine learning has an interwound story with statistics. It couldn't have been different! Statisticians have developed huge expertise in inferential processes based on samples of data, a topic that relies on methodologies that have become of central interest in current approaches to machine learning.

2.2.1 **Maximum likelihood estimation**

Suppose we are given a parametrical statistical model defined by the probability density $p_X(\theta, x)$. A likelihood for such a model is defined by the same formula as the density, but the roles of the data x and the parameter θ are interchanged, so that

$$L_x(\theta) = \pi_\theta(x) := p_X(x, \theta). \tag{1}$$

Hence there is a nice duality between the probability density π_θ and the likelihood L_x. While π_θ is a function of x with fixed θ, the likelihood L_x depends on θ, once x has been chosen. As such $L_x(\theta)$ expresses the likelihood that x is explained by the underlying probability distribution with the parameter θ. Likelihood is actually more general. If h is any positive function, then the function

$$L_x(\theta) = h(x)\pi_\theta(x) \tag{2}$$

is a generalized form of the likelihood associated with the statistical model π_θ. The method of maximum likelihood uses, as an estimator of the unknown true parameter value, the point $\hat{\theta}_x$ that maximizes the likelihood L_x. Hence we typically look for

$$\theta^* \in \arg\max_\theta L_x(\theta).$$

While maximization is what one typically bears in mind, as a matter of fact the method is rather vague. There are in fact cases in which the likelihood L_x needs not have a maximizer. In addition, if it does, the maximizer need not be unique. Instead of $L_x(\theta)$, the log-likelihood $l_x(\theta) = \ln L_x\theta$ is typically used. Clearly, the maximization of $L_x(\theta)$ yields the same estimation θ^* as $l_x(\theta)$. In Section 2.1.1, the notion of log-likelihood has been introduced when referring to loss functions for supervised learning. However, the principle of maximum likelihood is more general, since it involves any kind of probability distributions and therefore it can also be applied to unsupervised learning. There's more: One can think of likelihood of statistical models in functional terms, instead of considering a parametric model of θ, where the underlying model has a given probabilistic structure. Suppose we are given a probability distribution p_X which is sampled with values in $X = \{x_1, \dots, x_\ell\}$. We overload the symbol l by defining the log-likelihood as

$$l(p_X) := \sum_{\kappa=1}^{\ell} \log p_X(x_\kappa). \tag{3}$$

In case of i.i.d. variables

$$\max_{p_X} l(p_X) = \max_{p_X} \sum_{\kappa=1}^{\ell} \log p_X(x_\kappa) = \max_{p_X} \prod_{\kappa=1}^{\ell} p_X(x_\kappa) = \max_{P_X} L(p_X).$$

The maximum likelihood estimator (MLE) principle states that the samples in X of the probability distribution $p_X(\cdot)$ are explained by the specific probability distribution

$$p_X^* \in \arg\max_{p_X} L(p_X). \tag{4}$$

Hence, the term "likelihood" becomes clear, since the distribution p_X^* is the one that most likely explains X. As for supervised learning, p_X^* can be learned by minimizing the penalty function $-l$. The duality with supervised learning also involves the ill-position of the problem, which admits infinitely many solutions. As already stated by Eq. (2), the classic way of facing ill-position is that of formulating the maximum likelihood problem by a *parametric probability distribution*.

As an example, suppose that $p_X \equiv \mathcal{N}(\mu, \Sigma)$ is a normal distribution. In this case

$$\theta^* = \begin{pmatrix} \mu^* \\ \Sigma^* \end{pmatrix} \in \arg\max_{\theta} \sum_{\kappa=1}^{\ell} \ln \left(\frac{1}{\sqrt{(2\pi)^d \det \Sigma}} e^{-\frac{1}{2}(x_\kappa - \mu) \cdot \Sigma^{-1}(x_\kappa - \mu)} \right)$$

$$= \arg\min_{\theta} \left(\ln \det \Sigma + \sum_{\kappa=1}^{\ell} (x_\kappa - \mu) \cdot \Sigma^{-1}(x_\kappa - \mu) \right).$$

Let us consider the case $d = 1$ (univariate distribution).[4] We can promptly see that the maximum is achieved for $\hat{\mu}$ and $\hat{\sigma}^2$ such that

$$\hat{\mu} = \frac{1}{\ell} \sum_{\kappa=1}^{\ell} x_\kappa, \qquad \hat{\sigma}^2 = \frac{1}{\ell} \sum_{\kappa=1}^{\ell} (x_\kappa - \hat{\mu})^2.$$

The estimations $\hat{\mu} = \bar{x}$ and $\hat{\sigma}^2 = \sigma_s^2$ are referred to as the *sample mean* and the *sample variance*, respectively. Interestingly, they exhibit a different statistical behavior. It might be curious to know what happens when repeating this experiment based on samples of cardinality ℓ. In particular, it makes sense to analyze what are the expected values of $\hat{\mu}$ and $\hat{\sigma}^2$. We have

$$E\,\hat{\mu} = E\,\frac{1}{\ell} \sum_{\kappa=1}^{\ell} x_\kappa = \frac{1}{\ell} \sum_{\kappa=1}^{\ell} E\,x_\kappa = \mu,$$

where $x_1, \ldots, x_\ell$ are regarded as random variables distributed as $\mathcal{N}(\mu, \sigma)$. Hence, the estimator $\hat{\mu}$ satisfies the property $E\,\hat{\mu} = \mu$. Whenever this happens, we say that

---

[4] The general case of multivariate distribution is treated in Exercise 2.

the *estimator is not biased*. Now let us compute $E\hat{\sigma}^2$ for the sampled variance. We start noticing that

$$\sum_{\kappa=1}^{\ell}(x_\kappa - \mu)^2 = \sum_{\kappa=1}^{\ell}(x_\kappa - \hat{\mu})^2 + \sum_{\kappa=1}^{\ell}(\hat{\mu} - \mu)^2 + 2\sum_{\kappa=1}^{\ell}(x_\kappa - \hat{\mu})(\hat{\mu} - \mu).$$

Now we have

$$\sum_{\kappa=1}^{\ell}(x_\kappa - \hat{\mu})(\hat{\mu} - \mu) = \sum_{\kappa=1}^{\ell}x_\kappa\hat{\mu} - \mu\sum_{\kappa=1}^{\ell}x_\kappa - \ell\hat{\mu}^2 + \ell\mu\hat{\mu}$$

$$= \ell\hat{\mu}^2 - \ell\mu\hat{\mu} - \ell\hat{\mu}^2 + \ell\mu\hat{\mu} = 0.$$

Hence

$$\sigma_p^2 = \frac{1}{\ell}\sum_{\kappa=1}^{\ell}(x_\kappa - \mu)^2 = \sigma_s^2 + (\hat{\mu} - \mu)^2 \tag{5}$$

where σ_p is referred to as the *population variance*. Now, we can promptly see that the population variance is unbiased, since

$$E\sigma_p^2 = E\frac{1}{\ell}\sum_{\kappa=1}^{\ell}(x_\kappa - \mu)^2 = \frac{1}{\ell}\sum_{\kappa=1}^{\ell}E(x_\kappa - \mu)^2 = \frac{1}{\ell}\sum_{\kappa=1}^{\ell}\sigma^2 = \sigma^2.$$

On the other hand, from Eq. (5), for the sample variance we get

$$E\sigma_s^2 = E\sigma_p^2 - E(\hat{\mu} - \mu)^2 = \sigma^2 - E(\hat{\mu} - \mu)^2. \tag{6}$$

This states that the sample variance is in fact a biased estimate of the variance, and that the error corresponds with the variance of the mean $\hat{\mu}$. Now, because of the i.i.d. assumption, the variance of the mean is

$$E(\hat{\mu} - \mu)^2 = E\left(\frac{1}{\ell}\sum_{\kappa=1}^{\ell}x_\kappa - \mu\right)^2$$

$$= \frac{1}{\ell^2}E\sum_{i=1}^{\ell}\sum_{j=1}^{\ell}(x_i - \mu)(x_j - \mu)$$

$$= \frac{1}{\ell^2}E\sum_{\kappa=1}^{\ell}(x_\kappa - \mu)^2$$

$$= \frac{1}{\ell^2}\sum_{\kappa=1}^{\ell}E(x_\kappa - \mu)^2 = \frac{\sigma^2}{\ell}.$$

Finally, from Eq. (6) and the above equation on the variance of $\hat{\mu}$, we get

$$\mathrm{E}\sigma_s^2 = \frac{\ell - 1}{\ell}\sigma^2. \tag{7}$$

Hence the MLE principle yields a biased estimator that, however, achieves unbiasedness asymptotically ($\ell \to \infty$). Whenever this property holds, we say that the estimator is *consistent*. Notice that while the population variance cannot be computed numerically, since we don't know the precise value of μ, we can get an unbiased estimate from σ_s^2 by defining the following *unbiased sample variance*

$$s^2 = \frac{\ell}{\ell - 1}\sigma_s^2 = \frac{1}{\ell - 1}\sum_{\kappa=1}^{\ell}(x_\kappa - \hat{\mu})^2. \tag{8}$$

In this case of normal distribution, because of the relation between the population and the sample variance stated in Eq. (5), the log-likelihood can be reduced to

$$l(\mu, \sigma) \to l(\overline{x}, s) = \frac{1}{2}\ln s^2 + \frac{1}{2s^2}\left(\frac{\ell - 1}{\ell}s^2 + (\mu - \overline{x})^2\right). \tag{9}$$

This expression indicates that the log-likelihood, which in general depends on all the sample X, only depends on $\overline{x}$ and s^2. Hence, these two parameters are *sufficient* to represent the likelihood. Whenever this happens, we say that $\overline{x}$ and s^2 are sufficient statistics for the underlying parameters μ and σ^2.

EXERCISES

1. [*23*]   Suppose we are given an experiment that is repeated infinitely many times ("trials"), where the trials are independent. We also assume that the trials occur with an (unknown) probability p of success. Let us assume that we have r successes in m trials. Prove that the probability of a success on the next trial is $(r + 1)/(m + 2)$. Use this result to make a prediction of the event that "sun will rise tomorrow", when assuming that the age of the sun is 4.5 billion years.

2. [*18*]   Prove that in the case of multivariate normal distribution the MLE principle yields $\mu = (1/\ell)\sum_{\kappa=1}^{\ell} z_\kappa$ and $\Sigma = (1/\ell)(z_\kappa - \mu)(z_\kappa - \mu)'$.

3. [*17*]   Let $X^\sharp = \{x_1, \ldots, x_\ell\}$ be a sampling of X, and assume that the samples are uniformly distributed according to $p_X(x) = [x \in X]/\theta$. Use the MLE principle to estimate θ.

4. [*17*]   Let $X^\sharp = \{x_1, \ldots, x_\ell\}$ be a sampling of X, and assume that the samples are distributed according to

$$p_X(x) = \frac{1}{2\sigma}e^{-\frac{|x|}{\sigma}}.$$

Use the MLE principle to estimate σ.

5. [*17*]   Let $X^\sharp = \{x_1, \ldots, x_\ell\}$ be a sampling of X, and assume that the samples are distributed according to the Bernoulli distribution[5] $p_X(x) = p^x (1-p)^{1-x}$ where $x \in \{0, 1\}$. Use the MLE principle to estimate $\theta = p$.

2.2.2 Bayesian inference

Now we discuss the roots of the Bayesian approach to inference. As it will be clear soon, while there are some intriguing connections with MLE, the Bayesian approach is based on the fundamental assumption of considering the hypotheses to be inferred as random variables themselves.

Let us consider the case of classification, where an agent takes a decision on the basis of the input $x \in X \subset \mathbb{R}^d$. The decision is expressed by functions $f_i : X \to \{0, 1\}$ with $i = 1, \ldots, c$. Clearly, all f_i generate a discrete random variable I which takes on values $1, \ldots, c$. Let us define $\mathbb{D}_x = \{ y \in \mathbb{R}^d \mid x_i - \Delta/2 \le y_i \le x_i + \Delta/2, \quad i = 1, \ldots, d \}$, then we have

$$\Pr(I = i \mid X \in \mathbb{D}_x) \Pr(X \in \mathbb{D}_x) = \Pr(X \in \mathbb{D}_x \text{ and } I = i)$$
$$= \Pr(X \in \mathbb{D}_x \mid I = i) \Pr(I = i).$$

Now as Δ — the size of $\mathbb{D}$ — goes to 0, we recover the popular Bayes rule

$$\Pr(I = i \mid X = x) = \frac{p(x \mid I = i)}{p(x)} \Pr(I = i), \tag{1}$$

where $p(x \mid I = i) = p(x, i)/p_i$. Now the decision is taken based on maximization of the *posterior probability* $\Pr(I = i \mid X = x)$. Given x, one decides on its category by computing $\max_i \Pr(I = i \mid X = x)$. We have

$$i^* \in \arg\max_i \Pr(I = i \mid X = x) = \arg\max_i \frac{p(x \mid I = i)}{p(x)} \Pr(I = i)$$
$$= \arg\max_i p(x \mid I = i) \Pr(I = i), \tag{2}$$

so that $f_i^* = [i = i^*]$ and, of course, $i^* = \sum_i i \, [f_i^* = 1]$.

The Bayes rule states that the posterior probability $\Pr(I = i \mid X = x)$ can be determined once we know the *likelihood* $p(x \mid I = i)$ and the *prior probability* $\Pr(I = i)$. Clearly, the decision is independent of $p(x)$, which is referred to as *evidence*. We will now give the idea of how we can estimate the probability $\Pr(I = i \mid X = x)$ from observed data. Let L be a random variable which describes the unsupervised training set

$$\mathcal{L} = \bigcup_{j=1}^{c} \mathcal{L}_j = \bigcup_{j=1}^{c} \{x_{1(j)}, \ldots, x_{\ell(j)}\}.$$

---

[5] The Bernoulli distribution is typically associated with the process of tossing a coin. The random variable takes on values in $\{0, 1\}$ that are associated with heads and tails.

Eq. (2) can be reformulated by explicitly indicating the dependence on the learning, that is,

$$\Pr(I = i \mid X = x \text{ and } L_i = \mathcal{L}_i) = \frac{p(x \mid I = i \text{ and } L_i = \mathcal{L}_i)}{p(x)} \Pr(I = i). \qquad (3)$$

Here, there is the underlying assumption that $\Pr(I = i)$ is independent of observable data. Moreover, notice that we have assumed that learning is carried out independently for each class by using the associated training examples, so that the classification of category $j = 1, \ldots, c$ is independent of L_i for $i \neq j$. Hence, if we drop the category index $j = 1, \ldots, c$, and overload the symbol L to represent the generic class, we can simplify the notation. Hence $p(x \mid I = i \text{ and } L_i = \mathcal{L}_i)$ is simply rewritten as $p(x \mid L)$. The distinguishing feature of Bayesian learning is that the probability density is characterized by the parametrical form $p(x \mid \theta)$, where the parameter θ can be regarded as a random variable. Hence we have

$$p(x \mid L) = \int_{\mathcal{P}} p(x \mid \theta) p(\theta \mid L) \, d\theta, \qquad (4)$$

where

$$p(\theta \mid L) = \frac{p(L \mid \theta) p(\theta)}{\int_{\mathcal{P}} p(L \mid \theta) p(\theta) \, d\theta}. \qquad (5)$$

Here the probability density $p(x \mid L)$ is estimated on the basis of the observable data expressed by random variable L and involves the posterior $p(\theta \mid L)$, which is expressed by the Bayes rule in terms of the likelihood $p(L \mid \theta)$, the prior $p(\theta)$, and the evidence L. When using the classic i.i.d. assumption, the likelihood factorizes as

$$p(L \mid \theta) = \prod_{\kappa=1}^{\ell} p(x_\kappa \mid \theta), \qquad (6)$$

which dramatically simplifies computation. Eqs. (3), (4), (5), and (6) define the basic framework for Bayesian learning. Their solution is generally difficult. Interestingly, there are some intriguing connections with MLE. Suppose that $p(L \mid \theta)$ reaches a sharp peak at $\theta = \hat{\theta}$ and that the prior $p(\theta)$ takes on a value that only slightly changes in the neighbor of $\hat{\theta}$. Then from Eq. (4) we get $p(x \mid L) \simeq p(x \mid \hat{\theta})$, which corresponds with the result one would get from MLE.

2.2.3 Bayesian learning

A viable approach is that of solving Eq. 2.2.2–(4) recursively. Let $L_n = \{x_n\} \cup L_{n-1}$. Then we have

$$p(L_n \mid \theta) = p(x_n \mid \theta) p(L_{n-1} \mid \theta) = p(x_n \mid \theta) \frac{p(\theta \mid L_{n-1}) p(L_{n-1})}{p(\theta)}.$$

Plugging this equation into Eq. 2.2.2–(5), we get the recursive equation

$$p(\theta \mid L_n) = \frac{p(x_n \mid \theta) p(\theta \mid L_{n-1}) p(L_{n-1})}{\int_{\mathcal{P}} p(x_n \mid \theta) p(\theta \mid L_{n-1}) \, d\theta}, \tag{1}$$

$$p(x \mid L_n) = \int_{\mathcal{P}} p(x \mid \theta) p(\theta \mid L_n) \, d\theta. \tag{2}$$

A natural initialization is $p(\theta \mid L_0) = p(\theta)$, which tells us that the posterior probability $p(\theta \mid L)$ simply corresponds with the prior $p(\theta)$ at the beginning of learning. This equation offers immediately a computational scheme to determine the posterior $p(\theta \mid L_n)$. It is an online learning that is referred to as the *recursive Bayes learning*. There are, however, a couple of important issues to consider. First, the computation of the denominator of Eq. (1) can be very expensive. Second, we should understand whether the above recursive scheme converges, and, in that case, if it does converge to the optimal solution. Both these issues on recursive Bayes learning can nicely be grasped in Exercises 2 and 5, where we see how the denominator of Eq. (1) can be obtained by probabilistic normalization, and the process of convergence of $p(\theta \mid L_n)$ towards a delta distribution.

The structure of Bayesian learning suggests exploring the property of sufficient statistic that has already been investigated for the MLE (see Eq. 2.2.1–(9) for Gaussians). In general, the recursive Bayes learning computes $p(x \mid L_n)$ on the basis of all the previous information L_{n-1}. However, there are cases in which a sufficient statistics is enough for determining $p(x \mid L_n)$. A concise review of the literature is given in Section 2.5.

Bayesian recursive learning requires probabilistic normalization in Eq. (1), which might be expensive in general, especially if we consider that the posterior $p(\theta \mid L_{n-1})$ has a structure that is not the same as that of the prior (see Exercise 2). A classic approach to circumventing this problem is to use *conjugate priors*. It is often the case that the appropriate likelihood function is naturally suggested from the problem at hand, whereas the choice of the prior distribution is typically more subjective and arguable. However, once the likelihood function has been adopted, suppose that we choose the "right" prior, so that we needn't worry about the integral in the denominator. This can be done by introducing the notion of conjugate priors. Let $\mathcal{L}$ be a family of likelihood functions and $\mathcal{P}$ be a family of prior distributions. We say that $p \in \mathcal{P}$ is a conjugate prior to $\mathcal{L}$ if for any likelihood $l \in \mathcal{L}$, the corresponding posterior q remains in $\mathcal{P}$, that is, $q \in \mathcal{P}$. When choosing such a prior, the posterior belongs to the same family of distributions as the prior. Interestingly, the solution of Exercise 5 sheds light on the issue of conjugation between likelihood and prior. It is shown that, if we want to estimate the mean of a normal distribution, where the likelihood and the prior are both normal, then we end up with the conclusion that the posterior is normal, too. The case of Gaussians it somewhat special, since conjugation between the likelihood and prior also involves normal distributions. In the case of the uniform distribution discussed in Exercise 2, however, there is no conjugation between the likelihood and prior, since the posterior is not uniform! In Exercise 4, the case of Bernoulli random variables is discussed where the conjugation holds if we adopt *beta distribution* as prior. Like for the Gaussians, the conjugation strongly simplifies the MAP estimation.

EXERCISES

1. [*18*]   Suppose you are given a probability distribution that can be expressed as a mixture of Gaussians according to

$$p(x) = \sum_{j=1}^{2} \alpha_j \frac{1}{\sqrt{(2\pi)^d \det \Sigma}} e^{-\frac{1}{2}(z_\kappa - \mu_j) \cdot \Sigma^{-1}(z_\kappa - \mu)}.$$

Here we assume that the examples belong to two classes associated with the two Gaussians. Prove that the posterior probability of the class label c_κ of point x_κ is

$$\Pr(c = 1 \mid x_\kappa) = \frac{1}{1 + \exp(-\hat{w} \cdot \hat{x})}, \qquad \Pr(c = 2 \mid x_\kappa) = \frac{1}{1 + \exp(+\hat{w} \cdot \hat{x})}.$$

2. [*27*]   (Duda, Hart and Stork 2001.) Use recursive Bayes learning for estimating $p(x \mid L)$, in the case in which we believe that our one-dimensional samples come from the uniform distribution in $[0, \theta]$, $p(x \mid \theta) = [0 \le x \le \theta]/\theta$, with the additional assumption that $0 \le \theta \le 1$. Suppose that our prior, before any data arrive,[6] is $p(\theta \mid L_0) = p(\theta) = [0 \le x \le 10]/10$. Draw $p(x \mid L)$ for $L = \{4, 7, 2, 6\}$.

3. [*18*]   Given the Gaussian random variables $x_1 \sim \mathcal{N}(\mu_1, \sigma_1^2)$ and $x_2 \sim \mathcal{N}(\mu_2, \sigma_2^2)$, prove that

$$x_1 + x_2 \sim \mathcal{N}\left(\frac{\sigma_2^2 \mu_1 + \sigma_1^2 \mu_2}{\sigma_1^2 + \sigma_2^2}, \frac{\sigma_1^2 \sigma_2^2}{\sigma_1^2 + \sigma_2^2}\right), \qquad x_1 * x_2 \sim \mathcal{N}\left(\mu_1 + \mu_2, \sigma_1^2 + \sigma_2^2\right).$$

4. [*18*]   Consider a Bernoulli data distribution. Prove that the beta distribution is conjugate to the Bernoulli likelihood.

5. [*18*]   Use recursive Bayes learning for estimating $p(x \mid L)$, in the case in which we believe that our one-dimensional samples come from a normal distribution. In particular, assume that σ is given and that the only parameter to be estimated is $\theta = \mu$. We also assume that the prior is $\mathcal{N}(\mu_0, \sigma_0^2)$. Finally, extend the result to multivariate Gaussians.

2.2.4 Graphical modes

The described general framework of Bayesian learning may require a huge computational burden. In many real-word problems, the random variables have a network of dependencies that somewhat express the prior knowledge. An important step for the evolution of the Bayesian approach to machine learning is that of using a *probabilistic graphical model*, which is just a special way of representing a family of probability distributions.

---

[6] Here there's an overloading of symbol p, since we should have written $p_{\Theta|L}(\theta \mid L_0) = p_\Theta(\theta)$.

In the figure the five random variables A, B, C, D, and E are represented by a DAG where the vertexes correspond to random variables and the edges represent statistical dependencies between the variables. Graphs are an intuitive way of representing the relationships between many variables. Examples are, among others, neural networks, airplane connections, and different kinds of social network. A graph allows us to gain the abstraction connected with the conditional independence between the variables. Because of the expression of dependencies, we can compute the probability density of cooccurrence of the variable as follows:

$$p(a, b, c, d, e) = p(e \mid a, b, c, d) p(d \mid a, b, c) p(c \mid a, b) p(b \mid a) p(a)$$
$$= p(e \mid c) p(d \mid a, c) p(c \mid a, b) p(b) p(a).$$

In the second line of the above formula we have repeatedly used the properties of conditional independence between the random variables as described in the above DAG; for example, $p(e \mid a, b, c, d) = p(e \mid c)$ since the only parent of E is C. This can be generalized to any Bayesian network $B := \{V, A\}$, where V is the set of vertices and $A \subset V^2$ is the set of arcs. Now we assume that any vertex $v_i \in V$ is associated with the random variable V_i. We have

$$p(v_1, \ldots, v_\ell) = \prod_{i=1}^{\ell} p(v_i \mid v_{\mathrm{pa}(i)}). \tag{1}$$

Here, pa(i) is the set of parents of i. The previous graph gives an insight on the proof of this factorization (see Exercise 1). We notice in passing that a set of ℓ points generated according to a classic statistic models, like i.i.d. Gaussian distributions with mean μ and standard deviation σ, can be represented by the Bayesian network

$$B = \{(X_1, \mu), (X_1, \sigma), \ldots, (X_\ell, \mu), (X_\ell, \sigma)\}.$$

We say that we perform inference in Bayesian networks whenever we evaluate the probability distribution over some set of variables, given the values of another set of variables. As an example, still with reference to the above Bayesian network, we can compute $p(B \mid C = c)$, under the assumption that the variables are Boolean. We have

$$\mathrm{Pr}(BC = bc) = \sum_{ade=000}^{111} \mathrm{Pr}(ABCDE = abcde),$$

$$\mathrm{Pr}(C = c) = \sum_{b=0}^{1} \mathrm{Pr}(BC = bc),$$

$$\mathrm{Pr}(B = b \mid C = c) = \frac{\mathrm{Pr}(B = b, C = c)}{\mathrm{Pr}(C = c)},$$

where, for compactness, we have used the "random strings" ABCDE and BC, whose range are Boolean strings of length 4 and 2, respectively. This example immediately enlightens the complexity issues behind the inference. In particular, the first of these equations requires us to consider $8 = 2^3$ configurations that are associated with all the triples (A, B, E) that are combined with two configurations for defining any (B, C). Overall, there are 16 configurations. The structure of the computation, however, suggests that there is a curse of dimensionality associated with the marginalization of the variables.

A fundamental step towards the speed up of inference is based on the *belief propagation algorithm* that relies on the notion of *factor graph*. Insights on these concepts are given in Section 2.5. In this case, the theoretical and experimental advances are connected with the expression of special dependencies amongst the variables. For example, given three variables x_1, x_2, and x_3, we get the expression

$$X_1 \perp (X_2, X_3) \Leftrightarrow p(x_1, x_2, x_3) = p(x_1)p(x_2, x_3).$$

The network structure of the Bayesian network somewhat expresses the prior knowledge about the problem. Let us consider the classic example of document classification, in which each document is represented by a vector $x \in \{0, 1\}^d$. A generic component x_i represents the presence of a given feature of the document. In the simplest case one can think of representing the document as a bag of words, so as to associate x_i with the presence of the i-th keyword of the chosen dictionary in the document.[7] If we want to classify the document, we need a random variable Y to state whether the given document belongs to a certain class $i = 1, \ldots, c$. Now the classification task can be interpreted as an inferential process over the $d + 1$ random variables $(X_1, \ldots, X_d, Y)$. Suppose that for all i and j in $\{1, \ldots, d\} \subset \mathbb{N}$ we have $X_i \perp X_j$. Then

$$p(y, x_1, \ldots, x_d) := \Pr(Y = y, X_1 = x_1, \ldots, X_d = x_d) = q(y) \prod_{i=1}^{d} q_i(x_i \mid y),$$

where $q(y) := \Pr(Y = y)$ and $q_i(x \mid y) := \Pr(X_i = x \mid Y = y)$. Here $q(y)$ is the prior on the class, while $q_i(x_i \mid y)$ expresses the probability that feature i is present in the document, once we know that it belongs to the class defined by $Y = y$. Now we have

$$\arg\max_{y \in \{1, \ldots c\}} p(y, x_1, \ldots, x_d) = \arg\max_{y \in \{1, \ldots, c\}} q(y) \prod_{i=1}^{d} q_i(x_i \mid y).$$

A natural way to estimate $q(y)$ and $q_i(x \mid y)$ is to use the frequentist approach, which leads us to choose

$$\hat{q}(y) = \frac{1}{\ell} \sum_{\kappa=1}^{\ell} [y_\kappa = y], \qquad \hat{q}_i(x \mid y) = \frac{\sum_{\kappa=1}^{\ell} [x_{\kappa i} = x][y_\kappa = y]}{\sum_{\kappa=1}^{\ell} [y_\kappa = y]}. \qquad (2)$$

---

[7] This is in fact quite a poor representation. In classic information retrieval the term frequency tf is integrated with the inverse document frequency idf, and $x_i = tf \cdot idf$.

Interestingly (as shown in Exercise 2), the same conclusion on the estimation can be drawn by using MLE.

EXERCISES

  1. [*18*]   Prove Eq. (1) by induction on the number of the vertices.

  2. [*20*]   On the basis of the discussion on the naive Bayes classifier, prove Eq. (2) concerning the estimation of $q(y)$ and $q_i(x \mid y)$ by using MLE.

2.2.5 Frequentist and Bayesian approach

Frequentists and Bayesians come from a different school of thought. Bayesian inference uses probabilities to deal with both hypotheses and data. In addition, inferential processes depend on the prior and likelihood of observed data. The assumption on the prior, which is central to Bayesian inference, is typically regarded as subjective. It may be computationally intensive due to integration over many parameters.

By and large, a severe criticism of Bayesian inference is that prior is subjective! In many cases, it's typically difficult to choose a criterion for prior which is not arguable. Different people might come up with different priors and, therefore, they might arrive at different posteriors and conclusions. But, there's more: A fundamental philosophical objection concerns the assignment of probabilities to hypotheses. Basically, hypotheses do not constitute outcomes of repeatable experiments, and this is the source of discussion with frequentists, who state that there is no way of computing long-term frequency for the experimental validation. Strictly speaking, hypotheses are either true or false, regardless of whether one knows which is the case. Hence, one might be inclined to reject the principle of treating them as random variables and, consequently, of rejecting Bayesian inference. Hypotheses are not like data! A coin is either fair or unfair, and tomorrow it will be sunny, cloudy, or raining. Can we state something similar for hypotheses?

Frequentist inference never attaches probabilities to hypotheses, so that there are neither prior nor posterior probabilities. The inferential process depends on the likelihood for both observed and unobserved data, and it's typically more efficient than in the Bayesian framework. Frequentists believe that only data can be interpreted as a repeatable random sample. As a consequence, the frequency of occurrence is properly used to provide their probabilistic meaning. According to their view, we cannot do the same for the parameters, which are constant during repeatable processes of estimation.

According to the Bayesian viewpoint, parameters are unknown and are probabilistically described. We can think of the Bayesian approach as one which treats the probabilities as degrees of belief, rather than as frequencies generated by some unknown process. To sum up, in the Bayesian approach we assign probabilities to hypotheses, whereas the frequentist viewpoint is that of testing a hypothesis without an assigned probability.

Table 2.1 Meals at Berger's burgers.

Entrée	Cost [$]	Calories	Probability of arriving hot	Probability of arriving cold
Burger ($E = b$)	1	1000	0.5	0.5
Chicken ($E = c$)	2	600	0.8	0.2
Fish ($E = f$)	3	400	0.9	0.1

2.3 Information-based learning

Now we introduce information-based learning principles. They are inspired by the maximum entropy principle, which returns decisions that are consistent with constraints coming from the environment and turn out to be as unbiased as possible. There are intriguing connections with statistical mechanics, which contribute to shed light on the emergence of cognitive processes in both human and machines.

2.3.1 A motivating example

We start with a motivating example concerning strategic plans to be adopted in a fast-food restaurant. [This example has been proposed in the notes of the course on "Information, Entropy and Computation," by Seth Lloyd and Paul Penfield Jr., see http://www-mtl.mit.edu/Courses/6.050/2014/notes/chapter9.pdf.] At Berger's Burgers, a fast food restaurant, meals are prepared using high tech equipment to optimize the service. Three different meals are served. Burger Meal 1 (beef) costs 1 $, delivers 1000 calories, and has a probability of 0.5 of arriving cold. Burger Meal 2 (chicken) costs 2 $, has 600 calories, and a probability of 0.2 of arriving cold. Burger Meal 3 (fish) costs 3 $, has 400 calories, and has a 0.1 probability of being cold. This is summarized in Table 2.1.

There are several questions that can be asked, all involving the initial assumption about the buying habits, i.e., about the probabilities $\Pr(E = b) =: \Pr(b)$, $\Pr(E = c) =: \Pr(c)$, and $\Pr(E = f) =: \Pr(f)$ of each of the three meals being ordered. How can we discover these probabilities? Without any additional information, we could reasonably guess that $\Pr(b) = \Pr(c) = \Pr(f) = 1/3$, which is in fact the most unbiased decision. Suppose we are told that the average price of a meal is 1.75 $. This is clearly biasing the probability distribution of E. As an extreme case, if the average price is 3 $, it means that customers almost surely eat fish! (Formally, this means that customers eat fish with probability one.) Hence the information on the average price is definitely providing evidence on the probability distribution of serving meals.

It turns out that the above probabilities must satisfy the constraints:

$$\Pr(b) + \Pr(c) + \Pr(f) = 1, \quad \Pr(b) + 2\Pr(c) + 3\Pr(f) = 7/4. \tag{1}$$

When the information on the average price is not available, the choice of equal probabilities corresponds with the maximization of the entropy of the random variable E (entrée), which takes on the discrete values a, b, and c. If we keep the same principle, then from the observation of the average price, we can make a guess on $\Pr(b)$, $\Pr(c)$, and $\Pr(f)$ by maximizing

$$S(E) := -\Pr(b)\log\Pr(b) - \Pr(c)\log\Pr(c) - \Pr(f)\log\Pr(f),$$

under the constraints given by Eq. (1). Here we used base-e logarithm to define the entropy instead of the more common base-2 logarithm. Notice that the probabilities are not affected by this.

If we use the Lagrangian approach, we need to minimize the function

$$\begin{aligned}\mathcal{L} = &\Pr(b)\log\Pr(b) + \Pr(c)\log\Pr(c) + \Pr(f)\log\Pr(f) \\ &+ \lambda(\Pr(b) + \Pr(c) + \Pr(f) - 1) + \beta(\Pr(b) + 2\Pr(c) + 3\Pr(f) - 7/4).\end{aligned} \tag{2}$$

The minimization yields

$$\Pr(b) = \frac{e^{-\beta}}{e^{-\beta} + e^{-2\beta} + e^{-3\beta}}, \qquad \Pr(c) = \frac{e^{-2\beta}}{e^{-\beta} + e^{-2\beta} + e^{-3\beta}},$$

$$\Pr(f) = \frac{e^{-3\beta}}{e^{-\beta} + e^{-2\beta} + e^{-3\beta}},$$

where the satisfaction of the constraint $\Pr(b) + 2\Pr(c) + 3\Pr(f) = 7/4$ makes it possible to determine β as the solution of

$$-3e^{-\beta} + e^{-2\beta} + 5e^{-3\beta} = 0. \tag{3}$$

This yields $\beta = (1 + \sqrt{61})/6$ and, consequently $\Pr(b) = (19 - \sqrt{61})/24$, $\Pr(c) = (-4 + \sqrt{61})/12$, and $\Pr(f) = (13 - \sqrt{61})/24$. As a result, the entropy is $S(E) \simeq 1.05117$ nats $\simeq 1.51652$ bits. The knowledge of these probabilities makes it possible addressing a number of interesting questions. As an example, we can compute the average calorie count c, which turns out to be $c = 1000\Pr(b) + 600\Pr(c) + 400\Pr(f) = (25/3)(97 - \sqrt{61}) \simeq 743.248$. Likewise, the probability m_c of a meal being served cold is $m_c = (1/2)\Pr(b) + (1/5)\Pr(c) + (1/10)\Pr(f) = (46 - \sqrt{61})/120 \simeq 0.318$. What would be the solution of the problem, and particularly the value of Lagrangian multiplier β, in case of average price equal to 1 or 3?

2.3.2 Principle of maximum entropy

The estimation of the probabilities in the Berger's Burgers problem has been based on what is referred to as the *Principle of Maximum Entropy* (MaxEnt). It is based on the idea that the probability distribution of a random variable can be estimated in such a way to leave you the largest remaining uncertainty (i.e., the maximum entropy) consistent with your constraints. Berger's Burgers problem suggests a natural extension to the general case in which the environment requires us to enforce a set of n linear constraints. Interestingly, this problem is very well known in physics! The principle of maximum entropy offers a classic and elegant approach to connect "global" notions from thermodynamics, like temperature, to the energy of microscopic configurations of particles.

Let us replace the entrée random variable E with the generic state variable Y. In physics, we typically assume that y_κ, for $\kappa = 1, \ldots, n$, is the state associated with a particle, while p_κ and E_κ are the corresponding probability distribution and the energy of the state y_κ, respectively. Overall, the thermodynamic system is perceived as one with average energy

$$E = \sum_{\kappa=1}^{n} p_\kappa E_\kappa. \tag{1}$$

The maximum entropy principle states that the system will settle into configurations that maximize the entropy

$$S(p) = -\sum_{\kappa=1}^{n} p_\kappa \log p_\kappa, \tag{2}$$

where the probabilities satisfy the normalization condition $\sum_{\kappa=1}^{n} p_\kappa = 1$. There are intriguing connections with information entropy that can very well be captured when considering the Berger's Burgers example. If we associate the energy of the particles to the average meal price, then we end up with the same equations! When using the Lagrangian approach, this leads to the minimization of

$$\mathcal{L}(p) = \sum_{\kappa=1}^{n} p_\kappa \log p_\kappa + \beta \left(\sum_{\kappa=1}^{n} p_\kappa E_\kappa - E \right) + \lambda \left(\sum_{\kappa=1}^{n} p_\kappa - 1 \right).$$

Differentiation with respect to the variable p_i yields $\log p_i + 1 + \beta E_i + \lambda = 0$, from which we get

$$p_i = e^{-\lambda-1} e^{-\beta E_i}.$$

From the probabilistic normalization we get

$$e^{-\lambda-1} \sum_{\kappa=1}^{n} e^{-\beta E_\kappa} = 1,$$

and therefore

$$p_i^* = \frac{e^{-\beta E_i}}{\sum_{\kappa=1}^{n} e^{-\beta E_\kappa}} = \frac{1}{Z} e^{-\beta E_i}, \tag{3}$$

where the normalization term $Z = \sum_{\kappa=1}^{n} e^{-\beta E_\kappa}$ is referred to as the *partition function*. Now we need to determine the Lagrangian multiplier β, which can be obtained by imposing the other constraint,

$$E = \sum_{\kappa=1}^{n} p_\kappa E_\kappa = \frac{1}{Z} \sum_{\kappa=1}^{n} E_\kappa e^{-\beta E_\kappa}. \tag{4}$$

Given E, this is a nonlinear equation in β, which is in fact just a generalization of Eq. 2.3.2–(3). Interestingly, Lagrangian multiplier β has an intriguing meaning, which can be grasped when we start noticing that

$$E = -\frac{\partial}{\partial \beta} \log Z. \tag{5}$$

Now the maximum value of the entropy can be obtained when plugging Eq. (3) into the entropy definition, so that we get

$$
\begin{aligned}
S(p^*) &= -\sum_{\kappa=1}^{n} p_\kappa^* \log p_\kappa^* \\
&= -\frac{1}{Z} \sum_{\kappa=1}^{n} e^{-\beta E_\kappa} \log \frac{e^{-\beta E_\kappa}}{Z} \\
&= -\frac{1}{Z} \sum_{\kappa=1}^{n} \left(-\beta E_\kappa e^{-\beta E_\kappa} - \log Z e^{-\beta E_\kappa} \right) = (\beta E + \log Z).
\end{aligned}
\tag{6}
$$

Now let $T = 1/\beta$ and, moreover, define

$$F := -T \log Z. \tag{7}$$

Then, from Eq. (6), we get

$$E = ST + F.$$

Moreover, still from Eqs. (6) and (5), we have

$$\beta E + \log Z = -\beta \frac{\partial}{\partial \beta} \log Z + \log Z.$$

Then we get

$$S = -\frac{1}{T} \frac{\partial}{\partial T} \log Z \frac{\partial T}{\partial \beta} + \log Z = T \frac{\partial}{\partial T} \log Z + \log Z = \frac{\partial}{\partial T} T \log Z = -\frac{\partial F}{\partial T}.$$

This analysis attaches the meaning to the Lagrangian multiplier β, which is transferred to the temperature T.

EXERCISES

1. [17]   Consider the Berger's Burgers problem in the cases in which $E \to 1.00\,\$$ or $E \to 3.00\,\$$. Prove that $\beta \to +\infty$, that is, $T = 1/\beta \to 0$. Provide an interpretation of this "cold solution".

2. [20]   Consider a linear machine which performs supervised learning. Suppose we regard its weights as probabilities, so that $b + \sum_{\kappa=1}^{d} w_i = 1$. Formulate learning as MaxEnt, where the entropy is maximized under the constraints imposed by the training set. Discuss the corresponding solution. *Hint:* The constructed solution is somewhat opposite with respect to sparse solutions.

2.3.3 **Maximum mutual information**

The general framework of maximum entropy is not directly adequate to provide an appropriate interpretation of most interesting learning tasks, where the probabilities p_κ associated with the states y_κ also explicitly depend on the perceptual values x_κ. The entropy is defined for a discrete random variable Y, which expresses a compact symbolic description of the given perceptual data $x_\kappa \in X$. In machine learning we typically distinguish the cases in which random variable Y is supervised from the case in which there is no supervision.

We start considering the case of unsupervised learning, in which we can think of $(y_1, \ldots, y_n)$ as a vector with corresponding emission probabilities, associated with the encoding functions $f = (f_1, \ldots, f_n)'$. In order to enforce probabilistic normalization, we use a softmax function, so that for $i = 1, \ldots, n$

$$y_i = \Pr(Y = i \mid x, f) = \frac{e^{f_i(x)}}{\sum_{j=1}^{n} e^{f_j(x)}}. \tag{1}$$

As a consequence, the vector

$$y = ([y_1 > 0.5], \ldots, [y_n > 0.5])' \tag{2}$$

turns out to be the output code associated with the input x. Clearly, the softmax is enforcing solutions that are expected to approximate one-hot codes. Like in the previous discussion, we are expected to express a set of constraints on the probabilities which depend on the learning environment. Interestingly, even in the case in which there is no explicit constraint, we are interested in enforcing a special development, which promotes solutions that maximize the dependency of the output code on the inputs. We can express the conditional entropy of $Y(f)$ with respect to X as

$$S(Y \mid X)(f) = -\mathrm{E}_{XY}(\log p_{Y|X,F}(y \mid x, f))$$

$$= -\sum_{j=1}^{n} \int_X p_{XY|F}(j, x \mid f) \log p_{Y|X,F}(j \mid x, f) dx$$

$$= -\sum_{j=1}^{n} \int_X p_{Y|X,F}(j \mid x, f)) \log p_{Y|X,F}(j \mid x, f)) p(x) dx.$$

We notice in passing that, since we marginalize with respect to X only, the conditional entropy $S(Y \mid X)(f)$ can be regarded as a function of the element $f \in F$ chosen for modeling the probabilities defined by Eq. (1). The minimization of $S(Y \mid X)(f)$ corresponds with the enforcement of the dependency between random variable X and the attached symbol defined by random variable Y for a given $f \in F$. As an extreme interpretation, one might be willing to enforce the hard constraint $S(Y \mid X)(f) = 0$, but its soft-satisfaction plays a similar role in many real-world cases. The availability of finite unsupervised training set $X = \{x_1, \ldots, x_\ell\}$ makes it possible to replace the above conditional entropy with the corresponding empirical approximation

$$S(Y \mid X)(f) = -\frac{1}{\ell} \sum_{j=1}^{n} \sum_{\kappa=1}^{\ell} p_{Y \mid X, F}(j \mid x_{\kappa}, f)) \log p_{Y \mid X, F}(j \mid x_{\kappa}, f)). \tag{3}$$

We can easily realize that while the satisfaction of the constraint $S(Y \mid X)(f)$ is desirable in order to produce strong decisions, its satisfaction doesn't guarantee the discovery of appropriate solutions. Clearly, any trivial configuration like $Y = (1, 0, \ldots, 0)$ arising independently of X satisfies the constraint! The application of MaxEnt makes it possible to get rid of these trivial configurations. Given f, in order to apply the MaxEnt, we consider

$$Y(f) = \mathrm{E}_X(Y(X, F = f)) \simeq \frac{1}{\ell} \sum_{\kappa=1}^{\ell} p_{Y \mid X, F}(j \mid x_{\kappa}, f).$$

Then we maximize the entropy $S(Y)(f)$ under the constraint $S(Y \mid X)(f) = 0$. The maximization of the entropy $S(Y)(f)$ leads in fact to decisions that are as unbiased as possible. Let $0 \leq \mu \leq 1$. The soft interpretation of the constraint $S(Y \mid X)(f) = 0$ leads to the minimization of

$$D_\mu(X, Y)(f) := (1 - \mu)S(Y \mid X)(f) - \mu S(Y)(f). \tag{4}$$

Hence a collection of unsupervised data can be clustered into n groups by minimizing $D_\mu(X, Y)(f)$, which is in fact a soft-implementation of MaxEnt. Clearly, the value of μ may significantly affect the solution. Values of $\mu \simeq 1$ stress the discovery of strongly unbiased solutions that, however, tend to be independent of the random variable X. On the other hand, values of $\mu \sim 0$ are only enforcing the constraint $S(Y \mid X)(f) = 0$. However, as already pointed out, this ends up in trivial solutions that are, again, independent of X. Hence it is instructive to know more of $D_\mu(X, Y)(f)$ when μ provides an appropriate balance between the two terms of Eq. (4). We can provide a nice interpretation of $D_\mu(X, Y)(f)$ when rearranging D_μ as follows:

$$
\begin{aligned}
D_\mu &= (1 - \mu)S(Y \mid X)(f) - \mu S(Y)(f) \\
&= -\frac{1 - \mu}{\ell} \sum_{j=1}^{n} \sum_{\kappa=1}^{\ell} p_{Y \mid X, F}(j \mid x_{\kappa}, f)) \log p_{Y \mid X, F}(j \mid x_{\kappa}, f)) \\
&\quad + \frac{\mu}{\ell} \sum_{j=1}^{n} \sum_{\kappa=1}^{\ell} p_{Y \mid X, F}(j \mid x_{\kappa}, f)) \log \left(\frac{1}{\ell} \sum_{\kappa=1}^{\ell} p_{Y \mid X, F}(j \mid x_{\kappa}, f)) \right) \\
&= -\frac{1}{\ell} \sum_{j=1}^{n} \sum_{\kappa=1}^{\ell} p_{Y \mid X, F}(j \mid x_{\kappa}, f))) \log \frac{p_{Y \mid X, F}(j \mid x_{\kappa}, f))^{1-\mu}}{p_{Y \mid F}(j, f)^{\mu}}
\end{aligned}
$$

$$= -\sum_{j=1}^{n}\sum_{\kappa=1}^{\ell} \frac{p_{Y|X,F}(j\mid x_\kappa, f))}{\ell} \log \frac{p_{Y|X,F}(j\mid x_\kappa, f))^{1-\mu} p_{Y|X,F}(j\mid x_\kappa, f))^{\mu}}{p_{Y|\Phi}(j, f)^{\mu} p_{Y|X,F}(j\mid x_\kappa, f))^{\mu}}$$

$$= -\sum_{j=1}^{n}\sum_{\kappa=1}^{\ell} p_{XY|F}(x_\kappa, f, j) \log \frac{p_{XY|F}(x_\kappa, j, f) p_X(x_\kappa)^{-1}}{p_{Y|X,F}(j\mid x_\kappa, f))^{\mu} p_{Y|F}(j, f)^{\mu}}$$

$$= -D_{KL}(p_{XY|F=f} \| (p_{Y|F=f}\, p_{Y|X,F=f})^{\mu}\, p_X).$$

Here $D_{KL}(p_{XY|f} \| (p_{Y|f}\, p_{Y|X,f})^{\mu}\, p_X)$ is the *Kullback–Leibler divergence*[8] between distributions $p_{XY|F}$ and $(p_Y\, p_{Y|X,F})^{\mu}$. D_{KL} is also referred to as relative entropy; definitions and properties are discussed in Exercise 1. For $\mu = 1/2$, we have

$$D_{KL}(p_{XY|f} \| (p_Y\, p_{Y|X,f})^{1/2}\, p_X) = D_{KL}(p_{XY|f} \| (p_Y\, p_{Y|X,f}\, p_X)^{1/2}\, p_X^{1/2})$$

$$= D_{KL}(p_{XY|f} \| (p_{Y|f}\, p_{XY|F}\, p_X)^{1/2})$$

$$= \frac{1}{2} D_{KL}(p_{XY|f} \| p_{Y|\Phi}\, p_X) = \frac{1}{2} I(X, Y \mid f),$$

where $I(X, Y \mid f)$ is the mutual information between X and Y, conditioned on the choice of random variable $F = f$. Finally, for $\mu = 1/2$, the MaxEnt principle leads to discovering

$$f^* \in \arg\max_{f \in F} I(X, Y \mid f). \tag{5}$$

This can be given a direct interpretation, which is connected with the meaning of $I(X, Y \mid f)$. Whenever random variable Y depends on X, Eq. (5) prescribes the emission of symbols according to the probability distribution of Eq. (1), where f maximizes the mutual information $I(X, Y \mid f)$.

Now let us consider the case of supervised data. The previous analysis can still be carried out. The constraints that we impose in this case are explicit translation of the supervised pairs of the learning set. Hence, like for unsupervised learning, we can easily see that we need to impose soft-satisfaction of $S(Y \mid X)(f) := S(Y \mid X, F = f)$. Suppose that we dealing with two classes. In this case $p_{Y|X}$ follows the Bernoulli distribution,

$$p(y \mid x) = f(x)^{y}(1 - f(x))^{1-y} \tag{6}$$

and therefore

$$S(Y \mid X) = -\sum_{\kappa=1}^{\ell} \log p(y_\kappa \mid x_\kappa) = -\sum_{\kappa=1}^{\ell} (y_\kappa \log f(x_\kappa) + (1 - y_\kappa)\log(1 - f(x_\kappa))).$$

$$\tag{7}$$

---

[8] For the sake of simplicity, we replace $F = f$ with f.

The MaxEnt principle requires us to maximize $S(Y)$ under the soft-constraint $S(Y \mid X)$. The maximization of $S(Y)$ leads to configurations that are as unbiased as possible. The same weight given to $S(Y)$ and $-S(Y \mid X, f)$ leads to maximizing $I(X, Y, f) = S(Y) - S(Y \mid X, f)$. Basically, we ended up with the same principle of maximum mutual information stated by Eq. (5), already discussed for unsupervised learning.

This analysis crosses the discussion on loss functions for supervised learning carried out in Section 2.1.1. In particular, the relative entropy loss corresponds with Eq. (7). It's worth mentioning that while the relative entropy and the conditional entropy yield the same result in case of supervised learning, $S(Y \mid X, f)$ gives rise to more general conclusions, since it can be used also for unsupervised learning. Furthermore, we notice the common role of maximization of $S(Y)$ in supervised and unsupervised learning. However, in supervised learning the weight to be attached to $S(Y)$ depend on the way the training set is effectively representing the concept. Whenever it is abundant for capturing the underlying probability distribution, the term $S(Y)$ plays a minor role. This is not true for unsupervised learning, where $S(Y)$ does play a more crucial role. In particular, as already noticed, it prevents from creating trivial configurations. This discussion indicates that, even though $I(X, Y, f)$ gives rise to a general learning criterion, there's room for choosing information-based criteria that properly balance the terms $S(Y)$ and $S(Y \mid X, f)$.

EXERCISES

1. [*18*]   Given two probability distributions $p(x)$ and $q(x)$, their Kullback–Leibler distance is defined as $DL(p,q) := \mathrm{E}(p \log(p/q))$ Prove that $DL(p,q) \geq 0$. Suppose we define the symmetric function $S(p,q) := DL(p,q) + DL(q,p)$. Is this a metrics?

2.4 **Learning under the parsimony principle**

In this section, we introduce the parsimony principle and present its general philosophical nature according to the Occam razor. We discuss its role and different interpretations in science and, particularly, in machine learning.

2.4.1 **The parsimony principle**

In Section 1.2.1 we have discussed the puzzling nature of induction, which calls for well-posed explanations of the observed data. The parsimony principle (lex parsimoniae in Latin) is typically connected with classic Occam razor in philosophy, which states that *entities should not be multiplied beyond necessity*. Hence, whenever we have different explanations of the observed data, the simplest one is preferable. Clearly, this opens nontrivial issues on the correct interpretation of simplicity, an issue that, as pointed out in Section 1.2.1, might be controversial. The search for parsimony is a sort of universal feature pervading nearly any field of science. It provides a straightforward interpretation of many laws of nature (see Section 2.5 for a preliminary discussion) and it nicely drives decision process mechanisms. The unifying framework that arises is one in which an intelligent agent returns the simplest decision that is compatible with the environmental constraints. Unlike probabilistic and statistical inferences, such an agent decides on the basis of an appropriate notion of simplicity. The arbitrariness of any definition of simplicity seems to parallel related features, like elegance and beauty. While humans may easily disagree on the election and scoring of similar features, they are often attracted by the ambition of achieving their universal interpretation. This is especially interesting when the environment in which the agent is expected to live is formally described. Moreover, as it will be shown in the following, the parsimony principle shares borders with information theory and statistics, which allows us to introduce definitions of simplicity that exhibit interesting degrees of consistency with the solution that arises from these theories. Yet, we operate under a remarkably different principle that leads to different results. This raises the question of which principle we must obey. At a first look, the statistical framework is exactly what one is looking for, since it guarantees the optimal generalization behavior. Unfortunately, as better stated in Section 2.5, it relies on the arguable assumption of the *true state of nature*, which partly compromises the soundness of the related inferential mechanisms. On the other hand, the parsimony principle doesn't suffer from this lack of consistency with the hypotheses, but it is plagued by the arbitrariness of the formal translation of simplicity.

2.4.2 **Minimum description length**

A very natural way of using the parsimony principle is to assume a radically different philosophical viewpoint with respect to most statistical approaches to machine learning. We start to emphasize the importance of deciding among competing explanations of data. When introducing the general notion of induction, in Section 1.2.1, we touched some of its puzzling aspects, which can make the decision among competing explanations somewhat embarrassing. For example, why should we prefer Bob's interpretation of stylized clocks instead of Jack's? Clearly, not because Bob is in fact the supposed expert who's in charge of handling the test! In its apparent simplicity, the IQ example raises a very general problem that is of central importance to inductive and statistical inference. A possible way to face the problem is to use the minimum description length (MDL) principle. It is based on the idea that any regularity in the data can be used for compression, that is, we can use fewer symbols than those needed to describe the original source. The more regularities are captured, the more data can be compressed. The examples of Section 1.2.1 help understanding the deep meaning of capturing regularities. There are cases in which they are quite evident, in others it looks like data are truly random. Regardless of the tricky issues connected with the deep meaning of randomness, learning can properly be regarded as the act of discovering regularities. The following explicit case, which is often considered to introduce MDL, does help. Suppose we are in front of the sequence

$$
\begin{array}{llllllll}
n = & 1 & 2 & 3 & 4 & 5 & \cdots & 1000 \\
s_n = 01101 & 01101 & 01101 & 01101 & 01101 & \cdots & 01101
\end{array} \tag{1}
$$

which is an obvious example of data generated under a simple regularity. The prediction on what bits come next is easy: There is in fact an obvious periodic regularity. We can reproduce the sequence in a simple way by Algorithm S.

Algorithm S (*Periodic sequence generator*).   Print the sequence in Eq. (1).

S1. [Initialize.] Set $s \leftarrow 01101$ and $k \leftarrow 1$.

S2. [Print.] Print s. Terminate if $k = 1000$; otherwise $k \leftarrow k + 1$ and repeat this step.
∎

   We can promptly realize that there is a remarkable difference between the length of the original sequence and the length of the program that generates it. Suppose the bits are in fact coded just as any character by 1 byte. The length of the sequence (1) is about 5 Kbytes. The calculation of the length of the above generating program requires some assumptions. If we consider the two steps S1 and S2 and assume to count the characters composing the lines, as well as the control characters, then we find that the program is about 120 bytes long. While keeping the generating rule instead of the original sequence already yields a remarkable saving of space, we can immediately realize that as the length of the sequence increases, the saving increases dramatically. The symbolic structure of the program doesn't change, since the length of the sequence only affects step S2 (when we specify the bound on k), where the size of the period T is proportional[9] to $\log T$. To sum up, while the sequence length is $O(T)$, the length of the generating program is $O(\log(T))$. Of course, a similar difference involves the sequence length, since $n = 5T$ and therefore we keep the

---

[9] This is an asymptotic property of the general rule stating that the number of bits to store an integer m is $\lfloor 1 + \log m \rfloor$.

compression ratio $n/\log n$. This strong compression is the sign of a concept that has been strongly learned, one in which the regularities have been fully captured. Similar analyses can be drawn for the Fibonacci sequence 1.2.1–(1): Again, its computation by Eq. 1.2.1–(4) can be written by a program with constant length and, therefore, as $n \to \infty$, we only need a space proportional to $\log n$ to store the maximum number that must be generated. Interestingly, notice that this ratio $n/\log n$ holds regardless of the linguistic coding of the two programs involved in these two examples. The case in which a sequence is perfectly captured by a corresponding generating rule is pretty common in math; transcendental numbers like π and the Neper number can be expressed as series that characterize the numbers, just like a machine learning concept. In all these cases the program, which compresses the sequence, gains the maximum compression ratio $n/\log n$. It is a sort of *divine ratio*, which, to some extent, drives our ambition of disclosing what is behind an information source. As argued in Section 1.2.1, the elusive and puzzling nature of randomness leads us only to the partial discovery of hidden regularities.

Now, to formalize our ideas, we need to construct a description method, that is, a formal language to express the regularities to be captured in the data. The pseudocode used in this book, adopted in Algorithm S, is fine, but there are clearly infinitely many possibly different descriptions of the same rule. For example, Algorithm S could have been described using just one step:

S′. [Print the sequence.] Print 01101 1000 times.

While this description is shorter than S's, they share the same asymptotic property: The length of both of them is $O(\log n)$. Hence this extreme compression is not affected by the two different descriptions that have been used.

This discussion on randomness allows us to introduce the fundamental notion of Kolmogorov complexity. Suppose we are in a situation that has been widely discussed also in Section 1.2.1, in which a given string in $\{0, 1\}^*$ is interpreted by a hypothesis that predicts the future content of the sequence. Among the infinite number of possible hypotheses, which one should be preferred? Occam's razor suggests to *choose the simplest hypothesis that is consistent with the data*. Once we have chosen the formal language L, we can define the Kolmogorov complexity $C_L(x)$ of a string x as the length of the smallest program that outputs x. A problem with this definition is that complexity $C_L(x)$ is not only an inherent property of the information in x, but that it also depends on the language L. The previous remarks on the asymptotic behavior of the compression ratio suggest that the computational model might not affect the deep meaning of the definition.[10]

It can be proven that if we restrict the discussion between general-purpose (i.e. implementable by a universal Turing machine) computer languages, the complexities of a string x in the two languages L and M differ only by a constant c that does not

---

[10] This is quite a sophisticated issue that is not fully covered in this book. However, some additional details could be found in Section 2.5.

depend on the length of x but only on L and M:

$$|C_L(\mathsf{x}) - C_M(\mathsf{x})| \le c.$$

For this reason, it makes sense to elect any one of those languages as a representative of this asymptotic notion of complexity. We can also drop the dependence on L and refer to the *algorithmic complexity* $C(\mathsf{x})$ as an inherent property of the sequence. It's easy to realize that there exists a constant c, independent of x, such that $C(\mathsf{x}) < |\mathsf{x}| + c$.

Kolmogorov complexity offers a natural framework for describing the notion of *incompressible strings*, which are those for which $C(\mathsf{x}) \ge |\mathsf{x}|$. Incompressible strings have no redundancy, they are algorithmically random. But do incompressible strings exist? In Section 1.2.1, it has been shown that sequences that are apparently random are, on the other hand, driven by a simple generation rule. Bob and Jack were both missing a simple rule when looking at the stylized clocks! The pseudorandom generation rule 1.2.1–(9) clearly provides a sort of divine compression of a sequence that is apparently random! Couldn't it be the case that what we regard as random is simply the outcome of our lack of knowledge on the structure of the data? We can make a substantial step in understanding this question if we start considering collections of strings of a given length n, and analyze whether they can all be compressed. To make this question well-defined, suppose we consider the set S of all the 2^n binary strings; we want to see if it is always possible for all of them to be compressed. Let us assume that there exists a compression program p which can map any string of S to one of dimension $m < n$ in the following way: We map each string of S onto another binary string of length strictly less than n in such a way that this mapping is bijective. However, the number of all possible strings with any length less than n is

$$\sum_{\kappa=0}^{n-1} 2^\kappa = 2^n - 1 < 2^n = |S|.$$

The above inequality shows that there always exists at least one incompressible string! There is more: We can extend this counting argument to show that the vast majority of strings are mostly incompressible. Our strings are expected to be pretty big and, therefore, we are curious to see how the compression properties scale up. It makes sense to consider nearly incompressible (c-incompressible) strings that satisfy the property $|C(\mathsf{x})| \ge |\mathsf{x}| - c$, for some constant $c \in \mathbb{N}$. We have already seen that this holds for $c = 1$. The idea behind this definition is that the compression program p can only save a constant space c, which is irrelevant for long strings. If we repeat the previous counting argument, we have

$$|\overline{S}_c| \le \sum_{\kappa=0}^{n-c} 2^\kappa = 2^{n-c+1} - 1 < 2^n = |S|.$$

This allows us to calculate the fraction of c-incompressible strings. For $c = 10$ we have

$$\rho = \frac{|S| - |\overline{S}_c|}{|S|} = \frac{2^n - (2^{n-9} - 1)}{2^n} = 1 - \frac{1}{512} + \frac{1}{2^n}.$$

Clearly, this makes sense for large n, for which $\rho \simeq 0.998$. As a matter of fact, most long sequences of S are 10-incompressible! In this perspective, the presence of regularities can be interpreted as an exception. Only a few sequences enjoy the presence of a generative underlying rule; most long sequences cannot be usefully compressed. Notice that this says nothing about the randomness of single sequences. As already stated, randomness can be the result of a hidden structure. Hence almost all strings are random, but we cannot exhibit any particular string that is random. These properties on algorithmic complexity of sequences emerge in the framework of collections where one wants to capture the unifying regularities. Notice that the program p is expected to operate uniformly on the different sequences, just like machine learning programs which need to take decision on the basis of their input.

This framework of algorithmic complexity is very general, but it presents a couple of remarkable limitations. First, it can be shown that there exists no computer program that, for every string x, returns the shortest program that prints x. That is, the algorithmic complexity is uncomputable! Second, there is arbitrariness on the description method. While this issue is overcome asymptotically, in practice, we often deal with small data samples for which the invariance theorem does not produce any useful information. Then the hypothesis chosen by idealized MDL may have a nonnegligible impact on concrete examples.

Now we focus on the main topic of casting inferential processes as problems of discovering regularity in the data. MDL regards learning as data compression: It tells us that, for a given set of hypotheses $\mathcal{H}$ and data set X, we must find the hypothesis that compresses X most. This turns out to be a very general scheme of inductive inference, and it turns out to be adequate for dealing with model selection and overfitting.

Let us consider the problem of fitting the set of points indicated in Fig. 2.1. The experimental data come from measurements of the air pressure at different height. The fitting curves are constructed by using polynomial approximation according to the least squares regression. (Technical details on polynomial fitting can be found in Section 3.1.) As we can see in Fig. 2.1, there is a progressive increase of the quality of the data fitting until the degree reaches $m = 6$ (left). On the right, it is clear that for $m = 7$, while the fitting of the given data is very good, as height increases the approximation behavior becomes very bad! Basically, the natural laws suggest that the air pressure exhibits a strong decay towards zero, whereas polynomials can hardly fit similar behavior. Even in absence of knowledge of the laws that regulate the physical phenomenon, however, there is a good approximation for $m = 4$ and, especially, for $m = 6$. As the complexity of the model reaches $m = 7$, we experience a critical data overfitting.

This overfitting behavior is not a characteristic of this problem. As a consequence, model selection methods that are used in practice are required to choose a tradeoff

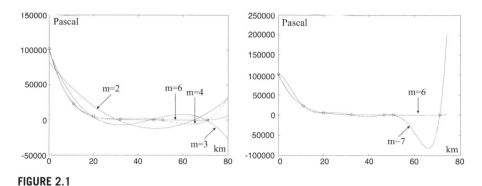

FIGURE 2.1

Dependence of the air pressure on the height. (Left) Progressive increase of the quality of data fitting as the degree of the polynomial grows, until $m = 6$. (Right) While from $m = 7$ the polynomial fits the experimental data very well, its behavior for height in the range of 60–80 km is very bad!

between goodness-of-fit and complexity of the models involved. Similar tradeoffs typically lead to better predictions of test data than one would get by adopting the "low" or "high" polynomial degree. We say that we are facing a *hypothesis selection problem* if we are interested in determining the optimal coefficients of a polynomial of a given degree, whereas in a *model selection problem* we are mainly interested in selecting the degree.

Now suppose that we are given a collection of candidate models $\mathcal{H} = \{H_1, \ldots, H_h\}$. In the previous case of polynomial data fitting, a candidate model is a polynomial of a certain degree, whereas different models correspond with the specific choice of the polynomial degree. Here any hypothesis is supposed to be described in a certain language and properly expressed by a corresponding program. In the following discussion we can interchange hypotheses and programs. As already noticed, the choice of the degree plays a fundamental role in the actual data fitting. Now we want to discover the hypothesis $H \in \mathcal{H}$ that better captures the hidden generating rule. First, it should explain the data as well as possible and, second, according to the Occam razor, it should also be as simple as possible. We can formally use the algorithmic complexity $C(H)$ to express the length of the hypothesis, so that we set $L(H) := C(H)$.

Suppose that the outcome of the measurement of a quantity y on some data $x_1, \ldots, x_\ell$ is described by the random variable $Y = (Y_1, \ldots, Y_\ell)$; on the same data we can compute the random variable $\hat{Y} = (\hat{Y}_1, \ldots, \hat{Y}_\ell)$ whose randomness is given by the multiple possible choices of H. We can say that $\hat{Y}$ is totally dependent on the discrete random variable whose possible values are $H_1, \ldots, H_h$. It is now natural to define the error $\Delta_i = \varepsilon(\hat{Y}_i, Y_i)$ (a natural choice is, for example, $\varepsilon(\hat{Y}_i, Y_i) = |\hat{Y}_i - Y_i|$) The encoding of this error given the model H, in bits, has a length denoted by $L(\delta)$, where $\delta = (\delta_1, \ldots, \delta_\ell)$, the extent of which depends on the way $\hat{Y}$ is correctly modeled by H. Notice that both $L(H)$ and $L(\delta)$ are description lengths expressed in bits.

The MDL principle prescribes that, amongst all possible hypotheses in $\mathcal{H}$, we choose

$$\hat{H} = \min_{H_\kappa \in \mathcal{H}} \left(L(H_\kappa) + L(\delta) \right). \tag{2}$$

In the problem of polynomial data fitting, $L(\delta)$ encodes the error between the given data y and the predicted value $\hat{y}$. Notice that the choice of high-order polynomials favors the reduction of this length, which can even reach the null value. As it will be shown in the next chapter, high order polynomial also suffers from ill-conditioning, which is just another face of the ill-position of learning with models with too many free parameters. However, the higher the degree of the polynomial, the higher the length $L(H)$ of its description. As already stated, a good balance between goodness-of-fit and complexity of the model is typically the winning solution.

In order to apply the MDL principle, according to Eq. (2), we need guidelines for expressing $L(H)$ and $L(\delta)$. As it will be seen in the following, the definition of $L(H)$ is more problematic, while $L(\delta)$ finds a good interpretation within the framework of the code theory. In order to provide a sound definition of the length of the correcting code $L(\delta)$, we need to assume that the application of the principle is associated with the knowledge of $\Pr(\Delta = \delta)$. In that case there exists an optimum correcting code with length

$$L(\delta) = -\log \Pr(\Delta = \delta), \tag{3}$$

that corresponds with the Shannon–Fano code.

2.4.3 MDL and regularization

Now let's go back to the polynomial fitting of functions of one variable, which is sketched in Fig. 2.1. Basically, we want to predict the function

$$f(x) = b + w_1 x + w_2 x^2 + \cdots + w_d x^d \tag{1}$$

on the basis of a training set of examples $\mathcal{L} = \{(x_1, y_1), \ldots, (x_\ell, y_\ell)\}$. Function f represents the hypothesis that has been denoted by H. More generally, this is a regression problem on $\mathcal{L}$ based on a function $f \in \mathcal{F}$. Specific assumptions on the structure of f, which allow us to collapse the optimization to a finite dimensional space, are clearly favoring the reduction of the computational complexity. However, at the moment, let's keep the more general assumption that naturally gives rise to the concept of *regularization*. We start by introducing the penalty index

$$E(f) = \frac{1}{\ell} \sum_{k=1}^{\ell} \left(y_\kappa - f(x_\kappa) \right)^2,$$

which is used to express the data fitting. According to the MDL principle, we need to provide a correcting code to express, for any $\kappa = 1, \ldots, \ell$, the error $y_\kappa - f(x_\kappa)$. Basically, we need to determine $\Pr(\Delta = \delta)$, which allows us to assess the corresponding code length. Here the hypothesis H_κ will be written as $F = f$. Of course,

$\Pr(\Delta = \delta) = \Pr(\hat{Y} = \hat{y}$ and $F = f)$. Then, under Gaussian probabilistic assumptions we have

$$\Pr(\hat{Y} = \hat{y} \text{ and } F = f) = \frac{1}{(2\pi)^{\ell/2}\sigma^\ell} \prod_{\kappa=1}^{\ell} e^{-\frac{(y_\kappa - f(x_\kappa))^2}{2\sigma^2}} = \frac{e^{-\sum_{\kappa=1}^{\ell} \frac{(y_\kappa - f(x_\kappa))^2}{2\sigma^2}}}{(2\pi\sigma^2)^{\ell/2}},$$

we get

$$L(\delta) = -\log \Pr(\Delta = \delta) = \frac{\ell}{2}\log(2\pi\sigma^2) + \sum_{\kappa=1}^{\ell} \frac{(y_\kappa - f(x_\kappa))^2}{2\sigma^2}.$$

Finally, according to Eq. 2.4.2–(2), the application of the MDL principle turns out to be translated into the optimization problem

$$f^* \in \arg\min_{f \in \mathcal{F}} \left(\sum_{\kappa=1}^{\ell} (y_\kappa - f(x_\kappa))^2 + 2\sigma^2 L(f) \right). \tag{2}$$

This equation offers an interpretation of learning, driven by MDL, which corresponds with the *regularization principle*. When adopting the framework of regularization, one looks for the optimization of an index that is composed of the *loss function* and the *regularization term*. Eq. (2) can be interpreted as a regularization problem with quadratic loss and with a regularization term that is specifically charged for discovering hypothesis with small description length. The loss term can be chosen as discussed in Section 2.1.1. However, while the quadratic choice has a direct probabilistic interpretation, for other loss functions such a link is not so explicit. Additional connections are established in Section 2.4.4. In regularization theory,[11] the term $L(f)$ expresses either the smoothness of f or a bound on appropriate norms in the vector space $\mathcal{F}$. Interestingly, both these ways of performing regularization can be interpreted as an algorithmic complexity measure.

In the case of polynomial, if we assume that each coefficient of Eq. (1) takes B bits, then a possible complexity measure is

$$L(f) = (d+1)B, \tag{3}$$

and therefore

$$(b, w_1, \ldots, w_d; d)^{*\prime} \in \arg\min \left(2B\sigma^2(d+1) + \sum_{\kappa=1}^{\ell} (y_\kappa - f(x_\kappa))^2 \right). \tag{4}$$

For any fixed d, this optimization problem corresponds with the classic Least Minimum Squares problem. The joint optimization with the dimension d, assumed as a variable, offers a solution to the problem of overfitting outlined in Fig. 2.1.

---

[11] More technical details on these two different views of regularization are given in Chapter 4.

While the solution of (4) relies on classic optimization in a finite dimensional space, the general statement for determining the solution f in a functional space is more involved (it will be discussed in Section 4.4.4). The drawback of (3) is that it is based on the direct representation of the hypothesis, while disregarding its true complexity. In some cases, two polynomials with the same number of weights can be described by codes having remarkably different length. One can in fact adopt a metalevel and think of the hypothesis f as a collection $\{(x_\kappa, f(x_\kappa))\}$ to be interpreted by MDL again. Clearly, smooth functions are more easy to predict, which suggests adopting description lengths reflecting this property. For single-dimensional inputs, one can define $L(f)$ as

$$L(f) = \int_X f'^2(x)\,dx = \|f\|^2 = \langle Df, Df \rangle, \tag{5}$$

which is a seminorm ($\|f\|^2 = 0$ for constant functions) that is clearly small for smooth functions. If f is a polynomial (we assume $w_0 := b$) $f(x) = \sum_{\kappa=0}^{d} w_\kappa x^\kappa$ then $f'(x) = \sum_{\kappa=1}^{d} \kappa w_\kappa x^{\kappa-1}$, and therefore we get

$$
\begin{aligned}
L(f) &= \int_X (f'(x))^2\,dx \\
&= \sum_{\kappa=1}^{d}\sum_{j=1}^{d} kj w_\kappa w_j \int_X x^{\kappa-1} x^{j-1}\,dx \\
&= \sum_{\kappa=1}^{d}\sum_{j=1}^{d} p_{\kappa j} w_\kappa w_j = \|w\|_p^2,
\end{aligned}
\tag{6}
$$

where $p_{\kappa j} := \kappa j \int_X x^{\kappa-1} x^{j-1}\,dx$ is positive semidefinite by construction and $\|w\|_p^2 := \sum_{\kappa,j} p_{\kappa j} w_\kappa w_j$. According to Eq. (6), we discover hypotheses with small description length whenever we choose polynomials with small coefficients, which lead to small $L(f) = \|w\|_p^2$. This way of expressing the parsimony of the decision can be generalized so we assume that

$$L(f) = \|w\|^p = \left(\sum_{\kappa=1}^{d} |w_\kappa|^p\right)^{1/p}. \tag{7}$$

The parsimonious discovery of f, discussed for $p = 2$, holds for other choices of p, but there are new intriguing properties that arise. In particular, when $p = 1$, an additional parsimonious discovery arises, which leads to sparseness in the space of the learning parameters.

2.4.4 Statistical interpretation of regularization

Suppose we are given a training set $\mathcal{L} = \{(x_1, y_1), \ldots, (x_\ell, y_\ell)\}$ that we want to approximate by the function $f : X \to \mathbb{R}$, so that $y_\kappa = \epsilon_\kappa + f(x_\kappa)$. The pairs (x_κ, y_κ)

are drawn from a known probability distribution, so that they can be regarded as instances of the random variable $L = (X, Y)$, where X is associated with x_κ and Y is associated with y_κ. The function f maps the random variable X to $F = f(X)$, which is expected to get as close as possible to Y. In the Bayesian statistical framework, we can relate the dependencies of these random variables. In particular, we look for the function f that maximizes $\Pr(F = f \mid L = \ell)$, which is the conditional probability of $F = f(X)$ given $L = (X, Y)$. In so doing, we provide a better explanation of random variable L in terms of dependencies between X and Y. Hence, following the Bayes rule, we have[12]

$$
\begin{aligned}
f^* &\in \underset{f \in \mathcal{F}}{\arg\max} \prod_{\kappa=1}^{\ell} p_F(f(x_\kappa) \mid (x_\kappa, y_\kappa)) \\
&= \underset{f \in \mathcal{F}}{\arg\max} \prod_{\kappa=1}^{\ell} \frac{p_L((x_\kappa, y_\kappa) \mid f(x_\kappa)) p_F(f(x_\kappa))}{p_L(x_\kappa, y_\kappa)} \\
&= \underset{f \in \mathcal{F}}{\arg\max} \prod_{\kappa=1}^{\ell} p_L((x_\kappa, y_\kappa) \mid f(x_\kappa)) p_F(f(x_\kappa)).
\end{aligned}
\tag{1}
$$

Now let us assume that

$$
p_L((x_\kappa, y_\kappa) \mid f(x_\kappa)) \propto \exp\left(-\frac{(y_\kappa - f(x_\kappa))^2}{2\sigma^2}\right),
\tag{2}
$$

$$
p_F(f(x_\kappa)) \propto \exp(-\gamma L(f)).
\tag{3}
$$

The first assumption corresponds to assuming that the random variable, available from the samples $\epsilon_\kappa = y_\kappa - f(x_\kappa)$, is normally distributed. Hypothesis (3) can be regarded as a model of the prior knowledge. Interestingly, when regarding $L(f)$ as the description length of the hypothesis f, the prior is coherent with the MDL: Small descriptions yield high prior density. From Eq. (1) we get

$$
\begin{aligned}
f^* &\in \underset{f \in \mathcal{F}}{\arg\max} \prod_{\kappa=1}^{\ell} \exp\left(-\frac{(y_\kappa - f(x_\kappa))^2}{2\sigma^2}\right) \exp(-\lambda L(f)) \\
&= \underset{f \in \mathcal{F}}{\arg\max} \exp\left(-\lambda\sigma^2 L(f) - \sum_{\kappa=1}^{\ell} (y_\kappa - f(x_\kappa))^2\right) \\
&= \underset{f \in \mathcal{F}}{\arg\min} \left(\lambda\sigma^2 L(f) + \sum_{\kappa=1}^{\ell} (y_\kappa - f(x_\kappa))^2\right),
\end{aligned}
$$

---

[12] While this is related to 2.2.2–(2), here we are not restricting the analysis to classification and, moreover, we formulate the problem of learning in the functional space $\mathcal{F}$.

where $\lambda = \gamma\sigma^2$. We can extend Eq. (2) by using the general definition based on loss 2.1.1–(1), so as get $p_L((x_\kappa, y_\kappa) \mid f(x_\kappa)) = \exp(-\lambda V(x_\kappa, y_\kappa, f))$. As a result, we get

$$f^* \in \arg\min_{f \in \mathcal{F}} \left(\lambda I_\cdot(f) + \sum_{\kappa=1}^{\ell} V(x_\kappa, y_\kappa, f) \right), \tag{4}$$

which nicely states the link between Bayesian decision and regularization.

2.5 Scholia

A remarkable topic that hasn't been covered in this book is the formulation of learning as a search problem, which has been made popular in T.M. Mitchell, *Artif. Intell.* **18** (1982), 203–226. Learning machines based on decision trees [see J.R. Quinlan, *Mach. Learn.* **1** (1986), 81–106, J.R. Quinlan *Programs for Machine Learning* (Morgan Kaufmann, 1993) and H. Blockeel, L. De Raedt, *Artif. Intell.* **101** (1998), 285–297] have been widely applied in real-world problems with remarkable achievements. In these cases we deal with discrete learning spaces where the search for the concepts is assisted by information-based criteria.

Section 2.1 Whenever we deal with sums of functions over the data, in statistics, we refer to M-estimators. Least Mean Square is a special case of M-estimators, but they have been the subject of extensive investigation in statistics [A. Shapiro, *Ann. Stat.* **28** (2000), 948–960]. Of course, *M*-estimators are at the basis of any risk function given as a sum of loss terms. An interesting discussion on loss function is given in L. Rosasco, E. De Vito, A. Caponnetto, M. Piana, A. Verri, *Neural Comput.* **16** (2004), 1063–1076.

Instead of following the formulation of risk minimization in finite dimensions, one can use variational calculus to search in infinite dimensional spaces. The learning agent is characterized by the function f which satisfies

$$\delta \int_{X \times Y} V(x, y, f) p_{X,Y}(x, y) \, dx dy = 0.$$

As already noticed, we can conveniently rewrite this equation as

$$\int_X \left(\delta \int_Y V(x, y, f) p_{Y|X}(y \mid x) dy \right) p_X(x) \, dx = 0.$$

Let us consider the case of single output models. Since $\forall x \in X$, $p_X(x) > 0$, we have

$$\int_Y \partial_f V(x, y, f) p_{Y|X}(y \mid x) \, dy = 0. \tag{1}$$

Now we follow again the path of discovering the optimal solution, in a more straightforward way, for regression in the case of L_2 and L_1 metrics. In case of quadratic loss $V(x, y, f) = (1/2)(y - f(x))^2$ we get

$$\int_Y (y - f(x)) p_{Y|X}(y \mid x) dy = 0 \rightarrow f(x) = \int_X y p_{Y|X}(y \mid x) \, dy = E_{Y|X}(y \mid x).$$

For the L_1 metrics, apart from singular points in which $y - f(x) = 0$, we have

$$\partial_f V(x, y, f) p_{Y|X}(y \mid x) = -[y - f(x) > 0] + [y - f(x) < 0].$$

Finally, from Eq. (1), we get

$$\int_Y [y - f(x) > 0] p_{Y|X}(y \mid x) \, dy = \int_Y [y - f(x) < 0] p_{Y|X}(y \mid x) \, dy,$$

which is satisfied by the conditional median 2.1.3–(5). Related analyses can be carried out for different values of the metric exponent p.

The discussion on bias variance dilemma reported in Section 2.1.4 concerns quadratic loss and is related to the seminal paper by S. Geman E. Bienenstock, R. Doursatet, *Neural Comput.* **4** (1992) 1–58. Studies for other loss functions can be found in J.H. Friedman, *Data Min. Knowl. Discov.* **1** (1997), 55–77 and R. Kohavi, D. Wolpert, *Proceedings of the Thirteenth International Conference on International Conference on Machine Learning* (1996), 275–283.

Maximum-likelihood estimation was discovered and widely popularized by Ronald Fisher, although previous studies by Carl Friedrich Gauss, Pierre-Simon Laplace, Thorvald N. Thiele, and Francis Ysidro Edgeworth contain remarkable traces of the basic idea [see J. Aldrich, *Stat. Sci.* **12** (1997), 162–176]. We can associate a score to observation x by defining the variable

$$v := \frac{\partial}{\partial \theta} \ln p_X(x, \theta).$$

We can promptly see that the first moment $\mathrm{E}(v \mid \theta)$ is null, since we have

$$
\begin{aligned}
\mathrm{E}\left(\frac{\partial}{\partial \theta} \ln p_X(x, \theta) \mid \theta \right) &= \int_X \frac{1}{p_X(x, \theta)} \frac{\partial p_X(x, \theta)}{\partial \theta} p_X(x, \theta) \, dx \\
&= \int_X \frac{\partial}{\partial \theta} p_X(x, \theta) \, dx = \frac{\partial}{\partial \theta} \int_X p_X(x, \theta) \, dx = \frac{\partial}{\partial \theta} 1 = 0.
\end{aligned}
$$

The second moment

$$\mathcal{I}(\theta) := \mathrm{E}(v^2 \mid \theta) = \mathrm{E}\left(\frac{\partial}{\partial \theta} \ln p_X(x, \theta) \right)^2 \tag{2}$$

is referred to as *Fisher information*. Since $\mathrm{E}_V(v) = 0$, the Fisher information is the variance of the score. Now suppose that $p_X(x, \theta)$ is twice differentiable with respect to θ. We have

$$
\begin{aligned}
\frac{\partial^2}{\partial \theta^2} \ln p_X(x, \theta) &= \frac{1}{p_X(x, \theta)} \frac{\partial^2}{\partial \theta^2} p_X(x, \theta) - \left(\frac{1}{p_X(x, \theta)} \frac{\partial}{\partial \theta} p_X(x, \theta) \right)^2 \\
&= \frac{1}{p_X(x, \theta)} \frac{\partial^2}{\partial \theta^2} p_X(x, \theta) - \left(\frac{\partial}{\partial \theta} \ln p_X(x, \theta) \right)^2.
\end{aligned}
$$

Now we have

$$
\mathrm{E}\left(\frac{1}{p_X(x,\theta)}\frac{\partial^2}{\partial\theta^2}p_X(x,\theta)\right) = \int_X \frac{1}{p_X(x,\theta)}\frac{\partial^2}{\partial\theta^2}p_X(x,\theta)p_X(x,\theta)\,dx
$$
$$
= \frac{\partial^2}{\partial\theta^2}\int_X p_X(x,\theta)\,dx = \frac{\partial^2}{\partial\theta^2}1 = 0.
$$

Finally, we get

$$
\mathcal{I}_X(\theta) := \mathrm{E}\left(\frac{\partial}{\partial\theta}\ln p_X(x,\theta)\right)^2 = -\mathrm{E}\left(\frac{\partial^2}{\partial\theta^2}\ln p_X(x,\theta)\mid\theta\right).
$$

This is another expression of Fisher information, which is a way of measuring the amount of information that an observable random variable X carries about the unknown parameter θ. This equation clearly indicates that the parameter θ yields high information whenever the probability distribution is concave with high second derivative, which corresponds with high sensitivity to the change of θ. Clearly, when considering i.i.d. variables, Fisher information is additive when considering any two independent variables, that is, $\mathcal{I}_{X,Y}(\theta) = \mathcal{I}_X(\theta) + \mathcal{I}_Y(\theta)$. However, it is worth mentioning that MLE and the associated definitions of score and Fisher information don't require the assumption of i.i.d. When dealing with large data collections, MLE is multivariate normal with mean θ (the unknown true parameter value) and variance $\mathcal{I}(\theta)^{-1}$. Note that in the multivariate case $\mathcal{I}(\theta)$ is a matrix, so the inverse Fisher information involves matrix inversion.

Now let us consider the perspective of information theory. Let us assume that the target $y_k \in Y$ and the predicted value $f(x_k)$, with $x_k \in X$, and X and Y being finite sets, are drawn from random variables with probability distributions $p_Y(\cdot)$ and $p_F(\cdot)$. A reasonable measurement of the disagreement can be given by the *Kullback–Leibler distance*, which was introduced by Kullback and Leibler in S. Kullback, R.A. Leibler, *Ann. Math. Stat.* **22** (1951), 79–86. Given two probability functions Pr_1 and Pr_2 and a random variable X, if we define $p(x) = \mathrm{Pr}_1(X = x)$ and $q(x) = \mathrm{Pr}_2(X = x)$, then the Kullback–Leibler (KL) distance between p and q can be defined as[13]

$$
D_{KL}(p\,\|\,q) := \sum_x p(x)\log\frac{p(x)}{q(x)}.
$$

This is a good way of assessing the agreement with the target, since $D_{KL}(p\,\|\,q)$ is nonnegative and $D_{KL}(p\,\|\,q) = 0$ iff $p \equiv q$. Now let $Y = \{0, 1\}$. Then the discrete probability distribution is only centered either on 1 or 0. Then the KL distance be-

---

[13] In case of infinite sets $D(p\,\|\,q) := \int_X p(x)\log(p(x)/q(x))\,dx$.

comes

$$D_{KL}(p \parallel q) = \sum_{x \in X} p_Y(x) \log \frac{p_Y(x)}{p_F(x)}$$

$$= \sum_{x \in X} p_Y(x) \log \frac{p_Y(x)}{p_F(x)} [Y(x) = 0] + \sum_{x \in X} p_Y(x) \log \frac{p_Y(x)}{p_F(x)} [Y(x) = 1]$$

$$= \sum_{x \in X} p_0 \log \frac{p_0}{p_F(x)} [Y(x) = 0] + \sum_{x \in X} p_1 \log \frac{p_1}{p_F(x)} [Y(x) = 1],$$

where p_0 and p_1 are respectively the probability of Y to be 0 and 1. Notice that in the case $p_0 = p_1 = 1/2$ we can further simplify the expression to

$$D_{KL}(p_Y \parallel p_f) = -\frac{1}{2} \sum_{x \in X} (\log p_F(x) + 1).$$

Moreover, under the assumption that $p_F(x) = f^y(1-f)^{1-y}$, the relative entropy assumes the form

$$-\frac{1}{2} \sum_{x \in X} \Big(y_x \log f(x) - (1 - y_x) \log(1 - f(x) + 1 \Big).$$

A related index can be defined when the target set is $Y = \{-1, 1\}$ (see Eq. 2.1.1–(6)). Notice that as $f(x) \to y_x$, we have $E(\hat{w}) \to 0$. We also notice that whenever an example is misclassified, the relative entropy penalty yields a strong penalization with respect to the quadratic and hinge functions introduced previously.

An excellent source of principles and technical developments on information-based learning is the book by J. Principle, *Information Theoretic Learning: Renyi's Entropy and Kernel Perspectives* (Springer, 2010).

Section 2.2 The maximum likelihood estimation (MLE) was introduced by and widely popularized by Ronald Fisher [see J. Aldrich, *Stat. Sci.* **12** (1997), 162–176]. It can be regarded as a special case of maximum a posteriori estimation (MAP) when we have a uniform prior. Bayesian inference and learning has a rich story of gradual achievements culminated in the composition of the framework of probabilistic reasoning and Bayesian networks [see e.g. J. Pearl, *Probabilistic Reasoning in Intelligent Systems: Networks of Plausible Inference* (Morgan Kauffman, 1988) and J. Pearl, *Causality: Models, Reasoning, and Inference* (Cambridge University Press, 2000)]. The book *Learning in Graphical Models* (MIT Press, 1999) by M.I. Jordan strongly contributed to establish the topics in the machine learning community. An in-depth analysis on the interplay between the frequentist and the Bayesian approach can be found in M.J. Bayarri, J.O. Berger, *Statist. Sci.* **19** (2004), 58–80.

Section 2.3 The principle of maximum entropy was introduced by E.T. Jaynes in two seminal papers in 1957 [*The Physical Rev.* **106** (1957) 620–630 and *The Physical Rev.*

108 (1957) 171–190], where he established a natural correspondence between statistical mechanics and information theory. When the constraints are expressed in terms of conditional entropy, the optimization of the entropy under their soft-satisfaction corresponds with the principle of maximum mutual information [see S. Melacci, M. Gori, *IEEE Trans. Neural Netw. Learning Syst.* **23** (2012), 1849–1861]. The principle of minimum description length is widely covered in *The Minimum Description Length Principle (Adaptive Computation and Machine Learning)* (The MIT Press, 2007) by P.D. Grünwald, along with interesting relationships with statistical learning.

Section 2.4 Occam's razor (lex parsimoniae in Latin) is a general principle of induction, attributed to William of Ockham,[14] stating that among competing hypotheses, the simplest should be selected. The preference for simplicity is mostly based on the *falsifiability criterion*; simpler theories are preferable to more complex ones because they are more testable [K.R. Popper, *The Logic of Scientific Discovery* (Routledge, 2002)]. While simplicity is clearly supported from an epistemological viewpoint, it does not necessarily yield general principles of science [A. Courtney, M. Courtney, *Phys. Can.* **3** (2012) 7–8]. There is a substantial agreement on the fact that the term Occam's razor was coined by Sir William Hamilton, centuries after William's of Ockham death. Apart from subtle reformulations, while the idea of the razor was surely inspired by William's of Ockham studies, it is also very well expressed by the popular statement "Entities must not be multiplied beyond necessity" that was formulated by the Irish Franciscan philosopher John Punch in 1639. Traces of the principle, however, have been found also in Aristotle and Ptolemy statements [J. Franklin, *The Science of Conjecture* (John Hopkins Press, 2015)]. Philosophers have been involved in a number of epistemological justifications of the razor. It has been argued that nature itself is simple and that simpler hypotheses about nature are more likely to be true. This has an aesthetic flavor and justifications are often drawn from teleological analyses on the apparent purpose of nature. In machine learning, the Occam's razor states that excessively complex models hardly generalize to new examples, whereas simpler models may capture the underlying structure better and may thus have better predictive performance. When carefully analyzing the razor principle, one ends up with the conclusion that it is not a primitive inference-driven model, but it is more a heuristic maxim for selecting the most appropriate inference!

Occam's razor is also used as a heuristics to construct models for interpreting laws of nature. A remarkable early example is the one coming from analytic mechanics, which is based on the application of the least action, extended later on, to quantum mechanics by the classic concept of path integral due to Feynman [R.P. Feynman Ph.D. thesis Princeton University, 1942]. J. Gleick on pages 60–61 of his book *Genius: The Life and Science of Richard Feynman* (Pantheon, 1992), says:

---

[14] Ockham of William (1287–1347), also Occam or Gulielmus Occamus in Latin, was an English Franciscan friar, scholastic philosopher, and theologian, who is believed to have been born in Ockham, a small village in Surrey.

Where Newton's methods left scientists with a feeling of comprehension, minimum principles left a sense of mystery. This thought is nicely reinforced in David Park's challenging question: *How does the ball know which path to choose?* The experimental validation is the only way to claim that the principle holds. As already pointed out, aesthetic considerations are not enough to claim that a ball or a planet is condemned to follow a predetermined path. This issue becomes central when trying to capture complex cognitive processes or to conceive models for decision making. We can think of cognitive processes that obey a sort of "cognitive law" based on the minimization of a generalized version of the Dirichlet integral in physics, which yields simplicity in its solution. Interestingly, within this context, the above Park's question on mechanics sheds light on the truly inductive nature of Occam's razor.

It is worth mentioning that in physics, more than minimizing the action, we look for stationary points! Hence we don't fully rely on a minimization principle, but we restrict to a sort of local analysis of the action. This holds for other disciplines in which related principles in which the "quest for optimality" has been posed with the purpose of deriving the governing laws [P.J.H. Schoemaker, *Behav. Brain Sci.* **14** (1991), 205–245]. Interestingly, the primary role of time leads to action principles where the quest for optimality is tacitly transformed into the search for stationary points.

An elegant way of formalizing the parsimony principle is to reduce it to the search for Minimum Description Length (MDL) representations of learning agents. A nice way of grasping immediately the essence of the MDL principle is that of underlying major difference with most methods of statistical machine learning. In the words of Rissanen [*Stochastic Complexity in Statistical Inquiry Theory* (World Scientific Publishing, 1989)], the widely recognized founders of MDL:

> *We never want to make the false assumption that the observed data actually were generated by a distribution of some kind, say Gaussian, and then go on to analyze the consequences and make further deductions. Our deductions may be entertaining but quite irrelevant to the task at hand, namely, to learn useful properties from the data.*

The Minimum Description Length Principle summarizes certain regularities in the data. Notice, however, that these regularities are present and meaningful independently of whether or not there is a *true state of nature*. In most statistical theories, one typically assumes that there is some probability distribution that generates the data, and then we regard the "noise" as a random quantity to interpret the data. In the MDL, what in statistics is called "noise" is in fact the discrepancy relative to the model, namely the residual number of bits needed for encoding. Thus, noise is not a random variable: It depends on the chosen model and on the observed data. Interestingly, in MDL the reference to *true state of nature* is not meaningful, simply because we are only looking for the shortest description length. It is important to realize that in many problems in which we use statistical machine learning we are making strong probabilistic assumptions that are unlikely to be verified in practice. Because of many very good achievements on new applications, we could claim that *those models are wrong, yet useful!* However, since they are often far away from modeling the true state of nature, one should always pay attention to inferential methods that rely on arguable assumptions. As pointed out by Peter Grünwald [*The Minimum Description Length Principle (Adaptive Computation and Machine Learning)* (The MIT Press, 2007)],

> *To be fair, we should add that domains such as spam filtering and speech recognition are not what the fathers of modern statistics had in mind when they designed their procedures ... they were usually thinking about much simpler domains ...*

It is worth mentioning that MDL guarantees the consistency whenever one knows the true probability distribution. However, this consistency is just an ideal statement for a method that does not make any distributional assumptions. MDL philosophy is quite agnostic about whether we have selected the right model that represents the true state of nature.

A central problem in MDL is that of expressing the codes $Y \mid H$ with the purpose of correcting the error with respect to the prediction supported by hypothesis H. At the dawn of information theory, the founders began to realize that one can construct codes for a given information source that might be significantly more efficient than the vanilla solution based on ordinary binary encoding. It was clear early that a substantial reduction of the average length of encoding can be achieved as soon as we realize that frequent symbols can be coded by brief codes, whereas long codes can be reserved for rare symbols.

The MDL principle was introduced by Jorma Rissanen in *Automatica* **14** (1978), 465–471. It is rooted in the theory of Kolmogorov on algorithmic complexity, developed in the 1960s by R.J. Solomonoff [*Inf. Control* **7** (1964), 224–254] and A.N. Kolmogorov [*Probl. Inf. Transm.* **1** (1965) 1–7]. Akaike's seminal paper on information criterion (AIC) was likely the first model selection method based on information-theoretic ideas, though it was proposed under a different context and philosophy [H. Akaike, *IEEE Trans. Autom. Control* **19** (1974), 716–723]. The consequences on appropriate design choices of the model in machine learning are typically reflected in corresponding problems of overfitting or underfitting. In the latter case, given the learning tasks, one needs to use more powerful models that can reasonably learn the training examples. In the second case, we need to reduce the number of free parameters, but data set augmentation and noise injection can also help to face the problem. For some machine learning tasks, especially in classification, it is reasonable to create new fake data. Noise injection, which is another viable approach to increase the dimension of the training set, was proven to be equivalent to Tikhonov regularization [C.M. Bishop, *Neural Comput.* **7** (1995), 108–116]. Ensemble methods, bagging, and boosting [T.G. Dieterich, *Proceedings of the First International Workshop on Multiple Classifier System* (2000), 1–15 and T.G. Dieterich, *Mach. Learn.* **40** (2000) 139–157] have produced remarkable results in improving generalization.

Linear threshold machines

3

For many men that stumble at the threshold
Are well foretold that danger lurks within.
WILLIAM SHAKESPEARE, in *King Henry VI* (1564–1616)

This book is about perceptrons—the simplest learning machines.
M. MINSKY and S. PAPERT, in Perceptrons, second printing (1988)

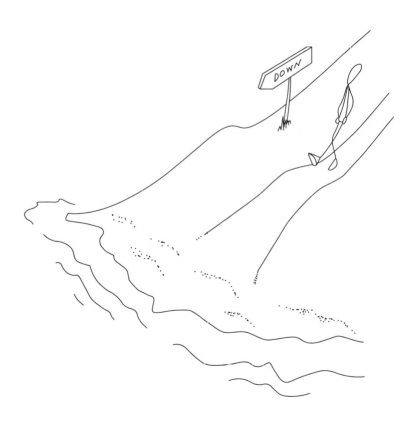

Machine Learning. https://doi.org/10.1016/B978-0-32-389859-1.00010-6

ONE OF THE simplest ways of modeling the interactions of intelligent agents with the environment is to expose them to a collection of pointwise constraints, namely to supervised pairs. This chapter is about the learning mechanisms that arise from the assumption of dealing with linear and threshold-linear machines. In most cases, the covered topics nicely intercept different disciplines, and are of remarkable importance to better grasp many approaches to machine learning.

3.1 **Linear machines**

EVERYBODY who is involved in experimental issues has likely heard of *Least Mean Square*. With many data available, one typically wants to best fit a model which is expected to describe the process under investigation.

Basically, given a collection of supervised pairs we are want to perform linear regression on the data so as to be able to make predictions on new points.

Two examples are shown in Table 3.1. In the first one, we want to predict the weight of an adult person from its height. The patterns are characterized by only one feature (the height), and the linear regression consists of finding w and b in $\mathbb{R}$ such that the prediction $f(x) = wx + b$ gets as close as possible to the target y, which represents the corresponding weight. A possible way of ensuring the fit is to weigh the errors by squaring them all, so as to minimize

$$E(w, b) := \frac{1}{2} \sum_{\kappa=1}^{\ell} (y_\kappa - wx_\kappa - b)^2, \tag{1}$$

where ℓ clearly denotes the number of examples. Now, this can be better re-written as

$$E(w, b) = b^2 \ell + \sum_{\kappa=1}^{\ell} y_\kappa^2 + w^2 \sum_{\kappa=1}^{\ell} x_\kappa^2 - 2w \sum_{\kappa=1}^{\ell} y_\kappa x_\kappa - 2b \sum_{\kappa=1}^{\ell} y_\kappa + 2bw \sum_{\kappa=1}^{\ell} x_\kappa. \tag{2}$$

Function (1) is quadratic in w and b. Any stationary point can be obtained when nullifying the gradient. We have

$$\frac{\partial E}{\partial w} = 2w \sum_{\kappa=1}^{\ell} x_\kappa^2 - 2 \sum_{\kappa=1}^{\ell} y_\kappa x_\kappa + 2b \sum_{\kappa=1}^{\ell} x_\kappa = 0;$$

$$\frac{\partial E}{\partial b} = 2\ell b - 2 \sum_{\kappa=1}^{\ell} y_\kappa + 2w \sum_{\kappa=1}^{\ell} x_\kappa = 0. \tag{3}$$

Table 3.1 Prediction tasks with more data available than unknowns.

(a)		(b)			
Height [m]	Weight [kg]	Type	Weight [kg]	Power [kW]	Time [s]
148	44.5	AUDI A3 2.0 TDI	1225	135	7.3
154	56.0	BMW S. 1 118D	1360	105	8.9
164	60.5	FIAT 500 L Pop Star	1320	77	12.3
174	63.5	FORD FOCUS	1276	74	12.5
182	74.5	RANGE ROVER SPORT	2183	375	5.3
194	84.5	FERRARI 458 S	1505	540	3.5
204	93.5	VOL. GOLF mpv	1320	63	13.2

The learning task on the left (a) refers to the prediction of the weight of a person on the basis of the height. The learning task on the right (b) consists of predicting the time it takes for a car to reach 100 km/h. The prediction is based on the mass of the vehicle and on the engine power.

Hence, the optimal solution w^* turns out to be

$$w^* = \frac{\begin{vmatrix} \sum_{\kappa=1}^{\ell} x_\kappa y_\kappa & \sum_{\kappa=1}^{\ell} x_\kappa \\ \sum_{\kappa=1}^{\ell} y_\kappa & \ell \end{vmatrix}}{\begin{vmatrix} \sum_{\kappa=1}^{\ell} x_\kappa^2 & \sum_{\kappa=1}^{\ell} x_\kappa \\ \sum_{\kappa=1}^{\ell} x_\kappa & \ell \end{vmatrix}} = \frac{\ell \sum_{\kappa=1}^{\ell} x_\kappa y_\kappa - \sum_{\kappa=1}^{\ell} x_\kappa \sum_{\kappa=1}^{\ell} y_\kappa}{\ell \sum_{\kappa=1}^{\ell} x_\kappa^2 - (\sum_{\kappa=1}^{\ell} x_\kappa)^2}$$

$$= \frac{\sum_{\kappa=1}^{\ell} x_\kappa y_\kappa / \ell - \overline{x} \cdot \overline{y}}{\sum_{\kappa=1}^{\ell} x_\kappa^2 / \ell - \overline{x}^2} = \frac{\hat{\sigma}_{xy}^2}{\hat{\sigma}_{xx}^2} \tag{4}$$

and

$$b^* = \overline{y} - \overline{x} \, w^*. \tag{5}$$

Remember that, given independent empirical observations $x_1, x_2, \ldots, x_\ell$, we define $E\,X \equiv \overline{x} := \sum_{\kappa=1}^{\ell} x_\kappa / n$. Then $\sigma_{xx}^2 := E\big((X - E\,X)^2\big)$ and $\sigma_{xy} := E\big((X - E\,X)(Y - E\,Y)\big)$ or, more explicitly,

$$\hat{\sigma}_{xx}^2 = \frac{1}{\ell} \sum_{\kappa=1}^{\ell} (x_\kappa - \overline{x})^2, \qquad \hat{\sigma}_{xy}^2 = \frac{1}{\ell} \sum_{\kappa=1}^{\ell} (x_\kappa - \overline{x})(y_\kappa - \overline{y}). \tag{6}$$

For a proof on the last step in Eq. (4), see Exercise 3.

As a matter of fact, the abundance of data leads to a single stationary point (w^*, b^*). Is it the global minimum? That is what we are really looking for, since we want our predictor to return values as close as possible to targets y. Of course, we have $E(w, b) \geq 0$ and $\lim_{w \to \infty} E(w, b) = \infty$. The same holds for the asymptotic behavior with respect to b. Now function E is in fact a paraboloid whose minimum corresponds with the vertex, and any optimization algorithm reaches its vertex. This suggests that instead of expressing E as a function of (w, b), one can change coordinates to $(\tilde{w}, \tilde{b})$, so that Eq. (1) can be more conveniently expressed by

$$\tilde{E}(\tilde{w}, \tilde{b}) = a_w(\tilde{w} - w^*)^2 + a_b(\tilde{b} - b^*)^2 + c. \tag{7}$$

In Exercise 4, we address the question on how this change of parameter takes place. The expression for the best fitting parameters, discovered by minimizing E, can be given a straightforward statistical interpretation. Suppose x and y are only marginally related, so that it is arguable to claim that there is any kind of hidden relation behind. This is reflected by $\sigma_{xy} \simeq 0$, which means that the best fitting only returns a constant, that is, $f(x) \simeq \bar{y}$. This makes sense: If there is no connection between x and y, it turns out that the best fitting reduces to $\bar{y}$. Something similar holds for very large σ_x^2. As σ_x gets larger and larger with respect to σ_{xy}, the relation between x and y is progressively blurred, so that we are back to the case of scarce correlation between x and y. What if we are, on the other hand, in front of a sort of deterministic process, so that $y_\kappa \simeq wx_\kappa + b$? As one expects, $\hat{w} \simeq w$ and $\hat{b} \simeq b$ (see Exercise 1).

Before going to the second case, one might wonder how accurate the discovered predictor is. We will see a trick to enrich this predictor, while using essentially the same idea. However, this very simple example enlightens an issue that is too often neglected. Is the available information sufficient to produce a good prediction? It is quite obvious that taller people are likely heavier than shorter people, but other input features are clearly important for more accurate predictions. The age and the race might play an important role in prediction. A sedentary person is more likely to be overweight than a sportsman, and other features can likely be proposed to construct a good feature representation. This is clearly the case in which we are using a poor pattern representation and, therefore, one should not expect miracles from math and algorithms! While this is quite clear in this simple case, in more complex learning tasks, where the appropriateness of the pattern representation cannot be trivially gained, it is not rare to see applications that rely on arguable beliefs about the actual capabilities of machine learning algorithms.

Now let us move to the second example of Table 3.1. Clearly, it can be treated in the same way, the only difference being that the prediction involves two inputs. That is, if we denote by x_1 the weight of the cars and by x_2 their engine power, then $f(x_1, x_2)$ predicts the time it takes to accelerate from 0 to 100 km/h, and the above quadratic index can still be used if we replace wx with $w_1x_1 + w_2x_2 + b$ or simply with $w \cdot x + b$ where x and w are now vectors in $\mathbb{R}^2$. Clearly, vectorial formulation unifies the two learning tasks. Unfortunately, in both cases, a linear model might not be accurate enough for our purposes and, therefore, one might be interested in exploring a richer model. We will see that the above mentioned linear vectorial representation can be conveniently used to model nonlinearity. In order to sketch the idea, let us go back to the problem of predicting the height of a person; the same analysis can be extended to higher-dimensional input spaces. If we choose the quadratic prediction $f(x) = w_2x^2 + w_1x + b$, we can still rely on a linear representation, since we can just pose

$$f(x) = w_2x_2 + w_1x_1 + b = w \cdot x + b, \tag{8}$$

which is just the same as the linear prediction model used to predict the acceleration time. Hence, given a learning task in a low-dimensional feature space (think of the height–weight learning task), the nonlinearity induced by a polynomial predictor corresponds somehow with a linear predictor in a high-dimensional space (think of the prediction of car acceleration). While this is a very nice way of involving nonlinearity in prediction, its concrete application might expose us to the perils of ill-conditioning! In particular, numerical problems arise as we start choosing high-order polynomials, since this significantly expands the range of the features. For example, when raising to power 10 the reals $x_1 = 0.1$ and $x_2 = 2$, we get $x_1^{10} \simeq 0$ and $x_2^{10} \simeq 1,000$, which shows how ill-conditioning arises (see Exercise 6 for details). While this feature enrichment can be very effective, one should not always expect miracles from such a general purpose pattern representation. This issue will be deepened in Chapter 4. In that case, it will become clear that the choice of the kernel does matter!

In addition, in both given examples, while the features are not very significant for the learning task, it is clear that we are missing relevant information to achieve very accurate predictions; the prediction of car acceleration time also misses very significant features (for example, the torque is much more informative than the engine power). The linear prediction can be especially effective when one has a good model of the process under investigation. For example, suppose one wants to predict the braking distance of a car running at a given velocity v. A good model is that of assuming uniform deceleration a, so as the braking distance s is $s = v^2/2a$. When posing $x = v^2$ and $w = 1/2a$, the prediction model turns out to be simply $f(x) = wx$, so that we are back to the above considered linear model, with the additional assumption that $b = 0$. If we are given a collection of pairs (v_κ, s_κ), we can find an LMS estimation of w, that is, of a. This is only formally similar to the two previous examples, since in this case the model of the braking distance has its roots in physical laws.

To sum up, it turns out to be very convenient to think of linear machines[1] in vector spaces, since different regression models can be converted into linear regression. This is formally covered in Section 3.1.1, where the general solution of the LMS problem is given in terms of *normal equations*. While there is nothing substantially new in treating vectorial learning tasks, playing with math enlightens some remarkable features of LMS and, most importantly, better links linear machines with neural nets and kernel machines.

So far, we have considered overdetermined problems, where the number of unknowns is smaller or equal to the number of examples. Clearly, in the opposite case, we cannot determine the unknown parameters, since there are infinitely many solutions. Hence, if the patterns belong to $X \subset \mathbb{R}^d$, the fitting is not well-posed in the case $\ell < 1 + d$. It looks like we need either additional examples or a different form of knowledge to provide a technically sound formulation of learning. As a matter of fact, this is not limited to the case $\ell < 1 + d$, but even in case of $\ell \geq 1 + d$, we can get

---

[1] Formally, they are affine models.

in trouble when using the previously suggested approach. This issue will be covered in Section 3.1.2 by a spectral analysis of the training data. However, in order to get a glimpse of the reason why the problem of learning can become ill-posed, we can just think of data which are somewhat dependent, so that there are examples which do not provide additional information or, at most, only add a very small novelty with respect to the previous data.

How can we face a problem with more unknowns than examples? What kind of information can be included to deal with such an ill-posed problem? The parsimony principle suggests a classic uniform solution: Amongst solutions which fit the given data, we want to select the one which offers somehow the simplest explanation. We can think of different parsimony criteria that are connected to the adopted notion of simplicity. Suppose that the examples $x \in X$ are samples of a probability distribution p_X, and regard simplicity just as of the smoothness of $f(x) = w^* x + b$. In so doing, a good *parsimony index* is

$$P(f) = \int_X \|\nabla f(x)\|^2 p_X(x)\, dx = w^2 \int_X p_X(x)\, dx = w^2, \tag{9}$$

where we have used the probabilistic normalization of p_X. Hence thinking of simplicity just as smoothness[2] of f leads to discovering solutions with small values of $\|w\|$. The new learning criterion can be based on discovering the minimum of

$$E(w, b) + \lambda w^2, \tag{10}$$

where the *regularization parameter* λ plays the important role of weighing the additional information coming from the above smoothness criterion with respect to the training data. An extreme formulation of learning is that of strictly enforcing the satisfaction of the constraints (perfect data fit), while minimizing w^2. As shown in Section 3.1.2, this corresponds with solving the problem of determining the pseudoinverse of a matrix. In Section 3.1.3, however, we provide evidence that the soft-constraining formulation turns out to be more adequate for learning tasks.

EXERCISES

1. [*M10*]   Given $f(x) = c \cdot x$ and $g(x) = x \cdot Ax$, then prove that $\nabla f(x) = c$ and $\nabla g(x) = (A + A')x$.

2. [*25*]   Let us consider the case of a linear machine with multi-dimensional output, that is one which based on the function $f(x) = Wx + b$, where $W \in \mathbb{R}^{d,n}$ and $b \in \mathbb{R}^n$. How can we extend the index of Eq. (1) to these vectorial functions? What are the normal equations in this extended case?

---

[2] While smoothness is a sound principle of simplicity, other principles are clearly possible. For example, one might want to provide selective weights for the different components of w. This is quite natural when we use polynomial regression, like in Eq. (8). In a similar case, one might prefer to penalize more the development of the second-order monomial (see also Exercise 7).

3. [*15*]   Using the definition of $\hat{\sigma}_{xx}^2$ and $\hat{\sigma}_{xy}^2$ prove that $\hat{\sigma}_{xx}^2 = \sum_{\kappa=1}^{\ell} x_\kappa^2/\ell - \bar{x}^2$ and $\hat{\sigma}_{xy}^2 = \sum_{\kappa=1}^{\ell} x_\kappa y_\kappa/\ell - \bar{x} \cdot \bar{y}$.

4. [*18*]   Given Eq. (1), let us consider the map $\mathcal{M}$ that changes coordinates as follows:

$$\begin{pmatrix} w \\ b \end{pmatrix} \xrightarrow{\mathcal{M}} \begin{pmatrix} \cos\phi & -\sin\phi \\ \sin\phi & \cos\phi \end{pmatrix} \begin{pmatrix} w \\ b \end{pmatrix}$$

What is the value of ϕ so that we can rewrite Eq. (1) as Eq. (7)? How can we determine a_w, b_w, and c?

3.1.1 Normal equations

We are given a linear map $f: X \subset \mathbb{R}^d \to \mathbb{R}$ that maps $x \mapsto w \cdot x + b$ with $w \in \mathbb{R}^d$ and $x \in X \subset \mathbb{R}^d$, a training set $\mathcal{L} = \{(x_\kappa, y_\kappa) \mid x_\kappa \in X, y_\kappa \in \mathbb{R} \text{ for } \kappa = 1, \ldots, \ell\}$, and the empirical risk $E_\ell(f)$, which measures the degree of match of f on the given training set. The pair (x_κ, y_κ) contains the generic κ-th input and the corresponding target y_κ. The targets are stacked into the vector $y \in \mathbb{R}^\ell$. The empirical risk function $E_\ell(f)$ can be written as

$$E_\ell(f) = E(w, b) = \frac{1}{2} \sum_{\kappa=1}^{\ell} (y_\kappa - f(x_\kappa))^2. \tag{1}$$

If we set $\hat{x} := (x_1, \ldots, x_d, 1)'$ and $\hat{w} := (w_1, \ldots, w_d, b)' \in W \subset \mathbb{R}^{d+1}$ then $f(x) = \sum_{i=1}^{d} w_i x_i + b = \hat{w} \cdot \hat{x}$. Now, let us introduce the notation

$$\hat{X} := \begin{pmatrix} \hat{x}_1' \\ \hat{x}_2' \\ \vdots \\ \hat{x}_\ell' \end{pmatrix} = \begin{pmatrix} x_{11} & x_{12} & \cdots & x_{1d} & 1 \\ x_{21} & x_{22} & \cdots & x_{2d} & 1 \\ \vdots & \vdots & & \vdots & \vdots \\ x_{\ell 1} & x_{\ell 2} & \cdots & x_{\ell d} & 1 \end{pmatrix} \in \mathbb{R}^{\ell, d+1}.$$

The matrix $\hat{X}$ summarizes the data available on the learning task; it is referred to as *information matrix*. We can rewrite Eq. (1) in a more compact form as follows:

$$E(\hat{w}) \equiv E(w, b) = \frac{1}{2} \|y - \hat{X}\hat{w}\|^2 = \frac{1}{2}(y - \hat{X}\hat{w})'(y - \hat{X}\hat{w})$$

$$= \frac{1}{2} y'y - y'\hat{X}\hat{w} + \frac{1}{2}\hat{w}'\hat{X}'\hat{X}\hat{w}. \tag{2}$$

Now we are ready to state the *least mean squares problem* as the one of determining

$$\hat{w}^* \in \underset{\hat{w} \in W}{\arg\min}\, E(\hat{w}).$$

This can be thought of a simplified form of supervised learning, in which we are only interested in data fitting. We can find stationary points by imposing $\nabla E(\hat{w}^*) = 0$.

Hence we have

$$(\hat{X}'\hat{X})\hat{w}^* = \hat{X}'y. \qquad (3)$$

These are referred to as the *normal equations*. It is easy to prove that any station-ary point $\hat{w}^*$ is necessarily an absolute minimum (see Exercise 2). In addition, we can also prove that $\hat{w}^*$ is *invariant under scaling of the inputs* (see Exercise 5). Depending on whether the condition $\det(\hat{X}'\hat{X}) = 0$ is met, we distinguish two re-markably different cases that are characterized by rank $\hat{X}$. When rank $\hat{X} = r < d + 1$, the given data do not bring enough information for uniqueness, since $\hat{X}$ is a *fat ma-trix* (more columns than rows). This case will be treated in the next section. When rank $\hat{X} = r = d + 1$, we are given at least $d + 1$ examples, which must be linearly independent. This corresponds with cases in which data bring enough information for yielding uniqueness of the solution, and $\hat{X}$ is a *skinny matrix* (more rows than columns). Of course the condition rank $\hat{X} = d + 1$ can hold also for square matrices ($\ell = d + 1$). Under this condition, there exists a unique solution $\hat{w}^*$ of the normal equations (3), that is,

$$\hat{w}^* = (\hat{X}'\hat{X})^{-1}\hat{X}'y, \qquad (4)$$

where the inversion of matrix $(\hat{X}'\hat{X})$ is guaranteed by the condition rank $\hat{X} = r = d+1$ (i.e., $\det \hat{X}'\hat{X} \neq 0$). An interesting property connected with the normal equations can be grasped by considering the matrix $P_\perp^d := \hat{X}(\hat{X}'\hat{X})^{-1}\hat{X}'$. We can promptly see that it is a *projection matrix*, since we have

$$(P_\perp^d)^2 = \hat{X}(\hat{X}'\hat{X})^{-1}\hat{X}'\hat{X}(\hat{X}'\hat{X})^{-1}\hat{X}' = P_\perp^d. \qquad (5)$$

Basically, $P_\perp^d$ projects $y \in Y \subset \mathbb{R}^\ell$ onto $\mathcal{R}(\hat{X})$. Here $\mathcal{R}(\hat{X})$ denotes the image space spanned by the columns of $\hat{X}$, while we will denote with $\mathcal{N}(\hat{X}')$ the kernel of $\hat{X}'$. Relatedprojection properties hold for $Q_\perp^d = \mathrm{Id} - \hat{X}(\hat{X}'\hat{X})^{-1}\hat{X}'$, which projects $y \in Y \subset \mathbb{R}^\ell$ onto the null space of $\hat{X}'$. This can be seen by noticing that

$$\hat{X}'Q_\perp^d y = \hat{X}'(\mathrm{Id} - \hat{X}(\hat{X}'\hat{X})^{-1}\hat{X}')y = 0,$$

that is, $Q_\perp^d y \in \mathcal{N}(\hat{X}')$.

$$E(\hat{w}^*) = \|Q_\perp^d y\|^2 \qquad (6)$$

the residual error corresponding with the optimal solution is mostly due to the spec-tral properties of $Q_\perp^d$. Clearly, $E(\hat{w}^*) \leq \|Q_\perp^d\|^2 \|y\|^2$. In addition, the *linear residual* $E(\hat{w}^*)$ also depends on y and, particularly, on its algebraic relation with $Q_\perp^d$, since the linear residual is null whenever $y \in \mathcal{N}(Q_\perp^d)$.

EXERCISES

1. [*18*] Let us assume that the data of a certain training set are generated by a nearly deter-minist stochastic process, for example, by measurements of a linear physical process such that $y_\kappa \simeq p x_\kappa + q$. Using normal equations, prove that $w^* \simeq p$ and $b^* \simeq q$.

▶ **2.** [*18*]   Prove that $HE(\hat{w}) = \hat{X}'\hat{X}$, where HE is the hessian matrix of E. Furthermore prove that HE is positive semi-definite. When does it happen that HE is positive definite?

3. [*M22*]   Given matrix $P_\perp^d = \hat{X}(\hat{X}'\hat{X})^{-1}\hat{X}'$, prove that:

i.   The eigenvalues of $P_\perp^d$ are either 0 or 1;

ii. $\mathrm{rank}(\hat{X}) = \mathrm{tr}(P_\perp^d)$.

What can we say about the spectrum and rank of $Q_\parallel^d$? Do these properties hold for $P_\perp^u$, $Q_\perp^u$ too?

▶ **4.** [*M22*]   Suppose you are given the solution $\hat{w}$ of the normal equations in a given reference. How does it change when roto-translating the reference?

▶ **5.** [*M24*]   Let $\hat{w}^*$ be the solution of the normal equations (3) for a given training set. How does it change when scaling the inputs?

6. [*M20*]   (Abnormal Equations.) The best fitting problem can be solved by using the following arguments: Given $\hat{X}\hat{w} = y$, if we left-multiply by matrix $M \in \mathbb{R}^{d+1,\ell}$ both sides, we get $M\hat{X}\hat{w} = My$, from which we get $\hat{w} = (M\hat{X})^{-1}My$. Any matrix M such that $(M\hat{X})^{-1}$ exists in the classic sense leads to these "abnormal equations" which solve the best fitting problem. Can you reconcile this paradox with normal equations?

7. [*M18*]   Let $\hat{w}^*$ be the solution of the normal equations (3) for a given training set. How does it change when using the preprocessing map $x \mapsto Tx$, where $T \in \mathbb{R}^{d\times d}$?

8. [*18*]   Give an example in which linear rescaling of the inputs is favorable.

▶ **9.** [*18*]   Let us consider the case $d = 1$. Then, beginning from the normal equations (3), prove that the optimal solution is given by Eq. 3.1–(4)–3.1–(5). *Hint:* Use the normal equations (3) for $d = 1$ and rearrange them by considering the definitions 3.1–(6).

10. [*16*]   Let us consider the simple classification problem where the training set is $\mathcal{L} = \{((0,0)',0), ((0,1)',1)\}$ and the classifier is a linear LMS. Determine the solution by using normal equations.

11. [*18*]   Let us consider the simple classification problem where the training set is $\mathcal{L} = \{((0,0)',0), ((0,1)',0), ((0,\alpha)',1)\}$ and the classifier is a linear LMS. Determine the solution by using normal equations and discuss the solution when changing $\alpha \in \mathbb{R}$.

▶ **12.** [*25*]   (Composition of experiments.) Suppose we have carried out two independent experiments involving the same probabilistic data distribution, which are summarized by the following (information matrix, target) $\{(X_1, y_1), (X_2, y_2)\}$. Each experiment is associated with the error function $E_i(\hat{w})$, $i = 1, 2$. Prove that on the composition of the two experiments, which consists of joining the data to get $\hat{X} = \binom{\hat{X}_1}{\hat{X}_2}$ and $\hat{y} = (\hat{y}_1', \hat{y}_2')'$, we have $E(\hat{w}) = E_1(\hat{w}) + E_2(\hat{w})$. Moreover let $\hat{w}_1^*$ and $\hat{w}_2^*$ be a solution of normal equation for experiment 1 and 2 respectively and $\hat{w}^*$ a solution for the joint problem then prove that $\hat{w}^* = (\hat{X}_1'\hat{X}_1 + \hat{X}_2'\hat{X}_2)^{-1}(\hat{X}_1'\hat{X}_1\hat{w}_1^* + \hat{X}_1'\hat{X}_2\hat{w}_2^*)$.

3.1.2 Undetermined problems and pseudoinversion

Now we consider the case rank $\hat{X} = r < d+1$, in which the problem is undetermined, and we assume that $\hat{X}$ is of full-rank, that is, $r = \ell < d+1$. In practice, there is not "enough" data available to get a unique solution. Basically, the equation

$$\hat{X}\hat{w} = y \tag{1}$$

admits $\infty^{d+1-\ell}$ solutions and, for each of those solutions $\hat{w}^*$, we have $E(\hat{w}^*) = 0$. The classic notion of matrix inversion can be extended also to include this case by considering *Moore–Penrose pseudoinversion* (see Section 3.5). In general, the above undetermined problem can be faced either by hard or soft-enforcing of the constraint (1).

In the first case, we can promptly check that

$$\hat{w}^* = \hat{X}'(\hat{X}\hat{X}')^{-1}y \tag{2}$$

is the *shortest length solution* that (strictly) satisfies the constraint (1). Notice that since we have assumed $\hat{X}$ to be a fat full-rank matrix, $\hat{X}\hat{X}'$ is invertible. In order to prove this property we begin noticing that the hard satisfaction can be immediately checked as follows:

$$\hat{X}\hat{w}^* = \hat{X}\hat{X}'(\hat{X}\hat{X}')^{-1}y = y.$$

Now let us consider any $\overline{w} \neq \hat{w}^*$ such that $\hat{X}\overline{w} = y$. Of course, we have $\hat{X}(\overline{w} - \hat{w}^*) = 0$. As a consequence, we have

$$(\overline{w} - \hat{w}^*) \cdot \hat{w}^* = (\overline{w} - \hat{w}^*) \cdot \hat{X}'(\hat{X}\hat{X}')^{-1}y = \big((\hat{X}(\overline{w} - \hat{w}^*)\big) \cdot (\hat{X}\hat{X}')^{-1}y = 0. \tag{3}$$

Finally, this yields

$$\|\overline{w}\|^2 = \|(\overline{w} - \hat{w}^*) + \hat{w}^*\|^2 = \|\overline{w} - \hat{w}^*\|^2 + \|\hat{w}^*\|^2 + 2(\overline{w} - \hat{w}^*) \cdot \hat{w}^* \geq \|\hat{w}^*\|^2.$$

Instead of the ordinary notation used for inversion, sometimes the following notation

$$\hat{X}^+ := \hat{X}'(\hat{X}\hat{X}')^{-1} \tag{4}$$

is used to explicitly refer to the *Moore–Penrose pseudoinverse* of $\hat{X}$. Interestingly, any $\overline{w}$ satisfies the orthogonality property (3), which clearly shows the importance of $\hat{w}^* = \hat{X}^+y$. Its associated minimality property can be grasped when studying $Q_\perp^u := I - P_\perp^u$, where $P_\perp^u = \hat{X}'(\hat{X}\hat{X}')^{-1}\hat{X}$. This projection property, like the one outlined in 3.1.1–(5), underlines another nice property that involves the *degree of freedom*, $\mathrm{df}(f)$, of function f, since it can be proven[3] (see Exercise 3) that

$$\mathrm{df}(f) = \mathrm{tr}(P_\perp) = \mathrm{rank}\,\hat{X}. \tag{5}$$

---

[3] The property holds for both $P_\perp^d$ and $P_\perp^u$.

As previously shown, $Q_\perp^u$ gives projection onto $\mathcal{N}(\hat{X})$. The property that $Q_\perp^u$ is a projection matrix can also be immediately drawn. For any $y \in Y \subset \mathbb{R}^\ell$, we have

$$(Q_\perp^u)^2 = (I - \hat{X}'(\hat{X}\hat{X}')^{-1}\hat{X}) \cdot (I - \hat{X}'(\hat{X}\hat{X}')^{-1}\hat{X})$$
$$= I - 2\hat{X}'(\hat{X}\hat{X}')^{-1}\hat{X} + \hat{X}'(\hat{X}\hat{X}')^{-1}\hat{X} \cdot \hat{X}'(\hat{X}\hat{X}')^{-1}\hat{X} = Q_\perp^u.$$

A straightforward way of capturing the meaning of the minimality property of $\hat{X}^+$ comes from posing the problem in the framework of constrained optimization. In that case, given $\mathcal{W}^+ = \{\hat{w} : \hat{X}\hat{w} = y\}$, one wants to discover

$$\hat{w}^* = \min_{\mathcal{W}^+} \|w\|^2.$$

As shown in Exercise 5, this can be attacked naturally by using the Lagrangian approach, which leads directly to conclude that $\hat{w}^* = \hat{X}^+ y$.

The problem of determining the minimal $\hat{w}$ that is consistent with constraint (1) can naturally be faced by the spectral analysis of $\hat{X}$, that is, by using the SVD decomposition. Let us assume that $\hat{X} = U\Sigma V'$, where $U \in \mathbb{R}^{\ell,\ell}$, $\Sigma \in \mathbb{R}^{\ell,d+1}$, $V \in \mathbb{R}^{d+1,d+1}$, and

$$\Sigma = \begin{pmatrix} \sigma_1 & & & 0 & \cdots & 0 \\ & \ddots & & & & \\ & & \sigma_r & & & \\ & & & \ddots & & \\ & & & 0 & \cdots & 0 \end{pmatrix}$$

with $r \leq \ell$. Now let $\xi := V'\hat{w}$. Using orthogonality of U and V, we have

$$\|y - \hat{X}\hat{w}\|^2 = \|y - U\Sigma V'\hat{w}\|^2 = \|U(U'y - \Sigma V'\hat{w})\|^2$$
$$= \|U\|^2 \|U'y - \Sigma V'\hat{w}\|^2$$
$$= \|U'y - \Sigma\xi\|^2 = \sum_{i=1}^{r}(\sigma_i\xi_i - u_i'y)^2 + \sum_{i=r+1}^{\ell}(u_i'y)^2, \tag{6}$$

where u_i is the ith column of U. Now we can promptly see that $\min_{\hat{w}} \|y - \hat{X}\hat{w}\|^2$ is reached for $\hat{\xi}^* = u_i'y/\sigma_i$ for $i = 1, \ldots, r$, while for $i = r+1, \ldots, d+1$ the value of $\hat{\xi}^*$ can be chosen arbitrarily. Using Eq. (6), this means that $\min_{\hat{w}} \|y - \hat{X}\hat{w}\|^2 = \sum_{i=r+1}^{\ell}(u_i'y)^2$. Finally, from the definition $V'\hat{w}^* = \xi^*$ we get $\hat{w}^* = V\xi^*$ and then

$$\hat{w}^* = \sum_{i=1}^{r}\frac{u_i'y}{\sigma_i}v_i. \tag{7}$$

Now this equation tells us that the optimal solution $\hat{w}^*$ is an expansion over the columns v_i of V that are associated with nonzero singular values σ_i. While this *spectral analysis* enlightens the structure of the solution, at the same time, it warns about numerical instability. If the original matrix has a null singular value then slight changes to the matrix on tiny positive numbers affect the pseudoinverse dramatically, since Eq. (7) requires computing the reciprocal of σ_i, which is critical for tiny numbers. We can come to the same conclusion exploiting a property of the pseudoinverse telling us that $(U \Sigma V')^+ = V \Sigma^+ U'$ (see Exercise 3) since the computation of Σ^+ is done by replacing nonzero σ_i with $1/\sigma_i$ on the diagonal of Σ. There are a number of other numerical methods to compute $\hat{X}^+$ (see, e.g., Exercise 6) but, in practice, instability is quite a common concern. For example, expression (4) of the pseudoinverse matrix also indicates the ill-conditioning that emerges when there are "nearly-dependent rows" in $\hat{X}$. Like for classic inversion of full-rank matrices, also pseudoinversion suffers of ill-conditioning. In the next section, we discuss how to face this critical issue.

EXERCISES

1. [*M14*]   For a general matrix A with complex entries, the pseudoinverse of A can be defined as the matrix A^+ that solves the following problem

$$P(A): \begin{cases} AA^+A = A, \\ A^+AA^+ = A^+, \\ (AA^+)^* = AA^+, \\ (A^+A)^* = A^+A. \end{cases}$$

Show that Eq. (4) solves such a problem.

2. [*HM26*]   (Penrose, 1955) Show that the problem $P(A)$ defined in Exercise 1 has a unique solution for any A.

3. [*16*]   Prove that, for every two unitary matrices U and V, $(UAV)^+ = V^*A^+U^*$.

4. [*M25*]   Something is wrong with the following property. Can you find where the mistake is? State the correct property, and explain why the proof is incorrect. "For any positive a and b the following SVD holds:

$$A = \begin{pmatrix} a & a \\ -b & b \end{pmatrix} = \begin{pmatrix} a & 0 \\ 0 & b \end{pmatrix} \begin{pmatrix} 1 & 1 \\ 1 & -1 \end{pmatrix}.$$

Proof. In order to find the decomposition $A = U \Sigma V'$, first of all let us compute

$$A'A = \begin{pmatrix} a & -b \\ a & b \end{pmatrix} \begin{pmatrix} a & a \\ -b & b \end{pmatrix} = \begin{pmatrix} a^2 + b^2 & a^2 - b^2 \\ a^2 - b^2 & a^2 + b^2 \end{pmatrix},$$

which has eigenvectors $(1/\sqrt{2})(1, 1)'$ with eigenvalue $2a^2$ and $(1/\sqrt{2})(1, -1)'$ with eigenvalue $2b^2$. Now because $A'A = V \Sigma' \Sigma V$ we have that

$$V = \frac{1}{\sqrt{2}} \begin{pmatrix} 1 & 1 \\ 1 & -1 \end{pmatrix}, \qquad \Sigma = \sqrt{2} \begin{pmatrix} a & 0 \\ 0 & b \end{pmatrix}.$$

Similarly, if we compute $AA' = U \Sigma \Sigma' U'$ as

$$AA' = \begin{pmatrix} a & a \\ -b & b \end{pmatrix} \begin{pmatrix} a & -b \\ a & b \end{pmatrix} = 2 \begin{pmatrix} a^2 & 0 \\ 0 & b^2 \end{pmatrix},$$

we can conclude that $U = I$. This ends the proof."

5. [*M20*]   Let us consider the problem of determining the shortest length solution of Eq. (1). Find the solution by using the Lagrangian approach.

6. [*M30*]   (Ben-Israel Cohen method) Prove that the following recursive scheme leads to the computation of $\hat{X}^+$

$$\hat{X}_{i+1}^+ = 2\hat{X}_i^+ - \hat{X}_i^+ \hat{X} \hat{X}_i^+,$$

whenever we start from $\hat{X}_0$ such that $\hat{X}_0 \hat{X} = (\hat{X}_0 \hat{X})'$. Notice that a possible initialization is $\hat{X}_0 = \alpha \hat{X}'$, but this is quite slow. A better computational choice is $\hat{X}_0 = (\mu I_d + \hat{X}' \hat{X})^{-1} \hat{X}'$.

7. [*18*]   Let us consider the case $d + 1 \le \ell$ along with the assumption of a skinny full rank $\hat{X}$. Prove that $P_\perp^d \in \mathcal{R}(\hat{X})$ (rank space generated by $\hat{X}$). Likewise, in the case of undetermined problems with a fat full rank matrix $\hat{X}$ prove that $P_\perp^u \in \mathcal{R}(\hat{X}')$.

8. [*M20*]   Prove that the spectrum of projection matrices $P_\perp^d$, $P_\perp^u$, $Q_\perp^d$, $Q_\perp^u$ is $\{0, 1\}$.

3.1.3 Ridge regression

An alternative way to deal with the lack of uniqueness of learning in the case $\ell < d + 1$ is to provide soft incorporation of the constraint 3.1.2–(1). We can formulate learning as the minimization of the index

$$E_r(\hat{w}) = \frac{1}{2} \| y - \hat{X}\hat{w} \|^2 + \frac{\mu}{2} \| \Pi_w \hat{w} \|^2, \tag{1}$$

where Π_w is the projection $\Pi_w \hat{w} = w \in \mathbb{R}^d$. The term $(\mu/2)w^2$ is motivated from the search of a parsimonious solution [see Eq. 3.1–(10)]. Predictions based on the minimization of this index are referred to as *ridge regression*. In this case, the stationary points of $E_r(\hat{w})$ satisfy

$$-\hat{X}'y + (\hat{X}'\hat{X})\hat{w} + \mu I_d \hat{w} = 0,$$

where $I_d = \Pi'_w \Pi_w$ We get

$$(\mu I_d + \hat{X}'\hat{X})\hat{w} = \hat{X}'y. \tag{2}$$

As proved in Exercise 1, we have $\det(\mu I_d + \hat{X}'\hat{X}) \ne 0$, so that the solution to Eq. (2) is

$$\hat{w} = (\mu I_d + \hat{X}'\hat{X})^{-1}\hat{X}'y. \tag{3}$$

It is interesting to notice that the scaling of the inputs has a relevant impact on the solution (see Exercise 3). Of course, for any value of the regularization parameter μ we get a different solution. It is instructive to study the limit behavior by considering the extreme cases $\mu \to 0$ and $\mu \to \infty$.

 Let us consider fat matrices. Like for the case of a hard constraint, treated in the previous section, it is instructive to carry out spectral analysis of $\hat{X}$, which clearly shows the role of the regularization parameter. Now we have

$$\mu I_d + \hat{X}'\hat{X} = \mu I_d + (U \Sigma V')' U \Sigma V' = \mu I_d + V \Sigma' U' U \Sigma V'$$
$$= \mu V I_d V' + V \Sigma' \Sigma V' = V \left(\mu I_d + \Sigma' \Sigma \right) V'.$$

Matrix V that is involved in the SVD of $\hat{X}$ is the one that diagonalizes $\mu I_d + \hat{X}'\hat{X}$. From Eq. (2) we get $V \left(\mu I_d + \Sigma' \Sigma \right) V' \hat{w}^* = (U \Sigma V')' y = V \Sigma' U' y$. If we left-multiply by $V' \in \mathbb{R}^{d+1,d+1}$ and remember that $\xi^* = V' \hat{w}^* \in \mathbb{R}^{d+1}$, we get $(\mu I_d + \Sigma' \Sigma) \xi^\star = \Sigma' U' y$, from which we can infer, in the same way we did in the previous section, that $\xi_i^* = \sigma_i u_i' y / (\sigma_i^2 + \mu)$ for $i = 1, \ldots, r$, and $\xi_i^* = 0$ if, instead, $i = r + 1, \ldots, d + 1$. From the orthogonality of V, we have $\hat{w}^* = V \xi^* = \sum_{i=1}^{d+1} \xi_i^* v_i$ and then

$$\hat{w}^* = \sum_{i=1}^{r} \frac{\sigma_i^2}{\sigma_i^2 + \mu} \frac{u_i' y}{\sigma_i} v_i, \tag{4}$$

where we can promptly see that the expansion along v_i (V's columns) are weighed by the *filtering parameter* $\phi_i(\mu) = (\sigma_i^2/(\sigma_i^2 + \mu))$. It turns out that, by an appropriate choice of μ, small singular values are filtered out, which is a desirable outcome in practice. Finally, we have practice. Finally, we have

$$\lim_{\mu \to 0} \hat{w}^*(\mu) = \lim_{\mu \to 0} \sum_{i=1}^{r} \phi_i(\mu) \frac{u_i' y}{\sigma_i} v_i = \sum_{i=1}^{r} \frac{u_i' y}{\sigma_i} v_i,$$

which corresponds with the solution under hard constraints given by Eq. 3.1.2–(7).

The solution (2) along with definition 3.1.2–(5) suggests extending the notion of degree of freedom to the case of ridge regularization by

$$\mathrm{df}(f) = \mathrm{tr}(P_\perp) = \mathrm{tr}\left(\hat{X}(\mu I_d + \hat{X}'\hat{X})^{-1} \hat{X}' \right).$$

Now from (4), we get

$$P_\perp = \hat{X} \sum_{i=1}^{r} \phi_i \frac{v_i u_i'}{\sigma_i} = U \Sigma V' \sum_{i=1}^{r} \phi_i \frac{v_i u_i'}{\sigma_i}$$
$$= U \Sigma \sum_{i=1}^{r} \frac{\phi_i (V' v_i) u_i'}{\sigma_i} \tag{5}$$
$$= U \sum_{i=1}^{r} \frac{\phi_i}{\sigma_i} (\Sigma e_i) u_i' = \sum_{i=1}^{r} \phi_i (U e_i u_i') = \sum_{i=1}^{r} \phi_i u_i u_i'.$$

Now notice that $u_i u_i'$ admits the only eigenvalue 1, since $(u_i u_i') u_i = u_i$. Moreover, from the property that the trace of a matrix is the sum of its eigenvalues we get

$$\mathrm{df}(f) = \mathrm{tr}(P_\perp) = \sum_{i=1}^{r} \frac{\sigma_i^2}{\mu + \sigma_i^2}. \tag{6}$$

Of course, $df(f) \rightarrow r$ as $\mu \rightarrow 0$, which corresponds with 3.1.2–(5). The degree of freedom decreases as μ gets higher and higher, and $df(f) \rightarrow 0$ as $\mu \rightarrow \infty$.

EXERCISES

▶ **1.** [*M28*] Let us consider the error function of ridge regression (1). Prove that in this case the hessian matrix $H E_r(\hat{w}) = \mu I_d + \hat{X}'\hat{X}$ and that $H E_r$ is invertible. This was the original motivation for ridge regression [see A.E. Hoerl, R.W. Kennard, *Technometrics* **12** (1970), 55–67].

2. [*30*] Let us consider the formulation of ridge regression, where one is looking for the minimization of the index (1), and replace w with $\hat{w}$ to minimize

$$E_r(\hat{w}) = \frac{1}{2}\|y - \hat{X}\hat{w}\|^2 + \mu\hat{w}^2.$$

Discuss the relationship of this new problem with the classic ridge regression.

3. [*20*] Discuss the role of α-scaling of the input $x \mapsto \alpha x$ in the solution of ridge regression given by Eq. (3).

▶ **4.** [*22*] A different formulation of ridge regression consists of minimizing the error with respect to the targets while enforcing a strict constraint on the norm of the weights. Hence, given $\omega > 0$, find $\hat{w}$ that solves

$$\min_{\hat{w} \in B_\omega(0)} \|y - \hat{X}\hat{w}\|^2,$$

where $B_\omega(0)$ is the ball of radious ω centered in the origin. Use the Lagrangian formalism to solve the problem.

▶ **5.** [*HM47*] Let us consider a one-layer classifier where in addition to the error function we introduce also a regularization term as in ridge regression:

$$E(\hat{w}) = \frac{1}{2}\sum_{\kappa=1}^{\ell}(y_\kappa - o(\hat{w}'\hat{x}_\kappa))^2 + \frac{1}{2}\mu w^2.$$

We know that the penalty term is local minima free in case of linearly separable examples. Does the property hold also for $E(\hat{w})$ after having introduced the regularization term?

6. [*25*] Let us consider the case of one-dimensional regression of Exercise 3.1.1–9 and suppose we preprocess the input x_κ so as to construct the feature space $\zeta_{i,\kappa} = x_\kappa^{i-1}$, $\forall i = 1,\ldots,d+1$. Discuss the solution by either using the normal equation, the pseudoinverse, and the ridge regression.

7. [*18*] Let us consider the quadratic regression expressed by Eq. (1) when $d = 2$. What is the effect on ill-conditioning of replacing the parsimony term $w_1^2 + w_2^2$ with $\alpha_1 w_1^2 + \alpha_2 w_2^2$ with different choices of α_1 and α_2?

3.2 **Linear machines with threshold units**

A linear machine can be used for both regression, i.e., finding a model that fits some data, and classification, i.e., constructing a function that correctly classifies inputs. Before reading this section, one might wonder whether there is any reason to introduce the nonlinearity of linear-threshold machines, when considering that linear machines can be used for regression, but they can also act as models for classification! As it will be clear in the end, this is not a trivial issue, and while the development of *linear-threshold machines* is intertwined with historical issues, it still plays a fundamental role, especially in neural networks. Furthermore, consider that — when dealing with classification — one might directly want the computational unit to return a Boolean to represent the membership of the input to the associated category. This is a good motivation to explore linear-threshold units but, as it will emerge from the rest of the section, there are other reasons that are intimately related to the joint choice of the unit and the loss function.

Formally, a linear threshold unit is a function that transforms an *activation* $a = \hat{w} \cdot \hat{x}$ onto an output using the rule

$$f(x) = \sigma(a) = \sigma(\hat{w} \cdot \hat{x}),$$

where σ can be chosen in different ways. Possible choices, amongst others, are

$$\begin{aligned}
&i. \ \ \sigma(a) = \text{sign}(a) \qquad &&ii. \ \ \sigma(a) = H(a), \\
&iii. \ \ \sigma(a) = \text{Th}(a) \qquad &&iv. \ \ \sigma(a) = \text{logistic}(a), \\
&v. \ \ \sigma(a) = (a)_+
\end{aligned} \qquad (1)$$

where $H(a) := [a \geq 0]$ is the Heaviside function, $\text{logistic}(a) := (1 + \exp(-a))^{-1}$ is referred to as the logistic function, and $(a)_+ := a\,[a \geq 0]$. Functions *i.* and *ii.* return Boolean-like variables, whereas the rest return real values. Clearly, in case of Boolean-like variables, gradient-based algorithms offer no heuristics to drive any learning process, whereas when turning to the continuum setting of computation, the gradient is a natural heuristic for adapting the parameters. While *iii.* and *iv.* (*squash functions*) are continuous approximations of *i.* and *ii.*, the *piecewise linear function* *v.*, which is referred to as a *rectifier*, is either inhibited or operates as a linear unit. Like for linear machines, a possible way to assess the error with respect to the target is to use the quadratic function 3.1.3–(1). However, other *loss functions* seem to play the right role in classification tasks. While the different functions f_i can be chosen as one of the linear-threshold machines given by (1), we might want to impose a *probabilistic normalization* on the output. One possibility is to use the *softmax function*

$$f_i(x) = \underset{i}{\text{smx}}(a_1, \ldots, a_c) := \frac{\exp(a_i)}{\sum_{j=1}^{c} \exp(a_j)}, \qquad (2)$$

normalized by construction, which also has the nice property that $\text{smx}_i \approx 1$ when $a_i = \max_{1 \leq k \leq c} a_k$ (see Exercise 2). Another remarkable property of softmax is that any translation $a_i \to a_i + \alpha$ of its argument leaves the output unchanged, since

$$\underset{i}{\mathrm{smx}}(a_1 + \alpha, \dots, a_c + \alpha) = \underset{i}{\mathrm{smx}}(a_1, \dots, a_c), \quad \text{for all } i = 1, \dots, c.$$

Instead of using the softmax (2), one could replace it with

$$f_i(x) = \frac{a_i}{\sum_{j=1}^c a_j},$$

which, however, does not possess the above softmax property. In this case, we get invariance when $x \to \beta x$.

We can promptly see that the nonlinearity of σ, as well as that of (2), is reflected in the above error index and, therefore, its stationary points cannot be expressed by normal equations any longer. Clearly, depending on the choice of σ, unlike the error function used for regression, $E(\hat{w})$ can be bounded, and the minimization can only yield an approximate value, since the targets are only reached asymptotically. The stationary points can be found using numerical algorithms; in most real-word applications involving many variables, gradient descent turns out to be adequate.

EXERCISES

1. [15]   Let $\hat{w}^*$ be the solution discovered for a threshold machine with a hard-limiting function. Is it still a solution when scaling the inputs?

2. [15]   Show that the softmax function defined in Eq. (2) is normalized to 1 (i.e. $\sum_{i=1}^c \mathrm{smx}_i = 1$) and that it indeed has the "softmax property": Suppose that there exists an a_i such that $a_i \gg a_j$ for all $j \neq i$, then $\mathrm{smx}_k \approx [k = i]$.

3.2.1 Predicate-order and representational issues

In this section, we want to give an insight on the connection between pattern recognition tasks and representational power of perceptrons. This is a really intriguing problem, since we are required to discover connections between human process of pattern interpretation and their input representation in perceptrons, which gives rise to fundamental questions also on their computational capabilities. Needless to say, there is a huge gap between abstract pattern notions and geometrical concepts inherited in the input space like, for instance, class separability! This is true when dealing with linguistic and visual patterns, as well as in most interesting learning problems. To give an insight on how we can fill the gap, we present the case of images. We are not interested in exploring natural images, since the essence of the difficulty of filling the mentioned gap can be appreciated also in the case of black and white images. Basically, we want to discuss predicates which take pictures X drawn in the retina R, and establish a given property of X. The retina $R \subset \mathbb{R}^2$ is where images are drawn. Formally, an image is a function $I : R \to \{0, 1\}$. A picture $X \subset R$ is defined as $X = \{x \in R : I(x) = 1\}$. For example, a given X on the retina R might be a circle, that is why we are interested in studying predicates like $[X \text{ is a circle}]$. Most of the

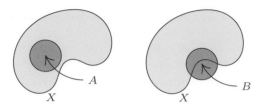

FIGURE 3.1

Mask predicates: $[A \subset X] = 1$ and $[B \subset X] = 0$.

times, it is more useful to consider a discrete retina $R^{\sharp}$. Consistently with the definition of X, a discrete figure $X^{\sharp}$ can be regarded as the set of all pixels x of the retina such that $I(x) = 1$. For example, a letter "c" on a 4×4 retina can be represented by a function I_c that gives the necessary information to build the image $X^{\sharp}$ as follows:

$$R^{\sharp} = \begin{array}{c}\boxplus\end{array} \xrightarrow{I_c} \begin{pmatrix} 0 & 1 & 1 & 0 \\ 1 & 1 & 0 & 0 \\ 1 & 0 & 0 & 0 \\ 1 & 1 & 1 & 0 \end{pmatrix} \equiv I_c(R^{\sharp}), \qquad X^{\sharp} = \textbf{C} \,.$$

Throughout this section, we will also use at our convenience both discrete and continuous representation of the retina. We are interested in properties of patterns that can reach high degree of abstraction. For example, we want to explore how it is possible to establish whether a given picture is convex, that is, whether for a given picture $[X$ is convex$]$ is true or false. The predicate can be computed by checking if for every p, q in X, and for every $\alpha \in [0 . . 1]$ we have that

$$r = \alpha p + (1 - \alpha) q \in X. \tag{1}$$

While this is clearly a geometrical property, it is not trivial to see how a perceptron could express such a property! The notion of connectedness of a given picture, defined by $[X$ is connected$]$ shares with $[X$ is convex$]$ the need for conquering abstraction, but it will be proven to be significantly harder for perceptron-based representations than convexity. On the other hand, there are some predicates that can be efficiently expressed by perceptrons. Let us consider

$$[p \subset X] \quad \text{and} \quad [A \subset X]. \tag{2}$$

The first case, $[p \in X]$, expresses membership of p to X. In the second case $[A \subset X]$ returns the true value (see Fig. 3.1) if and only if A, the *mask*, is a subset of X. The truth of predicate $[X$ is convex$]$ can be decided by checking the simpler predicate $[p \divideontimes q \in X]$ (here $p \divideontimes q$ is the midpoint between p and q) for all p and q in the figure, the simplicity of this predicate relying on the fact that it can be decided just by looking at three points.

Bearing this in mind, we can give a first definition of order of a predicate: We say that a predicate ψ is conjunctively local of order κ if there exist a set of predicates $\Phi = \{\varphi_1, \ldots, \varphi_D\}$ such that

$$\psi = \varphi_1 \wedge \varphi_2 \wedge \cdots \wedge \varphi_D = [\varphi_1 + \varphi_2 + \cdots + \varphi_D \geq D], \tag{3}$$

and each φ_i depends upon no more than κ points on the retina. For example, suppose that we are considering a two-dimensional retina, then all the possible predicates on this retina are the 16 Boolean functions of two variables; from the definition, the AND function $x_1 \wedge x_2$ is conjunctively local of order 1 since one can set $\varphi_1 = x_1$ and $\varphi_2 = x_2$, while, for example, the XOR function is conjunctively local of order 2 since $x_1 \oplus x_2 = (x_1 \vee x_2) \wedge (\bar{x}_1 \vee \bar{x}_2)$, so we can choose $\varphi_1 = (x_1 \vee x_2)$ and $\varphi_2 = (\bar{x}_1 \vee \bar{x}_2)$. The fact that the XOR predicate has order 2 corresponds with the impossibility of its realization by linearly separable functions, which can only attack predicates of order 1.

However, this is not a very good definition of order, since, as it stands, it still strongly depends on the particular choice of functions $\varphi_1, \ldots, \varphi_D$. In what follows, we will give a better definition of order for all those predicates that can be represented with a perceptron. The general perceptron ψ that we consider in this discussion is a linear threshold function with respect to Φ, meaning that there exist a number t and a set of weights $w_1, \ldots, w_D$ such that

$$\psi = [w_1\varphi_1 + w_2\varphi_2 + \cdots + w_D\varphi_D \geq t]. \tag{4}$$

Notice that φ emphasizes the peculiar preprocessing taking place on X along with its topological meaning.

Now let us come back to a general perceptron predicate, ψ, which operates on a certain retina R: Things are much more involved here. The problem is that we are not given a representation of the predicate, but only its abstract meaning, that can be complex to capture by LTU-based representations.

A sound definition of predicate order benefits from the notion of *support* of a predicate: Formally, the notion of support of φ is in fact used to capture the idea that, in general, the functions φ only depend on a limited portion of the retina. If we define $S = \{A \subset R : \varphi(X \cap A) = \varphi(X)\}$, then the support is the subset of the retina such that

$$\mathrm{supp}(\varphi) \in \arg\min_{A \in S} |A|.$$

We can promptly see that $\varphi(X \cap S) = \varphi(X)$ can easily be met for large S, whereas it might be difficult to verify the equality for "small" S. Here the function $|\cdot|$ for a continuous retina must be interpreted as the area function (Lebesgue measure). Intuitively, the support $\mathrm{supp}(\varphi)$ gives us the portion of the retina which is really relevant in the classification. In other words, the support contains the distinctive features, and it is clearly very important to express the degree of complexity of the predicate. It is in fact a useful concept to fully capture the notion of predicate order. Suppose that we can discover an upper bound on $|\operatorname{supp}(\varphi)|$. The order of a given predicate ψ is the smallest integer κ such that $|\operatorname{supp}(\varphi)| \le \kappa$ for all $\varphi \in \Phi$. The order $\kappa = \omega(\psi)$ of predicate ψ indicates that no decision on X can be taken with less than κ points which belong to the support S, which links this definition to Eq. (3). We can promptly see that the order of ψ is a property of ψ alone, and it is independent of the specific choice of Φ.

Let us consider the *mask predicate* $[A \subset X]$ associated with mask A. The concept of predicate order gives us an idea of its *locality*, that is, on how big the number of points to be jointly considered for constructing $\varphi_i \in \Phi$ is. It is quite easy to see that

$$\omega\big([A \subset X]\big) = 1. \tag{5}$$

The proof can be based on the construction of the set Φ corresponding to $[A \subset X]$. If we can find a set of predicates Φ through which we can express $[A \subset X]$ as a linear threshold function and such that $\operatorname{supp}(\varphi_i) = 1$ for all i, then this proves Eq. (5) since the predicate that we consider cannot have order smaller than 1. To show this, suppose that $A^\sharp = \{a_1, a_2, \ldots, a_{|A^\sharp|}\}$ and set $\varphi_i \equiv [a_i \in X^\sharp]$, then

$$[A^\sharp \subset X^\sharp] = \big[\varphi_1 + \varphi_2 + \cdots + \varphi_{|A^\sharp|} \ge |A^\sharp|\big]. \tag{6}$$

This comes from the fact that the above condition holds true when all points of the mask A are in X. Since φ_i depends only on one point of the retina, we have that $|\operatorname{supp}(\varphi_i)| = 1$. Notice that Eq. (6) is a possible description of an LTU to compute the mask predicate. Following the intuition that leads us to defining the order of a predicate, we will now show that the *convexity* predicate is convex a limited-order predicate. Now the verification of the condition in Eq. (1) requires three points, p, q, and r. Let us consider instead the following condition:

$$\forall p, q \in X : \; p \divideontimes q \in X. \tag{7}$$

Clearly, if any figure satisfies condition (1) then it also satisfies the above condition. Strictly speaking, the converse does not hold! One can in fact always choose $\alpha \in (0..1)$ such that $r = \alpha p + (1 - \alpha)q \in (\mathbb{R} \setminus \mathbb{Q})^2$. However, any such point r can be arbitrarily approximated by the dichotomic process

$$r_0 = p, \qquad r_{m+1} = \frac{r + r_m}{2},$$

which converges to r, that is, $r_m \longrightarrow r$. Hence, it turns out that Eq. (7) gives a weaker definition of convexity that, however, is fully acceptable for our purposes. Interestingly, the condition expressed by (7) is better suited for thinking about a perceptron-based representation. We can in fact express $[X^\sharp$ is convex$]$ by

$$[X^\sharp \text{ is convex}] = \left[\sum_{i=1}^{\binom{|X^\sharp|}{2}} \varphi_i \ge \binom{|X^\sharp|}{2} \right],$$

where the index i counts the pairs (p_i, q_i) of points on the figure and $\varphi_i = [p_i \divideontimes q_i \in X^\sharp]$ (here, of course, we have $w_i = 1$). Moreover, the support is $\operatorname{supp}(\varphi_i) = \{p_i, q_i, p_i \divideontimes q_i\}$, since $\varphi_i(X^\sharp \cap \text{supp}(\varphi_i)) = \varphi_i(X^\sharp)$, meaning in particular that $|\text{supp}(\varphi_i)| = 3$ for all $i = 1, \ldots, \binom{|X^\sharp|}{2}$. Hence we conclude that

$$\omega([X \text{ is convex}]) = |\text{supp}(\varphi_i)| = 3. \tag{8}$$

In order to grasp the connection between pattern recognition and representation in perceptrons, the introduction of conjunctively local predicates, along with the corresponding notion of predicate order, is very useful in understanding computational issues. However, the construction of predicates on images suggests that before thinking of any specific construction of features, one must decide which portion of the given picture to take into account. If we assume that the construction is crafted for the specific pattern recognition problem then visual features necessarily express local properties, so that for any pixel it is reasonable to isolate a pixel-centered portion of the picture to be used for feature extraction. Clearly, the portion cannot be too large, otherwise the amount of information which is involved will be of the same order as the information required for the decision itself! Strictly speaking, this assumption can be overcome if the features are learned, but also in this case the diameter limitation on pixel-centered portion of the picture is quite reasonable. The perceptrons which operate under this assumption are referred to as *diameter-limited perceptrons*. Interestingly, it will be shown that the computational limitations connected with the locality notion arising from the concept of predicate order are still there, and that they just assume a different form.

In order to become acquainted with the new concept of locality expressed by diameter-limited perceptrons, we go back to [X is convex]. This time, instead of convexity, we want to test a sort of *near convexity*. The idea is the following: Suppose you do want to accept as convex also figures that have a small hollow. This can be done by introducing a length scale r that fixes the degree of concavity that you are accepting. One way to tolerate such a violation is to define the set $p \mathrel{-\!\!\circ\!\!-} q$, that is, the set of points that have distance less or equal than r from the midpoint between p and q; in so doing, r naturally represents such scale. With this definition at hand, it is quite natural to say that a set C is *almost convex* if for every p and q in C we have that

$$p \mathrel{-\!\!\circ\!\!-} q \cap C \quad \text{is nonempty.} \tag{9}$$

Let us see now how we can implement the predicate of almost connectedness with a perceptron. First of all, notice that the condition of almost connectedness can fail because of two distinct reasons: Either there is a hole that is bigger than the set scale r or the figure is made up of more pieces that are well separated (at least by r); these two possibilities are displayed in Fig. 3.2. The condition expressed by Eq. (9) can be rewritten using the following perceptron LTU

$$[X \text{ is almost convex}] = \left[\sum_i [|p_i \mathrel{-\!\!\circ\!\!-} q_i \cap X| = 0] < 1 \right],$$

where the index i runs over all the couples of points of the image X. Notice that this predicates surely has finite order since $\varphi_i = [|p_i \mathrel{-\!\!\circ\!\!-} q_i \cap X| = 0]$ depends (of course, now thinking in terms of a retina with finite resolution) upon a limited number of points — in particular those inside the circle of radius r. In any case, both the locality analyses based on conjunctively local and on diameter-limited predicates lead to a promising conclusion about perceptron computational capabilities. Unfortunately, in many other interesting tasks, the order of the predicate explodes. The following analysis sheds light on fundamental limitations of perceptron-based

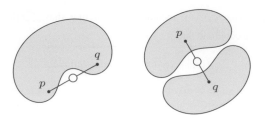

FIGURE 3.2

Failure of detection of almost convexity by diameter-limited perceptrons in a connected and nonconnected image.

representations. We prove that, unlike the predicates considered so far, [X is connected] is not conjunctively local of any order. Like for the case of convexity, we consider the weak definition, in which the property is analyzed for finite sets. The proof is given by contradiction and is based on considering the consequences of assuming that connectedness is a predicate of finite order κ. Let us consider the following three pictures:

$$X_1^\sharp = \rule{1cm}{0.8cm} \; , \qquad X_2^\sharp = \rule{1cm}{0.8cm} \; , \qquad X_2^\sharp = \rule{1cm}{0.8cm} \; ,$$

where all figures are composed of rows with $\kappa + 1$ pixels. Let us focus on $X_1^\sharp$ and $X_2^\sharp$, for which [$X_1^\sharp$ is connected] $= 1$ and [$X_2^\sharp$ is connected] $= 0$. Since [X is connected] is conjunctively local of order κ, it operates on $\varphi \in \Phi$ whose support is composed of κ points, that is, $|\operatorname{supp}(\varphi)| = \kappa$. Now all the κ points of $\operatorname{supp}(\varphi)$ must necessarily be located in the middle row of $X_1^\sharp$. This is in fact the only way of discriminating $X_1^\sharp$ from $X_2^\sharp$. Now, since there are $\kappa + 1$ pixels in the row, we can always find one of them which does not belong to $\operatorname{supp}(\varphi)$. The column j which identifies that pixel is used to construct figure $X_3^\sharp$, which is exactly the same as $X_2^\sharp$ apart from a new pixel p_{2j} located in column j. Of course, we have [$X_3^\sharp$ is connected] $= 0$ since $p_{2j} \notin \operatorname{supp}(\varphi)$, but this contradicts the fact that figure $X_3^\sharp$ is connected. Now one might suspect that this negative result is due to the lack of connection with the topology of the figures, and that diameter-limited perceptrons have a different behavior. On the opposite, this negative result still holds!

Let us consider any diameter-limited perceptron that is expected to discriminate whether a given picture is connected. We prove that no such perceptron exists by contradiction. Consider the following four figures:

$$X_1 = \rule{1cm}{0.4cm} \; , \qquad X_2 = \rule{1cm}{0.4cm} \; , \; X_3 = \rule{1cm}{0.4cm} \; , \qquad X_4 = \rule{1cm}{0.4cm} \; ,$$

suppose furthermore that, given the diameter of the perceptrons $2r$, the length of these images is at least 3 diameters. In this way it is always possible to distinguish three areas on each figure: The left and right edges and the center of the figure as displayed in Fig. 3.3. Now the hypothesis that connectedness can be established by means of a diameter-limited perceptron means that for a generic image X

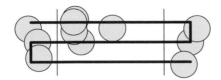

FIGURE 3.3

Inputs can be collected together into three groups according to where their support lies as shown here.

$$[X \text{ is connected}] = \left[\sum_i w_i \varphi_i \geq t \right],$$

or, in other words, that $\sum_i w_i \varphi_i - t$ must be nonnegative if the figure is connected and negative if the figure is nonconnected. The basic idea of the proof is to break the sum over i in the above expressions into three sums: One that runs only over the group of indices J of perceptrons that have support on the left, one over the indices K of perceptron with support on the center of the figure, and finally one sum over H of the indices of perceptrons with support on the right of the figure. In this way

$$[X \text{ is connected}] = \left[\sum_{j \in J} w_j \varphi_j + \sum_{k \in K} w_k \varphi_k + \sum_{h \in H} w_h \varphi_h \geq t \right].$$

Since X_2 is connected and X_1 is not, we surely have $\sum_i w_i \varphi_i(X_2) > \sum_i w_i \varphi_i(X_1)$; however, locally, X_1 and X_2 differ only on the right side, so we conclude that

$$\sum_{h \in H} w_h \varphi_h(X_2) > \sum_{h \in H} w_h \varphi_h(X_1). \tag{10}$$

Similarly comparing figure X_1 and X_3, we have

$$\sum_{j \in J} w_j \varphi_j(X_3) > \sum_{j \in J} w_j \varphi_j(X_1). \tag{11}$$

Locality tells us that $\sum_{h \in H} w_h \varphi_h(X_3) = \sum_{h \in H} w_h \varphi_h(X_1)$ and we also have

$$\sum_{h \in H} w_h \varphi_h(X_2) > \sum_{h \in H} w_h \varphi_h(X_3). \tag{12}$$

Then putting inequalities (10)–(12) together, knowing that

$$\sum_{j \in J} w_j \varphi_j(X_3) = \sum_{j \in J} w_j \varphi_j(X_4) \quad \text{and} \quad \sum_{h \in H} w_h \varphi_h(X_2) = \sum_{h \in H} w_h \varphi_h(X_4),$$

and observing that the sum over K does not change for any of the four images X_1–X_4, we finally have

$$\sum_i w_i \varphi_i(X_4) = \sum_{j \in J} w_j \varphi_j(X_3) + \sum_{k \in K} w_k \varphi_k(X_3) + \sum_{h \in H} w_h \varphi_h(X_2)$$

$$> \sum_{j \in J} w_j \varphi_j(X_3) + \sum_{k \in K} w_k \varphi_k(X_3) + \sum_{h \in H} w_h \varphi_h(X_2) = \sum_i w_i \varphi_i(X_3),$$

which is a contradiction since this last equation implies that X_4 is connected. This negative result suggests that diameter-limited perceptron fails capturing truly global properties. The impact of most of those critical results has been sometimes neglected, since in most cases their validity is not limited to single-layer perceptrons. For example, the result on connectedness holds for any diameter-limited perceptron, that is, it applies also for the multilayer networks discussed in Chapter 5.

EXERCISES

▸ **1.** [*10*]   Prove that the AND and OR functions are of order 1.

2. [*15*]   Generalize the result of the previous exercise and show that any *linearly separable* Boolean function $f(x_1, x_2, \ldots, x_n)$ is of order 1.

▸ **3.** [*25*]   Prove that the two-dimensional XOR is of order 2.

4. [*M32*]   (M. Minsky and S. Papert) Show that for a discrete retina $R^\sharp$, all the weights w_i and the threshold t in Eq. (4) can be confined to take only integer values.

5. [*20*]   Extend the result of Exercise 4 by showing that one can always get rid of the negative weights, so that the threshold function $[w_1 x_1 + w_2 x_2 + \cdots + w_D x_D \geq D]$ can be reduced to a positive threshold function with all the weights positive.

6. [*M22*]   (Retina graphs.) An interesting *retina graph* R_n can be built out of an $n \times n$ retina $R^\sharp$ in the following way: Draw a vertex inside each pixel and connect with edges the vertices of adjacent pixels. For example, from a 2×2 retina we obtain the graph $R_2 = $.

(a)   Show that in general a retina graph has n^2 vertices and $(n-1)(4n-2)$ edges.
(b)   Prove that for every $n \geq 1$ R_n is Hamiltonian.

7. [*M30*]   A set C of vertices in a graph is called convex if $[u \mathinner{.\,.} v] \subseteq C$ for all u and v in C and where we denote by $[u \mathinner{.\,.} v]$ the set of all points that belong to the shortest paths between u and v. This definition suggests a way to extend the notion of convexity for images on a discrete retina $R^\sharp$: We will say that an image $X^\sharp$ on an $n \times n$ retina is convex if the set $X^\sharp$ is convex in the retina graph R_n defined in Exercise 6. Is it possible to define midpoint convexity similar to that defined in Eq. (7) for the continuous case by introducing some *discrete midpoint function* $i \rightthreetimes j$?

3.2.2 Optimality for linearly separable examples

While we loose the linearity of normal equations, now we prove that under the hypothesis of *linearly separable* classes, the appropriate marriage of linear-threshold functions and loss function results in good formulations of learning in terms of computational complexity. In particular, we will see that some pairs lead to loss functions where there are no local minima different from the global minimum! Interestingly, it will be also shown that this is not always the case, since other pairs do not share this property. The analysis is carried out for the case of two classes, $c = 2$, but the same scheme can be adopted in the general case. Basically, in case of many classes ($c > 2$), each unit can be asked to discriminate a certain class with respect to the remaining ones.

Now let us determine the condition that any stationary point must satisfy, $\nabla E = 0$. Since the risk function accumulates the loss over all the examples, we have $E = \sum_\kappa e_\kappa$, and therefore the gradient can be calculated by

$$\frac{\partial E}{\partial \hat{w}_j} = \sum_{\kappa=1}^{\ell} \frac{\partial e_\kappa}{\partial \hat{w}_j} = \sum_{\kappa=1}^{\ell} \frac{\partial e_\kappa}{\partial a_\kappa} \frac{\partial a_\kappa}{\partial \hat{w}_j} = \sum_{\kappa=1}^{\ell} \hat{x}_{\kappa j} \delta_\kappa,$$

where $\delta_\kappa := \partial e_\kappa / \partial a_\kappa$ and $a_\kappa := \sum_j \hat{w}_j \hat{x}_{\kappa j}$. From now on, this is referred to as the *delta error* of the unit for the given example x_κ. Using vectorial notation, we have

$$\nabla E = \hat{X}' \Delta, \tag{1}$$

where $\Delta := (\delta_1, \ldots, \delta_\ell)'$. Of course, this makes sense only in points where the partial derivatives exist, and it is clear that δ_κ depends on the joint choice of the linear-threshold function and the loss function. For some of these choices, the delta error exhibits a nice sign structure that will be very useful in the following.

Sigmoidal Unit and Quadratic Loss. If we select the quadratic loss

$$e_\kappa = \frac{1}{2}(y_\kappa - \sigma(a_\kappa))^2$$

then we can promptly see that

$$\delta_\kappa = \sigma'(a_\kappa)(\sigma(x_\kappa) - y_\kappa). \tag{2}$$

As a consequence, Δ exhibits a sign structure which only depends on the category of the patterns. This is true for both the asymmetric and symmetric squash functions *iii.* and *iv.* of Eq. 3.2–(1), since $\sigma' > 0$.

Linear Unit and Hinge Function. Here, we couple the linear model (unit) with a hinge loss function, so that

$$e_\kappa = (1 - a_\kappa y_\kappa)_+.$$

Now, if there is strong sign agreement (i.e., if $a_\kappa y_\kappa \geq 1$), $e_\kappa \equiv 0$ and $\delta_\kappa = 0$, otherwise, $\delta_\kappa = -y_\kappa$. Written more concisely, we have $\delta_\kappa = -y_\kappa[1 - a_\kappa y_\kappa > 0]$. Once again Δ exhibits a sign structure which only depends on the category of the patterns.

In the case of the quadratic loss we can always sort the inputs in such a way that the Δ signs become

$$\text{sign}(\Delta) = (+ \cdots + - \cdots -)', \tag{3}$$

where $\text{sign}(x) = [x > 0] - [x < 0]$, and we have assumed that when applied to a vector v the sign function returns a vector of the same dimension with $(\text{sign}(v))_i =$

$\mathrm{sign}(v_i)$. The same holds for the hinge loss function, apart from the examples for which there is already strong sign agreement. Since the two classes are linearly separable, there exists $\alpha \in \mathbb{R}^{d+1}$ such that $\hat{X}\alpha$ gets the sign structure

$$\mathrm{sign}(\hat{X}\alpha) = (+ \cdots + - \cdots -)'. \tag{4}$$

Now notice that any solution of $\alpha'\hat{X}'\Delta = 0$ is also a solution of $\nabla E = \hat{X}'\Delta = 0$. Hence we have

$$(\alpha'\hat{X}')\Delta = (\hat{X}\alpha)'\Delta = 0 \Rightarrow \Delta = 0.$$

Now, in the case of the hinge loss, the condition $\Delta = 0$ means that we have discovered a configuration in which for all patterns there is strong sign agreement, namely a global minimum. This follows directly by contradiction. If there were no strong sign agreement for at least one pattern $\overline{\kappa}$, we would have $\delta_{\overline{\kappa}} = -y_{\overline{\kappa}} \neq 0$, so that we would end up with the contraction to $\Delta = 0$. The case of quadratic loss is more involved. Basically, $\Delta = 0$ cannot be reached, since the above vanishing of the gradient needs to be rewritten as $\nabla E \to 0 \Rightarrow \Delta \to 0$. Now the condition $\delta_{\kappa} = 0$ can correspond with both targets. However, we now prove that any numerical algorithm like gradient descent is attracted to the inferior $E = 0$, since $\partial^2 E / \partial \hat{w}_j^2 < 0$. This can be proven as follows:

$$\frac{\partial^2 E}{\partial \hat{w}_j^2} = \frac{\partial}{\partial \hat{w}_j} \sum_{\kappa=1}^{\ell} x_{\kappa j} \sigma'(a_\kappa)(f(x_\kappa) - y_\kappa)$$

$$= \sum_{\kappa=1}^{\ell} x_{\kappa j}^2 \sigma''(a_\kappa)(f(x_\kappa) - y_\kappa) + \sigma'^2(a_\kappa) \cdot x_{\kappa j}^2 \tag{5}$$

$$= \sum_{\kappa=1}^{\ell} x_{\kappa j}^2 \left(\sigma''(a_\kappa)(f(x_\kappa) - y_\kappa) + \sigma'^2(a_\kappa)^2 \right).$$

Now let us consider stationary points such that

$$(y_\kappa = 0 \wedge f(x_\kappa) \to 1) \vee (y_\kappa = 1 \wedge f(x_\kappa) \to 0). \tag{6}$$

Furthermore, if we assume that in those configurations $\sigma'^2/\sigma'' \to 0$, we conclude that

$$\frac{\partial^2 E}{\partial \hat{w}_j^2} \to \sum_{\kappa=1}^{\ell} x_{\kappa j}^2 \cdot \sigma''(a_\kappa)(f(x_\kappa) - y_\kappa) < 0,$$

since $\sigma''(a_\kappa)$ and $f(x_\kappa) - y_\kappa$ gets the opposite sign in the configuration associated with condition (6).

3.2.3 Failing to separate

In this section, we discuss the reason why linear-threshold units must be properly combined with a loss function, and show that no pair is satisfactory. What if we simply use linear units with the quadratic loss for classification? Unlike what one could expect from the optimality of normal equations, for classification problems such a choice is not necessarily adequate. In particular, we will show that there are cases in which we fail to separate linearly separable examples! To grasp the reasons of the failure, let us consider the following counterexample that clearly addresses what's wrong with that marriage.

We are given four examples, where the first and the fourth are repeated m times. This corresponds with attaching the weight m in the error function to these examples. In general, the new error function becomes

$$E(\hat{w}) = \frac{1}{2} \sum_{\kappa=1}^{\ell} c_\kappa (y_\kappa - \hat{w}'\hat{x}_\kappa)^2,$$ (1)

where c_κ can be regarded as a variable which counts the repetitions of examples $\hat{x}_\kappa$. As shown in Exercise 2, we can still use a slight modification of the normal equations, where

$$\hat{X} = \begin{pmatrix} 0 & 0 & m \\ -1 & 1 & 1 \\ 1 & 1 & 1 \\ 0 & 3/2m & m \end{pmatrix}, \qquad y = \begin{pmatrix} -m \\ -1 \\ -1 \\ m \end{pmatrix}.$$

Now, for $m = 1$ the solution is $\hat{w}' \propto (0, 10, -27/2)'$, which corresponds with the separation line $x_2 = 1.35$ (see the figure at the side) that correctly separates positive and negative examples. As $m \to \infty$, however, the solution is $\hat{w} \propto (0, 6, -9/2)$, which corresponds with the line $x_2 = 3/4$. Unlike the case $m = 1$, in this case LMS fails to separate the given examples. This undesirable behav-

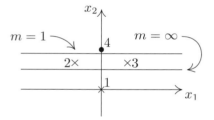

ior is essentially related to the nature of the Euclidian norm at the basis of LMS. This can promptly be understood in case of $d = 1$, where a threshold-like behavior that is desirable for classification would require using the supremum norm $\|\cdot\|_\infty$ instead of $\|\cdot\|_2$ (see Exercise 7 for further details).

The discussion in this section clearly indicates the need for LTU machines for classification. The desirable optimality that is gained by the normal equations for regression doesn't prevent from spurious solutions in classification tasks. Basically, the formulation by linear machines in classification might yield spurious configurations that are undesirable even if they correspond with the global minimum of the error function.

EXERCISES

▶ **1.** [20] Suppose we are given a training set with only two points, x_1 and x_2, and we consider a classification problem with $y = (1, -1)'$.

- (a) In the simple case $d = 1$ find an expression for the solution in terms of a linear machine by using normal equations, and prove that there exists only one solution.
- (b) For the general case $d > 1$ determine an expression of the solution under the assumption that $w = \alpha x_1 + \beta x_2$, with x_1 and x_2 not linearly dependent. In this case show that under the additional assumption that $x_1^2 = x_2^2$, $\hat{w}$ is an eigenvector of $\hat{X}'\hat{X}$.

▸ **2.** [*15*] Consider the weighed LMS problem defined with the error function in Eq. (1). Prove that the solution of this problem is still given by Eq. (4) after suitable redefinition of $\hat{X}$ and y.

3. [*17*] Start from a clustered pattern distribution with two spoiler examples and build the idealized representation choosing one representative point as follows:

with 1–4 being the points with coordinates $(0, 0)$, $(1, 0)$, $(1, 1)$, and $(0, 1)$, respectively. Assume for simplicity that the degeneracy of points 1 and 4 is m for both points and show that for any finite m the separation line obtained through the normal equations succeeds in separating the data, while for $m = \infty$ you don't achieve such separation as it is discussed in this section. What happens for $m = 0$? Interpret this result.

4. [*20*] Till now, we saw two examples (in Section 3.2.3 and in Exercise 3) in which the LMS solution fails to separate due to the presence of a degeneracy of some of the examples. Show, however, that there are examples in which even in the presence of this repetition there is no failure to separate. *Hint:* Consider a number of unknowns greater than the number of examples.

▸ **5.** [*M30*] Let us consider a threshold-linear neuron for classification and consider the case in which the targets are $y_\kappa \in \{\underline{d}, \overline{d}\}$, where $\underline{d} > 0$ or $\overline{d} < 1$. We can still argue that the error function is local minima free for linearly separable examples?

6. [*M30*] Let us consider the learning environment defined in Exercise 3 where the examples exhibit a natural clustering structure. The right-side learning environment presents a related case, where we dramatically simplify the basic idea and only consider four examples, where we are somehow representing the cluster centers. We also assume that 1 and 4 are repeated m times, where m is the cardinality of both clusters. Use the normal equations to prove that the solution is the one represented by the separation line, with positive slope, in between 1–3 and 4.

7. [*M25*] In this section, we discuss the separation failure when we combine linear units and the quadratic error. Now let us consider the generic p-norm

$$E(\hat{w}) = \left(\sum_{\kappa=1}^{\ell} (y_\kappa - f(x_\kappa))^p \right)^{1/p} .$$

What about the separation in the case $p \to \infty$?

▸ **8.** [*28*] Let us consider the classic seven segment display used to code 10 digits. Is the pattern set linearly-separable?

3.3 **Statistical view**

In this section we provide a statistical setting for linear machines. We will see that, under special assumptions on the probability distribution, they naturally arise from the Bayesian decision. Then we will reformulate LMS by assuming that data come from a certain probability distribution and, finally, we will explore the *shrinkage methods*.

3.3.1 **Bayesian decision and linear discrimination**

In Section 2.1.3, we have seen that once we choose the 0–1 loss function, the optimal decision which minimizes $\mathrm{E}\,L(Y, f(X))$ corresponds with Bayesian classifier, which returns the decision $f = (f_1, \ldots, f_c)$. Since for each fixed $i = 1, \ldots, c$ the probability $\Pr(I = i \mid X = x)$ is a function of x, the separation surface between two categories i and j is defined by the condition $\Pr(I = i \mid X = x) = \Pr(I = j \mid X = x)$. Using Eq. 2.2.2–(2) and assuming the case of multivariate Gaussian probability distribution

$$p(x \mid I = i) = \frac{1}{\sqrt{(2\pi)^d \det \Sigma_i}} e^{-\frac{1}{2}(x - \mu_i)' \Sigma_i^{-1}(x - \mu_i)},$$

we have that the separation surface is defined by the equation

$$(x - \mu_i)' \Sigma_i^{-1}(x - \mu_i) - (x - \mu_j)' \Sigma_j^{-1}(x - \mu_j) = \log \left(\frac{\det \Sigma_j}{\det \Sigma_i} \frac{P^2(I = i)}{P^2(I = j)} \right). \quad (1)$$

Hence, in general, Bayesian decisions under Gaussian assumption are quadratic surfaces in the input space. However, if $\Sigma_i = \Sigma_j = \Sigma$ then

$$2(\mu_j - \mu_i)' \Sigma^{-1} x + \mu_i' \Sigma^{-1} \mu_i - \mu_j' \Sigma^{-1} \mu_j + 2 \ln \frac{\Pr(I - j)}{\Pr(I = i)} = 0. \quad (2)$$

Hence, Bayes decision is the output of a linear machine where

$$w = \Sigma^{-1}(\mu_j - \mu_i), \qquad b = \frac{1}{2} \left(\mu_i' \Sigma^{-1} \mu_i - \mu_j' \Sigma^{-1} \mu_j \right) + \ln \frac{\Pr(I = j)}{\Pr(I = i)}. \quad (3)$$

Notice that if $\Sigma_i = \mathrm{diag}(\sigma_i)$ and $\Sigma_j = \mathrm{diag}(\sigma_j)$ then the separating plane is always orthogonal to the line which connects μ_i to μ_j. In the special case in which there is no prior, i.e., $\Pr(I = i) = \Pr(I = j)$, we can promptly check that the plane contains the middle point $x_m = (\mu_i + \mu_j)/2$. As we could expect, if one of the two classes has a higher prior then the plane moves toward the mean of that class.

3.3.2 **Logistic regression**

Suppose we classify examples in two categories by using LTU units. We want to discuss its statistical interpretations by determining $p(y \mid x)$, where $y \in \{0, 1\}$. We can promptly see that

$$p(y \mid x) = [y = 1]\sigma(a(x)) + [y = 0](1 - \sigma(a(x))) = (\sigma(a(x)))^y (1 - \sigma(a(x)))^{1-y} \tag{1}$$

which expresses a binomial distribution which in this case degenerates to Bernoulli distribution. According to the logistic hypothesis, we choose

$$\sigma(a(x)) = \frac{\exp(-a(x))}{1 + \exp(-a(x))}, \tag{2}$$

which, in turn, yields $1 - \sigma(a(x)) = 1/(1 + \exp(-a(x)))$. Notice that $1 - \sigma(x)$ is the classic logistic function.

We can easily generalize these equations to the case of $c > 2$ classes. In that case, softmax units are well-suited for guaranteeing the probabilistic normalization. This time the labels y_i, $i = 1, \ldots, c$ satisfy the condition $\sum_{i=1}^{c} y_i = 1$, and we still have $y_i \in \{0, 1\}$. We can express $p(y \mid x)$ by

$$p(y \mid x) = \prod_{i=1}^{c} [y_i = 1]\sigma(a(x)) = \prod_{i=1}^{c} (\sigma(a(x)))^{y_i}, \tag{3}$$

where

$$\sigma(a_i(x)) = \frac{\exp(a_i(x))}{\sum_{j=1}^{c} \exp(a_j(x))} = \frac{\exp(\hat{w}_i'\hat{x})}{\sum_{j=1}^{c} \exp(\hat{w}_j'\hat{x})}. \tag{4}$$

Notice that $p(y \mid x)$ in Eq. (3) follows a multinomial distribution. We can promptly realize that because of the probabilistic normalization, one of the c classes can always be considered as a reference, which suggests elaborating the softmax definition in such a way to save one out of c weight vectors. Let $i = c$ be the index of such a class. We can get rid of the parameters w_c of the class by noticing that

$$\sigma(a_i(x)) = \frac{[i = c]}{1 + \sum_{j=1}^{c-1} \exp(a_j(x) - a_c(x))} + \frac{[i \neq c]\exp(a_i(x) - a_c(x))}{1 + \sum_{j=1}^{c-1} \exp(a_j(x) - a_c(x))}$$
$$= \frac{1}{1 + \sum_{j=1}^{c-1} \exp(-\tilde{a}_j(x))}[i = c] + \frac{\exp(-\tilde{a}_i(x))}{1 + \sum_{j=1}^{c-1} \exp(-\tilde{a}_j(x))}[i \neq c], \tag{5}$$

where $\tilde{a}_i(x) := a_c(x) - a_i(x) = (\hat{w}_c - \hat{w}_i)\hat{x}$. We can easily verify[4] that for $c = 2$ this returns Eq. (1). An expressive interpretation of classifiers can be given by considering the *log-odds*, or *logit*, which is defined $\forall i = 1, \ldots, c - 1$ as

$$o_i(x) := \ln \frac{p(y_i \mid x)}{p(y_c \mid x)} = -a_i(x) = -w_i'\hat{x}. \tag{6}$$

---

[4] In the following, for the sake of simplicity, we replace $\tilde{a}_i(x)$ with $a_i(x)$.

It corresponds with the flipped sign output of the ith unit. In the literature we typically encounter $o_i = a_i(x)$, which is due to the different definition of the sigmoidal units that is typically used in neural nets.

Now we discuss parameter estimation by using MLE. Because of the multinomial distribution,

$$L = \prod_{\kappa=1}^{\ell}\left(1 - \sum_{j=1}^{c-1}\sigma(a_j(x_\kappa))\right)^{1-\sum_{j=1}^{c-1}y_{\kappa,i}} \prod_{i=1}^{c-1}\sigma\big(a_i(x_\kappa)\big)^{y_{\kappa,i}}.$$

The log-likelihood is

$$l = \sum_{\kappa=1}^{\ell}\sum_{i=1}^{c-1}\ln(\sigma(a_i(x_\kappa)))^{y_{\kappa,i}} + \sum_{\kappa=1}^{\ell}\ln(1 - \sum_{j=1}^{c-1}\sigma(a_j(x_\kappa)))^{1-\sum_{j=1}^{c-1}y_{\kappa,i}}$$

$$= \sum_{\kappa=1}^{\ell}\sum_{i=1}^{c-1}y_{\kappa,i}\ln\sigma(a_i(x_\kappa)) + \sum_{\kappa=1}^{\ell}(1 - \sum_{j=1}^{c-1}y_{\kappa,i})\ln(1 - \sum_{j=1}^{c-1}\sigma(a_j(x_\kappa))).$$

From Eq. (5), we get

$$l = \sum_{\kappa=1}^{\ell}\sum_{i=1}^{c-1}y_{\kappa,i}\ln\frac{\exp(-a_i(x_\kappa))}{1 + \sum_{j=1}^{c-1}\exp(-a_j(x_\kappa))}$$

$$+ \sum_{\kappa=1}^{\ell}\left(1 - \sum_{j=1}^{c-1}y_{\kappa,i}\right)\ln\frac{1}{1 + \sum_{j=1}^{c-1}\exp(-a_j(x_\kappa))}$$

$$= -\sum_{\kappa=1}^{\ell}\sum_{i=1}^{c-1}y_{\kappa,i}\,\hat{w}_i'x_\kappa - \sum_{\kappa=1}^{\ell}\ln(1 + \sum_{j=1}^{c-1}\exp(-\hat{w}_j'x_\kappa))).$$

Now $\forall h = 1,\ldots,c-1$ we have

$$\nabla_{\hat{w}_h}l = -\sum_{\kappa=1}^{\ell}y_{\kappa,h}\hat{x}_\kappa + \sum_{\kappa=1}^{\ell}\frac{\exp(-\hat{w}_h\hat{x}_\kappa)}{1 + \sum_{j=1}^{c-1}\exp(-\hat{w}_j'x_\kappa)}\hat{x}_\kappa$$

$$= -\sum_{\kappa=1}^{\ell}\left(y_{\kappa,h} - \frac{\exp(-\hat{w}_h\hat{x}_\kappa)}{1 + \sum_{j=1}^{c-1}\exp(-\hat{w}_j'\hat{x}_\kappa)}\right)\hat{x}_\kappa = -\sum_{\kappa=1}^{\ell}(y_{\kappa,h} - \sigma_h(x_\kappa))x_\kappa.$$

This equation for $\nabla_{\hat{w}_h}$ can be used for gradient-based learning algorithms since, unlike the case of linear units, here we can only provide a numerical solution. It's very instructive to compare this equation with Eq. 3.2.2–(2), which expresses the delta error in the case of quadratic loss. While both equations share the same basic structure, the presence of σ' in Eq. 3.2.2–(2) indicates a significant difference that arises during learning when the neurons end up in a saturated configuration. Unlike the case of quadratic loss, the logistic regression based on the maximization of the likelihood doesn't suffer from the same problem!

3.3.3 The parsimony principle meets the Bayesian decision

The Bayesian decision boundary that has been shown in the previous section is clearly robust with respect to points spread around the mean values. Interestingly, the regularization discussed for ridge regression seems to share a similar property. In order to give an insight on the effect of regularization, let us consider the case in which we are given a data distribution strongly centered around the two centroids. The case of Gaussian distribution treated in the previous case could be a good example in case $\det \Sigma \to 0$.

To make things easy, let us consider the idealized situation in which there are only two classes composed of a single labeled example, each one denoted by $(x_1, +1)$ and $(x_2, -1)$, respectively, so that $y = (-1, +1)'$. The normal equations turn out be

$$(\lambda I_d + \hat{x}_1 \hat{x}_1' + \hat{x}_2 \hat{x}_2')\hat{w} = (\hat{x}_1, \hat{x}_2)y.$$

Let us start considering the case in which there is no regularization, that is, $\lambda = 0$. If $d = 1$, the problem admits one solution (see Exercise 1–(a)). Now if $d > 1$, the problem admits infinitely many solutions. An interesting solution can be found as a linear combination of x_1 and x_2. When imposing $w = \alpha x_1 + \beta x_2$, we can easily check that (see Exercise 1–(b))

$$w = (2 + x_2^2 + x_1' x_2)x_1 - (2 + x_1^2 + x_1' x_2)x_2, \qquad b = x_2^2 - x_1^2. \tag{1}$$

Notice that, as one could expect, if $x_1^2 = x_2^2$ then $w \propto x_1 - x_2$ and $b = 0$. We can prove (see Exercise 1–(b)) that in this case the weight $\hat{w}$ is also an eigenvector of $\hat{X}'\hat{X}$, which is clearly related to Bayesian decision given by Eq. 3.3.1–(3).

Now let us go back to the original problem in which we have a distribution of points clustered around μ_i and μ_j. The singularity of the previous case gets the form of an ill-conditioned learning task. This critical condition can conveniently be understood by analyzing the spectral structure of matrix $\hat{X}'\hat{X}$. Because of the strongly clustered structure, the eigenvectors collapse to the nearly same value, which gives rise to ill-conditioning!

Once we use regularization, the solution is $\hat{w}^* = (\lambda I_d + \hat{X}'\hat{X})^{-1}\hat{X}'y$ where the inversion is now guaranteed by $\lambda \neq 0$. Now x_κ^i and x_κ^j denote the examples of two classes i and j. If the two classes have the same number of examples, that is, $\ell^i = \ell^j$, then the solution becomes

$$\hat{w}^* = (\lambda I_d + \hat{X}'\hat{X})^{-1}(x_1^i \dots x_{\ell_i}^i \mid x_1^j \dots x_{\ell_j}^j)' \cdot (+1 \dots +1 \mid -1 \dots -1)'$$

$$= (\lambda I + \hat{X}'\hat{X})^{-1}\left(\sum_{\kappa=1}^{\ell_i} x_\kappa^i - \sum_{\kappa=1}^{\ell_j} x_\kappa^j\right)$$

$$= (\lambda I_d + \hat{X}'\hat{X})^{-1}(\ell_i \overline{x^i} - \ell_j \overline{x^j})$$

$$\propto (\lambda I_d + \hat{X}'\hat{X})^{-1}(\overline{x^i} - \overline{x^j}).$$

This is related with Bayesian decision under Gaussian assumption in case $\Pr(I = i) = \Pr(I = j)$ and $\Sigma_i = \Sigma_j$. As the regularization parameter λ increases, the ill-conditioning disappears. As the regularization term starts to dominate, $[\lambda I_d + \hat{X}'\hat{X}]^{-1} \to \lambda I_d$ and, therefore, the separating plane becomes $(\overline{x^i} - \overline{x^j})'x = 0$.

3.3.4 LMS in the statistical framework

The formulation of LMS given is Section 3.1.1 consists of minimizing the error function which comes from joining the linear machine with a finite set of training data. Ideally, one would like to achieve a solution that, instead of using a sample of data, is based on the true data distribution — thus we regard the inputs and the targets as random variables with an associated distribution. Now, because of the assumption of using linear machines, the quadratic error can be rewritten[5] as

$$\mathrm{E}(Y - f(X))^2 = \int_{X \times Y} (y - \hat{w}'\hat{x})^2 p(x, y)\, dxdy.$$

We can follow an analysis which very much resembles that for normal equations. Let's consider the functional risk

$$E(\hat{w}) := \mathrm{E}(Y - f(X))^2 = \int_{X \times Y} \left(y^2 - 2y\hat{x}'\hat{w} + \hat{w}'\hat{x}\hat{x}'\hat{w} \right) p(x, y)\, dxdy$$
$$= \mathrm{E}(Y^2) - 2\,\mathrm{E}(XY)\hat{w} + \hat{w}'\,\mathrm{E}(XX')\hat{w},$$

where $\mathrm{E}(Y^2) := \int_{X \times Y} y^2 \cdot p(x, y)\, dxdy$, $\mathrm{E}(XY) := \int_{X \times Y} y\hat{x}' \cdot p(x, y)\, dxdy$, and $\mathrm{E}(XX') := \int_{X \times Y} \hat{x}\hat{x}' \cdot p(x, y)\, dxdy$. The minimum is reached when $\nabla E(\hat{w}) = 0$, that is, when

$$\mathrm{E}(XX')\hat{w} = \mathrm{E}(XY). \tag{1}$$

Hence, in the generalized Moore–Penrose sense, the solution can be written as $\hat{w}^* = \mathrm{E}(XX')^{-1}\mathrm{E}(XY)$. Notice that the above solution turns out to be the generalization of the one-dimensional case equations expressed by Eqs. 3.1–(4) and 3.1–(5). Let $\mu_X = \mathrm{E}\,X$ and $\mu_Y = \mathrm{E}\,Y$. Matrices $R_{XX} := \mathrm{E}(XX')$ and $R_{XY} := \mathrm{E}(XY)$ are related to the *covariance matrices* as follows:

$$\Sigma_{XX} := \mathrm{var}(X) = \mathrm{E}\left((X - \mu_X)(X - \mu_X)'\right) = \mathrm{E}(XX') - \mu_X \mu_X'$$
$$= R_{XX} - \mu_X \mu_X',$$
$$\Sigma_{XY} := \mathrm{covar}(X, Y) = \mathrm{E}\left((X - \mu_X)(Y - \mu_Y)'\right) = \mathrm{E}(XY) - \mu_X \mu_Y$$
$$= R_{XY} - \mu_X \mu_Y.$$

---

[5] We are considering the case of regression with one single output, but the extension to the case of multiple output is straightforward.

From an operational point of view, the natural question arises on how R_{XX} and R_{XY} can be computed. Notice that if we are given a training set composed of ℓ supervised pairs then R_{XX} can be expressed as

$$R_{XX} = \mathrm{E}(XX') = \int_X \hat{x}\hat{x}' p(x)\,dx = \int_X \sum_{\kappa=1}^{\ell} \delta(\hat{x} - \hat{x}_\kappa)\hat{x}\hat{x}'\,dx = \sum_{\kappa=1}^{\ell} \hat{x}_\kappa \hat{x}_\kappa' = \hat{X}'\hat{X},$$

$$R_{XY} = \mathrm{E}(XY) = \int_{X \times Y} \hat{x}\, y p(x, y)\,dxdy = \int_{X \times Y} \sum_{\kappa=1}^{\ell} \delta(\hat{x} - \hat{x}_\kappa, y - y_\kappa)\hat{x}\, y\,dxdy$$

$$= \sum_{\kappa=1}^{\ell} \hat{x}_\kappa y_\kappa = \hat{X}'y.$$

Hence, we are back to the matrix-based solution 3.1.1–(3) of normal equations for any finite collection of examples.

EXERCISES

▶ **1.** [*18*] Given a Gaussian probability density

$$p(x) = \frac{1}{\sqrt{2\pi}\sigma} e^{-\frac{(x-\mu)^2}{2\sigma^2}}$$

and a collection of measurements $\{(x_\kappa, p(x_\kappa)) : \kappa = 1, \ldots, \ell\}$, determine the LMS approximation for μ and σ.

2. [*15*] Suppose we are given a training set with two points only and consider the simple case $d = 1$. Consider a classification problem where $y = (+1, -1)$. Find an expression for the solution in terms of a linear machine by using normal equations and prove that there exists only one solution.

3. [*16*] Let us consider the problem stated in Exercise 2 for the general case of $X \subset \mathbb{R}^d$. Determine an expression of the solution under the assumption that $w = \alpha x_i + \beta x_j$.

▶ **4.** [*16*] Prove that the solution of the normal equations for the learning task composed of two points $\{(x_i, +1), (x_j, -1)\}$ is an eigenvector of $\hat{X}'\hat{X}$.

3.4 **Algorithmic issues**

So far, we have focused on foundations and general principles of learning. In this section we discuss how the developed methods naturally lead to learning algorithms. The focus on algorithms leads to the emergence of additional fundamental principles that are connected to the weight updating scheme, to the representational power, and to computational complexity. It has been shown that learning can be essentially regarded as an optimization process, but while following this path, we discover a somewhat different approach which is surprisingly related. We can conceive computational schemes to enable a human-inspired method that relies on the carrot and stick principle. Apparently, this is not closely related to optimization but, as it will be shown, there are in fact intriguing links.

3.4.1 **Gradient descent**

In this section we address algorithmic issues connected with learning the weights of LTU. The case of a pure linear machines is solved by normal equations that are based on specific numerical methods. Because of the nonlinearity induced by the threshold function, one must look for general numerical optimization methods. In most interesting applications, we need to face high-dimensional spaces, and therefore gradient methods are often preferred to second-order methods. Given the error function E, we need to discover its global minimum, which is in general hard! However, the case discussed in Section 3.2.2 shows that the marriage of LTU with appropriate error functions can lead to an error function where any local minimum is global. Algorithm G shows the general structure of any gradient descent learning scheme, where the optimization involves the overall error E, which comes from a given a finite learning set $\mathcal{L}$.

Algorithm G (*Gradient Descent*).   Given a training set $\mathcal{L}$, an error function E and a *stopping criterion C*, find the learned weight vector $\hat{w}_\kappa$ at κ and the index κ.

G1. [Initialize.] Compute the initial weights $\hat{w}_0^i$ and set $\kappa \leftarrow 0$, $\hat{w}_0 \leftarrow \hat{w}_0^i$, $g_0 \leftarrow 0$, $i \leftarrow 1$.

G2. [Compute the gradient.] Set $g_\kappa \leftarrow g_\kappa + \nabla e_i(\hat{w}_\kappa)$ and increase i by one.

G3. [More examples?] If $i \neq \ell$ go to G2, otherwise $i \leftarrow 1$ and continue to G4.

G4. [Update the weights.] Set $\hat{w}_{\kappa+1} \leftarrow \hat{w}_\kappa - \eta g_\kappa$.

G5. [Is C met?] If C is met, terminate the algorithm; the answer is $(\hat{w}_\kappa, \kappa)$. Otherwise, increase κ by one, $g_k \leftarrow 0$, and go back to step G2. ∎

Notice that, at any step κ of the weight updating, the gradient is accumulated by considering all the ℓ examples of the training set (step G2). Then the weights are updated according to the gradient descent heuristics (step G4), while the halt is checked by an appropriate *stopping criterion*. The learning rate η clearly controls the degree of weight updating and, depending on the problem, its choice might be indeed critical. The stopping criterion significantly affects the learning task. Clearly, the agent is expected to operate pretty well on the training set, but as we have seen in Chapter 2, one must avoid overtraining. The accumulation of the gradients of all the examples of $\mathcal{L}$ corresponds with the correct computation of $\nabla E(\hat{w})$ in any given configuration $\hat{w}$. Step G2 is in fact the direct consequence of applying the gradient operator ∇ to $E = \sum_i e_i(\hat{w})$. This updating based on the batch of the gradient is referred to as *batch mode*.

While regression can directly be approached by normal equations, when dealing with LTU, the gradient descent does require some additional analyses of the numerical behavior. In Section 3.2.2 we have analyzed the structure of the error function, which is local minima free under some conditions mostly connected with the linear-separability of the training set. Now we address the issue of convergence of Algorithm G, which clearly depends on the choice of the learning rate η. It is quite obvious that as η increases, we miss the gradient descent heuristics! Hence, while the algorithm is moving downhill from a single point, the steps of single weight updates cannot be too large so as not to loose the gradient heuristics. Clearly, we must expect that upper bounds on η depend on the pair $(\hat{X}, y)$, since the error function inherits the structure of the training set. Let us consider the case of linear prediction and quadratic error function. In this case the updating step G4 becomes

$$\hat{w}_{\kappa+1} \leftarrow \hat{w}_\kappa - \eta \hat{X}'(\hat{X}\hat{w}_\kappa - y) = (I - \eta \hat{X}'\hat{X})\hat{w}_\kappa + \eta y.$$

This offers a nice interpretation of the weight updating scheme, which evolves according to a linear system; it is asymptotically stable provided that the absolute value of the maximum eigenvalue of $I - \eta\hat{X}'\hat{X}$ is strictly less than 1. Now let us focus on the case $\ell \geq d + 1$, in which it makes sense to assume that $\hat{X}'\hat{X}$ is of full rank. If we use a feature transformation denoted by matrix P, we get $I - \eta\hat{X}'\hat{X} = PIP' - (P\Sigma P') = P(I - \eta\Sigma)P'$. Hence we can check the eigenvalues of $I - \eta\Sigma$ and impose their belonging to the unitary circle, that is, for every $i = 1, \ldots \ell$ we impose $|1 - \eta\sigma_i| < 1$, where $\Sigma = \text{diag}(\sigma_i)$. Now the upper bound on the learning rate turns out to be

$$\eta < \frac{2}{\sigma_{\max}}, \tag{1}$$

where $\sigma_{\max} = \max_i \sigma_i$. This has quite an intuitive meaning and suggests that convergence can always be achieved by either choosing a small learning rate, or by proper data normalization. However, as will be seen in Section 3.4.4, while this bound guarantees convergence, the computational complexity significantly depends on the overall spectral structure of $\hat{X}$, and particularly also on its minimum eigenvalue. Since the learning process is based on a batch of data, we might wonder what role in the convergence is played by the dimension ℓ of the training set. The condition expressed by Eq. (1) allows us to re-frame the problem into the analysis of the spectral structure of $\hat{X}'\hat{X}$. While bound (1) is related to ℓ, we cannot make any strong statements on such a relation. However, when data are remarkably clustered, we can approximate the process of learning within a reasonable degree of accuracy by an appropriate repetition of the clusters, which yields a bound on the dependency on ℓ of the maximum eigenvalue $\sigma_{\max}$. Notice that this analysis on the learning rate bound refers to the case $\ell \geq d + 1$. An extension to the case $\ell < d + 1$ is discussed in Exercise 7.

3.4.2 Stochastic gradient descent

When data come online, one might consider updating the weights by using directly the gradient associated with the incoming examples. In the above case of a finite learning set, we can change Algorithm G so as to synchronize the learning process with the occurrence of the examples. Intuitively, it seems reasonable to exploit information as it comes. Notice that, in general, the online learning scheme of Algorithm G′ might depart significantly from gradient descent.

Algorithm G′ — known in the literature as *Stochastic Gradient Descent* — is obtained from Algorithm G as follows:

Algorithm G′ (*Stochastic Gradient Descent*). Given a training set $\mathcal{L}$, an error function E and a *stopping criterion* C, find the learned weight vector $\hat{w}_\kappa$ at κ and the index κ.

G1′. [Initialize.] Compute the initial weights $\hat{w}_0^i$ and set $\kappa \leftarrow 0$, $\hat{w}_0 \leftarrow \hat{w}_0^i$, $i \leftarrow 1$.

G2′. [Update the weights.] Set $\hat{w}_{\kappa+1} \leftarrow \hat{w}_\kappa - \eta \nabla e_i(\hat{w}_k)$, $\kappa \leftarrow \kappa + 1$, and increase i by one.

G3′. [More examples?] If $i \neq \ell$ go to G2′, otherwise $i \leftarrow 1$, and continue to G4′.

G4′. [Is C met?] If C is met, terminate the algorithm; the answer is $(\hat{w}_\kappa, \kappa)$. Otherwise, increase κ by one and go back to step G2′. ∎

The gradient ∇e_i, which drives the online algorithm is also referred to as *stochastic gradient*. As $\eta \to 0$ and the dimension of the training set is kept small, stochastic gradient approximates with arbitrarily high degree the true batch-mode gradient heuristics. This can promptly be understood when considering the weight updating step G2′. If both η and ℓ are small, any loop G2′–G3′ updates the weights of small value $\Delta \hat{w}_\kappa$ and $\nabla E(\hat{w}_\kappa + \Delta \hat{w}) \simeq \nabla E(\hat{w}_\kappa)$. Hence the weight updating arising from G2′–G3′ closely approximates the weight updating of line step G4 in batch mode learning (Algorithm G).

Even though Algorithm G′ operates online, it can somehow be regarded as an approximation of the batch mode scheme, since it operates on a finite training set. A true online scheme is expected to process any sequence and returns the weights at any presentation of new examples, so that even the classic statistical distinction between learning and test set ceases to hold. Agent Γ describes how the previously formalized policy can be directly changed, so as to handle the presentation of an endless sequence of examples that do not necessarily exhibit a periodic structure.

Agent Γ (*SGD, endless sequences*). Given an *infinite* training set $\mathcal{L}$ and an error function E, update the learned weight vector $\hat{w}_\kappa$ at κ.

Γ1. [Initialize.] Compute the initial weights $\hat{w}_0^i$, set $\kappa \leftarrow 0$ and $\hat{w}_0 \leftarrow \hat{w}_0^i$.

Γ2. [Update the weights.] Return $\hat{w}_\kappa$ and set $\hat{w}_{\kappa+1} \leftarrow \hat{w}_\kappa - \eta \nabla e_\kappa(\hat{w}_k)$.

Γ3. [Go on.] Increase κ by one and go back to Γ2. ∎

While Algorithm G′ and Agent Γ share the same structure, learning tasks with no periodic assumption on the incoming sequence are significantly harder. Suppose that the supervisor presents random data to the agent. This is somewhat the opposite case of a periodic sequence, where the predictive behavior is based on processing the same bunch of data many times. In this case, we expect the agent to capture regularities that appear under more complex temporal structures.

Algorithm Γ is especially relevant in the case in which we consider linear units and quadratic error. In this case, the updating of the weights (step Γ2) turns out to be

$$\hat{w}_{\kappa+1} \leftarrow \hat{w}_\kappa + \eta(y_\kappa - \hat{w}'_\kappa \hat{x}_\kappa)\hat{x}_\kappa. \tag{1}$$

In order to understand the behavior of the algorithm and the conditions of its convergence, we consider the stochastic processes $\{X_\kappa\}$, $\{W_\kappa\}$, and $\{Y_\kappa\}$, and assume that $\{X_\kappa\}$ and $\{W_\kappa\}$ are independent. If we apply E on both sides, we have

$$\begin{aligned}
\mathrm{E}\, W_{\kappa+1} &= \mathrm{E}\, W_\kappa + \eta\, \mathrm{E}\left((Y_\kappa - W'_\kappa X_\kappa)X_\kappa\right) \\
&= \mathrm{E}\, W_\kappa + \eta\, \mathrm{E}(X_\kappa Y_\kappa) - \eta\, \mathrm{E}(X_\kappa X'_\kappa) \cdot \mathrm{E}\, W_\kappa \\
&= \mathrm{E}\, W_\kappa - \eta\left(\mathrm{E}(X_\kappa X'_\kappa) \cdot \mathrm{E}\, W_\kappa - \mathrm{E}(X_\kappa Y_\kappa)\right).
\end{aligned}$$

Eq. (1) offers an appealing connection. We have that $\mathrm{E}\, W_{\kappa+1} - \mathrm{E}\, W_\kappa = -\eta\nabla E(\hat{w})$. Then the associations $\hat{w}_\kappa = \mathrm{E}\, W_\kappa$, $\hat{x}_\kappa = \mathrm{E}\, X_\kappa$, and $y_\kappa = \mathrm{E}\, Y_\kappa$ yield

$$\hat{w}_{\kappa+1} \leftarrow \hat{w}_\kappa - \eta(\nabla E \hat{w} - y) = \hat{w}_\kappa - \eta(R_{XX}\hat{w}_\kappa - R_{XY}). \tag{2}$$

This statistical interpretation makes it possible to conclude that the bound in Eq. (1) holds also for the online Algorithms G′ and Γ when $\sigma_{\max} = 2R_{XX}^{-1}$.

3.4.3 The perceptron algorithm

So far learning as been regarded as an optimization problem. Now we explore a different corner of learning, which is perhaps more intuitive, since it is somehow related to the *carrot and stick principle*. One can regard learning as a process driven by the combination of rewards and punishment to induce the correct behavior. The mule moves towards the carrot because it wants to get food, and it does its best to escape the stick to avoid punishment. The same principle can be used for conquering the abstraction needed for learning a concept from a collection of labeled examples $\mathcal{L}$. We consider an online framework in which an LTU is updating its weights at any presentation of examples and continues by cycling on $\mathcal{L}$ until a certain stopping criterion is met.

We assume that the points belong to a sphere of radius R and they are *robustly separated* by a hyperplane, that is, $\forall(x_\kappa, y_\kappa) \in \mathcal{L}$

$$(i) \quad \exists R \in \mathbb{R} : \|x_\kappa\| \leq R, \qquad (ii) \quad \exists a \in \mathbb{R}^{d+1}, \exists \delta \in \mathbb{R}^+ : y_\kappa a \cdot \hat{x}_\kappa > \delta. \tag{1}$$

Notice that the robustness of the separation is guaranteed by the margin value δ.

Now let us consider Algorithm P, which runs until there are no mistakes on the classification of the training set.

Algorithm P (*Perceptron algorithm*).   Given a training set $\mathcal{L}$ with targets y_i taking values ± 1, find $\hat{w}$ and t such that the hyperplane perpendicular to $\hat{w}$ correctly separates the examples and t is the number of times that $\hat{w}$ is updated.

P1. [Initialize.] Set $\hat{w}_0 \leftarrow 0$, $t \leftarrow 0$, $j \leftarrow 1$, and $m \leftarrow 0$.

P2. [Normalize.] Compute R and for all $i = 1, \ldots, \ell$ set $\hat{x}_i \leftarrow (x_i, R)'$.

P3. [Carrot or stick?] If $y_j \hat{w}' \hat{x}_j \leq 0$, set $\hat{w} \leftarrow \hat{w} + \eta y_j \hat{x}_j$, $t \leftarrow t + 1$, $m \leftarrow m + 1$.

P4. [All tested?] Set $j \leftarrow j + 1$; if $j \neq \ell$ go back to step P3.

P5. [No mistakes?] If $m = 0$, the algorithm terminates; set $\hat{w} \leftarrow (w, b/R)$ and return $(\hat{w}, t)$.

P6. [Try again.] Set $m \leftarrow 0$, $j \leftarrow 1$, and go back to step P3.   ∎

Each of the ℓ examples is processed so as to apply the carrot and stick principle. Step P3 tests the condition under which the machine fails to separate. In case the classification is correct, there is no change of the weights, which already leads to successful separation. In the opposite case the weights are updated as described in step P3. The weight and the input vectors are properly rearranged as

$$\hat{w}_\kappa \equiv (w'_\kappa, b_\kappa/R)', \qquad \hat{x}_\kappa \equiv (x'_\kappa, R)',$$

where $R = \max_i \|x_i\|$, which corresponds with the definition given in Section 3.1.1 in case $R = 1$. This allows us to express $f(x) = w \cdot x + b = \hat{w} \cdot \hat{x}$.

Now we prove that if (1) holds then the algorithm stops in finitely many steps. The proof is based on following the evolution of the angle ϕ_κ between a and $\hat{w}_\kappa$ by means of its cosine

$$\varphi_\kappa = \arccos \frac{a \cdot \hat{w}_\kappa}{\|a\| \|\hat{w}_\kappa\|}. \tag{2}$$

We start by considering the evolution of $a \cdot \hat{w}_\kappa$. Clearly, it does not change until the machine makes a mistake on a certain example x_i. In that case the updating takes place according to step P3, so that $a \cdot \hat{w}_{k+1} = a \cdot (\hat{w}_\kappa + \eta y_i \hat{x}_i) = a \cdot \hat{w}_\kappa + \eta y_i a \cdot \hat{x}_i > \eta \delta$, where the last inequality follows from the hypothesis of linear-separability (1)–(ii). Hence, after t wrong classifications, since $w_0 = 0$ (step P1), we can promptly see by induction that

$$a \cdot \hat{w}_t > t \eta \delta. \tag{3}$$

Now for the denominator, we need to find a bound for $\|w_\kappa\|$, by using again the hypothesis of strong linear-separation. Again, in case there is a mistake on example x_i, we get

$$\|\hat{w}_{\kappa+1}\|^2 = \|\hat{w}_\kappa + \eta y_i \hat{x}_i\|^2 = \|\hat{w}_\kappa\|^2 + 2\eta y_i \hat{w}_\kappa \cdot \hat{x}_i + \eta^2 \|\hat{x}_i\|^2 \leq \|\hat{w}_\kappa\|^2 + 2\eta^2 R^2$$

After t mistakes we have

$$\|\hat{w}_t\|^2 \leq 2\eta^2 R^2 t. \tag{4}$$

Now we can always assume $\|a\| = 1$, since any two vectors a and $\check{a}$ such that $a = \alpha\check{a}$ with $\alpha \in \mathbb{R}$ represent the same hyperplane. Hence, when using the bounds (3) and (4), we have

$$\frac{t\eta\delta}{\sqrt{2\eta^2 t R^2}} = \frac{\delta}{R}\sqrt{\frac{t}{2}} \leq \cos\varphi_t \leq 1.$$

The last inequality makes it possible to conclude that the algorithm stops after t steps, which is bounded by

$$t \leq 2\left(\frac{R}{\delta}\right)^2. \tag{5}$$

This bound tells us a lot about the algorithm behavior. First, any scaling of the training set does not affect the bound. We can in fact say more about the scaling: Not only the bound is independent of scaling, but the actual number of steps needed to converge, as well as the whole algorithm behavior, does not change when replacing x_i with αx_i. In that case, the sphere which contains all the examples has radius αR, so that the previous scaling map yields $\hat{x}_i \rightarrow \alpha\hat{x}_i$. As a consequence, step P3 of the algorithm is unaffected, which means that the whole algorithm does not change. This property of scale invariance leads immediately to conclude that also the *learning rate* η does not affect the algorithm behavior, since it plays exactly the same role as the scaling parameter α. Clearly, this is also the conclusion we get from the expression of the bound, which is independent of η.

What if $w_0 \neq 0$? The previous analysis relies on the hypothesis $w_0 = 0$ in order to state the bounds (3) and (4), but we can easily prove that the algorithm still converges in case $w_0 \neq 0$. Exercise 9 proposes the formulation of a new bound which also involves w_0.

Now, when looking at Algorithm P and at the discovered bound on its convergence, one question naturally arises, especially if we start thinking of a truly online learning environment: What if the agent is exposed to an endless sequence whose only property is that its atoms (examples) are linearly separable? Notice that we need to rethink the given algorithmic solution, since we cannot cycle over the infinite training set! We can simply use the same carrot and stick principle so as to handle an infinite loop as shown in Agent Π.

Agent Π (*Online Perceptron*). Given an infinite training set $\mathcal{L}$ return $\hat{w}$ every time that it is updated and the total number of updates t that have occurred.

Π1. [Initialize.] Set $\hat{w}_0 \leftarrow 0$, $t \leftarrow 0$, and $j \leftarrow 1$.

Π2. [Normalize.] Compute R and for all $i = 1, \ldots, \ell$ set $\hat{x}_i \leftarrow (x_i, R)'$.

Π3. [Carrot or stick?] If $y_j\hat{w}'\hat{x}_j \leq 0$, set $\hat{w} \leftarrow \hat{w} + \eta y_j\hat{x}_j$, $t \leftarrow t + 1$; set $\hat{w} \leftarrow (w, b/R)$ and return $(\hat{w}, t)$.

Π4. [Go on.] Set $j \leftarrow j + 1$ and go back to step Π3. ∎

The algorithm is essentially the same, the only difference being that the principle is used for any of the incoming examples, which are not cyclic anymore. Unlike Algorithm P, in this case the weights are tuned whenever they are updated, since there is no stop. This time $\mathcal{L}$ is not finite, and therefore the above convergence proof does not hold.

However, we now show that the finiteness of $\mathcal{L}$ is not necessary, while it suffices to continue requiring conditions (1) to hold. In particular, (i) is needed in step $\Pi 2$, while (ii) gives crucial information for extending the proof. The strong linear separation means that there exist a finite set of examples $\mathcal{L}_s \subset \mathcal{L}$ such that $\forall(\hat{x}_j, y_j) \in \mathcal{L}_s$ and $\forall(\hat{x}_i, y_i) \in \mathcal{L} \setminus \mathcal{L}_s$,

$$y_j \hat{w}_j \hat{x}_j \leq y_i \hat{w}_i \hat{x}_i. \tag{6}$$

These examples completely define the separation problem, so that any solution on $\mathcal{L}_s$ is also a solution on $\mathcal{L}$. For this reason they are referred to as *support vectors*, since they play a crucial role in supporting the decision. Clearly, this holds also for a finite training set $\mathcal{L}$, but in this case the situation is more involved since we do not know in advance when the support vectors come. Hence, with no assumption on their occurrence, it is clear that the learning environment is not offering interesting regularities.

As it is discussed in Exercises 4 and 5, when one has an infinite training set, linear separability does not imply strong linear separability.

The discussion carried out so far has been restricted to considering linearly separable examples. On the other hand, when departing from this assumption, the perceptron cannot separate positive and negative examples, so that we are in front of representational issues more than of learning. However, one could be curious to know what happens when the previous learning algorithms on the perceptron are applied to nonlinearly separable training sets. Of course, the algorithm cannot end up with a separating hyperplane and the weights do not converge. Let us focus on Algorithm P, but the same conclusions can be drawn also in case of truly online examples. Suppose, by contradiction, that a certain optimal value $\hat{w}^*$ exists such that no change occurs after having presented all the ℓ examples. Clearly, this happens only for linearly-separable examples (see step P3), which contradicts the assumption. Interestingly, the behavior is *cyclic*, that is, after a certain transient the weights assume cyclic values, so that $\hat{w}_t$ is a periodic function with period ℓ.

While the perceptron algorithm exhibits a nice behavior for linearly separable examples, as we depart from this assumption, the cyclic decision is not very satisfactory. One would like a solution which separates as much as possible in any case! This can be achieved by a surprisingly simple change of the perceptron algorithm. Suppose we run the algorithm while keeping the best solution seen so far in a buffer (the *pocket*). Then the weights are actually modified only if a better weight vector is found, which gives rise to the name *pocket algorithm*.

3.4.4 Complexity issues

How expensive is learning with linear machines? As usual one can face this problem from two different corners. First, one can think of the complexity of the single algorithms; second the natural question arises on how complex is the learning problem itself. Any analysis aimed at understanding these issues cannot neglect that while Algorithm P and Agent Π carry out an inherently discrete updating scheme, all gradient-based algorithms only represent a discretization of an inherently continuous trajectory in the weight space.

In the first case, complexity analysis is quite simple, since we already know upper bounds on the number of mistakes in case of linearly separable patterns. Let us focus attention on Algorithm P and assume that R and δ are values independent of d and ℓ. We can promptly see that the algorithm is $O(d \cdot \ell)$, which is also a lower bound. Hence, the perceptron algorithm reaches the optimal bound $\Theta(d \cdot \ell)$. This holds also for Agent Π, which is protected by the upper bound on the number of mistakes. Of course, in both cases the optimality comes from the assumption that no example comes outside the bounded sphere of radius R or inside the hyperplane δ margin.

The analysis of gradient-based algorithms is more involved. Unlike the case of classic algorithms, technical instruments to capture the complexity of numerical algorithms are still on the frontier of research. However, there are some intriguing connections that nicely emerge when studying linear machines. Let us restrict the analysis to the case of the optimization of functions where any local minimum is also global (*local minima free*). When a proper formulation is given, this is in fact what linear machines have in common for both regression and classification tasks. Hence it is important to understand the cost of running Algorithm G. A straightforward abstraction consists of dealing with the related continuous system

$$\frac{d\hat{w}}{dt} = -\eta \nabla E(\hat{w}). \tag{1}$$

If E is local minima free, this suggests that the algorithm belongs to a certain complexity class. However, once we choose $\eta \in \mathbb{R}$, it is clear that the velocity of the above dynamics is strongly dependent on the steepness of E: While the dynamics is slow in plateaux, it is quick in ravines! Can we concretely discover a way to unify these dynamics? Suppose we adapt η during the gradient descent so that it increases in regions of small slope and decreases where the slope is high. Hence, let us replace Eq. (1) with

$$\frac{d\hat{w}}{dt} = -\eta_0 \frac{\nabla E}{\|\nabla E\|^2}. \tag{2}$$

Intuitively, this new learning rate $\eta = \eta_0 / \|\nabla E\|^2$ addresses very well the need to adapt the velocity of the dynamics. However, this transformation leads to a somewhat astonishing result! Let us analyze the temporal change of $\mathcal{E}(t) \equiv E(\hat{w}(t))$. We have

$$\frac{d\mathcal{E}}{dt} = (\nabla E) \cdot \frac{d\hat{w}}{dt} = -\eta_0 \frac{(\nabla E) \cdot \nabla E}{\|\nabla E\|^2} = -\eta_0.$$

Now we can always start from the same initial value $E_0 = E(\hat{w}_0) = \mathcal{E}(0) =: \mathcal{E}_0$, by normalizing the function. Then we have $\mathcal{E}(t) = \mathcal{E}_0 - \eta t$, and any global minima $(\mathcal{E} = 0)$ is reached after the finite time

$$\sigma = \mathcal{E}_0/\eta_0, \tag{3}$$

which leads us to conclude that Eq. (2) represents a terminal attractor dynamics. Now this suggests that, no matter what local minima free error function we are given, any gradient descent can be reached in the same finite time σ. Hence we are really in the presence of a class of problems that are characterized by the local minima free error functions, which are also referred to as *unimodal functions*. At a closer look, however, something strange emerges: The value of σ can be chosen arbitrarily small, since any value of η_0 is allowed. The differential equation (3) ends up in a *terminal attractor* in an arbitrarily finite time. The arbitrary definition of σ is in fact hiding a kind of complexity, which is absorbed by this terminal attractor computational model. There is in fact a singular point, which corresponds exactly with the terminal state, since $\nabla E \to 0$ leads to the explosion exactly as $E \to 0$. This explosion witnesses the singularity of the terminal configuration, and the tricky computational result on the arbitrary value of σ.

The terminal attractor computational model expressed by Eq. (3) can be given a nice interpretation by its discretization. Let us consider the weight updating for $t_\kappa = \tau\kappa$ and let

$$g_\kappa = -\nabla E(\hat{w}(t_\kappa)). \tag{4}$$

Then the variation of the error $\mathcal{E}(t_\kappa) = E(\hat{w}_\kappa) = E_\kappa$ can be calculated by using Taylor expansion with Lagrange remainder, that is,

$$E_{\kappa+1} = E_\kappa + g_\kappa \cdot (\hat{w}_{\kappa+1} - \hat{w}_\kappa) + \frac{1}{2}(\hat{w}_{\kappa+1} - \hat{w}_\kappa) \cdot \frac{\partial^2 E}{\partial \hat{w}^2}(\omega_\kappa)(\hat{w}_{\kappa+1} - \hat{w}_\kappa)$$

where ω_κ belongs to the line connecting $\hat{w}_\kappa$ and $\hat{w}_{\kappa+1}$. When considering the terminal attractor dynamics (2), we get

$$E_{\kappa+1} = E_\kappa - \eta\tau + \frac{1}{2}\left(\eta_0 \frac{g_\kappa \tau}{\|g_\kappa\|^2}\right) \cdot \frac{\partial^2 E}{\partial \hat{w}^2}(\omega_\kappa)\left(\eta_0 \frac{g_\kappa \tau}{\|g_\kappa\|^2}\right).$$

Hence, at any κ, the error with respect to the continuous dynamics can be bounded by imposing

$$|E_{\kappa+1} - E_\kappa| \le \frac{1}{2}\frac{\eta_0^2}{\|g_\kappa\|^4}\left|g_\kappa \cdot \frac{\partial^2 E}{\partial \hat{w}^2}(\omega_\kappa)g_\kappa \tau^2\right| \le \frac{1}{2}\frac{\tau^2\eta_0^2}{\|g_\kappa\|^2}\left\|\frac{\partial^2 E}{\partial \hat{w}^2}(\omega_\kappa)\right\|.$$

Now we can bound the maximum error by imposing

$$\frac{1}{2}\frac{\tau^2\eta_0^2}{|g_\kappa|^2}\left\|\frac{\partial^2 E}{\partial \hat{w}^2}(\omega_\kappa)\right\|\frac{\sigma}{\tau} < \epsilon_e. \tag{5}$$

This can be used to determine the quantization step to guarantee the error ϵ_e. Let us assume that the following bounds hold:

$$\|g_\kappa\| \geq \epsilon_g, \qquad \left\|\frac{\partial^2 E}{\partial \hat{w}^2}(\omega_\kappa)\right\| \leq H. \tag{6}$$

Then the bound (5) yields $\tau \eta_0^2 H \sigma/(2\epsilon_g^2) \leq \epsilon_e$, and finally it suffices to choose

$$\tau = 2\sigma \left\lfloor \frac{\epsilon_e \epsilon_g^2}{\mathcal{E}_0^2 H} \right\rfloor. \tag{7}$$

From this bound, we can calculate the number of steps to achieve an error below ϵ_e, that is, $\kappa^* = \lceil \mathcal{E}_0^2 H/(2\epsilon_e \epsilon_g^2) \rceil$. Now if we pose $\rho := \epsilon_e/\mathcal{E}_0$ then

$$\kappa^* = \frac{1}{2} \left\lceil \frac{\mathcal{E}_0 H}{\rho \epsilon_g^2} \right\rceil. \tag{8}$$

This bound offers relevant information in learning machines. The first application concerns the complexity of solving normal equations, including the case in which we introduce the regularization that yields ridge regression. To address those problems, one needs to know the computational complexity of determining the minimum of

$$E(\hat{w}) = \frac{1}{2}\|y - \hat{X}\hat{w}\|^2.$$

Now we have

$$\nabla E = -\hat{X}'y + \hat{X}'\hat{X}\hat{w}, \qquad H = \hat{X}'\hat{X},$$

and let us assume that $\det(\hat{X}'\hat{X}) \neq 0$. In order to impose condition (6), we express ∇E as follows:

$$\|\nabla E\| = \|\hat{X}'\hat{X}\hat{w} - \hat{X}'y\| = \frac{\|(\hat{X}'\hat{X})^{-1}\|\,\|\hat{X}'\hat{X}\hat{w} - \hat{X}'y\|}{\|(\hat{X}'\hat{X})^{-1}\|}$$

$$\geq \frac{\|\hat{w} - (\hat{X}'\hat{X})^{-1}\hat{X}'y\|}{\|(\hat{X}'\hat{X})^{-1}\|} = \frac{\epsilon_w}{\|(\hat{X}'\hat{X})^{-1}\|} = \epsilon_g.$$

Of course, the error on w can be related to the stopping criterion $\|y - \hat{X}\hat{w}\|^2 < \epsilon_e$, since $\epsilon_e = \beta \epsilon_w$ for an appropriate choice of $\beta > 0$. From Eq. (8) we get

$$\kappa^* = \frac{1}{2} \left\lceil \frac{\mathcal{E}_0 H}{\rho \epsilon_g^2} \right\rceil = \frac{1}{2} \left\lceil \frac{\mathcal{E}_0 \, (\|\hat{X}'\hat{X}\| \cdot \|(\hat{X}'\hat{X})^{-1}\|)^2}{\rho \qquad \epsilon_w^2} \right\rceil = \frac{1}{2} \left\lceil \frac{\beta^2}{\rho^2 \epsilon_e} \text{cond}^2(\hat{X}'\hat{X}) \right\rceil. \tag{9}$$

This equation expresses a fundamental principle on the complexity of learning with a linear machine! It shows that in case of quadratic error functions, the computational complexity measured by the discrete terminal attractor machines depends on the condition number of the Hessian of the error.

We notice in passing that while for quadratic error functions, or any other index for which the Hessian is generally non-null, the discovered bound makes sense in case of hinge functions the proposed analysis for discovering the bounds cannot be used straightforwardly. Notice that one cannot guarantee the gradient heuristics in those cases, since if we start from saturated configurations, there is no way to escape. As shown in Section 3.5, learning with linear machines can also be formulated by linear programming. This sheds additional light on complexity issues, since there's a huge literature in the field, along with software packages. One might also raise the question on whether the methods discussed in this chapter are worth mentioning when such a general framework is available. The complexity analysis of linear programming leads to a clear answers this question: There is no linear programming algorithm which exhibits the optimal complexity bound of the perceptron algorithm. The simplex method is generally efficient, but it explodes in the worst cases. The *Karmarkar algorithm* is polynomial, but it is sill far away from the optimality.

EXERCISES

1. [*15*] Consider Algorithm P and discuss why in step P5 we have inserted the assignment $\hat{w}_t \leftarrow (w_t, b_t/R)$. In particular, what happens if we directly return the weights $\hat{w}_t$ as computed by the loop P3–P4?

2. [*M15*] How does the bound in Eq. 3.4.3–(5) change if we assume that the examples x_i live in a general Hilbert space H with the norm $\|v\| = \sqrt{(v, v)}$ induced by a symmetric inner product.

▶ **3.** [*22*] Consider Algorithm P without the "normalization" step P2 and discuss what happens to the bound 3.4.3–(5).

4. [*15*] Consider a finite learning set $\mathcal{L}$ that is linearly separable. Can you conclude that they are also robustly separable, i.e., that Eq. 3.4.3–(1) holds?

▶ **5.** [*20*] Consider the same question of Exercise 4, this time for an infinite training set. What can you say about the convergence of the Agent Π?

6. [*20*] Suppose we set the bias b of $a = w \cdot x + b$ for an LTU using the Heaviside function to any $b \in \mathbb{R} \setminus \{0\}$. Can we always determine the same solution as in the case in which b is a free parameter? Does the perceptron algorithm separate a linearly-separable set under this restriction?

7. [*M25*] The convergence of batch-mode gradient descent established under condition 3.4.1–(1) is based on the assumption[6] $\ell \geq 1 + d$. What if this condition is violated? Propose a convergence analysis which extends the one proposed for condition 3.4.1–(1) by using the singular value decomposition of $(\hat{X}'\hat{X})$.

8. [*17*] Consider the problem of *pattern sorting*. Discuss its solution using linear machines. In particular, discuss the condition $y \in \mathcal{N}(Q^u_\perp)$.

9. [*20*] Consider the perceptron Algorithm P, where we start from $\hat{w}_o \neq 0$. Prove that the algorithm still halts with a separating solution and determine an upper bound to the number of mistakes.

---

[6] More precisely, we also need rank $\hat{X} = d + 1$.

10. [*17*]   Consider the problem of preprocessing in linear and linear-threshold machines. What is the role of α-scaling or of any linear map of the inputs? What happens when using ridge regression? *Hints:* Exercises 5, 7, 8, and 1 already provide a good basis for an in-depth analysis.

11. [*45*]   Suppose you are given a classification problem in which some of the features of the patterns are missing. Then assume that there is no pattern in the training set for which all features are missing. Analyze a solution based on the idea of expressing the inputs as the data generated by a linear machine which fits the available coordinates.

3.5 Scholia

Section 3.1 Linear and linear-threshold machines represent the foundations of many current machine learning models. The simplicity of the model leads to an in-depth understanding of many properties, some of which are also shared with more complex models.

The method of *Least Mean Squares* (LMS) is a way of determining an approximate solution of overdetermined systems, which is part of the background of everyone who needs to process experimental data. LMS is usually credited to Carl Friedrich Gauss, but antecedents to Gauss' publication were numerous, and in some cases controversial. There is evidence of earlier studies on determining the line of best fit for a set of points in the plane. Adrien-Marie Legendre published a paper on the method of least squares in 1805. His treatment, however, was not rooted in probability theory. That is what Gauss did in his 1809 publication, where he assumed that errors obey a normal distribution. A large variety of literature emerged early in the 1950s. For an in-depth discussion on the invention of LMS, see [S.M. Stigler, Ann. Stat. **9** (1993), 465–474].

Preliminary notions of pseudoinversion were put forward by Fredholm, Hilbert, and many others for integral and differential operators. However, the first specific work on matrix pseudoinverse can be traced back to Moore's studies the 1920 [E.H. Moore, *Bull. Am. Math. Soc.* **26** (1920), 394–395]. Later analysis by Roger Penrose in 1955 also led to the notion of matrix pseudoinverse using an axiomatic approach [*Proc. of the Cambridge Philos. Soc.* **51** (1955), 406–413]. At that time he was a graduate student at Cambridge and introduced the notion of generalized inverse operator $\mathcal{H}^{\#}$, which satisfied one or more of the following equations:

$$(i) \quad \mathcal{H}\mathcal{H}^{\#}\mathcal{H} = \mathcal{H} \qquad\qquad (iii) \quad (\mathcal{H}\mathcal{H}^{\#})^{+} = \mathcal{H}\mathcal{H}^{\#}$$

$$(ii) \quad \mathcal{H}^{\#}\mathcal{H}\mathcal{H}^{\#} = \mathcal{H}^{\#} \qquad\qquad (iv) \quad (\mathcal{H}^{\#}\mathcal{H})^{+} = \mathcal{H}^{\#}\mathcal{H}$$

In case $\mathcal{H}$ admits the ordinary inverse, it satisfies all the above equations. An operator which satisfies only (i), is called a 1-inverse of $\mathcal{H}$, one that satisfies (i) and (ii) is called a $(1, 2)$-inverse, one that satisfies (i), (ii), and (iii) is called a $(1, 2, 3)$-inverse. For matrices, a 1-inverse always exists and can be found by Gaussian elimination. The Moore–Penrose pseudoinverse, denoted $\mathcal{H}^{+}$, satisfies all four Penrose equations. Thus the Moore–Penrose pseudoinverse is a $(1, 2, 3, 4)$-inverse, just like for classic inversion. Hence, the Moore–Penrose pseudoinverse is equivalent to Moore's definition. A relevant study on generalized inverse in both theory and applications is [A. Ben-Israel, T.N.E. Greville *Generalized Inverses: Theory and Applications* (Wiley-Interscience, 1974)]. Moore–Penrose's pseudoinverse matrix seems to indicate a good direction to face undetermined learning tasks. The pseudoinverse $\hat{X}^{+}$ leads to the parsimonious solution $\hat{X}^{+}y$ of $\hat{X}\hat{w} = y$ with minimum value of $\|\hat{w}\|$. The solution of Exercise 3.1.2–5, where the adoption of the Lagrangian approach is proposed, enlightens a critical issue related to $\hat{X}^{+}y$, which makes it typically unsatisfactory in the framework of machine learning.

Basically, $\hat{X}^+ y$ is the solution of a hard-constrained problem, which means that $\hat{X}^+ y$ solves $\hat{X}\hat{w} = y$ perfectly, no matter what the data probability distribution is. Hence the beauty of the parsimony solution is blinded by the ill-position of the problem. The perfect fit might reveal a form of ill-conditioning (a related form of ill-conditioning can also affect problems with more examples than unknowns; as a matter of fact this is related to a drop of the matrix rank) that is clearly put forward by the spectral analysis of Section 3.1.2. This interesting issue can be covered from a different side. As pointed out in Section 3.4.4, the ill-conditioning of matrix $\hat{X}$ leads to an increase of the computational complexity of gradient descent learning algorithms. As this complexity appears, it looks like we are missing some relevant issue: Interestingly, in this case, it is the constrained-based formulation of learning that is wrong! The shift from hard to soft-constrained learning is the right direction to get rid of ill-conditioning. Early traces of this idea can be found in the principle of regularization. It was invented independently in different contexts, but it became widely known from its application to integral equations thanks to the studies of Andrey Tikhonov [see, e.g. A.N. Tikhonov, *Dokl. Akad. Nauk SSSR* **39** (1943), 195–198 and A.N. Tikhonov, V.Y. Arsenin, *Solution of Ill-Posed Problems* (Washington: Winston & Sons, 1977)]. Arthur E. Hoerl strongly contributed to popularizing ridge regression in the statistical community [A.E. Hoerl, R.W. Kennard, *Technometrics* **12** (1970), 55–67].

In real-world problems, especially in cases in which the patterns are represented by many coordinates, some of them might be missing. This frequently happens with clinical data: A profile is generally created on patients that is based on the availability of a number of exams, which provide the features. However, some of them are often missing, yet the decision might be immediately required. Learning with missing data is an active research topic, on which we have seen a remarkable number of contributions in the last few years. One straightforward approach is to guess the missing features by simply replacing them with their average. A rich survey on different approaches can be found in [J.W. Graham, *Annu. Rev. Psychol.* **60** (2009), 549–576], where there's a distinction between missing completely at random (MCAR), missing at random (MAR), and missing not at random (MNAR). A classic work is the one in [D.B. Rubin, *Biometrika* **63** (1976), 581–590], where it is argued that the process that causes missing data is very important in practical applications. An interesting way of attacking the problem is proposed in [Z. Ghahramani, M.I. Jordan, *Adv. Neural Inf. Process. Syst.* (1994), 120–127], which relies on the EM algorithm. Missing data can also be nicely handled in the framework of learning under the parsimony principle. A straightforward idea is the one proposed in Section 3.4, Exercise 3.4–11. As it will be shown in Chapter 6, the idea can be extended so as to benefit from rich descriptions of the environment to estimate missing data.

When relying on the parsimony principle, the choice of the regularization parameter is clearly very important. It turns out to be useful to analyze the way the regularization and the error terms are affected by λ. This can be done by injecting the solution 3.1.3–(4) into $\|w_\lambda\|^2$ and $\|y - \hat{X}\hat{w}_\lambda\|^2$. We have

$$\|w_\lambda\|^2 = \left\| \sum_{i=1}^{r} \phi_i \frac{u_i \cdot y}{\sigma_i} I_B v_i \right\|^2, \tag{1}$$

where

$$I_B = \begin{pmatrix} 1 & 0 & \cdots & 0 \\ \vdots & \ddots & \ddots & \vdots \\ 0 & \cdots & 1 & 0 \end{pmatrix}.$$

Now we have

$$\lim_{\lambda \to 0} \|w_\lambda\|^2 = \left\| \sum_{i=1}^{r} \frac{u_i' y}{\sigma_i} I_B v_i \right\|^2, \qquad \lim_{\lambda \to \infty} \|w_\lambda\|^2 = 0.$$

Likewise, when considering the expression of $P_\perp$ given by Eq. 3.1.3–(5), for the residual error we have

$$
\begin{aligned}
\|y - P_\perp y\|^2 &= \|y - \sum_{i=1}^{r} \phi_i [u_i u_i'] y\|^2 = \|y - \sum_{i=1}^{r} u_i \phi_i (u_i \cdot y)\|^2 \\
&= \|U'(y - \sum_{i=1}^{r} u_i \phi_i (u_i \cdot y))\|^2 = \|U' y - \sum_{i-1}^{r} e_i \phi_i (u_i \cdot y))\|^2 \\
&= \| \sum_{i=1}^{r} (u_i \cdot y) e_i - \sum_{i=1}^{r} e_i \phi_i (u_i \cdot y))\|^2 = \| \sum_{i=1}^{r} (u_i \cdot y) e_i (1 - \phi_i)\|^2 \\
&= \sum_{i=1}^{r} (u_i \cdot y)^2 (1 - \phi_i)^2.
\end{aligned}
\tag{2}
$$

Now we have $\lim_{\lambda \to 0} \|y - P_\perp y\|^2 = 0$, $\lim_{\lambda \to \infty} \|y - P_\perp y\|^2 = \sum_{i=1}^{r} (u_i' y)^2$. Eqs. (2) and (1) suggest a heuristic method to estimate the regularization parameter. Let us introduce $r(\lambda) := \ln \|w_\lambda\|^2$ and $e(\lambda) := \ln(\|y - P_\perp y\|^2)$. They are the parametric equation of a curve in the r–e plane defined by the condition $r = L(e)$. This curve looks like the character "L", which can be quickly seen when considering the limit conditions on $[0, \infty)$. In addition, there is a nice property which reminds us of the "L" structure, which is connected to the presence of a point at high curvature that might even look like a corner in certain circumstances. Let us consider the *corner point* $(e(\lambda^*), r(\lambda^*))$ defined according to the following conditions: (i) The tangent point in $e^* := e(\lambda^*)$ has slope -1, that is,

$$\frac{dL(e^*)}{de} = -1,$$

and (ii) the curve $\lambda \mapsto (e(\lambda), r(\lambda))$ is concave in a neighborhood of $e(\lambda^*)$. Let us assume that $e(\lambda)$ and $r(\lambda)$ are differentiable. Then we prove that these conditions correspond with the local minimization of

$$\mathcal{H}(\lambda) = \|y - P_\perp y\|^2 \hat{w}^2.$$

Indeed, we have

$$\mathcal{H}(\lambda) = \exp(e(\lambda) + r(\lambda)) = \exp(e(\lambda)) \exp(r(\lambda)).$$

Any local minima λ^* of $\mathcal{H}$ satisfies

$$\left(\frac{de(\lambda^*)}{d\lambda} + \frac{dr(\lambda^*)}{d\lambda}\right) \exp\left(e(\lambda^*) + r(\lambda^*)\right) = 0,$$

that is,

$$\frac{de(\lambda^*)}{d\lambda} + \frac{dr(\lambda^*)}{d\lambda} = 0,$$

which corresponds to condition (i). If in addition to this stationarity condition we consider the hypothesis that λ^* is a local minimum for $\mathcal{H}(\lambda)$, we get

$$e(\lambda) + r(\lambda) > e(\lambda^*) + r(\lambda^*),$$

which satisfies condition (ii) of concavity of the curve $\lambda \mapsto (e(\lambda), r(\lambda))$ in a neighborhood of λ^*. In the literature, this is referred to as the *L curve*. A good reference on L curve can be found in [V.N. Vapnik, *Statistical Learning Theory* (Wiley, 1998) section 13.2].

As it will be clear when dealing with kernel machines, the full exploitation of the parsimony principle, with the corresponding appropriate selection of the regularization parameter, can be achieved in the dual space, which is in fact a nice consequence of the linearity. A straightforward comparative complexity analysis of treating linear machines in the primal or in the dual space arises from the Woodbury equality. A dual form can be found also for the Rosenblatt perceptron. We can immediately express the dual form when noticing that the updating equation of Algorithm P can be better rewritten as

$$\hat{w}_{\kappa+1} \leftarrow \hat{w}_\kappa + \eta y_{i(\kappa)} \hat{x}_{i(\kappa)}$$

where we emphasize that the examples where the perceptron made mistakes are those indexed by $i(\kappa)$. From Algorithm P we can immediately conclude that the algorithm exits with

$$\hat{w}_{\kappa^*} = \eta \sum_{\kappa=1}^{\kappa^*} y_{i(\kappa)} \hat{x}_{i(\kappa)} \propto \sum_{\kappa=1}^{\kappa^*} y_{i(\kappa)} \hat{x}_{i(\kappa)}.$$

This is in fact the *dual representation* of the perceptron, which is the solution expressed as an expansion in the training set. Interestingly, the coefficients are either 1 or -1 and, not all inputs necessarily take part in the representation of the solution. Since $\kappa^* \leq 2(R/\delta)$, an upper bound on the number of terms in the above expansion is also available.

Section 3.2   The need to construct decision-based processes leads to the introduction of linear machines with threshold units (LTU). Their behavior is truly nonlinear, and they have been massively studied in the framework of computational geometry. In 1964, Thomas M. Cover completed his Ph.D. at Stanford University. He came up with the fundamental result [M. Cover Ph.D. thesis Stanford University (1964)] that LTU capacity is twice as much as the dimension of the input space. Among others, his studies early suggested that pattern classification in a high-dimensional space is

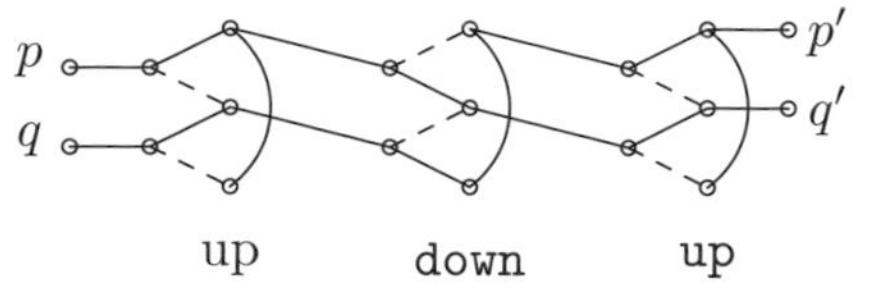

FIGURE 3.4

Link between the parity and connectedness predicates: The connectedness in the switching network is expressed by the predicate $[p$ is connected to $p']$, which is true if an even number of switches are in down position. In the figure we have two switches in down position and, consequently, $[p$ is connected to $p'] = 0$.

more likely to be linearly separable than in low-dimensional space. This topic will also be covered in Section 4.1.

Frank Rosenblatt, however, was the indisputably father of perceptron, which was conceived while working at Cornell Aeronautical Labs. Although early versions of software on the IBM 704 were soon available, the perceptron was intended to be a machine for image recognition, with weights encoded into potentiometers that were updated during learning by electric motors. Frank Rosenblatt's background on psychology conditioned significantly his view and the perspective of research in machine learning. In 1958, he published one of the most comprehensive contributions on perceptrons *Psychol. Rev.* **65** (1958), 368–408 that, however, is quite hard to read. He came up with a number of variants of perceptrons and learning rules, including the classic structure that has been the subject of massive investigation. Later on he made strong statements on the paradigm shift of perceptron, which led to controversial claims on its actual capabilities.

An in-depth critical analysis on representational issues in perceptrons, which created the basis for addressing those controversial claims, was carried out a few years after Rosenblatt's invention by Marvin Minsky and Seymour Papert in their seminal book *Perceptrons: An Introduction to Computational Geometry* (Cambridge: MIT Press, 1969). The analysis carried out in Section 3.2.1 is mostly based on the results of Minsky and Papert's view of perceptrons as abstract computational devices. The introduced concept of predicate order, along with the exemplification in predicates like mask, convexity, and connectedness, sheds light on what can be reasonably computed and on hard learning tasks. Interestingly, most predicates are hard to compute also for humans, unless we decide to use a sequential approach by patiently following the paths, which is in fact what perceptrons are unable and unexpected to do! It's important to realize that most of the properties exemplified in Section 3.2.1 concerning perceptron-like representations do apply to many other problems. There are in fact surprising links between tasks that look apparently very different. Fig. 3.4 offers a nice example of an intriguing link between parity and connectedness predicates. Under the finiteness assumption of $X^\sharp$, these apparently very different predicates are brought together thanks to appropriate constructions of *switching networks*. Intuitively, the connectivity of points in the network, which is established by $[X^\sharp$ is connected], is due to the appropriate selection of the switches, which is identified by the sequence S of up and down states.[7] We can promptly see that if $[S$ is even$] = 1$ (even number of down states) then $[X^\sharp$ is connected$] = 1$. These links the difficulty of computing connectedness with the difficulty of computing the parity of a string! Now we have seen that in case of two variables, the XOR is a predicate of order 2, that is, the predicate order is the same as the input dimension. Does it hold in general? It can be proven [see pages 56–57 of *Perceptrons: An Introduction to Computational Geometry* Expanded Edition (1987)] that the parity is indeed of order $|S|$, that is, its support is the whole sequence! Clearly, this establishes a strong link with the limitations behind the presentation of connectedness.

Minsky and Papert analyzed perceptrons from a formal point of view by using mostly topological issues and complexity analyses. They proved that the explosion of the predicate order is quite typical in a number of other topological predicates like, for instance, computing the number of holes in a picture. It is a common opinion that the results established in their book were responsible for shifting attention, during the 1970s and part of the 1980s, towards symbolic systems in the research in AI. Minsky and Papert added a chapter (epilogue) to their book in which they also address the relationship of their studies with the emerging studies on Parallel Distributed Processing [see *Nature* **323** (1986), 533–536], where they claimed that:

> ... *it was the limitation on what perceptrons could possibly learn that led to Minsky and Papert's (1969) pessimistic evaluation of the perceptron. Unfortunately, that evaluation has incorrectly tainted more interesting and powerful networks of linear threshold and other nonlinear units. As we shall see, the limitations of the one-step perceptrons in no way apply to more complex networks.*

The reply in the epilogue of Perceptrons' book opened a fundamental challenge for the years to come:

> *These critical remarks must not be read as suggestions that we are opposed to making machines that can "learn". Exactly the contrary! But we do believe that significant learning at a significant rate presupposes some significant prior structure. Simple learning schemes based on adjusting coefficients can indeed be practical and valuable when the partial functions are reasonably matched to the task* ...

In spite of the apparent simplicity, an in-depth analysis of perceptrons reveals that learning its weights might result in suboptimal solutions, even for linearly separable tasks. The issue of failing to separate linearly separable examples has been deeply covered by Eduardo D. Sontag and Héctor J. Sussmann [see *Neural Netw.* **4** (1991) 243–249] who nicely addressed an interesting counterexample given by M. Brady, R. Raghavan and J. Slawny in *IEEE Trans. Circuits Syst.* **36** (1989) 665–674. However, it early became clear that the failure arising in these conditions is essentially due to marrying neurons and error functions that are not compatible! This results in the

---

[7] In Fig. 3.4, we have $S = ($up, down, up$)$. In this case, $[S$ is even$] = 0$ and p is connected to q'.

emergence of *spurious local minima* that, for the given neuron, can be avoided by an appropriate choice of the error function [see M. Bianchini, M. Gori, *Neurocomputing* **13** (1996) 313–346 and M. Bianchini, P. Frasconi, M. Gori, M. Maggini, Neural Netw. Sys., Techn. and Appl. 1998, 1–51].

Section 3.3 Linear and linear-threshold machines have been the battlefield of statisticians who have dominated the subject for years. The important connections between the Bayesian decision and the linear discrimination have been early established and popularized in the 1960s, and it is now very well addressed in classic books [see, e.g. R.O. Duda, P.E. Hart, D. Stork, *Pattern Classification* second edition (New York: Wiley, 2001) and C.M. Bishop, *Pattern Recognition and Machine Learning* (Springer, 2006)]. The conditions in which linear machines can be jointly interpreted in the framework of statistics and of the parsimony principle have been the subject of discussion for years, and turn out to be related to linear discrimination. An in-depth analysis on logistic regression can be found in Applied Logistic Regression by D.W. Hosmer and S. Lemeshow second edition (Wiley, 2000). Both the statistical interpretation of learning and the one based on the parsimony principle lead to *shrinkage methods*, which are general techniques to improve LMS by adding constraints on the value of coefficients, an issue that has also been covered in the previous chapter. For linear machines a good coverage on shrinkage methods can be found in *The Elements of Statistical Learning* by T. Hastie, R. Tibshirani, J. Friedman (Springer, 2001).

Section 3.4 Unlike most of machine learning models, the assumption behind linear and linear-threshold machines leads to the identification of fruitful conditions on optimal convergence. While for linear machines the LMS formulation is inherently local minima free, it has been shown that the continuous optimization of error functions for LTU units can lead to spurious solutions that, however, can be properly avoided. When adopting online learning, a different perspective arises, that comes from the structure of the Widrow-Hoff algorithm [see B. Widrow, *Generalization and information storage in networks of adaline "neurons"* in *Self-Organizing Systems* edited by M.C. Yovits, G.T. Jacobi, G.D. Goldstein (Spartan, 1962) pages 435–461]. In the case of hard-limiting units — namely the classic perceptron — it has been shown that a different approach is required. The analysis on the perceptron algorithm described in Section 3.4.3 was given by A.B. Novikoff in *Symposium on the Mathematical Theory of Automata* **12** (1962), 615–622, who established the bound stated by Eq. 3.4.3–(5). The proof by contradiction makes use of the principle that the cosine of the angle ϕ between the normal vector of the separating hyperplane a and $\hat{w}$ cannot increase arbitrarily. The proof can be regarded as a convergence analysis of the algorithm, whereas the formulation of learning given in the remainder of the chapter is regarded as an optimization process. However, one can assume the same view while looking at Eq. 3.4.3–(2). Now, if we regard as an error function $E(\hat{w}) = -a'\hat{w}/\|\hat{w}\|$ then learning can be regarded as gradient descent along $E(\cdot)$. In so doing, we are back to the discussion on convergence when dealing with the continuous computational setting of complexity. We can promptly see that there are no false minima, that is, no local minima different from the global one. For this reason, Novikoff bound is not so surprising! An analysis based on the bound 3.4.4–(7) leads to an intriguing connection with Novikoff bound. Clearly, one cannot use a gradient descent algorithm based on $E(\hat{w}) = -a'\hat{w}/\|\hat{w}\|$, since a is unknown, but the analysis enlightens the reasons of the optimal bound.

The separation of positive a negative examples can also naturally be formulated as a liner programming problem. Given the training set $\mathcal{L} = \{(x_\kappa, y_\kappa) : \kappa = 1, \ldots, \ell\}$, $\forall \kappa \in 1, \ldots, \ell$ let us construct the map $\hat{x}_\kappa \mapsto \hat{x}_\kappa [y_\kappa = 1] - \hat{x}_\kappa [y_\kappa \neq 1]$. After this replacement the classification problem can compactly be written as

$$\min_{\hat{w}} 1 \quad \text{with } \hat{X}\hat{w} > 0, \tag{3}$$

which is a linear programming problem, for which there is huge literature of solving algorithms. Since the problem can be undetermined, one can look for a unique solution so as to face ill-conditioning. Hence it's natural to think of the principle used in Moore-Penrose pseudoinversion and in ridge regression, topics that will be mostly covered when discussing kernel machines in Chapter 4. In any case it's worth mentioning that while the simplex method might explode under some configuration, although polynomial, the Karmakar algorithm [see N. Karmakar, *Proc. of the Sixteenth Annual ACM Symposium on Theory of Computing* (1984), 302–311] doesn't achieve the optimal complexity bound of the perceptron algorithm.

The formulation of learning as continuous optimization shares the same structure of other apparently unrelated problems. One might wonder whether this ubiquitous formulation as continuous optimization is human trend to unify or is instead the inherent solution of most natural problems.

The notion of terminal attractor was introduced in Phys. Lett. A **133** (1988), 18–22, *Neural Netw.* **2** (1989), 259–274 by M. Zak and subsequently experimented in S.D. Wang, C.H. Hsu, *IJCNN* (1991), 183–189, M. Bianchini, M. Gori, M. Maggini, *ICANN* (1994), 26–29, and C.R. Jones, C.P. Tsang, *ICNN* (1993), 929–935. Its adoption to assess complexity in the continuous and discrete setting of computation has been proposed in M. Bianchini, M. Gori, M. Maggini, *ICANN* (1994), 26–29, M. Bianchini, S. Fanelli, M. Gori, M. Protasi, Neurocomputing **15** (1997), 3–13, P. Frasconi, S. Fanelli, M. Gori, M. Protasi, *Proceedings of International Conference on Neural Networks* (1997), 1240–1245, and in M. Gori *Continuous problem-solving and computational suspiciousness* in *Limitations and Future Trends in Neural Computation* (IOS Press, 2003), pages 1–22 edited by S. Ablameyko and M. Gori. The idea has also been used for addressing general optimization problems [see M. Bianchini, S. Fanelli, M. Gori, *IEEE Trans. Comput.* **50** (2001), 689–698]. The analysis enlightens the role of the circuit complexity and of the condition number, which depends on the degree of precision of the required solution. A general manifesto on the complexity in the continuous setting of computation is given in *Complexity and Real Computation* (Springer-Verlag, 1998) by L. Blum, F. Cucker, M. Shub, S. Smale, that also parallels some of the results given in M. Gori *Continuous problem-solving and computational suspiciousness* in *Limitations and Future Trends in Neural Computation* (IOS Press, 2003), pages 1–22 edited by S. Ablameyko and M. Gori. These studies offer a clear picture of the kind of complexity that arises when learning with neural networks. The circuit complexity is dependent on the number of connections, while the condition number of learning can explode when optimizing flat error functions.

The solution of a given problem in the framework of continuous optimization takes place by constructing a function that, once optimized, makes it possible to determine the solution of the problem. Basically, determining such function seems to be related to the creative process of designing algorithms in the classic discrete setting of computation. The elegance and generality of solutions based on continuous optimization, however, seems to represent also the main source of troubles that typically arise when approaching *complex* problems. The process of function optimization can either be hard because of the inherent complexity of the problem at hand or because of the way the problem is framed in the context of optimization. A wrong choice of the numerical algorithm may also affect the computational complexity significantly. The complexity of the optimization can be due to a *spurious* formulation of the problem, but can have also a *structural* nature, in the sense that the complexity can be inherently associated with the problem at hand. In the last case the problem gives rise to a sort of *suspiciousness* concerning the actual possibility to discover its solution under reasonable computational constraints. Whereas most practitioners use to accept without reluctance the flavor of suspiciousness arising from the approach and use to be proud of their eventual experimental achievements, one might be skeptical of problem solving by continuous optimization. As a matter of fact, the success of these methods is related to the problem at hand, and therefore one can expect an excellent behavior for a class of problems, whereas it can raise serious suspicions about the solution of others.

Gori and Meer [see M. Gori, K. Meer, *Math. Log. Q.* **48** (2002), 45–58] propose a general framework for problem solving using continuous optimization and give some theoretical foundations on the intuitive notion of suspiciousness by relating it to the theory of computational complexity. They introduce the concept of *action* as a sort of continuous algorithm running on an abstract machine, referred to as the deterministic terminal attractor machine (DTAM), which performs a terminal attractor gradient descent on the energy. This machine is conceived for running actions. For any instance of a given problem, the corresponding action runs on the DTAM and is guaranteed to yield a solution whenever the energy is local minima free. In this case the problem is called *unimodal*, and there is no need to initialize the DTAM. For complex problems one may require a *guessing module* for an appropriate initialization of the gradient descent. The corresponding machine is referred to as the nondeterministic terminal attractor machine (NDTAM), which suggests the introduction of nondeterministic unimodal problems. A discrete counterpart of these machines makes it possible to express the computational complexity in terms of the problem precision and of a properly defined condition number, which expresses for a given problem the degree of sensitivity of the input change. This is a generalization of the analysis carried out for linear machines in Section 3.4.3 (see, e.g., Eq. 3.4.4–(8) on the number of steps). A fundamental consequence of the proposed approach is that actions for unimodal problems can be conceived, which give rise to optimal algorithms with respect to the problem dimension. Examples to the problem of solving linear systems and to the problem of linear separation in computational geometry are presented, along with the corresponding complexity evaluation. As a consequence of the complexity bounds that are determined from the DTAM machine, it is shown that knowledge of lower bounds on the complexity of a given problem yields straightforward conclusions on its suspiciousness. Related studies in this direction are in M. Gori *Continuous problem-solving and computational suspiciousness* in *Limitations and Future Trends in Neural Computation* (IOS Press, 2003), pages 1–22 edited by S. Ablameyko and M. Gori, M. Bianchini, S. Fanelli, M. Gori, M. Protasi, *Solving Linear Systems by a Neural Network Canonical Form of Efficient Gradient Descent* in *Proceedings of the 1997 International Conference on Neural Information Processing and Intelligent Information Systems* (Springer, 1998) edited by N.K. Kasabov, R. Kozma, K. Ko, R. O'Shea, G.G. Coghill and Tom Gedeon and M. Bianchini, S. Fanelli, M. Gori, *IEEE Trans. Comput.* **50** (2001), 689–698.

Kernel machines

4

They could not see his face:
he was hooded, and above the hood he wore awide-brimmed hat,
so that all his features were overshadowed,
except for the end of his nose and his grey beard.
J.R.R. Tolkien, *The Lord of the Rings* (1954)

Here was a face with flashing eyes
and distorted features,
a face convulsed with hatred
and with the mad joy of
gratified revenge.
Arthur Conan Doyle, *The Lost World* (1912)

If you wish to put a finer tip on the pencil,
the device features
a serrated lip on its shavings drawer,
angled so that graphite swarf falls inside.
David Rees, *How to Sharpen Pencils* (2012)

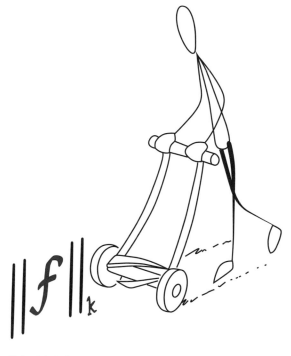

Machine Learning. https://doi.org/10.1016/B978-0-32-389859-1.00011-8

IN THE PREVIOUS chapter, we have discussed learning with linear machines by showing the fundamental role of the quadratic regularization term that gives rise to ridge regression. Amongst others, a couple of questions arise naturally: First, how can we perform regression for complex nonlinear maps or separate patterns that are nonlinearly separable? Second, how can we generalize to new examples even in the presence of small training sets? This chapter addresses both these two fundamental issues by introducing the elegant framework of kernel machines. It will be shown that when joining the need of constructing nonlinear machines, which requires an appropriate introduction to a rich feature space, with regularization requirements we naturally end up with the notion of kernel machine.

4.1 Feature space

In this section, we discuss the importance of feature enrichment in complex learning problems. This fundamental issue has already been addressed in Section 3.2.1 when discussing of the predicates on black and white pictures. The analysis has shown the importance of an appropriate feature selection for any decision process. This is in fact a fundamental topic in machine learning, which revolves around the never-ending discussion on whether appropriate features should be properly selected or learned from examples. Here, we cover the first case, which gives rise to the notion of kernel machines. While kernels can also be learned, most natural feature-learning takes place in connectionist models. In general, the features are determined by the *feature map*

$$\phi \colon X \subset \mathbb{R}^d \to H \subset \mathbb{R}^D,$$
$$x \to \phi(x). \tag{1}$$

In most cases $D \geq d$, and we often think of $D \gg d$; in the extreme case we will see that ϕ yields a feature space of infinite dimension. The analysis in Section 3.2.1 has involved the construction of functions φ that are essentially feature maps. The difference with respect to the above definition of ϕ is that φ operates on the retina, namely a two-dimensional structure.

4.1.1 Polynomial preprocessing

The linear machines discussed so far are limited either in regression or in classification. The linearity assumption in some real-world problems is quite restrictive. In addition, LTU machines can only deal with linearly-separable patterns. Interestingly, in both cases one can extend the theory of linear and LTU machines by an appropriate enrichment of the *feature space*. We focus attention on classification, but similar analysis can be drawn for regression tasks. We start by showing — by means of an example — how the linear separation concept can easily be extended.

Suppose we are given a classification problem with patterns $x \in X \subset \mathbb{R}^2$ and consider the associated feature space defined by the map $X \subset \mathbb{R}^2 \overset{\phi}{\to} H \subset \mathbb{R}^3$ such that $x \to z = (x_1^2, x_1 x_2, x_2^2)'$. Clearly, linear-separability in H yields a quadratic separation in X, since we have

$$a_1 z_1 + a_2 z_2 + a_3 z_3 + a_4 = a_1 \cdot x_1^2 + a_2 \cdot x_1 x_2 + a_3 \cdot x_2^2 + a_4 \geq 0.$$

It is obvious that ϕ plays a crucial role in the feature enrichment process; for example, in this case linear separability is converted into quadratic separability. This idea can be given a straightforward generalization by carrying out *polynomial processing* of the inputs. Let us consider the monomials coming from the raising to power p the sum of coordinates of the input as follows:

$$(x_1 + x_2 + \cdots + x_d)^p = \sum_{|\alpha|=p} \binom{p}{\alpha} x^\alpha,$$

where α is a multiindex, so that a generic coordinate in the feature space is

$$z_{\alpha,p} = \frac{p!}{\alpha_1! \alpha_2! \cdots \alpha_d!} x_1^{\alpha_1} x_2^{\alpha_2} \cdots x_d^{\alpha_d}, \tag{1}$$

and $p = \alpha_1 + \alpha_2 + \cdots + \alpha_d$. Any $z_{\alpha,p}$ is a pth order monomial; hence we can compose the general feature representation $z = \phi(x) = (z_{\alpha,p})$ where $p = 0, \ldots, p_m$ with monomials with order less or equal to p_m. While this space significantly increases the chance to separate the given classes, the problem is that the number of features explodes quickly! If we restrict to pth order monomials, we have

$$|H| = \binom{p+d-1}{p}, \tag{2}$$

which makes the computational treatment apparently unfeasible in high dimensional spaces. In Section 4.3, we will see that it is indeed possible to conceive machine learning schemes to deal even with infinite-dimensional spaces!

4.1.2 Boolean enrichment

While in case of real-valued inputs input enrichment is based on products of pairs of input, in case of Boolean-value inputs, the AND can be used with similar effect. To grasp the idea, let us consider the classification of the Boolean map defined by the XOR function

$$x_1 \oplus x_2 = (\bar{x}_1 \wedge x_2) \vee (x_1 \wedge \bar{x}_2). \tag{1}$$

Clearly, it is not linearly separable in the space (x_1, x_2). However, if we enrich the feature space by mapping $x = (x_1, x_2) \to (\bar{x}_1 \wedge x_2, x_1 \wedge \bar{x}_2)$, we can promptly see that we gain linear-separability, since the problem is reduced to dealing with the $\vee$ function, which is linearly-separable. This idea can be extended to any Boolean function.

The generalization involves the first canonical form — also referred to as Disjunctive Normal Form (DNF). Given any Boolean function f, it can be represented as

$$f(x_1, \ldots, x_d) = \bigvee_{j \in F} \bigwedge_{i=1}^{d} x_i^{\alpha_{ij}}. \tag{2}$$

Here, we use the assumption that $\alpha_{ij} \in \{-1, +1\}$, F is a set of indices and we pose

$$x^\alpha = [\alpha = 1]x \vee [\alpha = -1]\bar{x}. \tag{3}$$

In the previous case of XOR, $F = \{1, 2\}$, while α is

$$\alpha = \begin{pmatrix} 1 & -1 \\ -1 & 1 \end{pmatrix}.$$

Hence, the Boolean enrichment offered by the $\bigwedge_{i=1}^{d} x_i^{\alpha_{ij}}$ yields the natural feature representation of f, since "$\vee$" is linearly-separable. The features given by Eq. 4.1.1–(1), which were considered for the case of polynomial preprocessing, turn out to have a somewhat corresponding representation for Boolean functions, when choosing

$$z_{\alpha, j} = \bigwedge_{i=1}^{d} x_i^{\alpha_{ij}}. \tag{4}$$

Also in this case the number of features can easily explode. There are, however, some differences with respect to smooth functions. In this case, the number of features grows exponentially with the dimension of the input 2^d, whereas there is no superexplosion due to unlimited growth of the power. However, also in this case, we are in front of intractability issues that will be nicely covered by the introduction of kernel functions in Section 4.3. A different, yet equivalent, feature representation arises when considering CNF instead of DNF.

EXERCISES

1. [*15*] In Exercise 3.1.3–6, we have addressed the problem of dealing with polynomials of high degree, where we have seen the ill-conditioning connected with the Vandermonde matrix — see also Fig. 2.1, Section 2.4. Prove that, unlike for the case of Boolean functions, once the degree of the polynomial is chosen, the growth of the number of features with the input dimension doesn't result in combinatorial explosion.

▶ **2.** [*15*] Consider Eq. (4) for DNF-based feature representation of a Boolean function. Find a corresponding feature-based representation by using the CFN.

4.1.3 Invariant feature maps

Polynomial and Boolean preprocessing suggest that the hypothesis of linear separability can be overcome by appropriate construction of a new feature space. Clearly, these feature maps are only examples of clever choices for facing final learning tasks which are expected to benefit from an appropriate feature selection. There's more: Neither polynomial nor Boolean processing deal with crucial issues of *invariance* that arise in most interesting learning tasks. For example, the recognition of handwritten chars must be independent of translations and rotations. Moreover, it must be scale invariant. The ambition of extracting distinctive features from vision poses even more challenging tasks. While we are still concerned with feature extraction that is independent of the above classic geometric transformation, it looks like we are still missing the fantastic human skill of capturing distinctive features to recognize ironed and rumpled shirts! There is no apparent difficulty to recognize shirts by keeping the recognition coherence in case we roll up the sleeves, or we simply curl them up into a ball for the laundry basket. Of course, there are neither rigid transformations, like translations and rotation, nor scale maps that transform an ironed shirt into the same shirt thrown into the laundry basket. As pointed out in Section 5.4, the appropriate extraction of distinctive features is of crucial importance for visual tasks. The recognition and understanding of speech utterances is also strongly interwound with the problem of an appropriate extraction of distinctive features. This clearly requires us to determine features that are invariant under temporal translations, thus resembling related problems that arise for handwritten chars, where we need spatial (instead of temporal) translation invariance. A possible way of capturing feature invariance is to introduce a function which is the target of invariance. Let $\rho \colon X \to X$ be a map which transforms pattern x onto $\rho(x)$, so as both x and $\rho(x)$ share the same cognitive features. We can incorporate such a property when searching for features ϕ such that

$$\phi \circ \rho = \phi. \tag{1}$$

Since it must hold locally for all $x \in \mathbb{R}^d$, the above equation is equivalent to

$$J_\phi(x) \cdot \nabla \rho(x) = J_\phi(x),$$

where J_ϕ is the Jacobian of ϕ. This can be softly satisfied if we think of an agent which keeps small the penalty P given by

$$P(\phi) = \int_X \left(J_\phi(x) \nabla \rho(x) - J_\phi(x) \right)^2 dx.$$

This functional is hard to minimize, mostly because function ρ, which is expected to express the invariance, is not necessarily known. With reference to invariant features in computer vision, in Section 5.4, when discussing of convolutional networks, it will be shown that the above general invariant condition takes on a very expressive computational structure when we reduce the invariance over a temporal manifold, where we basically impose the principle of feature motion invariance.

4.1.4 Linear-separability in high-dimensional spaces

Linear and linear-threshold machines can deal with linearly-separable data. But is this hypothesis realistic? To ask this question, we must bear in mind that x is often regarded as an internal representation of a physical pattern. Hence linear-separability depends on the concrete physical structure of the pattern, as well as on the choice of its internal representation. While the first dependence results in a sort of structural complexity of the learning task, pattern representation offers a large degree of freedom, which can easily transform nonlinear separability into linear-separability.

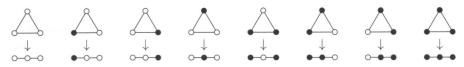

FIGURE 4.1

Three points in two (triangles) and one (lines) dimensional space. The one-dimensional points are constructed by projection. While in two dimensions all points configurations are linearly-separable, when projecting onto a single dimension, in two configurations the points are no longer linearly-separable.

In order to get an insight, let us consider the example shown in Fig. 4.1, where we are given three examples represented in one and two dimensional spaces. The representation onto a single dimensional space is simply created by a projection from the two-dimensional space. In two dimensions, all the eight dichotomies are linearly-separable, whereas this property is lost when projecting the examples onto a line. This holds under very general conditions: As the dimension of the space increases, the probability of linear-separation increases!

Suppose we are given ℓ training examples that are labeled as positive or negative. Hence $X \subset \mathbb{R}^d$ is partitioned by $X = X^+ \cup X^-$; in particular, we will say that X is *homogeneously linearly separable* if there exists a weight vector w such that $w \cdot x > 0$ if $x \in X^+$ and $w \cdot x < 0$ if $x \in X^-$. We further assume that the examples of the training set X are drawn in such a way that every subset of d or fewer vectors is linearly independent — whenever this happens, we say that the patterns are in *general position*. It is clearly a necessary condition for linear separability, to which it reduces whenever $\ell \le d$. We want to determine the number $c_{\ell d}$ of homogeneously linearly separable dichotomies that can be created in the d-dimensional space with ℓ examples. The key idea is gained by induction. Suppose we add one example $x_{\ell+1}$ to a collection of ℓ examples. Now let D be the number of linear dichotomies created by separating hyperplanes that contain the new point. If we move them slightly then any such dichotomy generates two new dichotomies $\{x_{\ell+1}\} \cup X^+$ and $\{x_{\ell+1}\} \cup X^-$. This can easily be proven. Let δ be the distance of the closest point from the generic dichotomy. Clearly, there are movements such that while $x_{\ell+1}$ is not on the hyperplane anymore, the categorization of the remaining points doesn't change. Hence $c_{(\ell+1)d} = (c_{\ell d} - D) + 2D = c_{\ell d} + D$. Now the problem is reduced to determining the number of dichotomies D. Since these dichotomies are created by hyperplanes that are constrained to contain $x_{\ell+1}$, they correspond with $c_{\ell(d-1)}$, since the constraint of passing for $x_{\ell+1}$, in case of patterns in general position, reduces of one the degree of freedom (see Exercise 5). Then we have

$$c_{(\ell+1)d} = c_{\ell d} + c_{\ell(d-1)}. \tag{1}$$

This recursive equation is paired with the initial condition $c_{1d} = 2$ and $c_{\ell d} = 0$ if $d \le 0$. Now we can induce the overall number of dichotomies by successively expanding Eq. (1). We have

$$
\begin{aligned}
c_{(\ell+1)d} &= c_{\ell d} + c_{\ell(d-1)} \\
&= c_{(\ell-1)d} + 2c_{(\ell-1)(d-1)} + c_{(\ell-1)(d-2)} \\
&= c_{(\ell-2)d} + 3c_{(\ell-2)(d-1)} + 3c_{(\ell-2)(d-2)} + c_{(\ell-2)(d-3)}
\end{aligned}
$$

$$\vdots$$

$$= \binom{\ell}{0} c_{1d} + \binom{\ell}{1} c_{1(d-1)} + \cdots + \binom{\ell}{\ell} c_{1(d-\ell)} = \sum_{i=0}^{\ell} \binom{\ell}{i} c_{1(d-i)}.$$

We have to be a little bit careful about the relative position of the integers i, d, and ℓ in order to correctly use the initial conditions. In particular, we have that

$$c_{(\ell+1)d} = \sum_{i=0}^{\ell} \binom{\ell}{i} c_{1(d-i)} \, [i < d] = 2 \sum_{i=0}^{d-1} \binom{\ell}{i},$$

since when $d > \ell$ and $i \in [\ell + 1 .. d - 1]$ one has $\binom{\ell}{i} \equiv 0$. So what we discovered is that

$$c_{\ell d} = 2 \sum_{i=0}^{d-1} \binom{\ell-1}{i} = 2^{\ell} - \binom{\ell-1}{d} F\left(\begin{matrix} 1, \ell \\ d+1 \end{matrix} \Big| \frac{1}{2} \right), \tag{2}$$

where $F\left(\begin{smallmatrix} a,b \\ c \end{smallmatrix} | z \right)$ is the hypergeometric function (see Exercise 6 for the details that lead to this last equality). From this last expression we see that whenever $d \geq \ell$ we have that $c_{\ell d} = 2^{\ell}$ since the coefficient in front of the hypergeometric function is identically zero, in all the other cases the numbers of homogeneously separable dichotomies are less that 2^{ℓ} because

$$F\left(\begin{matrix} 1, \ell \\ d+1 \end{matrix} \Big| \frac{1}{2} \right) = \sum_{k \geq 0} \frac{\ell^{\bar{k}}}{(d+1)^{\bar{k}}} 2^{-k}$$

is a positive number. Now we can determine the fraction of linearly-separable configurations over all possible 2^{ℓ} point configurations that arise from their labeling by "+" and "−":

$$p_{\ell d} := \frac{c_{\ell d}}{2^{\ell}} = 1 - \frac{\binom{\ell-1}{d}}{2^{\ell}} F\left(\begin{matrix} 1, \ell \\ d+1 \end{matrix} \Big| \frac{1}{2} \right). \tag{3}$$

Now consider the case $\ell = 2d$. As shown in Exercise 1, we have

$$p_{(2d)d} = \frac{1}{2}. \tag{4}$$

Hence, when the number of examples doesn't exceed the dimension, $p_{\ell d} = 1$. Interestingly, as we increase the number of examples to twice the critical number $\ell = d$, for which it is still $p_{\ell d} = 1$, we get $p_{(2d)d} = 1/2$. As $\ell \to +\infty$, the probability of linear separation goes to zero (see Exercise 1). This discussion leads naturally to the fundamental notion of *capacity*. Now let $\epsilon > 0$ and consider the largest number C such that if $\ell < (1 - \epsilon)C$ then $\forall \delta < 1, \forall \epsilon > 0$ there exists ℓ_0 such that for all $\ell > \ell_0$ the examples are linearly separable with $p_{ell,d} > 1 - \delta$. This value C is referred to as the capacity of the machine. For d sufficiently large we can prove that (see Exercise 7)

$$C \simeq 2d.$$

The definition of capacity finds its deep roots in the behavior of LTU in high dimensional spaces, where the value of $C = 2d$ represents a sort of cut-off on the number of examples.

There is a phase transition at C such that, beyond that value, there is no linear separation! A discussion on the effect of the dimensionality on $p_{\ell,d}$ is discussed in Exercise 8.

EXERCISES

1. [*12*]   Consider Cover's analysis on linear-separability in case $\ell = 2d$. Prove Eq. (4) and that $\lim_{\ell \to +\infty} c_{\ell d} = 0$.

2. [*10*]   Prove that $c_{\ell 1} = 2d$, without using Eq. (3).

3. [*M20*]   Give a formal proof by induction of Cover bound (3).

4. [*M25*]   Solve the recurrence relation in Eq. (3) using the theory of generating functions.

5. [*M30*]   Suppose that we are given patterns in *general position*. When referring to the recursive Eq. (1), prove that $D = c_{\ell(d-1)}$.

6. [*M25*]   Prove the second equality of Eq. (2).

▶ **7.** [*25*]   Prove that $C = 2d$.

8. [*22*]   Prove that as $d \to \infty$ we have $p_{\ell,d} \simeq [\ell < C]$.

4.2 Maximum margin problem

In this section, we show how an elegant robust expression of learning with enriched features of the input ends up in kernel machines. Formally, we introduce learning in the framework of computational geometry by formulating and solving the *Maximum Margin Problem* (MMP).

4.2.1 Classification under linear-separability

Let us consider a linear machine in the feature space that, as previously described in Section 4.1, is represented by

$$f(x) = w \cdot \phi(x) + b = \hat{w} \cdot \hat{\phi}(x), \tag{1}$$

where $\hat{\phi}(x) := (\phi_1(x), \ldots, \phi_D(x), 1)'$. We start considering a classification problem, where we are given a training set $\mathcal{L} = \{(x_\kappa, y_\kappa) : \kappa = 1, \ldots, \ell\}$ and $y_\kappa \in \{-1, +1\}$. Furthermore, let us assume that the feature space is chosen in such a way that the pairs of $\mathcal{L}$, once moved to the feature space, lead to the associated training set $\mathcal{L}_\phi = \{(\phi(x_\kappa), y_\kappa) : \kappa = 1, \ldots, \ell\}$, which is linearly separable. As pointed out in Section 4.1.4, this is indeed a reasonable assumption at high dimensions whenever $\ell \ll 2D$. Now let us consider the problem of determining $\hat{w}^\star$ such that

$$\hat{w}^* \in \underset{\omega \in W}{\arg\max} \left\{ \frac{1}{\|\Pi_w \omega\|} \min_\kappa \left(y_\kappa \omega \cdot \hat{\phi}(x_\kappa) \right) \right\}.$$

This problem has an interesting geometrical interpretation in the feature space, since the real $d(\kappa, \hat{w})$

$$d(\kappa, \hat{w}) := \frac{y_\kappa \hat{w} \cdot \hat{\phi}(x_\kappa)}{\|w\|} = \frac{|\hat{w} \cdot \hat{\phi}(x_\kappa)|}{\|w\|}$$

is the distance of $\phi(x_\kappa)$ to the hyperplane defined by $\hat{w}$ (that is, the set of points $x \in \mathbb{R}^d$ such that $\hat{w} \cdot \hat{x} = 0$). Now, for any $\alpha \in \mathbb{R}$, from $\hat{w} \cdot \hat{\phi}(x_\kappa) = 0$ we get $\alpha \hat{w} \cdot \hat{\phi}(x_\kappa) = 0$, which can be interpreted as the separation of the same points by the same hyperplane with parameters $\hat{w}_\alpha = \alpha \hat{w}$. Now we can always choose α such that for the closest point(s) $x_{\bar{\kappa}}$ to the separating hyperplane, we have $y_{\bar{\kappa}} \hat{w}_\alpha \cdot \hat{\phi}(x_{\bar{\kappa}}) = 1$. Hence the discussion on the given problem can equivalently be carried out by replacing $\hat{w}$ with $\hat{w}_\alpha$, so that for the sake of simplicity from now on we drop the index α and replace w_α with w. Based on these arguments, we can face an equivalent optimization problem that is based on the *parsimonious* satisfaction of a set of constraints that derive from the training set:

$$\begin{aligned} \text{minimize} \quad & \frac{1}{2}\|w\|^2 \\ \text{subject to} \quad & 1 - y_\kappa \hat{w} \cdot \hat{\phi}(x_\kappa) \le 0, \quad \kappa = 1, \ldots, \ell. \end{aligned} \tag{2}$$

The environmental constraint $1 - y_\kappa \hat{w} \cdot \hat{\phi}(x_\kappa) \leq 0$ expresses the strong sign agreement between the supervisor and the agent decisions. The problem can be approached by the Lagrangian formalism. In a sense, the problem reminds us the one stated in Exercises 3.1.2–5 to determine $\hat{w} = \hat{X}^+ y$. The two problems share the principle of discovering a parsimonious solution under a set of constraints deriving from the training set. In the case of Exercises 3.1.2–5, however, the constraint associated with each supervised pair (x_κ, y_κ) is bilateral, whereas the constraints (2) are unilateral. Moreover, while the pseudoinverse matrix emerges from the minimization of $\|\hat{w}\|^2$, in the maximum margin problem we adhere to the regularization framework, where the function to be minimized only involves w^2. An in-depth analysis on the relations between the two problems is proposed in Exercise 9. The minimization problem (2) requires us to determine stationary points $(\hat{w}^*, b^*, \lambda^*)'$ of

$$L(\hat{w}, \lambda) = \frac{1}{2}\|w\|^2 + \sum_{\kappa=1}^{\ell} \lambda_\kappa \left(1 - y_\kappa \hat{w} \cdot \hat{\phi}(x_\kappa)\right), \quad \text{with } \lambda \geq 0. \tag{3}$$

If we impose $\nabla L = 0$ then we have

$$\partial_w L(\hat{w}, \lambda) = w - \sum_{\kappa=1}^{\ell} \lambda_\kappa y_\kappa \phi(x_\kappa) = 0, \tag{4}$$

and

$$\partial_b L(\hat{w}, \lambda) = -\sum_{\kappa=1}^{\ell} \lambda_\kappa y_\kappa = 0. \tag{5}$$

These equations make it possible to separate the weight variables from the Lagrangian multipliers. In particular, since $w = \sum_{\kappa=1}^{\ell} \lambda_\kappa y_\kappa \phi(x_\kappa)$, we can re-write the Lagrangian as function of the Lagrangian multiplier only. Hence we have

$$\theta(\lambda) := \inf_{\hat{w} \in W} L(\hat{w}, \lambda)$$
$$= \frac{1}{2}\left\|\sum_{\kappa=1}^{\ell} \lambda_\kappa y_\kappa \phi(x_\kappa)\right\|^2 - \sum_{\kappa=1}^{\ell} \lambda_\kappa y_\kappa \left(\sum_{h=1}^{\ell} (\lambda_h y_h \phi(x_h)) \cdot \phi(x_\kappa) + b\right) + \sum_{\kappa=1}^{\ell} \lambda_\kappa$$
$$= \frac{1}{2}\sum_{h-1}^{\ell}\sum_{\kappa=1}^{\ell} \lambda_h \lambda_\kappa y_h y_\kappa \phi(x_h) \cdot \phi(x_\kappa) - \sum_{h=1}^{\ell}\sum_{\kappa=1}^{\ell} \lambda_h \lambda_\kappa y_h y_\kappa \phi(x_h) \cdot \phi(x_\kappa)$$
$$- b\sum_{\kappa=1}^{\ell} \lambda_\kappa y_\kappa + \sum_{\kappa=1}^{\ell} \lambda_\kappa = -\frac{1}{2}\sum_{h=1}^{\ell}\sum_{\kappa=1}^{\ell} \lambda_h \lambda_\kappa y_h y_\kappa \phi(x_h) \cdot \phi(x_\kappa) + \sum_{\kappa=1}^{\ell} \lambda_\kappa. \tag{6}$$

Again the replacement of $L((w^*, b^*)', \lambda)$ with $\theta(\lambda)$ is made possible by the specific structure of the problem, which allows us a complete change of the variables from the primal to the dual representation. Now let us define

$$k : X \times X \to \mathbb{R}, \qquad k(x_h, x_\kappa) := \phi'(x_h)\phi(x_\kappa). \tag{7}$$

It is referred to as a *kernel function*, which expresses a sort of similarity between any pair of points $(x_h, x_\kappa) \in X \times X$. From the previous analysis we promptly see that the maximum margin problem expressed in Eq. (2) is equivalent to the *dual optimization problem*:

$$\text{maximize} \quad \theta(\lambda) = \sum_{\kappa=1}^{\ell} \lambda_\kappa - \frac{1}{2} \sum_{h=1}^{\ell} \sum_{\kappa=1}^{\ell} k(x_h, x_\kappa) y_h y_\kappa \lambda_h \lambda_\kappa$$

$$\text{subject to} \quad \lambda_\kappa \geq 0, \quad \kappa = 1, \ldots, \ell, \tag{8}$$

$$\sum_{\kappa=1}^{\ell} \lambda_\kappa y_\kappa = 0.$$

This is a classic *quadratic programming* problem. Technical details on this issue are in Section 4.5, while a concise review on basic methods of constrained optimization at finite dimensions are given in Appendix A. The Lagrangian multipliers can be thought of as the *constraint reactions* to any single supervised pair. In this respect, learning can be regarded as a problem of determining nonnegative constraint reactions that maximize θ while respecting the balancing constraint $\sum_{\kappa=1}^{\ell} \lambda_\kappa y_\kappa = 0$. Intuitively, the maximum of $\theta(\lambda)$ indicates that while we keep minimizing with respect to $\hat{w}$ the Lagrangian $L(\hat{w}, \lambda) = \theta(\lambda)$, the minimum (w^*, b^*) is detected under the worst condition for the penalty term, which is driven by λ. Hence $(w^*, b^*, \lambda_\kappa^*)$ turns out to be a saddle point of L. As a result the optimal function turns out to be

$$f^*(x) := w^*\phi(x) + b^* = \sum_{\kappa=1}^{\ell} \left(\lambda_\kappa^* y_\kappa \phi(x_\kappa) \right) \cdot \phi(x) + b^* = \sum_{\kappa=1}^{\ell} y_\kappa \lambda_\kappa^* k(x_\kappa, x) + b^*. \tag{9}$$

This equation offers a dual representation of the solution of MMP. If we define[1] $\hat{\lambda} := (\lambda_1, \ldots, \lambda_\ell, b)'$ and $k_i(x) := k(x_i, x)$ then $f(x) = \hat{\lambda}' k(x)$. Hence primal and dual representations, $f(x) = \hat{w}'x = \hat{\lambda}' k(x)$, are just different expressions of the same function f, the only difference being the parameter space on which they operate. Primal $(\hat{w})$ and dual $(\hat{\lambda})$ parameters play their own role in decision in a truly complementary way. While primal parameters somewhat select most relevant coordinates in the feature space of single inputs, dual parameters select most relevant examples to carry out the decision. In this respect, the response to the input of pattern x depends on a sort of *similarity* between the pattern and any example x_κ of the training set, which is represented by $k(x_\kappa, x)$.

---

[1] In order not to overload the notation, we drop the "$\star$" to denote optimality.

Let $S_= := \{(x, y) \in \mathcal{L} : y\hat{w} \cdot \hat{\phi}(x) = 1\}$ and $S_> := \{(x, y) \in \mathcal{L} : y\hat{w} \cdot \hat{\phi}(x) > 1\}$. Now let us consider the discovered solution λ^* of the problem (8). The linear constraints guarantee the satisfaction of the KKT conditions which, along with the convex objective function of Eq. (2), make it possible to conclude that there is no duality gap. The KKT acronym stands for Karush–Kuhn–Tucker. These conditions are briefly surveyed in Section 4.5. Notice that while they are sufficient conditions to determine the optimal solution in the dual space as solution of the problem in Eq. (8), they might not be verified in some cases. This means that there is a duality gap and that $\sup_\lambda \theta(\lambda)$ doesn't return the optimal solution (see Exercises 1 and 2). The condition below emerges from the complementary slackness

$$\lambda_\kappa^*(y_\kappa f^*(x_\kappa) - 1) = 0, \quad \ell = 1, \dots, \ell.$$

Its satisfaction leads us to distinguish the following cases:

(i) $\lambda_\kappa^* = 0$. The stationarity condition is satisfied with an *interior coordinate*. In this case x_κ is called a *straw vector* and $y_\kappa f^*(x_\kappa) > 1$.

(ii) $\lambda_{\bar{\kappa}}^* > 0$. The stationarity condition is met on the border. In this case $x_{\bar{\kappa}}$ is called a *support vector*, and we have $y_{\bar{\kappa}} f^*(x_{\bar{\kappa}}) = 1$.

One might wonder whether straw vectors exist and under which circumstances. First, notice that the quadratic programming formulation leads us to conclude that the learning problem admits only one solution, that is, the primal, $\hat{w}$, and dual parameters are univocally determined, along with the maximum margin separating hyperplane. Notice that in some cases (see, e.g., Exercise 5 point 4.), the explicit expression of all the $|S_=|$ equations define univocally the separating hyperplane. Whenever this happens, we have $|S_=| \geq D = d$, that is, there are at least $D = d$ support vectors that freeze the separating configuration. Of course, $\lambda_{\bar{\kappa}} > 0$, since the constraint is active in $x_{\bar{\kappa}}$. On the other hand, if $y_\kappa f(x_\kappa) > 1$, we are in front of a straw vector, which clearly does not affect the solution of the problem, and for which $\lambda_\kappa = 0$. Another extreme case is when $|S_=| = 2$ (minimum number of supports), but we can also construct artificial problems for which the number of supports degenerates to the number of training examples (see Exercise 10).

Notice that in any feature space $\phi(X)$ that generates the kernel, we can determine the maximum margin hyperplane by passing from the dual to the primal solution, which can be done by using Eq. (4). To determine b^*, one could simply consider that $\forall \bar{\kappa} \in S_=$ we have

$$y_{\bar{\kappa}}\left(\sum_{\bar{h} \in S_=} \lambda_{\bar{h}} y_{\bar{h}} k(x_{\bar{\kappa}}, x_{\bar{h}}) + b^*\right) = 1, \tag{10}$$

from which we get

$$b^* = \frac{1 - y_{\bar{\kappa}} \sum_{\bar{h} \in S_=} \lambda_{\bar{h}} y_{\bar{h}} k(x_{\bar{\kappa}}, x_{\bar{h}})}{y_{\bar{\kappa}}}. \tag{11}$$

The problem with this solution is that it might be affected by numerical error. We can get a more reliable solution with a simple elaboration of (10). If we multiply by y_κ, we get

$$y_{\bar{k}}^2\left(\sum_{\bar{h}\in S_=}\lambda_{\bar{h}}y_{\bar{h}}k(x_{\bar{k}},x_{\bar{h}})+b^*\right)=y_{\bar{k}},$$

and if we accumulate all over the support vectors, we have

$$\sum_{\bar{k}\in S_=}\sum_{\bar{h}\in S_=}\lambda_{\bar{k}}y_{\bar{h}}k(x_{\bar{k}},x_{\bar{h}})+n_s b^*=\sum_{\bar{k}\in S_=}y_{\bar{k}},$$

where $n_s := |S_=|$. Finally, we determine b by

$$b^*=\frac{1}{n_s}\sum_{\bar{i}\in S_=}\left(y_{\bar{k}}-\sum_{\bar{h}\in S_=}\lambda_{\bar{k}}y_{\bar{h}}k(x_{\bar{i}},x_{\bar{h}})\right). \tag{12}$$

EXERCISES

1. [*M15*]　Let us consider the optimization problem $\min_{w\in\mathbb{R}^2}p(w)$, where $p\colon\mathbb{R}^2\to\mathbb{R}$, $(w_1,w_2)\mapsto w_1+w_2$ and $g_1(w_1,w_2)=w_1\le 0$, $g_2(w_1,w_2)=w_2\le 0$. Find the minimum by using the Lagrangian approach. Can you find the minimum directly, that is, without using the Lagrangian approach?

2. [*M23*]　Let us consider the optimization problem $\min_{w\in\mathbb{R}^2}p(w)$, where $p\colon\mathbb{R}^2\to\mathbb{R}:$ $(w_1,w_2)\mapsto w_1$ with constraints $g_1(w_1,w_2):=w_2-(1-w_1)^3\ge 0$ and $g_2(w_1,w_2):=-w_2\ge 0$. Prove that there is only one minimum and determine its value. What happens if we apply the KKT conditions? Is there a duality gap?

3. [*M25*]　Consider the minimization problem $\min_{w\in\mathbb{R}^2}p(w)$, where $p\colon\mathbb{R}^2\to\mathbb{R}:(w_1,w_2)\mapsto w_1^2+w_2^2$ with the constraint $h(w_1,w_2)=(x_1-1)^3-x_2^2=0$ and let $W:=\{(w_1,w_2)\in\mathbb{R}^2:h(w_1,w_2)=0\}$. Then consider the two following statements. Statement 1: "Let $B_R:=\{w\in\mathbb{R}^2:\|w\|\le R\}$. Then if one can find an $\bar{R}$ such that $W^c:=B_{\bar{R}}\cap W$ is nonempty, the given minimization problem is equivalent to $\min_{w\in W^c}p(w)$ since $p(w)=\|w\|^2$ and therefore for every point $y\in W\setminus W^c$ and $x\in W^c$ we have $p(y)>p(x)$. But if this is true, the problem always admits solution due to Weierstrass, and indeed in our example for every $\bar{R}\ge 1$, W^c is nonempty." Statement 2: "The constraint $h=0$ can be directly substituted in the objective function so that the problem reduces to the one-dimensional unconstrained minimization problem of $\pi(w)=w^2+(w-1)^3$. Since $\lim_{w\to-\infty}\pi(w)=-\infty$, the problem has no finite solution." Which of these two statements is wrong and why?

▶ **4.** [*M27*]　Let us consider the optimization problem $\min_{w\in\mathbb{R}^2}p(w)$, where $p\colon\mathbb{R}^2\to\mathbb{R}:$ $(w_1,w_2)\mapsto w_1w_2$ and with the following constraints

$$g_1(w_1,w_2)=w_1\ge 0,$$
$$g_2(w_1,w_2)=w_2\ge 0,$$
$$g_3(w_1,w_2)=-w_1^2-w_2^2\ge 1.$$

Determine the minima from direct analysis of the problem. Then use the Lagrangian approach. Is there a duality gap?

▶ **5.** [*25*]　Let us consider the learning problem defined by the set $\mathcal{L}=\{((0,0),+1),((1,0),-1),((1,1),-1),((0,1)-1)\}$.

1. Using the dual formulation, find $\hat{\lambda}$ and determine the support and straw vectors in the case of linear kernel.
2. Redo the computation by relabeling the points using a permutation of $(1, 2, 3, 4)$. Prove that the solution is not affected.
3. Redo the computation after having imposed a roto-translation to the training set. Prove that the solution is not affected and use general arguments to prove the generality of the invariance.
4. Solve the problem in the primal according to the formulation given by Eq. (2).

6. [*M20*] Prove that if a symmetric matrix is nonnegative definite, that is, $k \geq 0$, then its singular values are nonnegative.

7. [*20*] It has been proven that $|S| \geq D$, which corresponds with the number of conditions to determine univocally the separating hyperplane in the feature space. However, a different analysis seems to lead to a remarkably different conclusion on the number of support vectors. What's wrong with the following claim that in a D-dimensional feature space with randomly distributed examples only two support vectors exist?

Two support vector statement: Given any set of examples, we can always discover the closest pair with different labels by an exhaustive check, which takes $O(\ell^2)$. If the examples are drawn from a probabilistic distribution, there is only one such pair with probability one. Now let us denote this pair by (x^+, x^-) and consider the connecting segment $l((x^+, x^-))$. Then we construct the unique separating hyperplane as the one which is orthogonal to $l((x^+, x^-))$ and contains the middle point of the segment $l((x^+, x^-))$. This leads us to conclude that the pair (x^+, x^-) freezes the possible configurations and that the set of support vectors is $S = \{x^+, x^-\}$.

8. [*22*] Prove that when all the examples are support vectors, the Lagrangian multiples can be obtained by a linear equation.

▸ **9.** [*25*] Discuss the relations between the maximum margin problem, Moore–Penrose pseudo-inversion, and ridge regression with specific reference to the objective functions $\hat{w}^2$ and w^2. What happens in the maximum margin problem formulated according to Eq. (2) if we replace w^2 with $\hat{w}^2$?

10. [*30*] Construct two artificial examples in which the following extreme cases hold true: $|S_=| = 2$ and $|S_=| = \ell$.

11. [*15*] Give a geometrical interpretation of support and straw vectors concerning the solution of MMP.

12. [*22*] Provide qualitative arguments to explain the reason why the objective function in the primal optimization problem given by Eq. (3) is $1/2w^2$ instead of $1/2\hat{w}^2$.

13. [*15*] The MMP can be solved by quadratic programming either by solving optimization problem in Eq. (3) or in Eq. (8). Discuss complexity issues depending on the choice of ϕ.

14. [*15*] Let $k(x_h, x_\kappa) > 0$. Prove that function θ in Eq. (8) has only one critical point, which is a maximum. Then discuss qualitatively the meaning of the constraints.

15. [*20*] Discuss why the solution from b given by Eq. (12) is more stable numerically than that given by Eq. (11).

▸ **16.** [*M18*] Prove that $k(x_h, x_\kappa) \leq k(x_h, x_h) k(x_\kappa, x_\kappa)$.

4.2.2 **Dealing with soft-constraints**

The solution of MMP of the previous section offers an enlightening picture of the method that we can use for the perfect satisfaction of the supervision constraint under a parsimony criterion that corresponds with the maximization of the margin. Unfortunately, it relies on the critical assumption that the constraints can be perfectly satisfied — namely the patterns are assumed to be linearly separable. However, this is a risky guess! How can we really know whether the problem at hand is linearly-separable? Even though features that enlarge the dimension of the input space yield representations that Cover's theorem suggests to be separable with higher probability than the inputs, there's still a chance that we use quadratic programming on a set of hard constraints that are not verified. Here, we see how to extend the proposed optimization framework so as to relax the constraints.

Let us consider the function

$$E_q = \sum_{\kappa=1}^{\ell} V_q(y_\kappa f_q(x_\kappa) - 1) + \frac{1}{2}\|w\|^2 \tag{1}$$

where $V_q(\alpha) = q[\alpha < 0]$. We can think of the solution of MMP as the minimization of E_q as $q \to +\infty$. It is in fact easy to see that for any $q \in \mathbb{N}$ one gets an approximation of MMP, which arises from relaxing the constraints $y_\kappa f(x_\kappa) - 1 \geq 0$ (see Exercise 2).

Now, instead of dealing with the relaxed formulation (1), we discuss a different approach, which turns out to be very effective from a computational point of view. Suppose we introduce *slack variables* ξ_κ, $\kappa = 1, \ldots, \ell$, one for each example, which can be grouped in the space $\Xi \subset \mathbb{R}^\ell$. They are properly used for tolerating the violation of the constraints as follows:

$$\begin{cases} y_\kappa f(x_\kappa) > 1 - \xi_\kappa, \\ \xi_\kappa \geq 0. \end{cases} \tag{2}$$

Clearly, $\xi_\kappa = 0$ returns the previous MMP formulation. When $\xi \in (0, 1)$ we are still *inside the margin* and the solution is still correct, since there is sign agreement $y_\kappa f(x_\kappa) > 0$ with the target. The case $\xi = 1$ corresponds with $f(x_\kappa) = 0$, which is a clear instance of uncertain decision, while $\xi > 1$ represents the strongest constraint relaxation that might lead to errors. The constraints defined by Eq. (2) suggest us to define the following optimization problem:

$$\text{maximize} \quad \frac{1}{2}\|w\|^2 + C\sum_{\kappa=1}^{\ell} \xi_\kappa \tag{3}$$

$$\text{subject to} \quad y_\kappa f(x_\kappa) \geq 1 - \xi_\kappa, \quad \xi_\kappa \geq 0, \quad \kappa = 1, \ldots, \ell.$$

Here $C > 0$ is a proper parameter to express the degree of satisfaction of the corresponding constraints. Interestingly, while we allow their violation, this new optimization problem operates in a larger space including the slack variables, while imposing perfect satisfaction of (2). The introduction of the additional objective function (3) is motivated by the meaning that we can attach to the term $\sum_{\kappa=1}^{\ell} \xi_\kappa$. As already noticed, in case there is an error on x_κ, for the corresponding slack variable ξ_κ we have $\xi_\kappa > 1$. If $\mathcal{E}$ is the set of examples on which the machine takes the wrong decision, we have

$$\sum_{\kappa=1}^{\ell} \xi_\kappa \geq \sum_{\kappa \in \mathcal{E}} \xi_\kappa > |\mathcal{E}|,$$

and therefore $\sum_{\kappa=1}^{\ell} \xi_\kappa$ is an upper bound on the number of errors. Like for the case of hard constraints, we use the Lagrangian approach so that

$$L(\hat{w}, \xi, \lambda) = \frac{1}{2}\|w\|^2 + C\sum_{\kappa=1}^{\ell}\xi_\kappa - \sum_{\kappa=1}^{\ell}(y_\kappa f(x_\kappa) - 1 + \xi_\kappa)\lambda_\kappa - \sum_{\kappa=1}^{\ell}\mu_\kappa \xi_\kappa, \qquad (4)$$

which $\forall \kappa = 1, \dots, \ell$ can be paired with the KKT conditions:

$$\begin{aligned}
&\lambda_\kappa^* \geq 0, \quad y_\kappa f^*(x_\kappa) - 1 + \xi_\kappa^* \geq 0, \quad \lambda_\kappa^*\left(y_\kappa f^*(x_\kappa) - 1 + \xi_\kappa^*\right) = 0, \\
&\mu_\kappa^* \geq 0, \quad \xi_\kappa^* \geq 0, \quad \mu_\kappa^* \xi_\kappa^* = 0.
\end{aligned} \qquad (5)$$

In order to pass to the dual space, we determine the critical points of $L(\hat{w}, \xi, \lambda)$. We have

$$\partial_w L = 0 \Rightarrow w - \nabla_w \sum_{\kappa=1}^{\ell}\lambda_\kappa(y_\kappa(w'\phi(x_\kappa) + b) - 1 + \xi_\kappa) = w - \sum_{\kappa=1}^{\ell}\lambda_\kappa y_\kappa \phi(x_\kappa) = 0,$$

$$\partial_b L = 0 \Rightarrow \sum_{\kappa=1}^{\ell}\lambda_\kappa y_\kappa = 0,$$

$$\partial_{\xi_\kappa} L = 0 \Rightarrow C - \lambda_\kappa - \mu_\kappa = 0.$$

The last condition makes it possible to rewrite the Lagrangian (4) as

$$\begin{aligned}
L(\hat{w}, \xi, \lambda, \mu) &= \frac{1}{2}\|w\|^2 - \sum_{\kappa=1}^{\ell}\lambda_\kappa(y_\kappa \hat{w}'\hat{\phi}(x_\kappa) - 1) + \sum_{\kappa=1}^{\ell}(C - \lambda_\kappa - \mu_\kappa)\xi_\kappa \\
&= \frac{1}{2}\|w\|^2 - \sum_{\kappa=1}^{\ell}\lambda_\kappa(y_\kappa \hat{w}'\hat{\phi}(x_\kappa) - 1).
\end{aligned}$$

Interestingly, we get the same equation as the one of the primal formulation of MMP in case of hard constraints. When replacing $\hat{w}$ in $L(\hat{w}, \xi, \lambda)$, which is independent of

ξ and μ, we end up with the following optimization problem:

$$\text{maximize} \quad \theta(\lambda) = \sum_{\kappa=1}^{\ell} \lambda_\kappa - \frac{1}{2} \sum_{h=1}^{\ell} \sum_{\kappa=1}^{\ell} \lambda_h \lambda_\kappa y_h y_\kappa k(x_h, x_\kappa)$$

$$\text{subject to} \quad 0 \le \lambda_\kappa \le C, \quad \kappa = 1, \dots, \ell, \tag{6}$$

$$\sum_{\kappa=1}^{\ell} \lambda_\kappa y_\kappa = 0.$$

Surprisingly enough, we end up with an optimization problem which nicely matches the one for hard constraints, the only difference being that the nonnegativeness constraints on the Lagrange multipliers are turned into box constraints $0 \le \lambda_\kappa \le C$. This domain is referred to as a C-box. It's worth mentioning that as $C \to +\infty$ this soft-constraining problem is turned into the corresponding hard formulation, for which the box constraint simply becomes the nonnegativeness condition $\lambda_\kappa \ge 0$. Notice that there is no dependency on μ in the dual space. When considering the KKT conditions (5), we can promptly see that, whenever $\lambda_\kappa^* \ne 0$, we are in the presence of support vectors that are defined by

$$\xi_\kappa^* = 1 - y_\kappa f^*(x_\kappa). \tag{7}$$

If $\xi_\kappa^* \ne 0$, that is, if there is a violation of the associated hard constraint, then we have $\mu_\kappa^* = 0$. Hence in this case the optimization yields slack variables in the interior of Ξ, while solutions on the border ($\xi_\kappa^* = 0$) return hard satisfaction of the constraint. The results coming from the formulation in the dual space offer an interesting interpretation of the solution in the free space $W = \mathbb{R}^{d+1}$. The solution of the following optimization in the primal

$$\hat{w}^* \in \underset{\hat{w} \in W}{\arg\min} \left(\frac{1}{2} \|w\|^2 + C \sum_{\kappa=1}^{\ell} (1 - y_\kappa f(x_\kappa))_+ \right), \tag{8}$$

is equivalent to the minimization in the constrained optimization enriched space with the slack variables $(\hat{w}', \xi')'$. This is due to the special structure of the support vectors that is stated by Eq. (7), where we can see the equivalence between the generic slack variable ξ_κ and the value of the associated hinge function loss (see Exercise 1 for a detailed proof).

EXERCISES

1. [*M20*]   Prove that the optimization in the primal over the free domain of function in Eq. (3) is equivalent to the solution of the constrained optimization stated by Eq. (3).

2. [*20*]   Prove the claimed statement concerning the equivalence of MMP with the optimization of Eq. (1). Discuss the solution of the problem arising from any V_q function.

▶ **3.** [*15*]   Consider the soft-constrained formulation of MMP and discuss the solution as $C \to 0$.

4.2.3 Regression

In this section, we use the previously discussed kernel approach for regression. As usual we need to express the environmental constraints that in this case involve tracking tasks. Basically, the supervisor provides pairs (x_κ, y_κ) where this time $y_\kappa \in \mathbb{R}$. Now we wonder what is a natural translation of the strong sign agreement constraint stated by 4.2.1–(2). Let $\epsilon > 0$ and consider the constraint $|y_\kappa - f(x_\kappa)| \le \epsilon$. Clearly, for "small" ϵ this enforces the development of solutions $f(x_\kappa) \simeq y_\kappa$, which is exactly the purpose of regression. The constraint tells us sometime more: Its perfect fulfillment corresponds with an ϵ-insensitive condition, since the supervisor is tolerant with respect to errors until we cross the bound defined by the threshold ϵ. Like for classification, we can introduce slack variables. This time, for each example, the environmental constraints are given by

$$[y_\kappa - f(x_\kappa) \ge 0](y_\kappa - f(x_\kappa) \le \epsilon + \xi_\kappa^+)$$
$$+ [f(x_\kappa) - y_\kappa < 0](f(x_\kappa) - y_\kappa \le \epsilon + \xi_\kappa^-), \tag{1}$$

$$\xi_\kappa^+ \ge 0, \quad \xi_\kappa^- \ge 0, \tag{2}$$

where as usual $f(x) = w \cdot \hat{\phi}(x)$. Hence we formulate the regression problem as optimization of

$$E = \frac{1}{2}\|w\|^2 + C \sum_{\kappa=1}^{\ell} (\xi_\kappa^- + \xi_\kappa^+)$$

under the constraints (1) and (2). The Lagrangian is

$$L = \frac{1}{2}\|w\|^2 + C \sum_{\kappa=1}^{\ell}(\xi_\kappa^- + \xi_\kappa^+) + \sum_{\kappa=1}^{\ell} \lambda_\kappa^+ (y_\kappa - \hat{w}'\hat{\phi}(x_\kappa) - \epsilon - \xi_\kappa^+)$$

$$+ \sum_{\kappa=1}^{\ell} \lambda_\kappa^-(\hat{w}'\hat{\phi}(x_\kappa) - y_\kappa - \epsilon - \xi_\kappa^-) - \sum_{\kappa=1}^{\ell} \mu_\kappa^+ \xi_\kappa^+ - \sum_{\kappa=1}^{\ell} \mu_\kappa^- \xi_\kappa^-.$$

In order to pass to the dual space, we determine the critical points of L with respect to $\hat{w}$ and ξ. We have

$$\partial_w L = 0 \Rightarrow w - \sum_{\kappa=1}^{\ell} (\lambda_\kappa^+ - \lambda_\kappa^-)\hat{\phi}(x_\kappa) = 0,$$

$$\partial_b L = 0 \Rightarrow \sum_{\kappa=1}^{\ell}(\lambda_\kappa^+ - \lambda_\kappa^-) = 0,$$

$$\partial_{\xi_\kappa^+} L = 0 \Rightarrow C - \lambda_\kappa^+ - \mu_\kappa^+ = 0,$$

$$\partial_{\xi_\kappa^-} L = 0 \Rightarrow C - \lambda_\kappa^- - \mu_\kappa^- = 0.$$

If we eliminate $\hat{w}$ and ξ from the Lagrangian, we get

$$\theta(\lambda^+, \lambda^-) = -\frac{1}{2} \sum_{h=1}^{\ell} \sum_{\kappa=1}^{\ell} (\lambda_h^+ - \lambda_h^-)(\lambda_\kappa^+ - \lambda_\kappa^-) k(x_h, x_\kappa)$$

$$- \epsilon \sum_{\kappa=1}^{\ell} (\lambda_\kappa^+ + \lambda_\kappa^-) + \sum_{\kappa=1}^{\ell} y_\kappa (\lambda_\kappa^+ - \lambda_\kappa^-), \tag{3}$$

where $k(x_h, x_\kappa) = \langle \hat{\phi}(x_h), \hat{\phi}(x_\kappa) \rangle$. Like for classification, θ turns out to be independent of the Lagrangian multipliers μ^+ and μ^-. Moreover, since $\mu_\kappa^+ \geq 0$ and $\mu_\kappa^- \geq 0$, function $\theta(\lambda^+, \lambda^-)$ must be maximized over the domain defined by

$$\sum_{\kappa=1}^{\ell} \lambda_\kappa^+ = \sum_{\kappa=1}^{\ell} \lambda_\kappa^-, \quad 0 \leq \lambda_\kappa^+ \leq C, \quad 0 \leq \lambda_\kappa^- \leq C. \tag{4}$$

The role of balancing constraints is that of properly checking the departure from the targets of the function on the positive and negative side. The box constraints play the same role as in classification: Big values of C result in strong enforcement of the constraints, which involve a large range of variation of the reaction constraints λ_κ^+ and λ_κ^-.

Like for classification, we can provide a corresponding interpretation in the primal space by free optimization. In this case, the ϵ-*insensitive* condition $|f(x_\kappa) - y_\kappa| \leq 0$ can be implemented by using an ϵ-*insensitive loss function*, which doesn't return any loss in case of "small" departures from the target. In particular, if we use the *tub loss* $V(x, y, f) := [|y - f(x)| \geq \epsilon](|y - f(x)| - \epsilon)$ and the same parsimonious term as for classification then minimization of

$$E = C \sum_{\kappa=1}^{\ell} [|y_\kappa - f(x_\kappa)| \geq \epsilon](|y_\kappa - f(x_\kappa)| - \epsilon) + \frac{1}{2} \|w\|^2 \tag{5}$$

yields the same solution as minimization of (3) under the constraints (4) (see Exercise 1 for a proof). Notice that for $\epsilon = 0$ this function degenerates to $V(x, y, f) = |y - f(x)|$. This means that this loss becomes sensitive to any error with respect to the target. Notice that the loss $V(x, y, f) = |y - f(x)|$ corresponds with the soft-implementation of the pointwise bilateral constraints $y_\kappa = f(x_\kappa)$. Similar analyses hold for other any p-loss $V(x, y, f) = |y - f(x)|^p$. The case of quadratic loss is analyzed in Exercise 2.

EXERCISES

1. [20]   Prove that the solution in the dual space of regression by the minimization of (3) under the constraints in Eq. (4) corresponds with the minimization in the primal space of (5) in $\mathbb{R}^d$ (risk based on the hub loss).

2. [25]   Based on the kernel framework presented for ϵ-insensitive loss functions, formulate and solve regression for the quadratic loss.

3. [20]   The formulations corresponding to the hub loss produces representations of the solution in the dual space with support vectors. Why doesn't this happen for the square loss? Discuss the effect of ϵ in case of hub loss on the number of support vectors.

4. [15]   Discuss qualitatively the presence of support vectors depending on the choice of the kernel and of the loss function. Are there cases in which there are no straw vectors?

▸ **5.** [20]   Two students use the same software package for quadratic programming, and work on the same learning task by using also the same kernel. However, when reporting their results, while the performances are quite similar, they exhibit a report in which they indicate a different number of support vectors. How can this be possible?

4.3 **Kernel functions**

In the previous section, we have seen that the solution of MMP naturally leads to the notion of kernel. Interestingly, the presence of kernel functions emerge also in case in which we relax the constraints and also for regression when using ϵ-insensitive constraints. In this section we shed light on the notion of kernel and on its relations with the feature map.

4.3.1 **Similarity and kernel trick**

Let us restate the definition of kernel according to Eq. 4.2.1–(7) in a more general form

$$k \colon X \times X \to \mathbb{R}, \quad (x, z) \to \langle \phi(x), \phi(z) \rangle = \phi(x) \cdot \phi(z), \tag{1}$$

which clearly states that a kernel function plays the role of returning a *similarity* measure between any two points in the input space. Here $\langle \cdot, \cdot \rangle$ denotes the inner product in the feature space. It's important to point out a crucial property of kernels, which took some time to be recognized in most applications in machine learning: The input space X doesn't require any special structure. In particular, it is not necessarily a vector space; it's simply a set, which could also be finite. To attach a more natural meaning to this notion of similarity, we can replace $\langle \phi(x), \phi(z) \rangle$ with

$$k_\varphi(x, z) := \frac{\langle \phi(x), \phi(z) \rangle}{\|\phi(x)\| \cdot \|\phi(z)\|} = \cos \varphi(x, z), \tag{2}$$

where $\varphi(x, z)$ is the angle between x and z in the ϕ space. Of course, we have $\forall z \in X$, $k_\varphi(x, z) \leq k_\varphi(x, x)$, which states that no pattern is more similar to x than itself and the kernel is a symmetric function, that is, $k_\varphi(x, z) = k_\varphi(z, x)$. As discussed in Exercise 1, there is no transitivity. Suppose we establish the similarity by choosing a threshold s such that $x \sim z$ (x is similar to z) iff $k_\varphi(x, z) \leq s$. The violation of transitivity means that $(x \sim y) \wedge (y \sim z) \Rightarrow x \sim z$ doesn't hold true. The interpretation of the value returned by a kernel as a similarity measure has an immediate consequence on the meaning of the dual representation of f by Eq. 4.2.1–(9). The learning process returns appropriate values of the constraint reactions so as to select the inputs that are more similar to the incoming pattern x to be processed.

There are at least two classic ways of returning a similarity measure in a vector space. Some kernels arise when reducing k to a function depending on one variable only, $K \colon X \to \mathbb{R}$. Two noticeable cases are when $K(x - z) = k(x, z)$ and $K(x \cdot z) = k(x, z)$. The strong property of kernels is that they return a similarity measure between any two inputs which is based on their mapping to the feature space, without its direct involvement in the computation. This is referred to as the *kernel trick*. The following example highlights this issue. Here we have $X = \mathbb{R}^2$, $H = \mathbb{R}^3$, and

$$\begin{pmatrix} x_1 \\ x_2 \end{pmatrix} \xrightarrow{\phi} \begin{pmatrix} x_1^2 \\ \sqrt{2}x_1 x_2 \\ x_2^2 \end{pmatrix}. \tag{3}$$

In this case, we have

$$\begin{aligned} k(x_h, x_\kappa) &= (x_{h1}^2, \sqrt{2}x_{h1}x_{h2}, x_{h2}^2) \cdot \begin{pmatrix} x_{\kappa 1}^2 \\ \sqrt{2}x_{\kappa 1}x_{\kappa 2} \\ x_{\kappa 2}^2 \end{pmatrix} \\ &= x_{h1}^2 x_{\kappa 1}^2 + 2x_{h1}x_{h2}x_{\kappa 1}x_{\kappa 2} + x_{h,2}^2 x_{\kappa,2}^2 \\ &= (x_{h1}x_{\kappa 1} + x_{h2}x_{\kappa 2})^2 \\ &= \langle x_h, x_k \rangle^2. \end{aligned} \tag{4}$$

This equation indicates that, while $k(x_h, x_\kappa)$ is based on computation in the feature space, it can be equivalently expressed in the input space. This is especially convenient when the dimension of the feature space becomes huge, or even infinite. We can quickly realize that the extension to any input dimension and any polynomial preprocessing leads to a fast growth of the feature space dimension.

EXERCISES

▶ **1.** [*16*] Prove that the transitivity $(x \sim y) \wedge (y \sim z) \Rightarrow x \sim z$ doesn't hold for the similarity defined by Eq. (2).

▶ **2.** [*14*] Let us consider the polynomial kernel $k(x, z) = \langle x, z \rangle^2$. Show that kernel factorization is not unique.

  3. [*16*]   Given the function $k(x, z) = (c + \langle x, z \rangle)^2$ with $c > 0$, show that it is a kernel by exhibiting a feature representation.

4.3.2 Characterization of kernels

In this section, we discuss a fundamental property which allows us to characterize kernels. In order to grasp the idea, we benefit significantly from restricting the analysis on kernel k to the corresponding Gram matrix,

$$K(X_\ell^\sharp) = \begin{pmatrix} k(x_1, x_1) & \dots & k(x_1, x_\ell) \\ \vdots & & \vdots \\ k(x_\ell, x_1) & \dots & k(x_\ell, x_\ell) \end{pmatrix} \in \mathbb{R}^{\ell \times \ell},$$

which is a structured organization of the image of k over a sampling $X_\ell^\sharp = \{x_1, x_2, \dots, x_\ell\}$ of X. In a sense, the Gram matrix is a sort of picture of the kernel function at a certain resolution, which is connected with the cardinality of $X_\ell^\sharp$.

The Gram matrix of a kernel is then a simplification over a certain sample of the inputs. It allows us to replace functional analysis on the kernel with linear algebra on the associated Gram matrix, which contributes to simplifying the discussion, while retaining the essence of the properties.

So far, kernels are in fact functions for which there exists a feature map ϕ such that they can be expressed by Eq. (1).

Now we will see that kernels can also be characterized by the fundamental property of nonnegativeness of the associated Gram matrices on any finite sampling of X. We start by proving that for any such sampling $X_\ell^\sharp$ we have $K(X_\ell^\sharp) \geq 0$. Let us consider the transformation of $X \in \mathbb{R}^{\ell \times d}$ onto the feature space H, which yields

$$\Phi_\ell := \begin{pmatrix} \phi'(x_1) \\ \vdots \\ \phi'(x_\ell) \end{pmatrix} \in \mathbb{R}^{\ell \times D}.$$

From the definition of kernel, we get

$$K(X_\ell^\sharp) = \Phi_\ell \Phi_\ell', \tag{1}$$

which is clearly a nonnegative matrix. This can promptly be seen when considering that $\forall u \in \mathbb{R}^\ell$, if we pose $h_\ell := \Phi_\ell \cdot u$, we have

$$u \cdot K_\ell u = u \cdot \Phi_\ell \Phi_\ell' u = (\Phi_\ell' u)' \cdot (\Phi_\ell' u) = h_\ell^2 \geq 0.$$

In the following, for the sake of simplicity, we drop the dependency on $X_\ell^\sharp$ and use simply the notation K_ℓ for Gram matrices. Now we address the opposite question. Let us assume that a function k evaluated on a given training set $X_\ell^\sharp$ leads to a nonnegative definite Gram matrix. We want to prove that there exists a feature space and a proper feature map ϕ such that the factorization (1) holds true. This is equivalent to stating that such a function, when restricted to $X_\ell^\sharp$, is in fact a kernel. Notice that the factorization that will be constructed is not unique, and that for a given kernel, one can typically associate different feature spaces (see Exercise 2). In particular, we will construct a special feature space with $D = \ell$. We start by noting that $K_\ell = K_\ell'$ is symmetric and, since $K_\ell \geq 0$, it has nonnegative singular values (see Exercise 4 for a reminder). As a consequence, it is similar to the diagonal matrix of singular values $\Sigma = \Sigma^{1/2} \cdot \Sigma^{1/2}$, and if we use the spectral decomposition, and consider that k is also symmetric, we have

$$K_\ell = U \Sigma^{1/2} \cdot \Sigma^{1/2} U' = \Phi_\ell \Phi_\ell', \tag{2}$$

where $\Phi_\ell = U \Sigma^{1/2}$ and U is a square matrix whose columns are eigenvalues of K_ℓ. As a consequence, any coordinate of the Gram matrix can be written as

$(K_\ell)_{h\kappa} = k(x_h, x_\kappa) = \langle \phi(x_h), \phi(x_\kappa) \rangle$, where

$$\phi(x_i) = \begin{pmatrix} \sqrt{\sigma_1} u_{1i} \\ \sqrt{\sigma_2} u_{2i} \\ \vdots \\ \sqrt{\sigma_\ell} u_{\ell i} \end{pmatrix} \in \mathbb{R}^\ell, \quad i = 1, \ldots, \ell. \tag{3}$$

The discovered feature space $\Phi_\ell = U \Sigma^{1/2}$ is derived from the eigenvectors u of the given matrix K_ℓ. Hence the kernel can also be rewritten more explicitly as

$$k(x_h, x_\kappa) = \sum_{j=1}^{\ell} \sigma_j u_{hj} u_{\kappa j} = \phi(x_h) \cdot \phi(x_\kappa) = \langle \phi(x_h), \phi(x_\kappa) \rangle. \tag{4}$$

We can also define the feature map simply by posing $\Phi_\ell = U$, so that the ℓ features associated with any input correspond with the eigenvectors of K_ℓ. In so doing, we can rewrite Eq. (4) as

$$k(x_h, x_\kappa) = \langle \phi(x_h), \phi(x_\kappa) \rangle_\sigma, \tag{5}$$

where $\langle \cdot, \cdot \rangle_\sigma$ is the scalar product on H induced by the matrix Σ. To sum up, if we analyze kernels by the simplified picture offered by the Gram matrix, then we end up with the conclusion that they can be characterized by nonnegativeness. Notice that this is an algebraic property induced by the finiteness of the analysis on samples $X^\sharp$ of X. The kernel theory can in fact be applied in a very general case without making the assumption that X is a vector space. When working on any set $X^\sharp$ the properties of kernels assume a truly algebraic flavor; interestingly, they hold regardless of the sample dimension. This leads us to suspect that the condition of kernel nonnegativeness is a spectral property of the kernel itself also on vector spaces. While the nonnegativeness of k_ℓ on a single sample $X^\sharp$ of X cannot say too much about the function defined on $X \times X$, the uniform nonnegativeness for any sample is clearly a much stronger property! When restricting to the Gram matrix, Exercise 1 suggests the construction of this feature space for a polynomial kernel.

We can go one step further by considering infinite-dimensional feature spaces $(D \to +\infty)$. The spectral analysis on K_ℓ means that $D = \ell$, and therefore the exploration for infinite-dimensional feature in this case is connected with the corresponding analysis on infinite-dimensional input spaces X. Clearly, this is what is mostly interesting in machine learning. Apart from *transductive environments*, our intelligent agents are asked to process patterns that have never been seen before; they need to generalize to new examples. We can deal with enumerable sets X by thinking of a map that associates $\forall \ell \in \mathbb{N}$ the sample $X_\ell^\sharp$ to the corresponding Gram matrix $K(X_\ell^\sharp)$. In this case, any $x \in X$ can be regarded as an index to univocally identifying the feature vector

$$\phi(x) = (\phi_1(x), \phi_2(x), \ldots)' \in \mathbb{R}^\mathbb{N}, \tag{6}$$

which is an eigenvector of K_∞. This corresponds with the kernel factorized as

$$k(x, z) = \langle \phi'(x), \phi(z) \rangle_\sigma = \sum_{i=1}^{\infty} \sigma_i \phi_i(x) \phi_i(z), \tag{7}$$

where $\sigma_i \geq 0$, $x, z \in \mathbb{R}^d$ is any pair of inputs, and $\langle \cdot, \cdot \rangle_\sigma$ is the extension of the inner product defined by Eq. (5) in the space of sequences ℓ^2. An interesting example of the representation given by Eq. (7) is suggested in Exercise 8.

The extension of this spectral analysis to the continuum setting comes out naturally by introducing the functional operator

$$\mathcal{T}_k u(x) = \int_X k(x, z) u(z) \, dz, \tag{8}$$

which replaces the Gram matrix at finite dimension. Of course, we need to extend the associated notion of nonnegative definiteness that will be compactly written as $\mathcal{T}_k \geq 0$. This means that $\forall u \in L^2(X)$,

$$\langle u, \mathcal{T}_k u \rangle = \int_X (\mathcal{T}_k u(z)) \cdot u(z) \, dz = \int_X \left(\int_X k(z, x) u(x) \, dx \right) \cdot u(z), dz$$

$$= \int_{X \times X} k(x, z) u(x) u(z) \, dx dz \geq 0.$$

We can use arguments similar to the case finite dimensions to connect the nonnegative definiteness with the feature factorization. If we plug the representation (7) into the above quadratic form, we get

$$\langle u, \mathcal{T}_k u \rangle = \sum_{i=1}^{\infty} \sigma_i \int_X \int_X \phi_i(x) \phi_i(z) u(x) u(z) dx dz$$

$$= \sum_{i=1}^{\infty} \sigma_i \int_X \phi_i(x) u(x) \, dx \int_X \phi_i(z) u(z) \, dz = \sum_{i=1}^{\infty} \sigma_i \zeta_i^2 \geq 0,$$

where $\zeta_i := \int_X \phi_i(x) u(x) \, dx$. Now we show a constructive method to find a feature factorization of the kernel, once we are given a functional $\mathcal{T}_k \geq 0$. To get an insight on the idea, we remind that in case of finite dimensions, a nonnegative definite Gram matrix, which is analogous to having $\mathcal{T}_k \geq 0$, was factorized using spectral analysis. Now we follow the same principle by looking for eigenvalues and eigenfunctions of the functional equation

$$\mathcal{T}_K \phi_i = \sigma_i \phi_i, \quad i \in \mathbb{N}.$$

The above spectral equation is just the continuous counterpart of $K_\ell u_i = \sigma_i u_i$, which generalizes the already seen feature construction in the discrete setting, the only difference being that while $u_i \in \mathbb{R}^\ell$, ϕ_i belongs to a functional space. The equivalence $\phi_i \sim u_i$ arises whenever we think of sampling function ϕ_i using ℓ examples.

Hence we can construct the infinite dimensional feature vector of Eq. (6) by extending Eq. (3) according to

$$\phi(x_i) = \begin{pmatrix} \sqrt{\sigma_1} u_{1i} \\ \sqrt{\sigma_2} u_{2i} \\ \vdots \\ \sqrt{\sigma_\ell} u_{\ell i} \end{pmatrix} \rightarrow \phi(x) = \begin{pmatrix} \sqrt{\sigma_1} \phi_1(x) \\ \sqrt{\sigma_2} \phi_2(x) \\ \vdots \\ \sqrt{\sigma_\ell} \phi_\ell(x) \\ \vdots \end{pmatrix}. \tag{9}$$

This representation is a well-known result in functional analysis (Mercer's theorem).

Let us assume $k \colon X \times X \to \mathbb{R}$ is a continuous symmetric nonnegative definite function. Then there exists an orthonormal basis $\{\psi_i\}_{i=1}^{\infty}$ of X consisting of the eigenfunctions of k such that the corresponding sequence of eigenvalues $\{\sigma_i\}_{i=1}^{\infty}$ is nonnegative. The eigenfunctions ψ are continuous on X and k admits the representation

$$k(x, z) = \sum_{i=1}^{\infty} \sigma_i \psi_i(x) \psi_i(z).$$

EXERCISES

1. [20] Consider a function k defined on the set $X_4^{\sharp} = \{(0, 0), (1, 0), (1, 1), (0, 1)\}$, which is the same set considered in Exercise 4.2.1–5, and assume that its corresponding Gram matrix is

$$k(X_4^{\sharp}) = \begin{pmatrix} 0 & 0 & 0 & 0 \\ 0 & 1 & 1 & 0 \\ 0 & 1 & 4 & 1 \\ 0 & 0 & 1 & 1 \end{pmatrix}$$

Does this matrix come from a kernel? In that case, what is its Mercer's feature space?

2. [M20] Gram matrices coming from kernels are symmetric matrices. Prove that the eigenvalues of symmetric matrices are real.

3. [M20] Prove that Gram matrices have orthogonal eigenvectors.

4. [M20] Prove that a symmetric matrix is nonnegative definite iff its eigenvalues are non-negative.

5. [M22] Prove Cauchy–Schwarz inequality $|\langle x, z \rangle| \le \|x\| \|z\|$

▶ **6.** [M25] The notion of kernel involves similarity, just like the inner product, which is in fact a kernel in which the feature map is linear, $\phi = \mathrm{id}$. Prove that even though kernels are based on nonlinear feature maps, they still satisfy Cauchy–Schwarz inequality

$$\forall (x, z) \in X \times X, \quad k(x, z) \le \sqrt{k(x, x) k(z, z)}.$$

7. [M18] Discuss the feature extracted when regarding Eq. (2) as the Cholesky factorization of K_ℓ.

▶ **8.** *[23]* Let us consider a kernel $k : \mathbb{R} \times \mathbb{R} \to \mathbb{R}$, which satisfies the *translation invariance* condition $k(x, z) = K(x - z)$. Let $u = x - z$ and assume that

$$K(u) = \sum_{n=0}^{\infty} a_n \cos(nu),$$

where $a_n \geq 0$. Notice that $K(\cdot)$ is an even function, that is, $K(u) = K(-u)$. Prove that $k(\cdot, \cdot)$ is a kernel.

▶ **9.** *[15]* Suppose you are given a collection of feature maps $\{\phi_i : i = 1, \ldots, n\}$, where each of them is associated with a correspondent kernel $k_i = \langle \phi_i, \phi_i \rangle$. Then consider the linear combination defined by the feature map $\phi = \sum_{i=1}^{n} \alpha_i \phi_i$ along with the correspondent kernel $k = \langle \phi, \phi \rangle$. Prove that if $\langle \phi_i, \phi_j \rangle = \delta_{i,j}$ then $k = \sum_{i=1}^{n} \mu_i k_i$, where $\mu_i = \alpha_i^2$.

▶ **10.** *[15]* Let $x, z \in \mathbb{R}$ and $\delta > 0$. Prove that $k(x, z) := [|x - z| \geq \delta]$ is a kernel.

11. *[13]* The solution of the MMP leads to the functional representation given by Eq. 4.2.1–(9). Throughout this section, however, the bias term b of Eq. 4.2.1–(9) is ignored. Explain why we can always ignore the bias term by an appropriate definition of the kernel function.

▶ **12.** *[25]* Once one becomes acquainted with Gram matrices, it is immediately evident that they offer a natural and compact way of computing the outputs over all the training set. We have $f(x) = K_\ell \alpha$, where $\alpha, f(x) \in \mathbb{R}^\ell$. Suppose we have $K_\ell > 0$ then $\det K_\ell \neq 0$ and we can pose learning as the problem of data fitting which satisfies $k_\ell \alpha = y$, where $y \in \mathbb{R}^\ell$ is the vector which collects the supervised values. One can simply find

$$\alpha = K_\ell^{-1} y,$$

which perfectly fits the training set. This circumvents the problem of lack of training data that has been pointed out when using normal equations. Why should we bother with quadratic programming and more complex optimization techniques instead of just using the more straightforward solution of a linear equation?

4.3.3 The reproducing kernel map

In the previous section, we have discussed nonnegativeness of $\mathcal{T}_k$ as a fundamental characterization of kernels. A relevant byproduct is the discovery of a way of creating a feature map for any kernel that, according to Mercer's theorem, is based on the spectral analysis of $\mathcal{T}_k$. In this section we show another classic construction of a feature space associated with a given kernel, namely the one based on the reproducing kernel map. The idea is simple and elegant. In order to construct a feature representation of any pattern $x \in X$, we consider its similarity, as defined by the kernel k, with respect to all other patterns of X. Hence, while Mercer's features are eigenvectors of $\mathcal{T}_k$, in this case, we construct features by invoking the similarity as a central concept. Hence, for any $x \in X$, we construct its associated feature by introducing the following function:

$$\phi_x : X \to \mathbb{R}, \quad z \mapsto \phi_x(z) := k(x, z). \tag{1}$$

Notice that, on any finite set of patterns $X_\ell^\sharp$, this function is a collection of ℓ reals that express the similarity of x_i with respect to all the other patterns of $X_\ell^\sharp$. In this case it makes sense to regard $\phi_{x_i} \in \mathbb{R}^\ell$ as a finite-dimensional Euclidean vector. When looking at the intimate connection between this feature vector and the corresponding kernel, we use the notation $\phi_x(z) = k(x, z)$, while when regarding $k(x, y)$ as a function of y only we pose $k(x, y) = k_x(y)$. Like Mercer's features, the ones defined by Eq. (1) are in infinite-dimensional spaces. However, we now wonder whether this similarity-based feature a true kernel feature. In the affirmative case, we must discover an inner product such that $k(x, z) = \langle \phi_x, \phi_z \rangle_k$; this inner product can be chosen to be

$$\langle \phi_x, \phi_z \rangle_k^0 := k(x, z). \tag{2}$$

Clearly, because of the special definition of Eq. (1), the search for kernel features results in the above condition that only involves the kernel. Hence, the similarity-based features of Eq. (1) involve the *reproducing Hilbert space condition* stated by Eq. (2).

We say that a Hilbert space H of real-valued functions on X is a *reproducing kernel Hilbert space* (RKHS) if it has a function $k\colon X \times X \to \mathbb{R}$ with the following properties: First, for every $x \in X$, k_x belongs to H; second, for every $x \in X$ and for every $f \in H$, $\langle f, k_x \rangle_H = f(x)$. The second property is the reproducing property and k is called a reproducing kernel of H. Now, we want to understand more about $\langle \cdot, \cdot \rangle_k$, which does require us to gain an overall understanding of the space

$$H_k^0 = \left\{ f(x) = \sum_{\kappa=1}^\ell \alpha_\kappa k(x, x_\kappa) : \ell \in \mathbb{N}, \alpha_\kappa \in \mathbb{R} \text{ and } x \in X \right\}. \tag{3}$$

Given $u, v \in H_k^0$, expanded by $u(x) = \sum_{h=1}^{\ell_u} \alpha_\kappa^u k(x, x_h^u)$ and $v(x) = \sum_{\kappa=1}^{\ell_v} \beta_\kappa^v k(x, x_h^v)$, we define $\langle \cdot, \cdot \rangle_k^0$ by

$$\langle u, v \rangle_k^0 := \sum_{h=1}^{\ell_u} \sum_{\kappa=1}^{\ell_v} \alpha_h^u \alpha_\kappa^v k(x_h^u, x_\kappa^v). \tag{4}$$

Notice that if $u = k_x$, $v = k_z$ then we restore definition (2), which allows us to keep the notation $\langle \cdot, \cdot \rangle_k^0$. The question whether $\langle \cdot, \cdot \rangle_k^0$ is an inner product is covered in Exercise 1. It is proven that it possesses symmetry, bilinearity, and positive definiteness, which qualifies it as an inner product. Hence H_k^0 is a *pre-Hilbert space*. A distinctive property of this inner product arises when considering $\langle f, k_x \rangle_k^0$ for any $f \in H_k$. We have

$$\langle f, k_x \rangle_k^0 = \left\langle \sum_{\kappa=1}^\ell \alpha_\kappa k_{x_\kappa}, k_x \right\rangle_k^0 = \sum_{\kappa=1}^\ell \alpha_\kappa \langle k_{x_\kappa}, k_x \rangle_k^0 = \sum_{\kappa=1}^\ell \alpha_\kappa k(x, x_\kappa) = f(x). \tag{5}$$

This reproducing property attaches a special meaning to the defined inner product: For any point x the inner product $\langle f, k_x \rangle_k$ acts like an *evaluation* of f in x. Exercise 2 proposes an example to become acquainted with this inner product, while Exercise 3 goes one step further by proposing a general scheme for creating orthonormal functions in H_k. Now we show how we can build — starting from H_k^0 with $\langle \cdot, \cdot \rangle_k^0$ — the unique RKHS space associated with the kernel k exploiting the following two properties: First (P1), the evaluation functional δ_x (defined by $\delta_x(f) = f(x)$) is continuous on H_k^0; second (P2), all Cauchy sequences that converge to 0 pointwise also converge to 0 in the norm of H_k^0. Exercise 4 shows that properties P1 and P2 hold true. We can then define H_k as the set of functions over X for which there exists a Cauchy sequence $\langle f_n \rangle$ in H_k^0 that converges pointwise to f; of course, $H_k^0 \subset H_k$. Because of the above properties we can find a well-defined inner product on H_k and prove that with this product H_k is indeed the RKHS associated to the kernel k. First of all, let us define the inner product on H_k: Let f and g be two functions in H_k and let $\langle f_n \rangle$ and $\langle g_n \rangle$ be two H_k^0 Cauchy sequences converging pointwise to f and g, respectively. Then we set

$$\langle f, g \rangle_k := \lim_{n \to +\infty} \langle f_n, g_n \rangle_k^0.$$

This product is well defined in the sense that the limit exists and does not depend on the choice of the approximating sequences $\langle f_n \rangle$ and $\langle g_n \rangle$ but only on f and g. Also as one can expect, H_k^0 is dense in H_k. Exercise 5 shows that H_k is complete and therefore is a Hilbert space. Then at last we can show that H_k is indeed an RKHS with the same kernel k that we have used to build H_k^0 since for any $f \in H_k$ if we take $\langle f_n \rangle$ in H_k^0 that converges pointwise to f, we have

$$\langle f, k_x \rangle_k = \lim_{n \to +\infty} \langle f_n, k_x \rangle_k^0 = \lim_{n \to +\infty} f_n(x) = f(x).$$

Notice that H_k is not only an RKHS with kernel k, but it is also the only RKHS with that kernel: Every RKHS with reproducing kernel k must contain H_k^0 since for every $x \in X$, $k_x \in H_k$. Conversely, since H_k^0 is dense in H_k, we have that H_k is the only RKHS that contains H_k^0.

EXERCISES

1. [*M23*]   Prove that Eq. (4) defines an inner product on H_k.

2. [*25*]   Suppose we are given the kernel $k(x, z) = (x \cdot z)^2$. Give two functions in the RKHS H_k of k that are orthonormal.

3. [*M30*]   Extend the results of Exercise 2 to get an orthonormal set starting from the ℓ functions $f_\kappa(x) := x_\kappa \cdot x, \kappa = 1, \ldots, \ell$.

4. [*M27*]   Prove that the space H_k^0 defined in Eq. (3) satisfies the two properties (P1) and (P2).

5. [*M30*]   Prove that H_k is complete.

6. [*M30*]   Prove that H_k is separable.

4.3.4 Types of kernels

Now we discuss most popular kernels and analyze their associated feature space for either finite or infinite dimensions.

Clearly, the simplest kernel is the one which arises when posing $\phi = \text{id}$, so as $k(x, z) = x \cdot z$. For obvious reasons, this is referred to as the *linear kernel*. Working with linear kernels is equivalent to working with linear machines, as they inherit all their features. We can promptly see that learning with linear kernels is related to ridge regression. The perfect match arises when assuming the soft-constraint formulation with the quadratic loss.

The kernel in Eq. 4.3.1–(4) is one of the simplest examples of *polynomial kernels*. When $x, z \in \mathbb{R}^d$, the extension of that example can be compactly expressed by the multiindex notation and the Hadamard's componentwise product (Hadamard componentwise product of vectors $u, v \in \mathbb{R}^d$ is $(u_1 v_1, \ldots, u_d v_d)'$) as follows:

$$
\begin{aligned}
k(x, z) &= \left(\sum_{i=1}^{d} x_i z_i \right)^p \\
&= \sum_{|\alpha|=p} \frac{p!}{\alpha!} (x \circ z)^\alpha \\
&= \sum_{|\alpha|=p} \frac{p!}{\alpha!} \prod_{i=1}^{d} (x_i z_i)^{\alpha_i} \\
&= \sum_{|\alpha|=p} \left(\frac{p!}{\alpha!} \right)^{1/2} \prod_{i=1}^{d} (x_i)^{\alpha_i} \cdot \left(\frac{p!}{\alpha!} \right)^{1/2} \prod_{i=1}^{d} (z_i)^{\alpha_i} \\
&= \left\langle \left(\frac{p!}{\alpha!} \right)^{1/2} \prod_{i=1}^{d} (x_i)^{\alpha_i}, \left(\frac{p!}{\alpha!} \right)^{1/2} \prod_{i=1}^{d} (z_i)^{\alpha_i} \right\rangle_{|\alpha|=p}.
\end{aligned}
\tag{1}
$$

Where here α is a d-dimensional multiindex, that is, a d-tuple $\alpha = (\alpha_1, \ldots, \alpha_d)$ of nonnegative integers for which we use the notations $|\alpha| = \alpha_1 + \cdots + \alpha_d$, $\alpha! = \alpha_1! \cdots \alpha_d!$, and $x^\alpha = x_1^{\alpha_1} \cdots x_d^{\alpha_d}$. The feature vector for the unlabeled example u turns out to be

$$
\phi(u) = \left(\frac{p!}{\alpha!} \right)^{1/2} \prod_{i=1}^{d} (u_i)^{\alpha_i}
\tag{2}
$$

and the cardinality of the feature space is

$$
D = |\{ \alpha : |\alpha| = p \}| = \binom{p + d - 1}{d},
$$

which grows quickly with both d and p.

Gaussian functions offer another classic example of a kernel. Unlike polynomial kernels, only infinite-dimensional feature representations are known. We start proposing one of those representations by using Taylor's expansion in the simple case of $X = \mathbb{R}$. We have

$$
e^{-\gamma(x-z)^2} = e^{-\gamma x^2 - \gamma z^2} \left(1 + \frac{2\gamma xz}{1!} + \frac{(2\gamma xz)^2}{2!} + \cdots \right)
$$

$$= e^{-\gamma x^2 - \gamma z^2} \left(1 \cdot 1 + \frac{\sqrt{2\gamma}}{1!} x \cdot \frac{\sqrt{2\gamma}}{1!} z + \sqrt{\frac{(2\gamma)^2}{2!}} x^2 \cdot \sqrt{\frac{(2\gamma)^2}{2!}} z^2 + \cdots \right).$$

Now if we define

$$\phi(y) = e^{-\gamma y^2} \left(1, \sqrt{\frac{2\gamma}{1!}} y, \sqrt{\frac{(2\gamma)^2}{2!}} y^2, \dots \sqrt{\frac{(2\gamma)^i}{i!}} y^i, \dots \right)' \tag{3}$$

then we have $k(x, z) = G(x, z) = \langle \phi(x), \phi(z) \rangle$. This interpretation of Gaussian kernels makes it possible to think of the corresponding primal form as

$$f(x) = \sum_{i=0}^{\infty} w_i \phi_i(x) = \sqrt{2} \sum_{i=0}^{\infty} w_i \frac{\gamma^{i/2}}{i!} \frac{x^i}{e^{\gamma x^2}}.$$

Interestingly, in this simplified case of one-dimensional inputs, we can regard the Gaussian kernel as one which is generated by polynomial input preprocessing, where the inputs x^i are properly normalized by the factor $e^{\gamma x^2} i!/(\gamma^{i/2})$, which, for any i, yields its vanishing as $x \to \infty$.

Most of the kernels that are of interest in applications obey restrictions on the general structure of k that are already evident in the definition of linear, polynomial, and Gaussian kernels. In particular, linear and polynomial kernels are instances of a general class of kernels that are referred to as *dot product kernels*, which satisfy the condition

$$k(x, z) = K(\langle x, z \rangle) \tag{4}$$

The geometric intuition behind dot product kernels is that the similarity they induce depends on the angle between the vectors x and z, as well as on their magnitude. We can restrict the dependency to the angle when replacing $\langle x, z \rangle$ with $\langle x, z \rangle / (\|x\| \|z\|)$. This kind of similarity is very well suited to deal with information retrieval. In particular, the similarity on any two textual documents x and z, based on vector-based representations that use *tf-idf* (term frequency, inverse document frequency), can naturally be expressed by $\langle x, z \rangle$, as well as on more general functions $K(\langle x, z \rangle)$. The simplification behind dot product kernels is that the functional structure k is broken into two different steps: First, x and z are given the similarity measure $\langle x, z \rangle$ by the dot product and then such a similarity is enriched by the composition with function K. In so doing, we dramatically reduce the complexity of establishing whether k is a kernel, since we are reduced to considering K. This issue is covered in Exercise 1.

Another fundamental restriction on k is that of assuming

$$k(x, z) = K(x - z). \tag{5}$$

Gaussian kernels are a classic instance of this general notion of translation invariant kernels. Notice that one can go one step further by restricting k to obey the condition

$k(x, z) = K(\|x - z\|)$. In this case we say that $k(x, z)$ is a *radial kernel*. Exercise 8 invites the reader to discuss an interesting property of K which guarantees that k is a kernel. Like for dot product kernels, the underlying idea of splitting the computation of function k into two steps reduces the complexity. However, it's important to point out that both these simplifications make strong assumptions on the emergence of the appropriate feature representation. In particular, in case of radial kernels (see, e.g., Gaussian kernels) we are still in front of the curse of dimensionality issues that have been pointed out in Section 1.1 [see Eq. 1.1.4–(2)].

Another classic example of translationally invariant kernels are the B_n-splines, where

$$k(x, z) = B_{2p+1}(\|x - z\|), \quad \text{with} \quad B_n(u) := \bigotimes_{i=1}^{n} [|u| \le 1/2], \qquad (6)$$

where $\bigotimes_{i=1}^{n}$ is the n-fold convolution of the characteristic function of the interval $[-1/2, 1/2]$, and $\bigotimes_{i=1}^{0} [|u| \le 1/2] := [|u| \le 1/2]$. Here n denotes the number of times the convolution operator is repeated in the n-fold convolution. It can be proven that $k(\cdot, \cdot)$, defined by (6) is in fact a kernel (see Exercise 3). As pointed out in Exercise 4, B-spline kernels approximate Gaussian kernels as $n \to +\infty$.

The analysis of translation invariance and on dot product kernels indicates methods for creating kernels by appropriate composition. This is clearly part of a more general picture that gives insight on *kernel design*. It's quite easy to see that, given $\alpha > 0$, any two kernels k_1, k_2, then

$$k(x, z) = k_1(x, z) + k_2(x, z), \quad k(x, z) = \alpha k_1(x, z), \quad k(x, z) = k_1(x, z) \cdot k_2(x, z),$$
$$(7)$$

are also kernels. The proof is given in Exercise 5.

EXERCISES

1. [*M30*]   Consider the function $k(x, z) = K(\langle x, z \rangle)$. Give conditions on K which guarantee that k is a kernel.

2. [*M30*]   Let k be a translation invariance kernel such that $k(x, z) = K(x - z)$. Prove that if $\hat{K}(\omega) \ge 0$ then k is a kernel.

3. [*M22*]   Prove that k defined by Eq. (6) is a kernel.

4. [*M25*]   Prove that the function g defined by

$$g(u) := \lim_{n \to +\infty} B_n(u)$$

is a Gaussian.

5. [*M23*]   Prove compositional properties of kernels stated by Eq. (7).

6. [*HM45*]   The formulation of learning with SVM, which corresponds with the maximum margin problem, leads to an optimization problem that is local minima free. Discuss the complexity of learning under the terminal attractor model described in Section 3.4.

4.4 Regularization

In this section we introduce learning as a regularization problem and show that its solution is given by a kernel expansion on the training set. While the basic idea behind regularization has been already spread throughout the book, here we present a more detailed view of regularization which begins with remarks on the compactness of the class of solution functions $H = \{f : X \to \mathbb{R}\}$ which characterize our intelligent agents. Among others, we met regularization issues in linear machines when discussing the different behavior of normal equations depending on the relation between the number of examples ℓ and the input dimension d. In the case $d + 1 > \ell$ normal equations admit $\infty^{d+1-\ell}$ solutions, which corresponds with an ill-position of learning. Clearly, the linear machine is characterized by a function defined $\forall x \in X set$ as $f(x) = \hat{w} \cdot \hat{x}$, which is not in a compact set, since $\hat{w}$ with arbitrary large values are possible! This is in fact what we must avoid: The values of the weights must not be too big. This idea also emerged in Section 3.1 when we began promoting the parsimony principle by introducing the regularized risk [see Eq. 3.1–(10)]. When weights are kept small, we implicitly enforce stability, so that small changes of the inputs yield small changes on regression. The boundedness on $\|w\|$ was also the outcome of a simplest reformulation of the maximum margin problem in Section 4.2 [see Eq. 4.2.1–(2)]. Hence parsimony and stability issues both suggest boundedness of the weights, which is also the outcome of the geometrical robustness principle of MMP.

The formulation of learning problems as simple pointwise constraint satisfaction may give rise to ill-conditioning when we don't possess enough information. This results in noncompact H with infinite solutions, whereas one is typically interested in the definition of a well-defined behavior of the agent. How can we address this problem? Of course, one could search the solution in a compact set by imposing the bounds $\|f\| \leq F$, with $F > 0$. This constraint could be added to the others deriving from the learning environment to contribute towards the definition of a unique solution of optimization. However, it's quite easy to realize that F would be quite an arbitrary value and that the adoption of the parsimony principle, which is translated by the minimization of the objective function $\|f\| \leq F$, under the environmental constraints definitely offers a more sound framework for learning.

4.4.1 Regularized risks

In Section 2.4, we have already introduced the regularized risk by connecting its definition to the MDL principle. It became clear that we can translate the parsimony into an appropriate selection of metrics in the functional space H. When regarding this space as the RKHS of a given kernel then it makes sense to define $\|f\|_k$ as the measure of parsimony. Hence, given a training set of supervised pairs, learning can be regarded as the problem of minimizing $\|f\|_k$ under the soft-constraints deriving from the training set. Formally,

$$f^* \in \arg\min_{f \in H_k} \left(E_{\text{emp}}(f) + \mu P(f) \right) \tag{1}$$

Here, $E_{\text{emp}}(f)$ is the empirical risk, μ is the *regularization parameter*, while $P(f)$ is the *parsimony term*. It's quite obvious that there are in fact many degrees of freedom in this problem coming from both the choice of the loss function V and of the parsimony term. We have already addressed different choices of $E_{\text{emp}}(f)$, which comes from the selection of the corresponding loss function in Section 2.1.1. Now we discuss two possible choices for $P(f)$.

First, we start considering H_k as the ambient for the solution of (1). Basically, this means that we believe in a given kernel k, along with its connected issues on similarity and feature map, and we explore solutions in the dual space, that is, in the reproducing Hilbert space of kernel k. If $f(x) = \sum_{\kappa=1}^{\ell} \alpha_\kappa k(x, x_\kappa)$ then the measurement of parsimony in H_k is

$$\|f\|_k^2 = \sum_{\kappa=1}^{\ell} \sum_{h=1}^{\ell} k(x_h, x_\kappa) \alpha_h \alpha_\kappa. \tag{2}$$

The parsimony term $P(f) = \bar{P}(\|f\|_k^2)$ — where $\bar{P}$ is a monotone function — keeps the penalization of nonparsimonious behavior as $\|f\|_k^2$ increases. An interesting issue related to the presence of the bias term b in the kernel expansion is proposed in Exercise 1. It's worth mentioning that while the regularization term in the primal space only involves $\|w\|^2$, in the dual space all unknown quadratic terms $\alpha_h \alpha_\kappa$ are involved. That is, while in the primal the bias term doesn't contribute to composing the parsimony term, in the dual space all the constraint reactions are involved. An in-depth analysis on the connection between the solution of the MMP in case of soft-constraints and the minimization of $\|f\|_k^2$ according to Eq. (2) is discussed in Exercise 2. The connection arises when posing $\alpha_\kappa := y_\kappa \lambda_\kappa$, so that $\|f\|_k^2 = -\theta(\lambda)$, that is the (sign-flip) objective function of the quadratic programming problem for the solution of MMP according to the formulation given in Eq. (8). Notice that this notion of parsimony is interwound with the assumption that $f(x) \in \text{span}\{k(x, x_1), \ldots, k(x, x_\ell)\}$.

In Section 2.4, we have introduced a different notion of parsimony which is connected with the smoothness of the solution f. The idea is that of thinking in terms of description length of f. It was pointed out that it is easy to predict f when it doesn't change, that is, when $Df = 0$. The parsimony term can be regarded as a sort of description length $P(f) = L(f) = \langle Df, Df \rangle$. A closer look at this definition immediately suggests that it doesn't fully capture other relevant information about f. Clearly, Df only captures an important issue of f, but all derivatives of any order are clearly important — including $D^0 f = f$. This remark suggests replacing Df with richer m-th order differential operators $Pf = \sum_{i=0}^{m} \pi_i D^i$, where $\pi_i \in \mathbb{R}$. These operators are referred to as *regularization operators*. As shown in Appendix B, the order m can become infinite. In so doing, we can consider the regularization term

$$\|f\|_P^2 := \langle Pf, Pf \rangle, \tag{3}$$

which can replace $\|f\|_k$ in Eq. (1). We are in front of another functional problem, where again the unknown is a function.

The regularization problems which arise from these two different definitions of the parsimony term are treated in Sections 4.4.2 and 4.4.3 for parsimony terms based on the norm in RKHS, and Section 4.4.4 for parsimony terms based on regularization operators.

4.4.2 Regularization in RKHS

The formulation of learning according to Eq. 4.4.1–(1) requires us to discover the solution on H_k, which is a functional space. This is significantly different with respect to what we have seen so far! Linear and linear-threshold machines are characterized by a finite set of unknown learning parameters, while their structure is fixed in advance. The search in H_k is much more ambitious, since the underlining assumption is that the functional space derives from the generating kernel, but the functions can be generated by with $f(x) \in \text{span}\{k(x, x_1), \ldots, k(x, x_\ell)\}$, where there is no restriction on ℓ. This shouldn't be confused with the fact that we formulate supervised learning over a finite set of supervised pairs. Strictly speaking, while the mathematical and algorithmic framework that has been used so far relies on the discovery of a vector of finite dimension, here we look for functions, which can be thought of as elements of infinite-dimensional spaces. Surprisingly enough, the discovery of f^*, according to Eq. (1), needn't involve classic variational methods, since we can reduce the problem to finite dimension.

In order to see how this simplification arises, let $f \in H_k$ and consider the learning process over a training set of ℓ examples. Then we can construct the kernel expansion over the examples

$$f_\|(x) = \sum_{i-1}^{\ell} \alpha_i k(x, x_i).$$

While changing the coefficients α_i, we generate the space H_k^ℓ, while its orthogonal complement $H_k^\perp = H_k \setminus H_k^\ell$ is composed of functions $f_\perp$ which are orthogonal to any $f_\|$. Hence, we can decompose f as $f = f_\| + f_\perp$. In order to calculate the regularized risk, we first need to determine the loss, which requires us to express $f(x_\kappa)$. We have

$$f(x_\kappa) = \langle f, k_{x_\kappa} \rangle = \langle f_\| + f_\perp, k_{x_\kappa} \rangle = \langle f_\|, k_{x_\kappa} \rangle + \langle f_\perp, k_{x_\kappa} \rangle = f_\|(x_\kappa), \qquad (1)$$

where in order to simplify the notation, we dropped H in the inner product $\langle \cdot, \cdot \rangle_H$. For the regularization term we have $\|f\|^2 = \|f_\||^2 + \|f_\perp\|^2$ and therefore

$$P(f) = \bar{P}(\|f\|_k^2) = \bar{P}\left(\left\|\sum_{i=1}^{\ell} \alpha_i k_{x_i}\right\|^2 + \|f_\perp\|^2\right). \qquad (2)$$

Here, as stated when defining the regularization term, we assume that $\bar{P}(\cdot)$ is a monotone function. Now we are ready to determine the minimum of the regularized risk.

From Eqs. (1) and (2), we get

$$
f^* \in \underset{f \in H_k}{\arg\min} \left(\sum_{\kappa=1}^{\ell} V(x_\kappa, y_\kappa, f(x_\kappa)) + \mu \bar{P}(\|f\|_k^2) \right)
$$

$$
= \underset{f \in H_k}{\arg\min} \left(\sum_{\kappa=1}^{\ell} V(x_\kappa, y_\kappa, f_\|(x_\kappa)) + \mu \bar{P}(\|f_\|\|^2 + \|f_\perp\|^2) \right)
$$

$$
= \underset{f \in H_k^\ell}{\arg\min} \left(\sum_{\kappa=1}^{\ell} V(x_\kappa, y_\kappa, f_\|(x_\kappa)) + \mu \bar{P}(\|f_\|\|^2) \right) \ni f_\|^*.
$$

Hence the minimization of the regularized risk is represented by

$$
f^*(x) = f_\|^*(x) = \sum_{\kappa=1}^{\ell} \alpha_\kappa^* k(x - x_\kappa). \tag{3}
$$

The representation of the solution f^* turns out to be simply an expansion of kernel k at the points of supervision, a property that naturally arises because of the reproducing property of k and of the way we measure the parsimony. Notice that in order to state this property we didn't make any special assumption on the empirical risk. For instance, the kernel expansion property holds also for nonconvex risk functions! While this is a strong property, the role of convexity emerges early when we start thinking about the related algorithmic framework (see Section 4.4.3).

The kernel expansion in Eq. (3) is of crucial importance from a computational point of view, since, as shown in the next section, we can reformulate learning in $\mathbb{R}^\ell$, which is finite-dimensional. The collapse of dimensionality is a fundamental property that arises in the presence of a pointwise constraint, and plays a crucial role in speeding up learning algorithms.

4.4.3 Minimization of regularized risks

The collapse of dimensionality that arises from the kernel expansion stated by Eq. 4.4.2–(3) offers a straightforward scheme for proposing a learning algorithm. If we plug f^* as given by Eq. 4.4.2–(3) into the regularized risk defined by Eq. 4.4.1–(1), we get

$$
E(f) = E_{\text{emp}}(f) + P(f) = E_{\text{emp}}^\alpha(\alpha) = V(\alpha) + \mu \bar{P}(\alpha' K \alpha),
$$

where V is defined by $V(\alpha) = \sum_{\kappa=1}^{\ell} V(x_\kappa, y_\kappa, \sum_{h=1}^{\ell} \alpha_h k(x_\kappa - x_h))$ and $P = \mu \bar{P}$, being μ the regularization parameter ($\mu = C^{-1}$ see Eq. 4.2.2–(3)). This equation makes it possible to convert the problem of learning $f \in H_k$ into the finite-dimensional optimization problem

$$
\min_\alpha \left(V(\alpha) + \mu \bar{P}(\alpha' K \alpha) \right). \tag{1}
$$

As already pointed out, the solution of this problem can be hard if E^{α}_{emp} is not convex. In case of quadratic loss and $\bar{P} = \text{id}$ we have $V(\alpha) = (1/2)\|y - K\alpha\|^2$. The optimal value can be determined by imposing $\nabla_{\alpha} E^{\alpha}_{\text{emp}}(\alpha) = 0$, from which we derive

$$\alpha^* = (\mu \, \text{Id} + K)^{-1} y. \tag{2}$$

When restricting to linear kernels, this is exactly the same result discovered in Chapter 3, when we introduced the notions of primal and dual spaces.

In general, we cannot express explicitly the solution as in Eq. (2) and therefore we need numerical solutions. Since, as usual, most interesting problems in machine learning involve very high dimension, gradient descent is typically the most common approach for determining α^*. Notice that the updating $\alpha \leftarrow \alpha - \nabla_{\alpha} E^{\alpha}_{\text{emp}}(\alpha)$ is a truly batch mode scheme, and it can be carried out according to ordinary guidelines in numerical analysis, as well as by following the analysis carried out for linear threshold machines in Section 3.4. However, when paralleling this learning scheme with what has been discussed for linear threshold machines, one becomes early curious to explore the new face of online learning corresponding with dual representations. If we strictly follow the dual representation of f, we get in trouble early! The problem is that the memory for the accumulation of data explodes, and therefore we need to think of updating mechanisms to avoid memorizing of the inputs. This is reasonable, since it is quite obvious that we can get rid of many examples that are not crucial for learning. This surely happens when the loss function generates support vectors (see Exercises 4.2.3–4 and 4.2.3–5). While there is already a rich literature on online learning for kernel methods, namely functions in dual representations (see Section 4.5), we notice that primal representations offer a more natural structure for online updating. There's more! The interpretation of kernels given in the following section suggests that, at least for an important class of kernels, we need to reformulate kernel methods in a more general framework, which is suitable for online learning even in the perspective of processing of temporal information. (See Sections 1.1, 1.2, and 1.4 to appreciate the dichotomy between online learning and temporal learning.)

4.4.4 Regularization operators

In Section 3.1, we have introduced a parsimony term which is based on the principle of preferring a smooth solution. This idea has been developed in Section 4.4.1 by the introduction of differential operators to implement minimum description length solutions [see Eq. 4.4.1–(3)]. This opens the door to a new framework of supervised learning that is based on discovering functions that minimize

$$E(f) = \sum_{\kappa=1}^{\ell} V(x_{\kappa}, y_{\kappa}, f(x_{\kappa})) + \frac{1}{2}\mu\langle Pf, Pf \rangle. \tag{1}$$

The minimization of E is a way of implementing the general principle of discovering a parsimonious solution under environmental soft-constraints. The parsimony is guaranteed by the objective functions, while the constraints are enforced, with weight $C = 2\mu^{-1}$, by the empirical risk. The wide range of loss functions paired with the many possible choices of the regularization operator P makes the minimization of E quite a complex problem to handle — though all those choices are still framed under the same general principles. We notice in passing that this formulation is an amenable formulation for incorporating any eventual boundary constraints on f. This is something new with respect to what has been covered so far, yet it's of fundamental importance in the real world. In fact, one might be interested in discovering bounded solutions in any point of the domain (see, e.g., the discussion in Section 5.3.3). It's quite clear that as we add boundary constraints, we restrict the class of admissible solutions, which results in richer interpretations of the learning environment. An intriguing example related this discussion, from a truly analytical point of view, is given in Exercise 3.

Now let us consider the variation of the regularized risk (1). In a sense, this is a sort of gradient computation for infinite dimensions. We consider the variation of the risk and of the regularization terms separately. Function $f : X \to \mathbb{R}$ is replaced with $f + \epsilon h$, where $\epsilon \in \mathbb{R}$, while h is a variation. Because of linearity

$$\delta \sum_{\kappa=1}^{\ell} V(x_\kappa, y_\kappa, f) = \sum_{\kappa=1}^{\ell} \delta V(x_\kappa, y_\kappa, f),   \tag{2}$$

which reduces to the calculation of $\delta V(x_\kappa, y_\kappa, f) = V'_f(x_\kappa, y_\kappa, f)\epsilon h$. This can be given the following distributional interpretation:

$$\delta \sum_{\kappa=1}^{\ell} V(x_\kappa, y_\kappa, f) = \int_X \sum_{\kappa=1}^{\ell} V'_f(x_\kappa, y_\kappa, f(x))\delta(x - x_\kappa)\epsilon h(x)\, dx$$
$$= \epsilon \left\langle \sum_{\kappa=1}^{\ell} V'_f(x_\kappa, y_\kappa, f)\delta_{x_\kappa}, h \right\rangle,   \tag{3}$$

where $\delta_{x_\kappa}(x) = \delta(x - x_\kappa)$. Now we calculate $\delta\langle Pf, Pf \rangle$ and get

$$\delta\langle Pf, Pf \rangle = \langle P(f + \epsilon h), P(f + \epsilon h) \rangle - \langle Pf, Pf \rangle = 2\epsilon\langle Pf, Ph \rangle + o(\epsilon).   \tag{4}$$

While the risk term is factorized with an explicit h term, the regularized term contains Ph. This is common with these functionals, but we can get around this problem by using a classic trick, which arises from by part integrations — in case of $P = D$ (reduction of the general differential operator P to prime derivative). Here, things are more involved, but the unified notion of adjoint operator comes to help. We have

$$\langle Pf, Ph \rangle = \langle P^*Pf, h \rangle   \tag{5}$$

provided that

$$\forall \kappa = 0, 1, \ldots m - 1, \quad [x \in \partial X]h^{(\kappa)}(x) \equiv 0.   \tag{6}$$

Now if we define $L = P^*P$ then we can express the condition $\delta E(f) = 0$ by using Eqs. (3)–(5). If we use the fundamental lemma of variational calculus, then we have

$$Lf(x) + \frac{1}{\mu}\sum_{\kappa=1}^{\ell} V'_f(x_\kappa, y_\kappa, f(x))\delta(x - x_\kappa) = 0. \tag{7}$$

In this differential equation, the agent computes a value $f(x)$ whose match on supervised pairs (x_κ, y_κ) is measured by $V'_f(x_\kappa, y_\kappa, f(x_\kappa))$. We can think of the input as a distribution $\delta(x - x_\kappa)$, $\kappa = 1, \ldots, \ell$ on points x_κ.

Let $g(x, x_\kappa)$ be the Green function of differential operator L, that is,

$$Lg(x, x_\kappa) = \delta(x - x_\kappa).$$

Whenever[2] $g(x, x_\kappa) = g(x - x_\kappa)$, we say that the Green function is *translation invariant*. Since the equation is linear, we can use the superposition principle to state that the function

$$f_{\parallel}(x) = \sum_{\kappa=1}^{\ell}\alpha_\kappa g(x, x_\kappa) \tag{8}$$

is a solution of (7). Now we can parallel the analysis of Section 4.4.3. When plugging $f_{\parallel}$ into Eq. (7), we get

$$\sum_{\kappa=1}^{\ell}\left(\alpha_\kappa + \frac{1}{\mu}V'_f(x_\kappa, y_\kappa, f(x))\right)\delta(x - x_\kappa) = 0.$$

This is always satisfied for $x \neq x_\kappa$. The satisfaction on the training set yields $\alpha_\kappa + (1/\mu)V'_f(x_\kappa, y_\kappa, f(x_\kappa)) = 0$. When using Eq. (8), finally, $\forall \kappa = 1, \ldots, \ell$, we get

$$\alpha_\kappa + \frac{1}{\mu}V'_f\left(x_\kappa, y_\kappa, \sum_{h=1}^{\ell}\alpha_h g(x_\kappa, x_h)\right) = 0. \tag{9}$$

If the loss function is quadratic, that is, $V(x_\kappa, y_\kappa, f(x_\kappa)) = (1/2)(y_\kappa - f(x_\kappa))^2$ then $V'_f(x_\kappa, y_\kappa, f(x_\kappa)) = (f(x_\kappa) - y_\kappa)$. When plugging into Eq. (9), we get

$$\mu\alpha_\kappa + \sum_{h=1}^{\ell}\alpha_h g(x_\kappa, x_h) = y_\kappa. \tag{10}$$

Interestingly, once we rewrite this equation by using matrix notation, we realize that it is the same as Eq. 4.4.3–(2), since we get $(\mu\,\mathrm{Id} + G)\alpha = y$, where G is the

---

[2] For sake of simplicity, we overload the symbol g to represent two different functions.

Gram matrix of the Green function g. Of course, in order to get the unique solution $\alpha = (\mu\,\mathrm{Id}+G)^{-1}y$, we need the nonsingularity of matrix $\mu\,\mathrm{Id}+G$. In Section 3.1.3, a related problem has been proposed for in the primal form for ridge regression (see Exercise 3.1.3–1). Because of the same algebraic structure, we conclude that the non-negativeness of G is sufficient to guarantee the invertibility. In addition, it becomes clear that Eqs. (2) and (10) are the same whenever $G = K$, that is, when the kernel and Green function coincide. The reduction to the solution of a system of equations is another instance of collapse of dimensionality that clearly holds also for nonquadratic loss functions, since the property is generally stated in Eq. (9). However, when abandoning quadratic loss, Eq. (9) becomes nonlinear and therefore its solution can be remarkably more complex. Of course, like for the analysis with kernels, we can use the principle of collapsing of dimensionality by plugging Eq. (8) into the regularized risk definition given by Eq. (1). While we can replace directly the expression of f in the loss, we need to express $\langle Pf, Pf \rangle$ directly as a function of the finite-dimensional vector α. To this purpose, notice that

$$\langle Pf, Pf \rangle = \langle P \sum_{h=1}^{\ell} \alpha_h g(x - x_h), P \sum_{\kappa=1}^{\ell} \alpha_\kappa g(x - x_\kappa) \rangle$$

$$= \sum_{h=1}^{\ell}\sum_{\kappa=1}^{\ell} \alpha_h \alpha_\kappa \langle Lg(x, x_h), g(x, x_\kappa) \rangle = \sum_{h=1}^{\ell}\sum_{\kappa=1}^{\ell} g(x_h, x_\kappa)\alpha_h\alpha_\kappa$$

and therefore we end up with the finite-dimensional optimization problem

$$\alpha^* \in \arg\max_{\alpha} \left(\sum_{\kappa=1}^{\ell} V\left(x_\kappa, y_\kappa, \sum_{h=1}^{\ell} \alpha_h g(x_\kappa - x_h)\right) + \mu\alpha'G\alpha \right). \tag{11}$$

The connection between kernels and Green function has its roots in the links between the representations of f given in Eqs. (3) and (8). In order to better grasp the connections between kernels and Green functions related to supervised learning, we need to shed light on an issue that has been purposely ignored so far. First, notice that the analysis relies on the strong boundary assumption stated by (6). Second, even though we restrict to this assumption, there are uniqueness issues that are still open. Can we have different solutions of Eq. (7) under the boundary conditions (6)? If two functions, f_1 and f_2, exist, then we must have $L(f_2(x) - f_1(x)) = Lu(x) = 0$ where $u(x) := f_2(x) - f_1(x)$. Hence the unicity comes out when the kernel of L only contains the null function, that is $\mathcal{N}L = \{0\}$.

EXERCISES

▶ **1.** [*18*] Consider the definition of $\|f\|^2$ given in Eq. 4.4.1–(2) and the case in which $f(x) = \sum_{\kappa=1}^{\ell} \alpha_\kappa k(x, x_\kappa) + b$ with $b \neq 0$. How can we redefine $\|f\|^2$?

2. [*22*] Discuss the relation between the formulation of learning in Eq. 4.4.1–(1), where $\|f\|_k$ is defined according to Eq. 4.4.1–(2), with MMP with soft-constraints when choosing the hinge loss function.

3. [*M25*]   Consider the minimization of the functional

$$E(f) = \int_{-1}^{+1} f^2(x)\big(1 - f'(x)\big)^2 dx$$

where the function $f : [-1, 1] \to \mathbb{R}$ is expected to satisfy the boundary conditions $f(-1) = 0$ and $f(0) = 1$. Prove that there is no solution in $C^1([-1, 1]; \mathbb{R})$, whereas, when allowing one discontinuity of f', the function $f(x) = x[x \geq 0]$ minimizes E.

▶ **4.** [*M26*]   Let us consider the case of functions $f : \mathbb{R} \to \mathbb{R}$ and the differential operator

$$L = \sum_{\kappa=0}^{+\infty} (-1)^\kappa \frac{\Sigma^{2\kappa}}{\kappa!2^\kappa} \frac{d^{2\kappa}}{dx^{2\kappa}},$$

which is associated with the Gaussian kernel. Prove that $\lim_{x \to 0} Lg = \infty$.

5. [*M15*]   Prove that for any given polynomial kernel $k(x - z)$, there is no regularization operator $L = P^*P$ such that $Lk(\alpha) = \delta(\alpha)$, that is, there is no associated Green function to $k(x - z)$. The proof can promptly be given by contradiction. Let L be any candidate regularization operator of the polynomial Green function k. Clearly, because of the regularities of k, we have $Lk \neq \delta$.

6. [*M16*]   Consider the perceptual space $X = \mathbb{R}$ and the regularization operator $P = D^2$. Prove that $L = D^4$ and that $g(x) = |x|^3$

7. [*M23*]   Given $L = (\Sigma^2 \operatorname{Id} - \nabla^2)^n$ prove that its Green function is a spline.

4.5 Scholia

Section 4.1 Kernel machines strongly emerged in mid-1990s as an alternative to the connectionist wave raised in mid-1980s. Amongst others, reasons for sustaining this alternative approach were rooted in the more solid mathematical foundations that characterize kernel machines. The efficiency of learning, along with the generalization capabilities due to the regularization framework in which they are defined, raised attention in the machine learning community, but also in related fields where connectionist models were massively applied. The field of kernel machines has been the subject of a systematic description also in textbooks [see *An Introduction to Support Vector Machines: And Other Kernel-Based Learning Methods* (Cambridge University Press, 2000) and *Kernel Methods for Pattern Analysis* (Cambridge University Press, 2004) by N. Cristianini, J. Shawe-Taylor and *Learning with Kernels: Support Vector Machines, Regularization, Optimization, and Beyond, Adaptive Computation and Machine Learning* (MIT Press, 2002) by B. Schölkopf, A.J. Smola].

The idea of feature enrichment by polynomial preprocessing has been proposed early in the 1960s by a number of people [see, e.g., the seminal books by N.J. Nilsson *Learning Machines* (McGraw-Hill, 1965) and the book by R.O. Duda and P.E. Hart, subsequently published also with the contribution of D.G. Stork *Pattern Classification* (Wiley, 2001)]. The proposal of the Boolean counterpart of feature enrichment was implicitly advocated mostly in seminal book *Perceptrons: An Introduction to Computational Geometry* (Cambridge: MIT Press, 1969) by M. Minsky and S. Papert at the end of the 1960s.

Feature enrichment is in fact stimulated by the limitation of the linear separability assumption. However, its concrete impact in real-world learning tasks is sometimes neglected — some learning tasks in applications are in fact linearly or nearly-linearly separable. The analysis that leads to determining the probability of linear separation was proposed by Thomas Cover in his PhD thesis *Geometrical and Statistical Properties of Linear Threshold Devices* (Stanford Electronics Labs., 1964). Issues on invariance in feature extraction has been the subject of attention for years, since people early realized their role in the complexity of learning. Like in most signal processing tasks, in speech recognition, it was early clear that classifiers could have hard time in directly processing the input signal. The act of speech generation is in fact controlled by the brain which sends appropriate signals to the speech articulators. The message to be decoded is in fact related to this sequence of commands, whereas signal production involves the physical process of signal transmission. The short-time Fourier transform and the linear prediction filter techniques create a sequence of frames — every 10–20 ms — that represent features with a remarkable degree of invariance with respect to all the utterance information apart from the corresponding textual transcription. Most challenging issues of invariance, however, can be found in computer vision, where most of the emphasis has been placed on translation, rotation, and scale invariance. The seminal work by Lowe on Scale Invariant Feature Transform (SIFT) [D.G. Lowe, *Int. J. Comput. Vis.* **60** (2004), 91–110] offers one the most effective methods to extract useful features for object recognition. A more general view on invariance, that is framed in group theory, is given in the technical report *On Invariance and Selectivity in Representation Learning* (MIT, 2015) by F. Anselmi, L. Rosasco, T. Poggio. In order to provide an assessment of the generalization to new examples, a good way of expressing the capacity is to rely on the notion of *VC dimension*. It is based on the following idea. Suppose we are given a collection of examples and consider a function f taken from a certain class, which separates the patterns in a precise way, thus yielding a corresponding labeling. Of course, there are 2^ℓ different ways of labeling the given ℓ examples. A machine which supports a very rich class of functions might be able to implement all the possible 2^ℓ separations, in which case it is said to *shatter* all the ℓ points. In general, however, not all examples are shattered. The VC dimension of a certain class of function is the largest ℓ such that there exists a set of ℓ points that can be shattered by the class of function — $VC = \infty$ in case no such ℓ exists. The VC dimension is useful to make predictions on the generalization to new examples. Suppose that the examples in the training and in the test set are drawn with the same probability distribution — no restrictive assumptions are given on the distribution. If $h < \ell$ is the VC dimension of the given class of functions (machine) then the structural risk $E(f)$ and the empirical risk $E_{\text{emp}}(f)$ differ by at most $\beta(h, \ell, \delta)$, that is, $E(f) \leq E_{\text{emp}}(f) + \beta(h, \ell, \delta)$ where

$$\beta(h, \ell, \delta) = \sqrt{\frac{1}{\ell}\left(h\left(\ln\frac{2\ell}{h} + 1\right) + \ln\frac{4}{\delta}\right)}.$$

Here, we promptly see how the increment of the dimension ℓ of the training set allows us to squeeze the term $\beta(h, \ell, \delta)$, as well as the special role of the VC dimension h [see, e.g. B. Schölkopf, A.J. Smola, *Learning with Kernels: Support Vector Machines, Regularization, Optimization, and Beyond, Adaptive Computation and Machine Learning* (MIT Press, 2002)].

Section 4.2 Although a number of ingredients have already been put forward, the first clear formulation of the maximum margin problem along with its solution was given in B.E. Boser, I.M. Guyon, V.N. Vapnik, [*Proc. of the 5th Annu. Workshop on Comput. Learn. Theory* (1992), 144–152], while the soft margin version was introduced a few years later. The paper defined the basic elements of support vector machines that are characterized by using kernels, absence of local minima, sparseness of the solution, and very good generalization — which is in fact the outcome of the maximization of the margin.

The introduction of SVM makes it possible to look at learning from a different perspective with respect to most classic neural networks models that rely on computation in the primal space $X \subset \mathbb{R}^d$. The representation of the optimal solution as a kernel expansion in the training set is in fact the consequence of the classic dual formulation of optimization problems, which is concisely described in Appendix A. The classic book by on convex optimization by S. Boyd and L. Vandenberghe [*Convex Optimization* (Cambridge University Press, 2004)] is a rich source of optimization methods whose computational structure resembles that used in SVM. The sparseness of the solution, which corresponds with the emergence of support vectors, is the outcome of the formulation of the maximum margin problem, whose solution in the dual space yields no duality gap.

The kernel methods discussed in the book assume that the function maps to a real-valued domain, which is used to model a single task. In classification problems one typically assumes that any class is modeled by a single SVM, so that multiple classification is carried out by training each SVM to discriminate their associated class with respect to all the patterns of all the others. This approach is a mixed blessing. While its simplicity is definitely attractive, the corresponding classifiers yield an output which is not probabilistically normalized. Moreover, while the kernel captures the smoothness of the single functions, the regularization doesn't capture any useful relative information among the different tasks. The extension of kernels to a multitask environment has been systematically carried out in T. Evgeniou, C.A. Micchelli, M. Pontil, [*J. Mach. Learn. Res.* **6** (2005), 615–637].

The described quadratic programming formulation of learning is inherently batch-mode, namely it requires us to provide the training set at the beginning. When the agent is expected to learning online, kernel machines, like any dual representation of the task f, face the problem of guessing where to expand the functions. When thinking of a truly online framework, there is limitation on the length of the sequence in advance, which means that one could expect that the optimal representation of the task corresponds with an expansion over infinitely many points.

A number of methods have been proposed in the literature to deal with on-line data. In some papers [see, e.g., J. Thorsten, *Making large-scale support vector machine learning practical* edited by C.J.C. Burges, A.J. Smola (MIT Press, 1999), 169–184, E. Osuna, F. Freund, F. Girosi, *Neural Netw. for Signal Process.—Proc. of the IEEE Workshop* (1997), 276–285 and K. Crammer, Y. Singer, *J. Mach. Learn. Res.* **2** (2002), 265–292] we can see the unifying idea of decomposing the problem into manageable subproblems over part of the data. A different approach, proposed by T. Frieß, N. Cristianini and C. Campbell [*Proc. of the Fifteenth Int. Conf. on Mach. Learn.* (1998), 188–196], relies on componentwise optimization. A different idea is pursued by G. Cauwenberghs, T. Poggio [*Advances in Neural Inf. Process. Syst.* (2000), 388–394], where the solution is constructed recursively, one point at a time. Basically, the solution on $\ell + 1$ training examples is expressed in terms of that found for ℓ examples.

Section 4.3 It has been pointed out that kernels are essentially a way of expressing similarity between two patterns. The feature function ϕ doesn't necessarily map a vectorial space X to the feature space $F \in \mathbb{R}^D$; we don't require that the input space X posses any special structure, the only condition being transferred to the feature space. The idea of the kernel trick, which definitely reveals the importance of the kernel concept, was introduced in M.A. Aizerman, E.A. Braverman, L. Rozonoer, *Automat. and Remote Control* **25** (1964), 821–837.

Among others, a strong property of kernels is that which leads to a straight-forward approximation of $\| f \|_k$ once the solution α^* of a learning problem has been discovered. If K is the corresponding Gram matrix, it has been shown that $\|f^*\|_k = \alpha \cdot K\alpha \geq 0$. The nonnegativeness of k is in fact the basic requirement to measure the norm of f, which expresses the degree of smoothness of the solution. The extension of the spectral analysis of K to the function domain leads to Mercer's features. The corresponding theorem was proven by J. Mercer [*Philos. Trans. R. Soc. Lond.* **209** (1909) 415–446]. A systematic treatment can be found in volume 1 of the book *Methods of Mathematical Physics* (Wiley, 1989) by R. Courant, D. Hilbert. Mercer features show the power of kernel approaches. They are very general since they construct rich frequency features, but unfortunately, this generality is also indicating the weakness of the methods: We need features that are invariant with respect to relevant transformation (see, e.g., the requirements in speech, language, and vision).

Reproducing Kernel Hilbert Spaces (RKHS) were formally introduced by N. Aronszajn [*Trans. Am. Math. Soc.* **68** (1950), 337–404], and soon became the subject of attention in data analysis by Parzen and Wahba, who adopted the idea for dealing with data smoothing by using spline models [see, e.g., the seminal book by G. Wahba [*Spline Models for Observational Data*, (SIAM, 1990)]. They also proposed the following enlightening introduction to the idea of kernel. Let $f: [0, 1] \to \mathbb{R}$ be a smooth function in the neighborhood of $x = 0$. We can write $f(x)$ in terms of $f(0)$ by:

$$f(x) = f(0) + \int_0^x f'(u)\, du$$

$$= f(0) + \int_0^x D_t(u - x) f'(u)\, du$$

$$= f(0) + (u - x) f'(u)\big|_0^x - \int_0^x (u - x) f''(u)\, du$$

$$= f(0) + \frac{f'(0)}{1!} x + \int_0^1 (x - u)_+ f''(u)\, du.$$

Now if we apply integration by parts again, we get

$$f(x) = f(0) + \frac{f'(0)}{1!} x + \frac{f''(0)}{2!} x^2 + \int_0^1 (x - u)_+^2 f'''(u)\, du.$$

We can easily verify by induction that

$$f(x) = \sum_{\kappa=0}^{m-1} \frac{f^{(\kappa)}(0)}{\kappa!} x^\kappa + \int_0^1 \frac{(x - u)_+^{m-1}}{(m-1)!} f^{(m)}(u)\, du.$$

Now let us consider the class of functions B_m such that $\forall \kappa = 0, \ldots, m - 1$, $f^{(\kappa)(0)} = 0$. If $f \in B_m$ then

$$f(x) = \int_0^1 \frac{(x - u)_+^{m-1}}{(m-1)!} f^{(m)}(u)\, du = \int_0^1 g_m(x, u) f^{(m)}(u)\, du,$$

where

$$g_m(x, u) = g_m(x - u) = \frac{(x - u)_+^{m-1}}{(m-1)!}.$$

Now denote by W_m^0 the collection of functions $f \in B_m\}$ such that $\forall \kappa = 1, \ldots, m-1$, we have that $f^{(\kappa)}$ is absolutely continuous while $f^{(m)} \in L_2$. We can prove that W_m^0 is a RKHS with kernel [G. Wahba, *Spline Models for Observational Data*, (SIAM, 1990)]

$$k(x, z) = \langle g_m(x, \cdot) g_m(z, \cdot) \rangle = \int_0^1 g_m(x, w) g_m(z, w) \, dw. \tag{1}$$

This expresses factorization with feature map $\phi(v) := g_m(v, \cdot)$.

Kernels are just the outcome of the choice of the feature map ϕ. Its choice does define the corresponding kernel, while a given kernel is susceptible to many different feature map representations. Kernel methods are inherently limited by the need to choose a kernel. This is in fact a strong limitation with respect to learning by artificial neural networks, where the purpose of learning is also that of detecting the best features. When choosing a kernel we somewhat bias the solution of the problem, since we are implicitly making assumptions on the features that are relevant for the task. While many feature maps are in fact possible interpretations of the computation carried out with a given kernel, all of them are somehow equivalent and provide quite a significant bias to the solution. As a matter of fact, as pointed out in many real-world applications [see, e.g., A. Moschitti, S. Quarteroni, R. Basili, S. Manandhar, *Proc. of the 45th Annual Meeting of the Assoc. of Comput. Linguistics* (2007), 76–783], when carrying out experiments, the appropriate selection of kernels turns out to be crucial.

A possible way of dealing with this limitation is that of choosing a kernel function which is a linear combination of a given collection of kernels $\{k_i : i = 1, \ldots, n\}$, that is, $k = \sum_{i=1}^n \alpha_i k_i$, where $\mu_i \geq 0$ in order to guarantee the nonnegativeness of k. In so doing, we regard the kernel as a conic combination of kernels of the given collection. This has been proposed in G.R.G. Lanckriet, T. De Bie, N. Cristianini, M.I. Jordan, W.S. Noble, *Bioinformatics* **20** (2004), 2626–35, for applications to bioinformatics. It is proven that we can determine the general solution, which includes the discovery of the weights μ_i, by quadratically constrained quadratic programming (QCQP). Unfortunately, this approach is somewhat limited because of its computational requirements. In F.R. Bach, G.R.G. Lanckriet, M.I. Jordan, *Proc. of the Twenty-First Int. Conf. on Mach. Learn.* (2004), 6 this limitation is significantly reduced by posing the problem as a second-order cone programming. Another remarkable computational improvement of the method proposed in [G.R.G. Lanckriet, T. De Bie, N. Cristianini, M.I. Jordan, W.S. Noble, *Bioinformatics* **20** (2004), 2626–35] is given in M. Gönen, E. Alpaydín, *J. Mach. Learn. Res.* **12** (2011), 2211–2268.

Section 4.4 While the foundations on support kernel machines were independently pioneered by a number of people, the seminal book on statistical learning theory by V.N. Vapnik [*Statistical Learning Theory* (Wiley-Interscience, 1998)] likely represents one of the first significant contributions. Early studies on the principle of regularization were carried out by A.N. Tikhonov and V.Y. Arsenin [see *Solution of Ill-Posed Problems* (Winston & Sons, 1977)].

The framework of regularization based on differential operators described in Section 4.4.4 was introduced in machine learning by T. Poggio and F. Girosi in the technical report *A Theory of Networks for Approximation and Learning* (MIT Press, 1989), and was subsequently elaborated in related papers [F. Girosi, M. Jones, T. Poggio, *Neural Comput.* **7** (1995), 219–269 and F. Girosi, M. Jones, T. Poggio, *Adv. Comput. Math.* **13** (2000), 1–50]. The intriguing connection between the Green function of L and the kernel comes also with important differences. While the kernel theory, as proposed in the framework of the RKHS, disregards the behavior on the frontier of the perceptual space X, the theory based on regularization operators yields a solution which is significantly affected by the conditions that are given on ∂X. An in-depth analysis on this issue is given in G. Gnecco, M. Gori, M. Sanguineti, *Neural Comput.* **25** (2013), 1029–1106. While the kernel approach has been spread in the machine learning community, only a few people have focused attention on regularization operators. As shown in Section 4.4.4, also in this case, we look for a representation theorem that turns out to be the outcome of the superposition principle. It is in fact this representation that, like in the framework of the RKHS, gives rise to the efficient quadratic programming algorithm. One might wonder why we do not directly integrate numerically the Euler-Lagrange equations 4.4.4–(7), instead of relying on the Green function. Apart from the presence of the distribution, we can immediately see that such an approach is not viable because of the curse of dimensionality in the perceptual space! Hence, while the partial differential equations 4.4.4–(7) represent a formulation of supervised learning also in presence of boundary conditions, no numerical method can reasonably return its solution. This limitation seems to be a sort of warning on the appropriateness of interpreting learning as a parsimonious best fitting of the constraints associated with the supervised pairs. Something similar happens when more complex constraints are involved. On the other hand, in nature, any learning process is framed in the temporal domain, so one might suspect that capturing the regularities in non-temporally ordered collections is remarkably more difficult than online learning. Interestingly, while the emphasis on given learning collections seems to be very well-suited to deal with statistical methods, it might lead to problems that are intrinsically more difficult than the corresponding problems faced in nature. When data are indexed by time, the laws of learning that are expressed by partial differential equations in the feature space — for which we face the intractability due to the curse of dimensionality — could be translated into ordinary differential equations, which correspond to moving on a temporal manifold of the feature space. Preliminary studies on this idea can be found in M. Gori, M. Maggini, A. Rossi, *Neural Netw.* **81** (2016) 72–80.

Deep architectures

I call architecture frozen music
J.W. Goethe, *Letter to Eckermann* (1829)

Deeper and more profound,
The door of all subtleties!
Lao-tzu, *The Way* (c. 604–c. 531 B.C.)

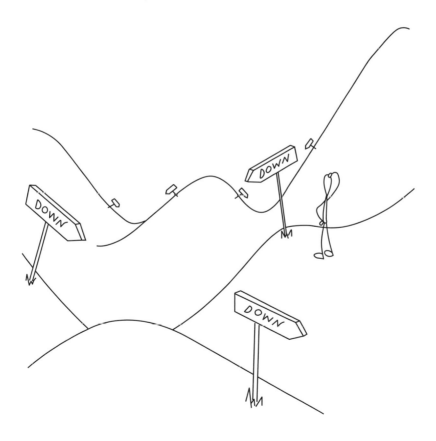

Machine Learning. https://doi.org/10.1016/B978-0-32-389859-1.00012-X

217

IN THIS CHAPTER, we give foundations on feedforward neural networks and incorporate some new recent developments on deep learning. This is a central topic in machine learning both for foundational and applicative implications. From the foundational side, it has been the subject of a multifacet analysis involving, amongst others, computational geometry, circuit theory, circuit complexity, approximation theory, optimization theory, and statistics. The impact that feedforward networks have been experiencing, however, is likely to be due mostly to their application in an impressive number of different domains.

5.1 Architectural issues

While linear and linear-threshold machines construct a map from the input to the output, without providing any internal representation, feedforward neural networks enrich the computation by hidden neurons. They construct an internal representation of the input that is based on input features in the spirit of what has been mostly described in Section 4.1. Once the hidden neurons are organized in layers, it becomes clear that higher degree of abstraction is gained as the output layer is approached. When regarding the hidden neurons as units that support appropriate features it becomes very important to focus on their computation to gain more information on the extracted features. One relevant difference concerns the pattern of connections. In the case of fully connected units, one expects the neurons to construct a very large class of features, since each one processes all the inputs for carrying out the decision. However, the discussion in Section 4.1 has already suggested the importance of gaining invariant properties. In the case of handwritten character recognition, like in any image processing task, one very much would like neurons to detect features that are invariant under scale and roto-translation. (Similar requirements are typically imposed in computer vision, but as pointed out in Section 5.4 and 5.7, there are more natural invariant properties to impose that involve motion.) No matter what kind of feature are extracted by the neurons, the case of full connections typically comes with the tacit assumption that the neurons are all different, so that the corresponding features are not invariant. Fully-connected units process the whole set of input coordinates, so if they share the same weights then they always return the same output.[1] Duplication and, more generally, replication of the units become very interesting whenever we abandon fully connected architectures. The case of handwritten character recognition and, more generally, that of image processing naturally lead to the selection of neurons acting on proper *receptive fields*. Neurons that are replicated over receptive fields extract properties that are invariant under translation. Hence weight sharing in neurons operating on receptive fields yields translational invariance features.

---

[1] There is not much to gain from the perfect duplication of two hidden neurons, apart from fault-tolerance. However, this is not a trivial issue. The solution of Exercise 1 sheds light on the role of neuron duplication.

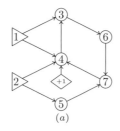 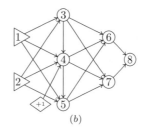

(a)   (b)

FIGURE 5.1

(a) The case of cyclic graphs: If there is an ordering conflict (see the loop $3 \longrightarrow 6 \longrightarrow 7 \longrightarrow 4 \longrightarrow 3$) then the static neural computation of linear threshold units leads to inconsistent computations (see vertex 4, where the explicit bias link is also drawn). (b) The partial ordering of Directed Acyclic Graphs (DAG) gives rise to a consistent computation on all the vertices. In both drawings the triangular vertices are inputs and the rectangular ones are biases; furthermore, given any two vertices with labels u and v, the weight attached to the arch $u \longrightarrow v$ is w_{uv}, while the bias relative to v is b_v.

5.1.1 Digraphs and feedforward networks

The discussion on LTU, especially on the representational issues given in Section 3.2.1, provides motivation to go beyond computational models based on single units. Any of those units can be regarded as vertices of a graph which carries out collective computation. If the neuron is a classic LTU then we can promptly realize that the only consistent computational mechanism that can be constructed is necessarily based on Directed Acyclic Graph (DAG) $\mathcal{G} = (V, A)$, where V is the set of vertices and A is the set of arcs (multiset of ordered pairs of elements of V). This can be seen in Fig. 5.1 (notice that we only represent the bias for unit 4, whereas the other bias links are not drawn for the sake of simplicity). On the left a cyclic graph gives rise to inconsistent computation because of a conflict in the ordering on neuron 4. Clearly, the inconsistency arises because of the presence of units in a loop. On the other hand, whenever we are given a partial ordering relation defined on the vertexes, the LTU-based computation is consistent and takes place according to a *data flow* computational scheme.

In particular, the data flow scheme for the network of Fig. 5.1–(b) can be expressed by the following partially ordered set:

$$S = \{\{1, 2\}, \{3\}, \{4\}, \{5\}, \{6, 7\}, \{8\}\} . \tag{1}$$

Whenever the network is based on a DAG, we say that it has a *feedforward structure*. In particular, a *feedforward neural network* is a DAG $\mathcal{G}$ with $V = I \cup H \cup O$, together with the following computational structure:

$$x_p = v_p \, [p \in I] + \sigma \left(\sum_{q \in \mathrm{pa}(p)} w_{pq} x_q + b_p \right) [p \in H \cup O]$$

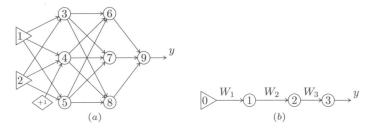

FIGURE 5.2

Feedforward network with multilayered structure. (*a*) The neurons are organized into two *hidden layers* $(3, 4, 5)$ and $(6, 7, 8)$, and there is an output layer composed of neuron 9 only. Clearly, there is a partial ordering of the vertices required by feedforward networks, with the additional layered structure; there is no connection within the layers. (*b*) Compact representation of the above multilayered network with layers and interconnection matrices. Here matrix W_l is associated with the pair of layers $l - 1$ and l, so that its generic coordinate is $w_{i(l), j(l-1)}$.

where w_{pq} and b_p are real numbers (see the caption of Fig. 5.1). The activation relative to the vertex p will therefore be defined as $a_p = \sum_{q \in \mathrm{pa}(p)} w_{pq} x_q + b_p$. In this case, there is a partial ordering since no time constraint in the computation involves sets $\{1, 2\}$ and $\{6, 7\}$, whereas the total ordering would require subsets with a single vertex only. A very interesting special case is the one in which the feedforward structure is due to a *multilayered* organization, where the units are partitioned into *ordered layers* with no internal ordering. Fig. 5.2 gives an example of a hierarchical organization based four layers. In this case we have

$$S = \{\{1, 2\}, \{3, 4, 5\}, \{6, 7, 8\}, \{9\}\}. \tag{2}$$

Here, we have the total ordering $\{1, 2\} \prec \{3, 4, 5\} \prec \{6, 7, 8\} \prec \{9\}$, whereas there is no ordering inside the layers.

The layered structure allows us to express in a compact form the data flow propagation of the input. When referring to Fig. 5.2, we can promptly see that the weights associated with a layer can compactly be represented by a corresponding matrix, so the output turns out to be

$$y = \sigma(W_3 \sigma(W_2 \sigma(W_1 x))),$$

where $\sigma := (\sigma_1, \ldots, \sigma_m)$ is a vectorial function constructed by squashing each unit of $W_l x_{l-1}$. In general we have

$$x_l = \sigma(W_l x_{l-1}) \tag{3}$$

which is initialized with $x_0 := x$. Of course, the role of σ is crucial in the neural network behavior. It is worth mentioning that in case of linearity, a feedforward network of L layers just collapses to a single layer! This can promptly be seen, since in that case we have $\sigma := \text{id}$ and therefore we conclude that $y = \prod_{l=1}^{L} W_{\ell} = Wx$, where $W := \prod_{l=1}^{L} W_{\ell}$.

The computational collapse of layers is a rare property. When σ is not linear, the recursive layer-based computation in Eq. (3) typically enriches significantly one of single layers! Let $\sigma_{W_l}(x) := \sigma(W_l x_{l-1})$. In general, there is no matrix W_3 such that $\sigma_{W_2}(\sigma_{W_1}(x)) = \sigma(W_3 x)$. As it will be shown in the following, this is nicely exploited in the construction of feedforward neural networks with rich computational capabilities. Notice that the neurons that have been considered so far determine the output according to

$$y = g(w, b, x) = \sigma(w'x + b).$$

They are referred to as *ridge neurons*. Another classic computational scheme is based on

$$y = g(w, b, x) = k(\|x - w\|/b),$$

where k is usually a bell-shaped function, which are called *radial basis function* neurons.

EXERCISES

1. [*17*]   Suppose you are given a feedforward neural network with one hidden layer where two neurons are equal. Prove that there exists an equivalent network where the two equal units can be replaced with one of them.

2. [*M21*]   Let us consider a three layer linear feedforward neural network, where the matrices W_1 and W_2 are simultaneously diagonalizable.[2] Prove that W_1 and W_2 are commutative matrices and therefore $y = W_2 W_1 x = P \operatorname{diag}(\omega_{2,i}\omega_{1,i})P^{-1}$. In the case W_1 and W_2 are invertible, prove that $y = (W_1^{-1} + W_2^{-1})^{-1}(W_1 + W_2)x$.

5.1.2 Deep paths

In order to get an insight on the role of deep structures, let us consider the extreme case of a cascade of two units. For the sake of simplicity, let us also assume that the bias in null, that is, $b_1 = b_2 = 0$. The output is simply $y = \sigma(w_2\sigma(w_1 x))$. In this case, if $\sigma = \text{id}$, that is, $\sigma(w_2 z)$ and $\sigma(w_1 x)$ are linear, then, in addition to the collapsing to linearity, we also gain the commutative property $y = \sigma(w_2(\sigma(w_1 x))) = \sigma(w_1(\sigma(w_2 x))) = w_1 w_2 x$. This does not hold in general in the multidimensional case (see Exercise 2 for additional details). Most importantly, the collapsing property does not hold for other most interesting functions. Here are a few examples.

5.1.2.1 Heaviside function

Let us consider the case of the hard-limiting computation that is based on the Heaviside function $\sigma(a) = H(a)$, where as usual $a = wx + b$. The two-unit cas-

---

[2] Matrices W_1 and W_2 are simultaneously diagonalizable, whenever there exists P such that $W_1 = P \operatorname{diag}(\omega_{1,i})P^{-1}$ and $W_2 = P \operatorname{diag}(\omega_{2,i})P^{-1}$.

cade computation yields $y = H(w_2 H(w_1 x + b_1) + b_2)$. Then we can promptly see that $H(w_i x + b_i) = [x \geq -b_i/w_i]$, i.e., the characteristic function of the interval $[-b_i/w_i, \infty)$.

Now we want to see whether there exists (w_3, b_3) such that

$$H(w_2 H(w_1 x + b_1) + b_2) = H(w_3 x + b_3);$$

that is to say, we want to see whether it is possible to find w_3 and b_3 such that $f(x) = [w_2[w_1 x + b_1] + b_2]$ and $g(x) = [w_3 x + b_3]$ return 1 for the same values of x. Clearly, $g(x)$ is 1 if and only if $x \geq -b_3/w_3$; on the other hand, we have that $f(x)$ is 1 for $x \geq -b_1/w_1$ when $w_2 + b_2 \geq 0$ and for $x < -b_1/w_1$ whenever $b_2 \geq 0$. This means that $f(x)$ can assume four different forms depending on the values of w_2 and b_2: $f(x) = 1$ if $w_2 + b_2 \geq 0$ and $b_2 \geq 0$; $f(x) = 0$ if $w_2 + b_2 < 0$ and $b_2 < 0$; $f(x) = [x \geq -b_1/w_1]$ if $w_2 + b_2 \geq 0$ and $b_2 < 0$; and finally, $f(x) = [x < -b_1/w_1]$ if $w_2 + b_2 < 0$ and $b_2 \geq 0$.

All this means that when $f(x) = 1$ or $f(x) = 0$ for all values of x, we cannot find w_3 and b_3 such that $f = g$; however, this is the case when the neural network returns the same value regardless of the input, so for what concerns machine learning this is not a relevant problem. The same conclusion obviously holds for the case in which $f(x) = [x < -b_1/w_1]$. On the other hand, when $f(x) = [x \geq -b_1/w_1]$, we can set $b_3/w_3 = b_1/w_1$ to achieve $f = g$. This means that collapsing from two Heaviside neurons to one is possible only under some restrictions on the values of w_2 and b_2.

As we consider $d > 1$ then the horizontal growth of the neural network, joined with the cascade of Heaviside LTU, yields functions that make the situation even worse. An in-depth analysis of this case will be given in Section 5.2.1 which concerns Boolean functions.

5.1.2.2 Rectifier

Like for the Heaviside function, a chain of two rectifiers does not necessarily collapse to a rectifier. Here is an example which also nicely shows the links between rectifiers and sig-moidal functions. Let the output $y = \left(1 - (1 - x)_+\right)_+$ be a cascade of two equal units with $\sigma(a) = \sigma(wx + b) = (1 - x)_+$, where $w = -1$ and $b = 1$. We can promptly see (as it is shown in the picture on the right) that

$$y = \left(1 - (1 - x)_+\right)_+ = x[0 \leq x \leq 1] + [x > 1] \equiv s(x).$$

A good approximation of the rectifier is $(a)_+ \simeq \ln(1 + \beta e^a)$ with $\beta > 0$. In Exercise 1, it is proven that if $\beta = e/(e - 1)$ then the above approximation of a cascade of two rectifiers yields a good approximation of the logistic sigmoid logistic$(a) = 1/(1 + e^{-a})$.

5.1.2.3 Polynomial functions

Another example of enrichment that does not collapse to single unit computation arises in case of polynomial functions.

Let us consider the case $y = \sigma(a) = a^2$, where $a = wx$. We can easily see that the cascade of two units does not collapse. We have $y = (w_2(w_1x)^2)^2 = w_2^2 w_1^4 x^4$ and there is no $w_3 \in \mathbb{R}$ such that $\forall x \in \mathbb{R}$ we have $w_3^2 x^2 = w_2^2 w_1^4 x^4$. Clearly, polynomial functions significantly enrich the input. If $\sigma(a) = \sum_{\kappa=0}^{m} \alpha_\kappa a^\kappa$, the output of the first unit is

$$y_1 = \sigma(a) = \sum_{i=0}^{m} \alpha_i (wx+b)^i = \sum_{i=0}^{m} \alpha_i \sum_{\kappa=1}^{i} \binom{i}{\kappa} w^\kappa b^{i-\kappa} x^\kappa$$

$$= \sum_{\kappa=0}^{m} \left(\sum_{i \geq \kappa} \alpha_i \binom{i}{\kappa} w^\kappa b^{i-\kappa} \right) x^\kappa = \sum_{\kappa=0}^{m} \beta_\kappa x^\kappa.$$

Since $y_1 = \sum_{\kappa=0}^{m} \beta_\kappa x^\kappa$, we have

$$y_2 = \sum_{\kappa=0}^{m} \alpha_\kappa \left(\sum_{h=0}^{m} \beta_h x^h \right)^\kappa = \sum_{\kappa=0}^{m} \alpha_\kappa \sum_{|j|=\kappa} \frac{\kappa!}{j_1! \cdots j_m!} \prod_{h=0}^{m} (\beta_h x^h)^{j_h}$$

$$= \sum_{\kappa=0}^{m} \sum_{|j|=\kappa} \prod_{h=0}^{m} \alpha_\kappa \frac{\kappa!}{j_1! \cdots j_m!} \beta_h^{j_h} x^{h+j_h}.$$

Hence the cascade of two units of mth degree doubles its degree.

5.1.2.4 Squash functions

The case in which $y(x) = (1+e^{-x})^{-1}$ leads to constructing cascading of units that are different from the original squash function. We can promptly see this property by looking for a solution of

$$y(x) = \frac{1}{1 + e^{-\frac{1}{1+e^{-x}}}} = \frac{1}{1 + e^{-wx-b}}.$$

The solution of this equation would yield a squash function with weights (w, b) that is equivalent to the cascade of two units both with weights $w = 1$ and $b = 0$. We can promptly see that this equation does not admit solutions for any x, which shows that also the squash functions do not collapse in the cascade of units. However, the qualitative behavior of $\sigma(w_2 \sigma(w_1 x + b_1) + b_2)$ does not change significantly. In particular, like the single unit, $\sigma(w_1 x + b_1)$, this cascade is still a monotonic function (see Exercise 4).

5.1.2.5 Exponential functions

Finally, let us consider the case the exponential functions $y = e^a$. Like in the previous cases, there is no w_3 such that $y = e^{w_2 \cdot e^{w_1 x}} = e^{w_3 x}$. This can promptly be seen when considering that this equation is equivalent to $w_2 \cdot e^{w_1 x} = w_3 x$. This holds true in the general case of computation in $\mathbb{C}$, which corresponds with real exponentiation and with sinusoidal functions in the case of values in $\mathbb{R}$. More details on this issue are covered in Exercise 3.

From this preliminary analysis it is clear that the cascading of units typically enlarges the space of functions. Interestingly, the way the space is enriched very much depends on the choice of σ. Exercise 5 proposes a nice example that involves polynomials, where we can clearly see the limitation of the constructed space.

EXERCISES

1. [20] Let us consider the cascade of two rectifiers, each defined by $\sigma(a) = \ln(1 + \beta e^a)$, where[3] $\beta := (e - 1)/e$. Prove that

$$\ln\left(1 + \beta \exp\left(1 - \ln\left(1 + \beta \exp\left(1 - x\right)\right)\right)\right) \simeq \frac{1}{1 + e^{-x}}$$

2. [26] Let us consider the cascade of the rectifiers $\rho(x, x_1)) := (x - x_1)_+$ and $\rho_2(x, x_2) := (x - x_2)_+$. Prove that

$$\forall x \in X, ((x - x_1)_+ - x_2)_+ = ((x - x_2)_+ - x_1)_+$$

Does this commutativity property of the composition, i.e. $\rho_1(\rho_2(x, x_1)) = \rho_1(\rho_2(x, x_2))$, hold for general rectifiers?

3. [26] Let us consider the simple case of $d = 1$ and $y = \sigma(a) = e^{wx}$, where $w, x \in \mathbb{C}$. Prove that there is no collapse in the cascade $y = e^{w_2 \cdot e^{w_1 x}}$. Once the results have been established, provide its interpretation in the real field $\mathbb{R}$ by showing that there is no collapse in the cascade of sinusoidal functions.

4. [18] Let us consider the cascade of two sigmoidal functions, where $\sigma(a) = 1/(1 + e^{-a})$ and $a = wx + b$. Prove that $\sigma(w_2\sigma(w_1 x + b_1) + b_2)$ is monotone.

5. [16] Let us consider the simple case of $d = 1$ and $y = \sigma(a)$, where $a = wx + b$ and $\sigma(a) = a^2 + c$. Given such units, analyze the computational capabilities of a cascade of n units. Prove that, no matter how deep we choose the network, there are polynomials that cannot be realized.

5.1.3 From deep to relaxation-based architectures

The discussion on deep paths of the previous section suggests considering the extreme case of neural networks with infinite depth. For this to make sense, in general, we need to provide a rule to describe how the weights change along the paths. A simple rule is that of assuming that there is a layered structure that represents a motif to be repeated. In so doing, we can construct a network of arbitrary depth! Of course, as will become more clear in the following, while the layers of a deep network can play a different role, the repeating of a motif leads to a different notion of depth. In a sense, the repeating of the motif corresponds with an extreme interpretation of deep networks that can be regarded as recurrent networks.

---

[3] They are regarded as an approximation of $(1 - (1 - x)_+)_+$.

Let us assume that this motif is represented by $\mathcal{N} = (V, A)$, so that $x = f(w, u)$, where the input u and the output x are both vectors of $\mathbb{R}^c$. The repeating of the motif can be expressed by stating that for every fixed positive ε, there is an index $\bar{\imath} \in \mathbb{N}$ such that for $t > \bar{\imath}$,

$$x_{t+1} = f(w, u_t), \quad \|x_{t+1} - x_t\| < \varepsilon, \tag{1}$$

whereas $u_0 := u$ is the input that is fed at the beginning of the iteration. If the dynamics evolves according to a relaxation process towards an equilibrium point, then the map f is such that the equation $x_{t+1} = f(w, x_t)$ admits a fixed-point x^* that satisfies

$$x^* = f(w, x^*).$$

At a first look, the solution of this equation does not seem to depend on the input u, but only happens on whenever f only admits one fixed point, regardless of $u_0 = u$. In that case the parametric dependence is well expressed by the matrix $\hat{W} \in \mathbb{R}^{c \times c}$, so that $f(x) = \hat{W}' \hat{x}$. Hence discovering the fixed point corresponds to finding the eigenvector of W associated with the eigenvalue $\lambda = 1$, that is, to finding $\hat{x}$ such that $(W - \text{Id})\hat{x} = 0$. Hence either $\hat{x} = 0$ or we have infinitely many solutions. Clearly, this is not a desirable behavior since, as proven in Exercise 1, the linearity leads to conclude that $\hat{x}^* \propto u$. In case of nonlinear units, if there is convergence to a fixed point, we can experience a more interesting dynamics.

A particular case that has been the subject of in-depth investigation is the *Hopfield neural network*. In that case a single layer of neurons is used where $\sigma(a) = \text{sign}(a)$ and matrix W is symmetric with $w_{ii} = 0$. The corresponding model is expressed by

$$x_{i(t+1)} = \text{sign}\left(\sum_j w_{ij} x_{jt} + b_i\right).$$

In Section 6.3.3, these neural networks will be discussed in detail in the framework of recurrent neural networks and neurodynamics.

EXERCISES

1. *[18]* Let us consider the relaxation-based network defined by Eq. (1) in the case of linear f. Prove that, unless $\hat{x} = 0$, Eq. (1) converges to $\hat{x} \propto u$ (the output is proportional to the input).

5.1.4 Classifiers, regressors, and auto-encoders

Feedforward neural network is used for classification and regression, as well as for pattern encoding. In the first case, the network is expected to return a value $z = f(w, x)$ which is as close as possible to the target y.

In the second case, the target becomes the input itself (as it is shown in Fig. 5.3), so that the network is expected to minimize $V(x, y, F(w, x))$. In the case of classifiers the output contains a code of the class of the input. In the simplest case, one is interested simply in the decision on the membership of x to a certain class $\mathcal{C}$. Hence

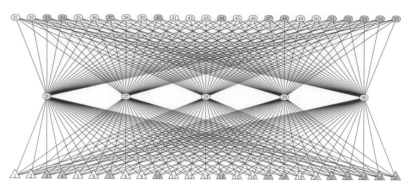

FIGURE 5.3

A feedforward neural network that encodes the handwritten character "2" that we used in Section 1.1.3.

if $x \in \mathcal{C}$ then the target is $y = 1$, otherwise $y = 0$. Multiclassifiers can be constructed in different ways. We must use an output layer with a number of units that is enough to code the class. In Exercise 1, a discussion is suggested on different class encodings. In particular, the one-hot encoding is compared with the Boolean encoding. The analysis returns a result which explains why most of the real-world experiments have been using one-hot encoding instead of more compact output representations.

In order to deal with multiclassification, we can either use

   (1)

The left-hand side network is a modular architecture where each of the three classes is connected to three distinct hidden neurons. On the other hand, the right-hand side configuration in (1) is a fully-connected network that leads to a richer classification process, in which all the hidden units contribute jointly to define the classes. Of course, in this case the network optimizes the discrimination capabilities, since the competition amongst the classes takes place by using all the features extracted in the hidden neurons. However, the left-side network presents the remarkable advantage of being modular, which favors the gradual construction of the classifiers. Whenever we need to add a new class, the fully-connected network does require a new training, whereas the modular one only requires training the new module. Most of the issues raised in classification also hold for regression. However, it is worth mentioning that the output neurons are typically linear in regression tasks, since there is no need to approximate any code.

The encoding architecture is an extension of matrix factorization in linear algebra. In that case we are given a matrix T and we want to discover factors W_1, W_2, so $T = W_2 W_1$. The process of encoding consists of mapping $x \in \mathbb{R}^d$ to a lower dimension

$\bar{y}$, which is the number of hidden units. One would like the network to return $z = f(w, x) \simeq x$, so that the output of the hidden neurons h can be regarded as a code of the input x.

EXERCISES

1. [*21*]   Let us consider one hidden layer networks for classification into 4 categories. Discuss the different behavior of two networks with the same number of inputs and hidden layers ($|I| = 6$ and $|H| = 5$) but with a number of output units $|O| = 4$ and $|O| = 2$, respectively. In the first case, *one-hot encoding* is used, that is,

$$\mathcal{C}_1 \sim (1, 0, 0, 0), \mathcal{C}_2 \sim (0, 1, 0, 0), \mathcal{C}_3 \sim (0, 0, 1, 0), \mathcal{C}_4 \sim (0, 0, 0, 1),$$

while in the second case

$$\mathcal{C}_1 \sim (0, 0), \mathcal{C}_2 \sim (0, 1), \mathcal{C}_3 \sim (1, 0), \mathcal{C}_4 \sim (1, 1).$$

Provide arguments to conclude that learning with two neurons is more difficult than learning with four neurons. *Hint:* In order to define a degree of difficulty, discuss the different separation surfaces in the two learning tasks.

5.2 Realization of Boolean functions

The understanding of classic AND–OR Boolean circuits offers important insights to capture the structure of feedforward neural networks. We will see that the traditional design methods used in switching theory can profitably be used to expose LTU-based realizations, though similar realizations are typically far from representing optimal solutions from a circuit complexity point of view.

5.2.1 Canonical realizations by AND–OR gates

The role of function σ has been shown to be very important even in the case of paths of single units. However, linear units have shown collapsing behavior, which does not hold for other σ functions. In addition, linear units also collapse in multidimensional spaces, whereas, as it will be shown in this section, LTUs like the Heaviside function give rise to a rich computational behavior. In order to understand this important property, we explore where a cascade of LTU units leads in the case of Boolean functions.

Here we will use the following notation for two-dimensional Boolean functions: Assuming that 1 corresponds to "true" and 0 to "false," we will say that the sequence of the four values $f(0,0)\,f(0,1)\,f(1,0)\,f(1,1)$ is the *truth table* of the Boolean function f. For example, the truth table of the AND function is 0001, while that of the OR function is 0111. Of course, this notation can be readily generalized for a Boolean function of n variables $f(x_1, x_2, \ldots, x_n)$; in this case the truth table will be the sequence of the 2^n numbers

$$f(0,0,\ldots,0,0)\,f(0,0,\ldots,0,1)\,f(0,0,\ldots,1,0)\ldots f(1,1,\ldots,1).$$

Let us consider Heaviside linear-threshold units and start with the AND function $\wedge$. We want to realize the truth table 0001 by $x_1 \wedge x_2 = [w_1 x_1 + w_2 x_2 + b \geq 0]$. Now let $\mathcal{W}_\wedge$ be the set of vectors $(w_1, w_2, b)'$ in $\mathbb{R}^3$ such that

$$(b < 0) \wedge (w_2 + b < 0) \wedge (w_1 + b < 0) \wedge (w_1 + w_2 + b > 0). \tag{1}$$

Notice that each of the propositions that are AND-ed in the definition of $\mathcal{W}_\wedge$ are a direct translation of the truth table. We can promptly see that $(w_1, w_2, b) = (1, 1, -3/2) \in \mathcal{W}_\wedge$ is a possible solution. In addition, the solution space $\mathcal{W}_\wedge$ is convex (see Exercise 1 for additional details and for the proof of convexity).

Likewise, the $\vee$ Boolean function can be implemented by a linear-threshold function $[w_1 x_1 + w_2 x_2 + b \geq 0]$. In particular, we can prove that one solution is $(w_1, w_2, b) = (1, 1, -1/2)$ and that, like for $\mathcal{W}_\wedge$, the solution $\mathcal{W}_\vee$ space is convex (see Exercise 2).

Now suppose we are given the exclusive OR function

$$x_1 \oplus x_2 = \neg x_1 \wedge x_2 \vee x_1 \wedge \neg x_2. \tag{2}$$

Unlike in the case of $\wedge$ and $\vee$, the set

$$\mathcal{L} = \{((0,0),0), ((0,1),1), ((1,0),1), ((1,1),0)\} = \square$$

is clearly not linearly separable. Formally, this comes out directly when considering that for any candidate separation line, the following proposition must hold:

$$(b < 0) \wedge (w_2 + b) > 0 \wedge (w_1 + b) > 0 \wedge (w_1 + w_2 + b < 0).$$

Now it is easy to see that there is no solution. If we sum up the second and the third inequalities, we get $w_1 + w_2 + 2b > 0$. Likewise, if we sum up the first and the fourth inequalities, we get $w_1 + w_2 + 2b < 0$, so we end up with a contradiction. Hence, we conclude that $\mathcal{W}_\oplus = \emptyset$. A nice graphical interpretation of $\mathcal{W}_\oplus = \emptyset$ is given in Exercise 4.

The above discussion essentially proves that we cannot compute the XOR function using a single LTU. We will now show that instead there are many ways to represent the XOR using a multilayered network Fig. 5.4.

FIGURE 5.4

Network for the evaluation of the XOR.

Looking at Fig. 5.4, we immediately realize that input x_1 and x_2 must be mapped by the hidden layer to x_3 and x_4 such that it can be linearly separated by the neuron 5. For example, in Fig. 5.5–(a) it is shown how this can be done using a "geometrical" approach; here the two evenly dashed lines have equations $x_1 + x_2 + 1/2 = 0$ and $x_1 + x_2 + 3/2 = 0$. Neurons 3 and 4 classify the points of the Boolean square according to the rule $x_3 = [x_1 + x_2 - 1/2 \geq 0]$ and $x_4 = [-x_1 - x_2 + 3/2 \geq 0]$; in this way, as one can see from Fig. 5.5–(a), the inputs are mapped into a separable configuration.

Another way to implement the XOR function can be done by noting that both $\neg x_1 \wedge x_2$ and $x_1 \wedge \neg x_2$ can be represented by an LTU with the Heaviside function. This is a straightforward consequence of the above discussed representations of $\wedge$ and $\vee$ by threshold functions. We can promptly see that a function for the realization of $\oplus$ can be constructed by using the canonical representation $x_1 \oplus x_2 = (\neg x_1 \wedge x_2) \vee (x_1 \wedge \neg x_2)$.

Now let us begin with the construction of $(\neg x_1 \wedge x_2)$ and $(x_1 \wedge \neg x_2)$. When thinking of the $\wedge$ and $\vee$ realization, we can promptly realize that the solution is similar, since any min-term is linearly separable. In Fig. 5.5–(b) we can see the lines corresponding with the two min-terms and the mapping of each example onto the hidden layer representation. The line corresponding to the neuron 3 has equation $-x_1 + x_2 - 1/2 = 0$, while the one that corresponds to 4 has equation $x_1 - x_2 - 1/2 = 0$; in fact, we have $\neg x_1 \wedge x_2 = [-x_1 + x_2 - 1/2 \geq 0]$ and $x_1 \wedge \neg x_2 = [x_1 - x_2 - 1/2 \geq 0]$. Like in Fig. 5.5–(a), we can see that such a representation is linearly-separable. Because of the first canonical expression of XOR, the output unit 5 acts as an OR, which again is linearly separable.

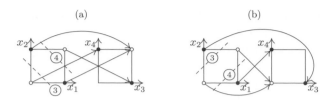

FIGURE 5.5

(a) The examples are not linearly separable when represented at the input layer (units 1, 2). Their mapping to the hidden layer (units 3, 4) yields a linearly-separable representation. (b) Solution based on the first canonical form of XOR. Units 3 and 4 detect the min-terms, $x_1 \wedge \neg x_2$ and $\neg x_1 \wedge x_2$, respectively. On top of the representation created by those units, neuron 5 acts as an OR.

It is worth mentioning that the solution given in Fig. 5.5–(b) can also be given a related interpretation in terms of the second canonical form. We have

$$x_1 \oplus x_2 = (x_1 \vee x_2) \wedge (\neg x_1 \vee \neg x_2).$$

In this case, max-term $x_1 \vee x_2$ is realized by unit 3, while max-term $\neg x_1 \vee \neg x_2$ by unit 4. This time the output neuron 5 acts as an AND.

Interestingly, as discussed in Exercise 5, XOR can be realized by saving one neuron with respect to the solution discussed herein.

The analysis of the XOR function sheds light on the general structure of the solution space concerning any Boolean function. Let us focus on the representation based on the first canonical form. The solution space $\mathcal{W}_\oplus$ is defined by:

$$
\begin{aligned}
&00 \quad b_3 < 0, b_4 < 0, b_5 < 0, \\
&01 \quad w_{32} + b_3 \geq 0, w_{42} + b_4 < 0, w_{54} + b_5 \geq 0, \\
&10 \quad w_{31} + b_3 < 0, w_{41} + b_4 \geq 0, w_{53} + b_5 \geq 0, \\
&11 \quad w_{31} + w_{32} + b_3 < 0, w_{41} + w_{42} + b_4 < 0, w_{53} + w_{54} + b_5 \geq 0.
\end{aligned}
$$

Now, following arguments similar to those used for $\mathcal{W}_\wedge$ and $\mathcal{W}_\vee$, we conclude that this is a convex set. The same holds true if we consider the realization of XOR based on the second canonical form. Basically, we have identified the two convex sets $\mathcal{W}_\oplus^I$ and $\mathcal{W}_\oplus^{II}$ that correspond to the first and second canonical representation of the solution. Of course, these sets are not connected since $\mathcal{W}_\oplus^I \cap \mathcal{W}_\oplus^{II} = \emptyset$.

This analysis for the $\oplus$ function can be extended to any Boolean map by relying on canonical forms. If we use the first canonical form, we have

$$f(x) = \bigvee_{j=1}^{m} \bigwedge_{k=1}^{s_j} u_{jk} = (u_{11} \wedge \cdots \wedge u_{1s_1}) \vee \cdots \vee (u_{m1} \wedge \cdots \wedge u_{ms_m}), \qquad (3)$$

where u_{ij} are literals, which mean either a variable x_i or its complement. Clearly, functions $\wedge$ and $\vee$ can be extended by using associativity. We have that

$$(x_1 \wedge x_2) \wedge x_3 = x_1 \wedge (x_2 \wedge x_3).$$

This cascade computation can be also carried out by a single threshold-linear unit with three inputs, that is to say, one

instead of,

where $x_{5'} = [w_1 x_1 + w_2 x_2 + w_3 x_3 + b \geq 0]$ with the weights satisfying the inequalities

$$
\begin{array}{lll}
000 & \circ & b < 0, \\
001 & \circ & w_3 + b < 0, \\
010 & \circ & w_2 + b < 0, \\
011 & \circ & w_2 + w_3 + b < 0, \\
100 & \circ & w_1 + b < 0, \\
101 & \circ & w_1 + w_3 + b < 0, \\
110 & \circ & w_1 + w_2 + b < 0, \\
111 & \bullet & w_1 + w_2 + w_3 + b \geq 0.
\end{array}
$$

Like $\wedge$, the $\wedge_3$ logical function is linearly separable, and therefore a single linear-threshold unit can be used for its implementation. If we map the truth table of the multidimensional $\wedge_m$, $m \in \mathbb{N}$ onto the Boolean hypercube then we can promptly see that the property of linear separability still holds (see Exercise 3). As a consequence, any function written as a conjunction of literals can be realized by a linear-threshold unit when associating $\neg x_i \to 1 - x_i$. Finally, since also the multidimensional $\vee_i$ can be realized by a single linear-threshold unit, we conclude that *any Boolean function can be given the two-layered representation* based on the conjunctive normal form in Eq. (3). A related realization can be given by using the second canonical form of a Boolean function (see Exercise 7).

An interesting question arises concerning the replacement of the Heaviside function with the sign function. It can be proven that any feedforward neural network based on Heaviside function neurons can be mapped to a corresponding network with neurons whose activation is nonlinearly transformed by the sign function (see Exercise 8).

EXERCISES

1. [*M15*]   Prove that the set $\mathcal{W}_\wedge$, corresponding to the AND function, is convex.

2. [*M18*]   Given a neuron, determine the set of weight solutions S for the implementation of the $\vee$ function and prove that $w_u = w_v = 1, b = -1/2$ is a solution.

3. [*M20*] Let us consider the multidimensional AND

$$\check{z} = \bigwedge_{\kappa=1}^{m} \check{x}_{\kappa}.$$

Construct a linear-threshold function which implements $\bigwedge$. Let $\rho < 1/2$ be a threshold value which enables the definition of the true $1 - \rho$ and false ρ values. Formally, if z is the neural representation of Boolean $\check{z}$ then $z > 1 - \rho$ is associated with $\check{z} = \top$ and $z < \rho$ is associated with $\check{z} = \mathsf{F}$. Given ρ determine the limit dimension m_ρ such that the linear-threshold map is possible [see also P. Frasconi, M. Gori, M. Maggini, G. Soda, *Mach. Learn.* **23** (1996), 5–32].

4. [*M16*] Construct the separation surfaces of the XOR function once you have determined a solution by using the rectifier nonlinearity.

5. [*17*] Prove that the neural network in the figure on the side can realize the XOR by giving the corresponding values of the weights.

6. [*25*] Construct a neural network with 4 units (2 inputs, 3 hidden, one output) such that its separation surface is open. Choose different weight configurations to construct open separation surfaces, and also one case in which they are closed.

7. [*18*] Find the linear-threshold representation of a given Boolean function by using the second canonical form. This representation is the dual of Eq. (3).

8. [*18*] Let us consider the class of Boolean function and the corresponding realization by linear-threshold unit feedforward architectures given in Section 5.2.1 by using the Heaviside function. Given any such realization, can we always construct a corresponding realization based on the sign function? In the affirmative case show the constructive scheme to convert the given network to the equivalent sign-based architecture.

9. [*18*] Let us consider the following Boolean function:

$$\langle \mathsf{x}, \mathsf{y}, \mathsf{z} \rangle := (x \wedge y) \vee (x \wedge z) \vee y \wedge z.$$

Prove that it returns the *majority bit*, namely the bit which is more frequent in the string x, y, z. Is it a linearly separable function?

5.2.2 Universal NAND realization

First and second canonical forms are only two of the many possibilities of expressing Boolean functions. In order to appreciate the range of different realizations, let us consider the following example that nicely shows two extreme representations. Suppose we want to realize the function $f(x) = \overline{x_1 \cdot x_2 \cdot x_3}$. When using De Morgan laws, we get $y_3 = \overline{x}_1 + \overline{x}_2 + \overline{x}_3 = \overline{x_1 \cdot x_2 \cdot x_3}$. Clearly, this holds for any number of variables, that is,

$$f(x) = \bigvee_{i=1}^{d} \neg x_i = y_d$$

where y_d is recursively determined by $y(i) = y(i-1) \vee \neg x_i$ and $y(0) = 0$. While $\bigvee_{i=1}^{d} \neg x_i$ is the most shallow representation, the equivalent recursive computation based on y_d fully expands in depth!

This example gives insights on the important role of the NAND operator. It is clearly a linearly separable function, one which can be implemented by an LTU. Interestingly, this holds in any dimension. It is well-known that the NAND operator possesses the universal property of representing any Boolean function. The idea of pretty simple. The first canonical representation requires constructing monomials by using the AND and NOT operators that are accumulated by the OR. Now, to conclude on the claimed universal property, we simply need to prove that AND, OR, and NOT can be realized by the NAND. This comes from elementary properties of Boolean algebra, namely

$$\bar{x} = \overline{x \cdot x} = \text{nand}(x, x),$$

$$x \cdot y = \overline{\overline{x \cdot y}} = \text{nand}(\text{nand}(x, y), \text{nand}(x, y)),$$

$$x + y = \overline{\bar{x} \cdot \bar{y}} = \text{nand}(\text{nand}(x, x), \text{nand}(y, y)).$$

For example, the XOR function in two dimensions becomes

$$x_1 \oplus x_2 = x_1 \cdot \bar{x_2} + \bar{x_1} \cdot x_2 = \overline{\overline{x_1 \cdot \bar{x_2}} + \overline{\bar{x_1} \cdot x_2}} = \overline{\overline{x_1 \cdot \bar{x_2}} \cdot \overline{\bar{x_1} \cdot x_2}}.$$

Its circuit representation indicates that we need five NAND operators. When replacing the AND with the corresponding LTU unit, we can promptly see that the resulting architecture contains more neurons than that which is simply based on the first-order canonical representation. In addition, while the depth of the network of Fig. 5.4 is two, in this case the depth is three! This is not surprising: The universal property of the NAND does not necessarily result in efficient realizations of the functions.

5.2.3 Shallow vs deep realizations

The discussion carried out in the previous section has enlightened important issues on the difference between shallow and deep realizations of the same function. The multidimensional XOR is a nice example to appreciate realizations. It can be proven that any depth-2 AND–OR circuit computing the XOR requires at least $2^{d-1} + 1$ gates (see Section 5.7), which clearly raises a warning on the effectiveness of shallow architectures. Basically, depth-2 AND–OR circuit allows us to construct functions according to the first (second) canonical form. The problem of computational explosion arising in functions like XOR is due to the construction of min-terms (max-terms), whose number explodes with the dimension of the function. Hence any model which looks at Boolean function from the corner of min-terms (max-terms) bounces against classic issues of intractability of the satisfiability problem.

Deep circuits turn out to be a very natural way of circumventing this complexity problem. Because of the associativity of XOR, for $d \geq 2$, we can write

$$\text{xor}(x_1, \ldots, x_d) = \bigoplus_{i=1}^{d} x_i = y_d,$$

$$y(i) = y(i-1) \oplus x_i, \quad y(1) = x(1).$$

(1)

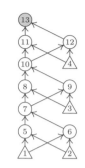

FIGURE 5.6

Deep realization of the XOR function.

The above neural solution, which is based on the XOR realization discussed in Section 5.2.1, is a straightforward translation of Eq. (1). In the figure a stream of bits is applied at node 1 by the *forward connections* used so far. However, the connection which links neuron 5 to neuron 4, indicated in gray, is not a usual synaptic connection, it simply returns y_5 delayed by the same time that synchronizes the input stream. Basically, $y_4(i) = y_5(i-1)$. Vertex 5 is the one returning the XOR output. Its value, once properly delayed, is fed to neuron 4, which, along with the input x_i, is used to compute $y(i)$. This architecture is referred to as a *recurrent neural network*, since the output is computed by the recurrence equations (1). Its computational scheme can nicely be described for any limited sequence by *time unfolding* of the recurrent network, which is shown for the case $d = 4$ in Fig. 5.6. This deep network presents an impressive computational advantage with respect to those coming from the canonical forms. Since we need to select half of the min-terms, the number of units grows exponentially with d. In the case of the above deep network the number of units is only proportional to d. This strong circuital computational complexity difference between shallow and deep networks will be discussed intensively in this book. However, the XOR function is a nice example to show that circuit complexity issues are quite involved, since there are also cases in which appropriate shallow realizations can break the border of exponential explosion. Interestingly, this can be gained when shifting attention towards LTU-based realizations.

So far the realization of Boolean functions has been driven by Boolean algebra. However, LTU-based realizations can be found that directly express the given function. As an example, we further investigate the multidimensional XOR. For the sake of simplicity, let us assume $d = 4$. Since the XOR function corresponds with the parity of the corresponding string, we can construct pairs of neurons devoted to detect the presence of an even number of 1 bits. In this case two pairs of neurons are devoted to detect the presence of 1 or 3 bits equal to 1, respectively. Now we construct a multilayer network with one hidden layer where each neuron can realize the $\leq$ and $\geq$ relations, while the output neuron accumulates all the hidden values and fires when the accumulated value exceeds a threshold which corresponds with the dimension $d = 4$. Hence the neurons are fired according to

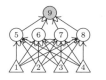
$$x_5 = H(\sum_{\kappa=1}^{4} x_\kappa - 1)$$
$$x_6 = H(1 - \sum_{\kappa=1}^{4} x_\kappa)$$
$$x_7 = H(\sum_{\kappa=1}^{4} x_\kappa - 3)$$
$$x_8 = H(3 - \sum_{\kappa=1}^{4} x_\kappa)$$
$$x_9 = H(\sum_{\kappa=5}^{8} x_i - 3)$$

FIGURE 5.7

Shallow realization of the XOR function. Unlike the solution based on canonical representations of Boolean functions, this realization has circuit complexity $O(d)$.

$$\sum_{i=1}^{4} x_i \geq 1 \rightarrow x_5 = 1, \quad \sum_{i=1}^{4} x_i \leq 1 \rightarrow x_6 = 1,$$

$$\sum_{i=1}^{4} x_i \geq 3 \rightarrow x_7 = 1, \quad \sum_{i=1}^{4} x_i \leq 3 \rightarrow x_8 = 1, \quad \sum_{i=5}^{8} x_i \geq 3 \rightarrow x_9 = 1.$$

Now let us analyze the incoming sequences depending on their parity. In case the parity is odd, since we have a specific pair of units which is fired, that part contributes 2 to the sum accumulated for the output unit 9. Basically, in this case, for one of the firing pairs $\sum_{i=1}^{4} x_i$ is equal to 1 or 3, and therefore for that unit the output is $H(0) = 1$. The other nonspecific pairs contribute 1, since only one of the two neurons of the pair turns out to be fired. Hence $\sum_{i=5}^{8} x_i = 3$, and therefore $x_9 = 1$. In case the parity is even there is no specific pair which reacts. All pairs of units exhibit uniform firing in this case, since only one of the units of a pair is fired. Hence $x_9 = \mathcal{H}(2-3) = 0$. Clearly, this can be generalized to the d-dimensional XOR, where we need d hidden units devoted to spot the odd numbers $2\kappa + 1 \leq d$. Again, in case of odd parity, the pairs of neurons corresponding to the odd number contribute 2 to the output, whereas all remaining pairs contribute 1. Finally, in case of even parity only half of the hidden neurons are fired.

Hence we conclude that there is a shallow depth-2 threshold architecture which realizes the XOR with $O(d)$ neurons. While both the neural networks depicted in Figs. 5.6 and 5.7 are efficient $O(d)$ realizations of the XOR, there is, however, an important difference: Notice that the solution based on Fig. 5.6 is robust with respect to the change of the weights, whereas the computation based on Fig. 5.7 is clearly sensitive to any violation of the comparison conditions. However, we can easily see that an opportune definition of the threshold confers robustness also to the shallow architecture (see Exercise 3. This robustness issue will be faced in a systematic way in the next section by the generalizing the realization technique proposed herein.

EXERCISES

1. [*16*]   Given $f(x) = \overline{x_1 \cdot x_2 \cdot x_3} = \text{nand}(x_1, x_2, x_3)$, consider the following transformation into a deep net:

$$y_3 = \text{nand}(y_2, x_3),$$
$$y_2 = \text{nand}(y_1, x_2),$$
$$y_1 = \text{nand}(y_0, x_1),$$
$$y_0 = 1$$

This transformation into a deep architecture makes use of nand associativity. Is it correct?

2. [22] Consider the following neural network:

with given weights $w_{32} = w_{41} = w_{63} = w_{64} = w_{65} = 1$, $w_{31} = 0$, $w_{42} = w_{51} = w_{52} = -1$ and with biases $b_3 = -1$, $b_4 = 1/2$, $b_5 = 9/2$, $b_6 = -5/2$. Assume also that $\sigma(x)$ is the sign function. Prove that $X = \{x \in \mathbb{R}^2 : x_6 = f(w, x) \geq 0\}$ is the region drawn in the above figure. How many different regions can be created?

3. [18] Let us consider the XOR realization of Fig. 5.7. Propose a design solution for the implementation of the four units of the hidden layer to face robustness issues. The proposal must simply determine the appropriate values of the threshold of neurons. Prove that such a choice yields robustness to face noisy input and, at the same time, it is also robust with respect to the choice of the weights. What is the space of the weights which guarantees the solution? Is it convex?

5.2.4 LTU-based realizations and complexity issues

The last example on the XOR realization indicates the superiority of LTU w.r.t. AND–OR gates. Intuitively, the additional efficiency that arises when dealing with LTU realization is due to the expression by real-valued weights instead of simple Boolean variables. In Section 5.8 the role of real-valued weights will be briefly discussed; in the case of Boolean functions, it will be shown that we can restrict the weight space to integer values without loosing computational power.

The feedforward network of Fig. 5.6 also opens the question on whether we should prefer shallow or deep networks for the realization of Boolean functions.[4]

However, before facing this fundamental problem, we want to show that the basic ideas behind the depth-2 construction proposed for the XOR in Fig. 5.7 can be extended to the class of *symmetric functions*. A Boolean function $f : \{0, 1\}^d \to \{0, 1\}$ is said to be symmetric provided that $f(x_1, \ldots, x_d) = f(x_{(1)}, \ldots, x_{(d)})$, where $(x_{(1)}, \ldots, x_{(d)})$ is any of the $d!$ permutations of $(x_1, \ldots, x_d)$. The XOR and the EQUIV functions are examples of symmetric functions. Other examples are given in Exercise 1.

---

[4] In the following, this issue will be covered also in the case of real-valued functions that are more interesting in machine learning.

Now we illustrate a depth-2 construction for the class of symmetric functions. The idea behind this construction is pretty simple, and it is very similar to construction schemes that are used also for real-valued functions. Intuitively, we want our network to spot the input configurations for which it is expected to return 1. In the case of Boolean functions, that is covered herein, the problem is hard in general, since those configurations cannot be efficiently determined. In case of symmetric functions, however, the spotting configurations are characterized only by $\sum_{\kappa=1}^{d} x_\kappa$. Hence this makes the problem somewhat similar to the case of real-valued domains, which, in this case, becomes one-dimensional! The spotting of the input configurations that makes the Boolean function true can simply be based on the value of $\sum_{\kappa=1}^{d} x_\kappa$, which is what makes LTU networks very well-suited to compute symmetric functions. A depth-2 network can be constructed which identifies the firing configurations of the function by neurons in the hidden layer. This is essentially what is done in Fig. 5.7, where each of the two firing configurations, corresponding with $\sum_\kappa x_\kappa = 1$ and $\sum_\kappa x_\kappa = 3$, is properly detected by two pairs of LTU. This idea can in fact be extended and formalized as follows. Given the interval $[0, d]$, let us construct a family of disjoint intervals

$$K = \left\{ [\underline{k}_1, \overline{k}_1], \dots [\underline{k}_s, \overline{k}_s] \right\}, \tag{1}$$

where $\underline{k}_{i+1} > \overline{k}_i$. Now the given function is characterized by the condition

$$f(x_1, \dots, x_d) = 1 \Leftrightarrow \exists j \in [0, d] : \sum_{i=1}^{d} x_i \in [\underline{k}_j, \overline{k}_j]. \tag{2}$$

Let us construct a depth-2 network, whose layers are composed as follows:

(*i*) *First layer* (configuration-spotting pairs)

$$\underline{y}_{k_j} = H\left(\sum_{i=1}^{d} x_i - \underline{k}_j\right), \quad \overline{y}_{k_j} = H\left(\overline{k}_j - \sum_{i=1}^{d} x_i\right) \tag{3}$$

(*ii*) *Second layer* (accumulation)

$$y(x_1, \dots, x_d) = H\left(\sum_{j=1}^{s} (\underline{y}_{k_j} + \overline{y}_{k_j}) - s - 1\right) \tag{4}$$

We want to prove that this network returns $y(x_1, \dots, x_d) = f(x_1, \dots, x_d)$. Let us consider Eq. (2), which characterizes the given function. We distinguish the two cases, $\sum_{i=1}^{d} x_i \notin [\underline{k}_j, \overline{k}_j]$ and $\sum_{i=1}^{d} x_i \in [\underline{k}_j, \overline{k}_j]$. In the first case, for $j = 1, \dots, s$, either $\underline{y}_{k_j}$ or $\overline{y}_{k_j}$ are true, that is, $\underline{y}_{k_j} + \overline{y}_{k_j} = +1$. Hence $y(x_1, \dots, x_d) = H((\sum_{j=1}^{s} 1) - s - 1) = H(s - s - 1) = 0 = f(x_1, \dots, x_d)$. In the other case, for $j = 1, \dots, s$, both $\underline{y}_{k_j}$ and $\overline{y}_{k_j}$ are true, that is, $\underline{y}_{k_j} + \overline{y}_{k_j} = 2$ for some $j = 1, \dots, s$, whereas, for $i \neq j$, we have $\underline{y}_{k_i} + \overline{y}_{k_i} = 1$. As a result we have $y(x_1, \dots, x_d) = \text{sign}(1 + s - s - 1) = \text{sign}(0) = +1 = f(x_1, \dots, x_d)$.

Let us use this construction for the previous example of the XOR with $d = 4$. The choice $s = 2$, $\underline{k}_1 = \overline{k}_1 = 1$ and $\underline{k}_2 = \overline{k}_2 = 3$ clearly characterizes the XOR with four bits. Then we can promptly check that the network construction based on Eqs. (3) and (4) corresponds with the realization of Fig. 5.7. Exercise 2 proposes examples of symmetric functions to be realized by the above design scheme.

This LTU-based realization is an example of the significant benefit of using LTU instead of AND–OR gates! Here we need at most $1 + 2\lceil d/2 \rceil$ units, that is, we achieve the upper bound $O(d)$. This shows the impressive improvement with respect to the $O(2^d)$ bound coming from AND–OR realizations.

While this analysis seems to indicate that also shallow networks can conquer efficient realizations, now we show that, still for symmetric functions, the construction of deep nets gives rise to significant reduction of circuit complexity. There are many techniques for exploring efficient realization that are briefly summarized in Section 5.8. Here, we present a construction which is based on *telescopic series*. We will present the idea for the construction of a depth-3 circuits for a symmetric function, with the purpose of seeing the fundamental complexity gain with respect to the previous techniques. The construction method is still based on an appropriate family of intervals like Eq. (1), but this time $K = \{[\underline{k}_0, \overline{k}_0], \dots [\underline{k}_s, \overline{k}_s]\}$, where $\underline{k}_0 = 0$. Moreover, we impose that $\overline{k}_i < \underline{k}_{i+1}$ and use the configuration spotting principle stated by Eq. (2). The realization is also based on an appropriate division of $[0, d]$ into another family of subintervals, where each element $[\underline{k}_{i_j}, \overline{k}_{i_j}]$ is defined by the two-dimensional index i_j, where $1 \leq j \leq l$ and $1 \leq i_j \leq r$. For reasons that will become clear when applying the telescopic technique, for all $j = 1, \dots, l$, we set $\underline{k}_{0_j} = \overline{k}_{0_j} = 0$. The subintervals are chosen in such a way that $[0, d] \subset \mathbb{N}$ is split into.

$$[\underline{k}_{1_1}, \underline{k}_{2_1} - 1], [\underline{k}_{2_1}, \underline{k}_{3_1} - 1], \dots, [\underline{k}_{r_1} - 1, d] \tag{5}$$

where each subinterval, apart at most for the last one, contains the same number l of elements $\underline{k}_s$ and $\overline{k}_s$. Moreover, we have $\underline{k}_{i_1} < \overline{k}_{i_1} < \underline{k}_{i_2} < \overline{k}_{i_2} < \cdots < \underline{k}_{i_l} < \overline{k}_{i_l}$. The network is constructed as follows:

(i) *First layer* (configuration spotting neurons). Let $i := i_1$. For all $i = 1, \dots, r$, the first layer of neurons carries out the computation

$$z_i = H\left(\sum_{j=1}^{d} x_j - \underline{k}_{i_1}\right) \tag{6}$$

(ii) The *second layer* of neurons is identified by the telescopic series. Then, for all $h = 1, \dots, l$, we have

$$\overline{t}_h = \overline{k}_{1_h} z_1 + (\overline{k}_{2_h} - \overline{k}_{1_h}) z_2 + \cdots + (\overline{k}_{r_h} - \overline{k}_{(r-1)_h}) z_r = \sum_{i=1}^{r} (\overline{k}_{i_h} - \overline{k}_{(i-1)_h}) z_i,$$

$$t_h = \underline{k}_{1_h} z_1 + (\underline{k}_{2_h} - \underline{k}_{1_h}) z_2 + \cdots + (\underline{k}_{r_h} - \underline{k}_{(r-1)_h}) z_r = \sum_{i=1}^{r} (\underline{k}_{i_h} - \underline{k}_{(i-1)_h}) z_i.$$

(7)

These telescopic series are used to construct the second layer, which is based on the computation of the following $2l$ hidden units:

$$\overline{q}_h = H\left(\overline{t}_h - \sum_{j=1}^{d} x_j\right), \quad \underline{q}_h = H\left(\sum_{j=1}^{d} x_j - \underline{t}_h\right)$$

(*iii*) The output comes from the accumulation of the telescopic series at the third layer according to

$$f(x) = H\left(\sum_{h=1}^{l} 2(\overline{q}_h + \underline{q}_h) - 2l - 1\right)$$

(8)

Notice that at the first layer the computation is driven by $\underline{k}_{j_1}$. Now we prove that this depth-3 neural network computes the symmetric function defined according to the characterization condition (2). Suppose that $\sum_{j=1}^{d} x_j \in [\underline{k}_{m_1}, \underline{k}_{(m+1)_1} - 1]$. The first hidden layer simply returns the vector[5]

$$z = (\underbrace{1, 1, \ldots, 1}_{\text{up to } m}, 0, 0, \ldots, 0)' \in \mathbb{R}^r.$$

The computation at the second layer involves the telescopic series (7). Due to the discovered structure of z computed at the first layer, the propagation to the second layer yields

$$\overline{t}_h = \sum_{i=1}^{r} (\overline{k}_{i_h} - \overline{k}_{(i-1)_h}) z_i = \sum_{i=1}^{m} (\overline{k}_{i_h} - \overline{k}_{(i-1)_h}) z_i = \overline{k}_{m_h} - \overline{k}_{0_h} = \overline{k}_{m_h}.$$

(9)

The same holds true for the other series, that is, $\underline{t}_h = \underline{k}_{m_h}$. Now, because of the characterization condition (2), condition $f(x) = 1$ means that we can always find $h \in [1, l]$ such that $\underline{k}_{m_h} \leq \sum_{j=1}^{d} x_j \leq \overline{k}_{m_h}$. Let us analyze the consequence of the processing of at the second layer. We have

$$\overline{q}_h = H\left(\overline{t}_h - \sum_{j=1}^{d} x_j\right) = H\left(\overline{k}_{m_h} - \sum_{j=1}^{d} x_j\right) = 1,$$

---

[5] Intuitively, if $\sum_{j=1}^{d} x_j$ belongs to the mth multiinterval, the first m out of r hidden neurons fire. Notice that the firing involves $\underline{k}_{i_1}$, that is, the index i_j with $j = 1$.

$$\underline{q}_h = H\left(\sum_{j=1}^{d} x_j - \underline{t}_h\right) = H\left(\sum_{j=1}^{d} x_j - \underline{k}_{m_h}\right) = 1.$$

Then

$$\overline{q}_h + \underline{q}_h = \begin{cases} 2; & \text{if } \sum_{j=1}^{d} x_j \in [\underline{k}_{m_h}, \overline{k}_{m_h}] \\ 1. & \text{if } \sum_{j=1}^{d} x_j \notin [\underline{k}_{m_h}, \overline{k}_{m_h}] \end{cases}$$

As a consequence the network returns, at the last layer the output

$$y = H\left(\sum_{h=1}^{l} 2(\overline{q}_h + \underline{q}_h) - 2l - 1\right) = H\left(\sum_{h=1}^{l} 2l + 2 - 2l - 1\right) = 1.$$

Similarly, if $f(x) = 0$ then there is no h such that $\sum_{j=1}^{d} x_j \in [\underline{k}_{m_h}, \overline{k}_{m_h}]$. As a consequence, $\forall j = 1, \dots, l$ we have $\overline{q}_h + \underline{q}_h = 1$. Then

$$y = H\left(\sum_{h=1}^{l} 2(\overline{q}_h + \underline{q}_h) - 2l - 1\right) = H\left(\sum_{h=1}^{l} 2l - 2l - 1\right) = 0.$$

Now let us analyze the circuit complexity of this realization. There is in fact a crucial step which leads to a dramatic complexity cut with respect to $O(d)$, which is related to the telescopic property stated by Eq. (9), which makes it possible to compute $\underline{q}_h$ and $\overline{q}_h$ with $O(1)$! This allows us to build a second layer of $2l$ neurons that carry out computation with complexity $O(l)$. Based on this premise, consider that the first layer requires r units, the second layer requires $2l \leq \lceil d/r \rceil$, while the output only requires one unit. Hence in total we need $r + \lceil d/r \rceil + 1$ units, which is minimum for $r = \sqrt{d}$. Finally, this results in a circuit complexity that is bounded by $1 + 2\sqrt{d}$, that is, $O(\sqrt{d})$. We can apply the presented realization technique to the XOR, since it is a symmetric function. We begin by determining the intervals that characterize the XOR. We choose $s = 7$ and set the family K of intervals by $\underline{k}_1 = \overline{k}_1 = 1, \underline{k}_2 = \overline{k}_2 = 3$, $\underline{k}_3 = \overline{k}_3 = 5, \underline{k}_4 = \overline{k}_4 = 7, \underline{k}_5 = \overline{k}_5 = 9, \underline{k}_6 = \overline{k}_6 = 11, \underline{k}_7 = \overline{k}_7 = 13, \underline{k}_8 = \overline{k}_8 = 15$ that clearly defines[6] odd parity. Then we need to construct the multiinterval $\underline{k}_{i_j}$ and $\underline{k}_{i_j}$. We pose $r = 4, l = 4$ and define

$$\underline{k}_{1(1)} = 1, \quad \underline{k}_{2(1)} = 5, \quad \underline{k}_{3(1)} = 9, \quad \underline{k}_{4(1)} = 13,$$
$$\underline{k}_{1(2)} = 3, \quad \underline{k}_{2(2)} = 7, \quad \underline{k}_{3(2)} = 11, \quad \underline{k}_{4(2)} = 15.$$

In so doing, each subinterval contains two $\underline{k}_i$. In particular,

$$\underline{k}_1 = 1, \quad \underline{k}_2 = 3 \in [\underline{k}_{1(1)}, \underline{k}_{1(2)}],$$

---

[6] Notice that, as already seen, this choice does not guarantee robustness. This can be achieved as we choose nonzero measure intervals $[\underline{k}_i, \overline{k}_i]$.

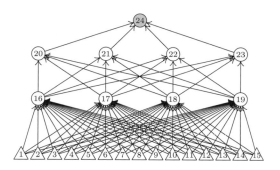

FIGURE 5.8

Realization of the XOR with $d = 15$.

$$\underline{k}_3 = 5, \quad \underline{k}_4 = 7 \in [\underline{k}_{2(1)}, \underline{k}_{2(2)}],$$
$$\underline{k}_5 = 9, \quad \underline{k}_6 = 11 \in [\underline{k}_{3(1)}, \underline{k}_{3(2)}],$$
$$\underline{k}_7 = 13, \quad \underline{k}_8 = 15 \in [\underline{k}_{4(1)}, \underline{k}_{4(2)}].$$

The corresponding neural network is shown in Fig. 5.8.

These results on symmetric functions show that a small increment in depth leads to a significant complexity cut. The parity of 1,000,000 variables can be realized with a network of about 1,000 LTU! The increment in depth has leads to even better results. For a class of symmetric functions with periodic structure (e.g., parity), there are depth-m networks such that the bound $O(\sqrt{d})$ can be reduced to $O(md^{1/m})$. Among the complexity requirements of good realizations, it is opportune to consider also the possible explosion of the weights of the developed solution. The symmetry of pictures offers an appropriate example to illustrate the issues. Symmetry can be formalized as an equality predicate between Boolean words, and it will be denoted as $\mathrm{simm}_d(x, y)$. For instance, checking symmetry for the pictures 010110011010 and 0111101011110 is reduced to checking the equality $x = y$:

$$\underbrace{010110}\underbrace{011010} \quad \underbrace{011110}1\underbrace{011110}$$
$$\underbrace{010110}_{x}\underbrace{010110}_{y} \quad \underbrace{011110}_{x} \underbrace{011110}_{y}$$

To face the problem, we introduce the comp_d (comparison) function, which is defined as follows:

$$\mathrm{comp}_d(x, y) = \begin{cases} 1 & \text{if } x \geq y \\ 0 & \text{if } x < y \end{cases} = H\left(\sum_{i=0}^{d-1} 2^i (x_i - y_i)\right) \tag{10}$$

We can promptly see that

$$\mathrm{simm}_d(x, y) = \mathrm{comp}_d(x, y) \bigwedge \mathrm{comp}_d(y, x).$$

Since $\wedge$ can be represented by a single LTU, a depth-2 neural network allows us to compute symmetry. As we can see from Eq. (10), however, this realization does require an exponential increment of the weights of the neurons of the hidden layer! However, we can easily circumvent this problem and compute $\mathrm{simm}_d(x, y)$ by the bitwise equality check

$$\mathrm{simm}_d(x, y) = \bigwedge_{i=1}^{\lfloor d/2 \rfloor} \neg(x_i \oplus y_i). \tag{11}$$

For the realization, notice that $\overline{x_i \oplus y_i} = H(x_i - y_i) + H(y_i - x_i) - 1$. Hence, while $\overline{x_i \oplus y_i}$ is a depth-2 circuit, since we need not accumulate it before its forwarding to the $\wedge$ unit; we can in fact send it directly to the unit so as $\mathrm{simm}_d(x, y)$ is realized itself by a depth-2 network. While this nicely addresses the issue of weight explosion in the computation of symmetry, the expression by Eq. (10) for the comparison function is plagued by representational problems of the weights. It is very instructive to see that the increment in depth offers possibilities to face this issue. To provide a realization with bounded weights, here we discuss a method that makes use of AND–OR gates. In particular, we give a depth-3 realization.[7] Let us consider the following recursive definition of $\mathrm{comp}_d(x, y)$:

(i) $\mathrm{comp}_1(x, y) = x_1 \vee \neg y_1$;
(ii) $\mathrm{comp}_d(x, y) = (x_d \vee \neg y_d) \vee \big((x_d \vee \neg y_d) \wedge \mathrm{comp}_{d-1}(x, y)\big)$.

Now let us define the following variables b_κ, $\kappa = 1, \ldots, d$:

$$b_1 = \bigwedge_{j=1}^{d} x_j \wedge \neg y_j,$$

$$b_k = x_\kappa \vee \neg y_\kappa \bigwedge_{j=k+1}^{d} x_j \wedge \neg y_j,$$

$$b_d = x_d \bigwedge \neg y_d.$$

We can easy see by induction that

$$\mathrm{comp}_d(x, y) = \bigvee_{k=1}^{d} b_k.$$

Hence $\mathrm{comp}_d(x, y)$ can be realized by a network where at the first hidden layer we compute $x_j \wedge \neg y_j$ and $x_j \vee \neg y_j$, while at the second hidden layer we compute b_k. Finally, $\mathrm{comp}_d(x, y)$ is computed by OR-ing the b_k at the output. Since AND–OR gates

---

[7] Another depth-2 realization of $\mathrm{comp}_d(x, y)$ is referred to in Section 5.8 which also gets rid of the problem of weight explosion.

Table 5.1 Depth-size tradeoffs for different problems and techniques.

Depth	Technique	Symmetric functions	Parity	Symmetry
2	AND–OR based	$O(2^d)$	$O(2^d)$	$O(d)$
2	Shallow LTU	$O(d)$	$O(d)$	–
3	Telescopic LTU	$O(\sqrt{d})$	$O(\sqrt{d})$	–
$\vdots$	$\vdots$	$\vdots$	$\vdots$	$\vdots$
m	Telescopic LTU	–	$O(md^{1/m})$	–

can be computed by an LTU, we conclude that a depth-3 feedforward neural network can compute comparison. This is very interesting, since it indicates that while the previous depth-2 network computes comparison at the expense of the weight explosion, just adding one more layer suffices to overcome this complexity issue!

The discussion in this section has raised many questions on efficiency concerning the tradeoffs involving depth and size in neural networks. The answer to some of them seems to indicate clearly that the exploration of *deep networks* is very important also in the restricted class of Boolean functions. Some of the results are summarized in Table 5.1, where we can see remarkable improvements arising from the computation with more layers. In addition, as indicated in the case of comparison function, the growth of the depth can help when facing the problem of exponential explosion of the weights. Additional issues, along with the historical framework, are discussed in Section 5.8.

EXERCISES

1. [*22*]   Provide an AND–OR realization of following symmetric functions:

$$\text{maj}_d(x) = \begin{cases} 1 & \text{if } \sum_{\kappa=1}^d x_\kappa \geq d/2, \\ 0 & \text{otherwise,} \end{cases} \qquad \text{ex}_i^d(x) = \begin{cases} 1 & \text{if } \sum_{\kappa=1}^d x_\kappa = i, \\ 0 & \text{otherwise.} \end{cases}$$

Then convert such a realization into one which is based on replacing the AND–OR gates with the corresponding LTU units.

2. [*18*]   Use depth-2 LTU-based realization schemes to construct circuits for functions defined in Exercise 1.

3. [*M20*]   When using telescopic techniques, we can prove that $\sum_{n=1}^\infty 0 = 1$. The proof is straightforward. We have

$$\sum_{n=1}^\infty 0 = \sum_{n=1}^\infty (1-1) = 1 + \sum_{n=1}^\infty (1-1) = 1.$$

This clearly contradicts the obvious $\sum_{n=1}^\infty 0 = 0$. Why?

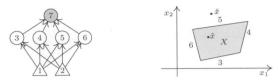

FIGURE 5.9

Classification in $\mathbb{R}^2$. The neural network with hard-limiting LTU returns $f(X) = 1$. Each neuron in the hidden layer $(4, 5, 6, 7)$ is associated with a corresponding line that together define the convex domain X.

5.3 Realization of real-valued functions

Real-valued functions can model both regression and classification problems. Interestingly, they share many common properties, some of them also related to Boolean functions. There are, however, important differences that clearly enable different structures in the neural realization.

5.3.1 Computational geometry-based realizations

When the input space is composed of real-valued features, new geometrical properties arise, but the overall computational scheme is closely related to what has been seen for Boolean functions. This is especially true in case of hard-limiting LTU. Let us begin with the example of Fig. 5.9, where a neural network with two inputs is expected to classify the patterns belonging to the convex domain X, drawn on the right. It is interesting to see the way any $x \in \mathbb{R}^2$ is mapped to the hidden layer $h \in \mathbb{R}^4$.

Let us consider the point $\hat{x} \in X$. We can promptly see that the weights of the four hidden units can easily be chosen in such a way that $x_3 = x_4 = x_5 = x_6 = 1$. We simply need to choose the weight vector w_i, $i = 1, \ldots, 4$ according to $w_i'\hat{x} > 0$. When we interpret the outputs of the hidden units as Boolean variables, the condition $\hat{x} \in X$ is clearly equivalent to stating the truth of $x_3 \wedge x_4 \wedge x_5 \wedge x_6$. If neuron 7 acts as an AND gate, we immediately conclude that $x_7 = \mathsf{T}$. Clearly this is due to the fact that $\hat{x} \in X$ and for any other $\check{x} \notin X$, on the other hand, $x_7 = \mathsf{F}$. Notice that, for the specific choice of $\check{x} \notin X$ of Fig. 5.9, we have $x_3 = x_4 = x_6 = 1$, but $x_5 = 0$, which immediately yields $x_7 = \mathsf{F}$.

Now we want to prove that any neural architecture with one hidden layer and hard-limiting LTU characterizes convex domains, just like in the example. Let $0 \le \alpha \le 1$ and suppose we have any two patterns $\hat{x}_1$ and $\hat{x}_2$ that belong to X, and consider any $\hat{x} = \alpha\hat{x}_1 + (1-\alpha)\hat{x}_2$. Since $\hat{x}_1 \in X$, we have $\hat{h}_1 > 0$. (Here we compactly express the componentwise condition $\hat{h}_i > 0$.) Likewise for $\hat{x}_2$ we have $\hat{h}_2 > 0$. Hence

$$\dot{h} = W_0\hat{x} = W_0(\alpha\hat{x}_1 + (1-\alpha)\hat{x}_2) = \alpha W_0\hat{x}_1 + (1-\alpha)W_0\hat{x}_2$$
$$= \alpha\hat{h}_1 + (1-\alpha)\hat{h}_2 > 0.$$

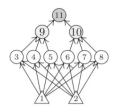
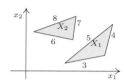

FIGURE 5.10

Classification in $\mathbb{R}^2$. The neural network with hard-limiting LTU returns $f(X) = 1$. The non-connected domain $X = X_1 \bigcup X_2$ is detected by a depth-3 neural network, where at the second hidden layer the convex domains X_1 and X_2 are isolated. Then, in the output, X is recognized by OR-ing the propositions that define X_1 and X_2.

Hence the neural network characterization property $\hat{h}_1 > 0$, for $\hat{x}_1 \in X$, and $\hat{h}_2 > 0$, for $\hat{x}_2 \in X$, is converted into $\hat{h} > 0$ for all $\hat{x} \in X$, which means that X is convex.

The characterization of convexity is a good property for the realization of more complex functions. Suppose we want to realize a neural network capable of recognizing nonconnected domains like that drawn on the right in Fig. 5.10. From the previous analysis we can promptly see that neurons 4, 5, 6 and 6, 7, 8 characterize the convex sets X_1 and X_2, respectively. In particular, their recognition is carried out by the neurons 9 and 10 that, according to what has been established in the previous analysis, must act as AND gates. Finally, $x \in X_1 \bigcup X_2$ holds true if and only if $x \in X_1$ or $x \in X_2$, which is in fact the OR operation that is carried out by the output neuron 11.

It is quite obvious that the construction shown in Fig. 5.10 holds for any nonconnected domain composed of convex parts. The construction also suggests how the network grows depending on the complexity of the given task. Clearly, the number of units in the first hidden layer depends on the complexity of the single convex domains, as well as on their number, which is the only parameter that, on the other hand, affects the second hidden layer.

The construction shown for nonconnected convex sets can be used to realize any concave set. The basic idea is shown in Fig. 5.11. The computational mechanism is exactly the same as that shown in Fig. 5.10 for nonconnected domains. This holds true since we can provide a partition of X by convex sets X_1 and X_2 that are detected according to the already shown scheme.

In Fig. 5.11, we show that there are neurons that are shared in the detection of the single convex parts. In the figure, neuron 5 is in fact participating in the construction of both convex sets. Notice that the shown realization does not require decomposing the given set according to a partition. Of course, if $X_1 \bigwedge X_2 \neq \emptyset$, the OR-ing of the neurons 8 and 9 still returns the correct response. In general, given X, we need to find a family of sets $F_X = \{X_i : i = 1, \ldots, m\}$ such that $\bigvee_{i=1}^{m} X_i = X$. This construction provides an insight on the complexity of detecting concave sets. In addition, it suggests that there are many different constructions that depend on the way X is covered by the family F_X. In Fig. 5.12–(a, b) we can see two nice examples to illustrate this

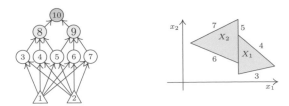

FIGURE 5.11

Classification in $\mathbb{R}^2$ of the concave set $X = X_1 \wedge X_2$. The processing is the same as that of Fig. 5.10, but in this case, one neuron (5) is shared in the two convex sets.

(a) (b) (c) (d)

FIGURE 5.12

Different domains that can be realized by a depth-3 neural network.

issue. In order to realize the "star" (a), first of all, we can look for an appropriate compositional representation that takes place by composing figures $X_\kappa \in F_X$ of the family $F_X \subset 2^X$. The composition takes place on 2^X by using ordinary set operators. The star can be composed as indicated below

$$\bigstar = \bigwedge \cup \bigtriangledown = \left(\bigwedge \setminus \bigtriangleup\right) \cup \left(\bigtriangledown \setminus \bigtriangledown\right) =: (X_1 \setminus X_2) \cup (X_3 \setminus X_4).$$

Notice that, for this to make sense, the set operators must act on elements of F_X that are properly located in the retina. We can promptly see that the four convex domains, X_1, X_2, X_3, and X_4, can be realized by three neurons each at the first hidden layer. Now $X_1 - X_2$ can be realized by considering a common neuron by five neurons, and the same holds for $X_3 - X_4$. Two units at the second layer, and one output, complete the architecture. Notice that the learning task of Fig. 5.12–(c) contains a hole. Domains with holes typically correspond with complex realizations; in this case we can promptly see that a depth-3 neural network can provide a realization by composing five convex sets of three units each. This example clearly shows that this decomposition can be carried out in many different ways. For example, another composition is the one suggested by Fig. 5.12–(c), where we consider the hole as one set, along with the other five triangles. Finally, the annulus of Fig. 5.12–(d) can be based on a depth-2 network with two radial basis function hidden units only.

The realization of the Mickey mouse domain of Fig. 5.12–(b) can be based on

$$\text{🐻} = \text{⬤} \cup \text{⚫} \cup \text{⚫}$$

In this case, the composition requires detecting circles, which can naturally be carried out by radial basis function units. Hence three units at the first layer and one output unit are enough for the realization. Notice that this depth-2 neural network cannot be realized by hard-limiting LTU, which can only approximate the circles. Better approximations are possible with smooth functions such as the hyperbolic tangent.

The depth-3 architecture used for nonconnectedness and concavity can be used for detecting pictures with holes.

To sum up, nonconnectedness and concave domains can be constructed by appropriate neural composition. These constructions very much resemble the *cover problem*. An insight on these connections will be given in Section 5.8, where we can capture important issues connected with the computational complexity of learning.

EXERCISES

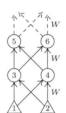

1. [*18*] Provide the realization for the "star" learning task of Fig. 5.12–(a) and (c).

2. [*20*] Consider the infinite network in the picture, where the weight layer matrix is independent of the layer. Compute its response once we are given W.

5.3.2 Universal approximation

In this section, we extend the discussion previously centered on geometric issues, and review basic ideas and result on universal approximation by discussing also the role of the neural nonlinearity. In addition to classification, we will discuss functions conceived for regression and their corresponding realization.

We start by presenting a general technique that relies on the principle of griding the function domain, so as to associate any single hypercube with a real value that provides an opportune approximation of the function. Without limitation of generality, the technique is proposed for functions $f : X \subset \mathbb{R}^d \to \mathbb{R}$. The basic idea is sketched in Fig. 5.13, where the input space X is partitioned into n hypercubes, which might not necessarily be of the same volume. Basically, we assume that $X = \bigcup_{i=1}^{n} B_i$ and $B_i \bigcap B_j = \emptyset$ whenever $i \neq j$.

For the sake of simplicity, in the figure, a $2D$ domain is gridded by uniform squares. Each of these boxes is represented by a corresponding node, numbered from 3 to 18, that is expected to contain the value of the function in the center of the corresponding box. The given function can be approximated on a certain box B_i by $f(x) \simeq f^*(x) = f_i[x \in B_i]$, where we assume that f_i takes the value of the center of box. Hence the approximating function f^* over X is

$$f^*(x) = \sum_{i=1}^{n} f_i[x \in B_i].$$

Now we need to find an appropriate realization of the characteristic $[x \in B_i]$. The idea can be grasped from Fig. 5.13, where each box is associated with a feedforward

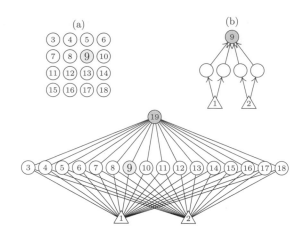

FIGURE 5.13

Approximation on a domain $X \subset \mathbb{R}$. A griding by 16 squares, numbered from 3 to 18, is used to associate each node with a hidden unit. The corresponding output connection sets the value of the function.

network (see, e.g., box B_9, which is associated with the network–(b)). In this $2D$ case every box B_i can be defined as

$$B_i = \{(x_{i,1}, x_{i,2}) \in \mathbb{R}^2 : (x_{i,1} \geq x_{i,1}^-) \wedge (x_{i,1} < x_{i,1}^+) \wedge (x_{i,2} \geq x_{i,2}^-) \wedge (x_{i,2} < x_{i,2}^+)\}. \tag{1}$$

Now a selective response to these boxes can be obtained from the neural net of Fig. 5.13–(b). Two pairs of hard-limiting hidden units detect the proposition stated by Eq. (1) that define B_i. Then, in the same figure, there are 16 similar networks used to select all the boxes that cover the domain X of f. Finally, the value of the function inside each domain is approximated by $f_i \cdot [x \in B_i]$.

It is worth mentioning that this constructive method of realizing a function is closely related to those proposed in Section 5.3.1. However, since the realization proposed in this section needs to construct functions that return real numbers, it is convenient to rely on the simplified assumption of covering X by a partition. In so doing, a depth-3 network provides a *universal approximation* by a very general scheme that holds also for regression. However, both realizations share the need to solve a *set cover* problem. It is quite obvious that the approximation of a given function increases when increasing the number n of approximating boxes. Exercise 1 proposes the analysis of the approximation of functions which satisfy Lipschitz condition. The main conclusion is that the number n of hypercubes to achieve a certain accuracy ϵ is exponential with the dimension d of X, that is, $n = \propto (1/\epsilon)^d$. This depth-3 network is a sort of *vanilla realization* that, clearly, is very expensive.

Now we show that we can provide realizations that, like those inspired by canonical representations of Boolean functions, only require one hidden layer. In order to grasp the idea, let use restrict the analysis to one-dimensional functions

$f: X \subset \mathbb{R} \to \mathbb{R}$. The domain can be partitioned into the collection of intervals $\{[x_{i-1}, x_i : i = 1, \ldots, n\}$, where $x_i^m = (x_{i-1} + x_i)/2$ is the coordinate of the middle point. Hence the function f can be approximated by

$$f^*(x) = \sum_{i=1}^{n} (f(x_i^m) - f(x_{i-1}^m)) H(x - x_i), \qquad (2)$$

where we assume $f^*(x_0^m) = 0$. As proposed in Exercise 2, we can prove that $f^*(x_i^m) = f(x_i^m)$ by checking the value in the middle of each interval, and that, under the condition that f satisfies the Lipschitz condition, the error is $O(1/n)$. Interestingly, this staircase approximation suggests the replacement of $H(x - x_i)$ with $\sigma(x - x_i)$, where $\sigma(\alpha) = 1/(1 + e^{-\alpha})$. It is easy to see that $H(x - x_i) \simeq \sigma(x - x_i)$ for large values of the weights. To sum up, the approximation based on Eq. (2) suggests a straightforward depth-2 feedforward neural network realization. Basically, we conclude that the universal approximation is gained by using one hidden layer only. This property fails not only for one-dimensional domains but, as it will be pointed out, it is in fact general.

Let $f: X \in \mathbb{R}^d \to \mathbb{R}$ be a Lebesgue integrable function, so that we can define its Fourier transform

$$\hat{f}(\xi) = \int_X f(x) e^{-2\pi i \xi \cdot x} \, dx.$$

When inverting $\hat{f}(\xi)$, we can express f as

$$f(x) = \frac{1}{(2\pi)^d} \int_{\mathbb{R}^d} f(\xi) e^{2\pi i \xi \cdot x} \, d\xi.$$

Now any function of practical interest typically gets a bounded spectrum, so as we can restrict the above integral over a bounded domain $\Xi \subset \mathbb{R}^d$. If we partition Ξ into n hypercubes B_i with the same volume, so $\Xi \subset \bigcup B_i$ and $B_i \cap B_j = \emptyset$ for $i \neq j$, then we can approximate f by means of

$$f(x) \simeq V \sum_{\kappa=1}^{n} f(\xi_\kappa) e^{2\pi i \xi_\kappa \cdot x} = \sum_{\kappa=1}^{n} \omega_\kappa e^{2\pi i \xi_\kappa \cdot x}$$

where $V = \mathcal{L}^d(B_i)/(2\pi)^d$ and $\omega_\kappa = V f(\xi_\kappa)$. Now let $\omega_\kappa = |\omega_\kappa|(\cos(\phi_\kappa) + i \sin(\phi_\kappa))$. Since f is a real-valued function, we only take $\Re\left(\omega_\kappa e^{2\pi i \xi_\kappa \cdot x}\right)$ and get

$$f(x) \simeq \sum_{\kappa=1}^{n} |\omega_\kappa| (\cos \phi_\kappa \cos 2\pi \xi_\kappa \cdot x - \sin \phi_\kappa \sin 2\pi \xi_\kappa \cdot x) \qquad (3)$$

This representation of f promptly suggests a realization by a depth-2 feedforward neural network. Let $w_\kappa = 2\pi \xi_\kappa$. We can associate any hypercube with the two hidden units which return $\cos w_\kappa \cdot x$ and $\sin w_\kappa \cdot x$, respectively. Interestingly, since by using the staircase expression (2), we gain universal approximation of any function of a single variable, then $\cos w_\kappa \cdot x$ and $\sin w_\kappa \cdot x$ can be expressed accordingly. Hence we conclude that a depth-2 network with sigmoidal functions provides universal approximation also in multidimensional spaces.

EXERCISES

1. *[M22]* Let us consider the *brute force* universal approximation scheme described for functions $f : X \subset \mathbb{R}^d \to \mathbb{R}$ in Section 5.3.2, where the boxes are hypercubes of dimension p. Prove that if f satisfies the Lipschitz condition than the approximation accuracy is exponential with d.

2. *[18]* Given $f : X \subset \mathbb{R} \to \mathbb{R}$, prove the stairwise representation f^* return f on the nodes of the grid, that is, $f(x_k^m) = f^*(x_k^m)$. Then prove that if f satisfies the Lipschitz condition then the accuracy is $O(1/n)$.

5.3.3 Solution space and separation surfaces

Most of the studies on feedforward nets have an experimental flavor: Once the weights of the nets have been learned, it is common not to inquire on the structure of the solution. In this section we shed light on a few interesting properties of the solution space of the learning problem, which provide some useful advices in the experimental setup. We begin pointing out a curious property of the solution space which, as we will see, is at the basis of the success and strong diffusion of this machine learning model.

Suppose we are dealing with feedforward nets having one hidden layer only. For the sake of simplicity, let us consider the same network used for the XOR predicate, which has two hidden units. We can easily see that, no matter what the input x is, if we permute the hidden units then we get the same output, that is,

where Υ returns the output of the network, once we apply a generic $x \in X$. The reason is that the permutation does not change the accumulation of the activations on units 5.

Of course, this property holds regardless of the number of hidden units. Let I and H denote the input and hidden layer. Then forward propagation yields

$$x_i = \sigma\left(b_i + \sum_{j \in H} w_{ij}\sigma\left(b_j + \sum_{\kappa \in I} w_{j,\kappa} x_\kappa\right)\right)$$

$$= \sigma\left(b_i + \sum_{j \in \mathrm{perm}(H)} w_{i,j}\sigma\left(b_j + \sum_{\kappa \in I} w_{j,\kappa} x_\kappa\right)\right).$$

This equation indicates that the computation is independent of the $|H|!$ different permutations of the neurons in the hidden layer. A network with as few as 10 hidden units, which is typically just a toy in most real-world experiments exhibits at least 3,628,800 solutions! This has a great impact on the likelihood of discovering the absolution minimum of the error function. Like for any other configuration, in this case there are nearly 4 million different solutions which all return the same absolute minimum. Clearly, in many real-world experiments, the number of these configurations is really huge and explodes with the cardinality of the hidden layer.

In case of deep nets, suppose the $h = |H|$ hidden units are split into p layers. We can promptly see that the number S of equivalent configurations shift from $h!$ to

$$S(h_1, \ldots, h_p) = \prod_{\sum_i h_i = h} h_i!. \tag{1}$$

Because of symmetry, maxima and minima of S correspond with equal values of $h_i = h/p$, that is,

$$S(h_1, \ldots, h_p) = \prod_{\sum_i h_i = h} h_i! \le ((h/p)!)^p. \tag{2}$$

Here, for the sake of simplicity, h is a multiple of p. In order to understand the effect of distributing the hidden units in different layers, we consider $\log S$, that is,

$$\log S(h_1, \ldots, h_p) = \sum_{\sum_i h_i = h} \log h_i! \le p \log (h/p)!.$$

The minimum is reached for $p = h$, namely when $\forall i = 1, \ldots, p : h_i = 1$. We restrict the analysis of the maximum to the case of large values of h_i. In order to see the role of p, from Stirling formula[8] we have

$$\ln S(h_1, \ldots, h_p) = p \left(\frac{h}{p} \ln \frac{h}{p} - \frac{h}{p} + O\left(\ln \frac{h}{p} \right) \right).$$

For big h and big h we get

$$\ln S(h_1, \ldots, h_p) \simeq h \ln \frac{h}{p} - h$$

that is maximized for $p = 1$ (OHL nets), which corresponds also with stating that

$$S(h_1, \ldots, h_p) = ((h/p)!)^p \le h! \tag{3}$$

---

[8] Since ln and log only differ by a factor, the analysis is the same. However, log is preferable for the following complexity analysis.

A general proof of this equation is proposed in Exercise 3. Eqs. (2) and (3) clearly indicate that the maximum number S of symmetric configurations arises in the case of OHL nets. This suggests that the discovery of absolute minima is strongly favored in this case, whereas the OHL assumption does not optimize the search for different configurations.

Interestingly, this nice symmetry is not the only remarkable property of the solution space. Suppose we use $\sigma = \tanh$. Then

where the gray level in the connections of the second network indicates that the weights are the same as the corresponding weights of the first network (black connections), with flipped signs. So, for example, connection $w_{3,1}$ in the first network becomes $-w_{3,1}$ in the second network. In general, whenever σ is an odd function, we have

$$\sum_{j \in H} w_{i,j} \sigma \left(\sum_{\kappa \in I} w_{j,\kappa} x_\kappa \right) = \sum_{i \in H} (-w_{i,j}) \sigma \left(\sum_{\kappa \in I} (-w_{j,\kappa}) x_\kappa \right).$$

This identity reveals the invariance, which is gain when replacing

$$(w_{i,j}, w_{j,\kappa}) \rightarrow (-w_{i,j}, -w_{j,\kappa}). \tag{4}$$

For any given configuration of weights $\mathcal{C}_w$, the above sign flip yields $\mathcal{C}_w \rightarrow \mathcal{C}'_w$, which is equivalent. The number of sign flip corresponds with 2^h. When considering Eq. (1), the overall number of configurations turn out to be

$$|\mathcal{C}_w| = \prod_{i=1}^{p} 2^{h_i} h_i!$$

Clearly, this is huge even for small networks. As will be seen later, this huge number of equivalent configurations is one the main reasons for the availability of efficient learning algorithms. Related analyses can be carried out concerning a comparison between shallow and deep nets when considering this additional symmetry.

When looking inside feedforward nets, there are other "math goodies" with strong practical impact. Let us analyze the frontier of the solution space. Given a multilayered network $\mathcal{N}$, which returns on the given $x \in X$ the single output $F(w, x)$, we call *separation surface* the set

$$S := \{x \in X \subset \mathbb{R}^d : F(w, x) = 0\}. \tag{5}$$

The separation surface depends on the architecture of the net as well as on the neuron nonlinearity. We separate the discussion between ridge and radial-basis function neurons. We mostly study ridge neurons equipped with the Heaviside function. Let us begin with XOR nets in the case of $\sigma :=$ sign and assume that the weights are

$$w_{3,1} = 1, \quad w_{3,2} = 1, \quad b_3 = -1/2,$$
$$w_{4,1} = 1, \quad w_{4,2} = 1, \quad b_4 = -3/2,$$
$$w_{5,3} = 1, \quad w_{5,4} = 1, \quad b_5 = -1/2.$$

Clearly, $\forall x \in X_{\oplus}$ defined by

$$X_{\oplus} = \{(x_1 + x_2 - 1/2 > 0) \wedge (x_1 + x_2 - 3/2 < 0)\}$$

we have $F(w, x) > 0$ and $S_{\oplus} = \partial X_{\oplus}$. If we consider two-dimensional spaces and with hard-limiting nonlinearities, like the Heaviside and sign functions, then the separation surface is the frontier of polygons. For $h = |H| = 2, 3$, and 4, we have

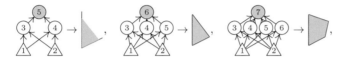

where each network is associated with the corresponding detected domain. As we can see, depending on the value of h, there is a fundamental difference between the cases $h = 2$ and $h > 2$. In the former, there are only two hidden units, which can only generate domains bounded by two separating lines. Hence the detected domain cannot be bounded. The above mentioned case of the XOR network belongs to this class; it is characterized by parallel separating lines that, however, still define an unbounded domain. For $h = 3$ and 4, the separating surfaces can define the above depicted bounded polytopes and, clearly, the same holds for higher values of h.

These examples suggest that in order to generate bounded domains, the network the value of h must be at least $d + 1$. In case of $d = 3$, the first polytope with this property is the tetrahedron, which has 4 faces ($h = 4$). Notice, however, that the condition

$$h = |H| > d \tag{6}$$

is only *necessary* for the domain to be bounded. For example, for $d = 3$, we can have

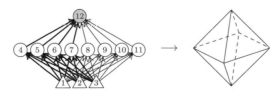

Of course, a network with 8 hidden neurons can generate a *diamond*-shaped bounded domain. However, in this case, if we use only the 4 units, represented by thick connections, then the generated domain is not bounded! The problem proposed in Exercise 6 gives additional insights on the nature of the problem.

This analysis can be extended to the case of monotonic σ (see Section 5.8). If σ is not monotonic then bounded domains can also be created with $h = |H| = d$ (see Exercise 1). In case of deep nets, the conclusions drawn for OHL nets are not remarkably different, since the first hidden layer plays the role of the single hidden layer in OHL nets. Notice that, as pointed out in Section 5.8, in no way the capabilities of constructing bounded domains result in the guarantee of their discovering!

In order to construct classifiers with guaranteed generation of bounded domains by ridge neurons architectures, we can adopt auto-encoders. The idea is that of modeling each class by a corresponding network, which is expected to create a more compact representation of the inputs. In this case an auto-encoder generates the separation surface

$$S = \{x \in X \subset \mathbb{R}^d : \| F(w, x) - x \| = \epsilon\} \tag{7}$$

where $\epsilon > 0$ is an appropriate threshold. We can see that if the output neurons of the auto-encoders are linear then the domain D, defined by the frontier S, is bounded. This can be gained by considering any $x \in D$. It satisfies $\| x - \hat{w}_2 \sigma(\hat{w}_1' \hat{x}) \| < \epsilon$. Now, if we consider any input scaled along x, we have

$$\| \alpha x - \hat{w}_2 \sigma(\hat{w}_1' \alpha \hat{x}) \| = |\alpha| \| x - \frac{1}{\alpha} \hat{w}_2 \sigma(\hat{w}_1' \alpha \hat{x}) \| < \epsilon.$$

Since σ is upper bounded, there exists $|\alpha| < A \in \mathbb{R}_+$ such that this condition is only be met for $|\alpha| < A \in \mathbb{R}^+$. Hence, no matter which direction defined by $x/\|x\|$ is considered, if x is auto-encoded, there always exists $|\alpha| > A$ such that αx is not auto-encoded, that is, D is bounded.

The analyses on the structure of the separation surfaces in the case of radial function neurons lead to very different results. The intuition is that radial basis function neurons are only active in a bounded region close to the center, which means that the issue of open separation surfaces doesn't hold anymore (see Exercise 2).

EXERCISES

1. [*M20*] As shown in Section 5.5.3, the analysis on separation surfaces using the sign or the Heaviside functions leads to conclude that bounded domains can only be generated whenever $h = |H| > d$. When restricting to $X \subset \mathbb{R}^2$, prove that in case of nonmonotonic functions, bounded domains can be generated also when $h = |H| \leq d$. Then generalize the property to any dimension.

2. [*18*] Prove that OHL nets with radial basis functions in the hidden layer always generate closed separation surfaces (bounded domains).

3. [*M22*] Prove the inequality (3) for any h and p.

5.3.4 Deep networks and representational issues

In this section, we show that, like for Boolean functions, the depth/size tradeoff is of fundamental importance also for real-valued functions. We start the discussion with the example shown in Fig. 5.14, where two networks with two-dimensional inputs

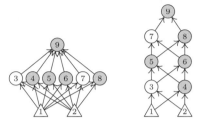

FIGURE 5.14

Shallow versus deep networks.

and six hidden neurons are compared. They are expected to be used as classifiers, so as only one output neuron is used. Furthermore, let us assume that the networks are composed of rectifier neurons.

Depending on the input, the hidden neurons are either activated (linear regime) or deactivated. Once a neuron is deactivated, it does not propagate its activation to forward paths. Hence, if we know which neurons are activated by the given input, we can directly compute the output. In the shallow net we have

$$
\begin{aligned}
x_9 = {}& (w_{9,4}w_{4,1} + w_{9,5}w_{5,1} + w_{9,6}w_{6,1} + w_{9,8}w_{8,1})x_1 \\
& + (w_{9,4}w_{4,2} + w_{9,5}w_{5,2} + w_{9,6}w_{6,2} + w_{9,8}w_{8,2})x_2 \\
& + b_9 + w_{9,8}b_8 + w_{9,6}b_6 + w_{9,5}b_5 + w_{9,4}b_4.
\end{aligned} \tag{1}
$$

We can easily see that the same analysis for the deep net (right-side in Fig. 5.14) yields

$$
\begin{aligned}
x_9 = {}& (w_{9,8}w_{8,5}w_{5,4}w_{4,1} + w_{9,8}w_{8,6}w_{6,4}w_{4,1})x_1 \\
& + (w_{9,8}w_{8,5}w_{5,4}w_{4,2} + w_{9,8}w_{8,6}w_{6,4}w_{4,2})x_2 + b_9 + w_{9,8}b_8 \\
& + w_{9,8}w_{8,6}b_6 + w_{9,8}w_{8,5}b_5 + (w_{9,8}w_{8,5}w_{5,4} + w_{9,8}w_{8,6}w_{6,4})b_4.
\end{aligned} \tag{2}
$$

From Eqs. (1) and (2), we can promptly induce the general rule: The inputs[9] x_1, x_2, 1 propagate their values to the output through all possible paths passing from active neurons. The processing is therefore restricted to the set A of active neurons. Let I be the set of inputs and denote by path$_j(i, 0)$ any path j which connects the input $i \in I$ to the single output o. In addition, if κ is any vertex in the graph, we denote by $\kappa \rightarrow$ a connected vertex according to the topological sort. Hence, in general, we have

---

[9] Notice that, as usual, we regard 1 as the virtual input to consider the bias contribution.

$$x_o = \sum_{i \in I} x_i \underbrace{\sum_{j=1}^{|\text{path}(i,o)|} \prod_{\kappa \in \text{path}_j(i,o)} w_{\kappa \to ,\kappa}}_{\overline{w}_{o,i}} + \underbrace{\sum_{\alpha \in A} b_\alpha \prod_{\kappa \in \text{path}_j(\alpha,o)} w_{\kappa \to ,\kappa}}_{\overline{b}}$$

$$= \sum_{i \in I} \overline{w}_{o,i} x_i + \overline{b}.$$

(3)

This indicates a reduction to the linear regime and the corresponding equivalent weights $\overline{w}_{o,i}$ and $\overline{b}$.

Let us compare the shallow and deep nets. In both the shallow and the deep net $A = \{4, 5, 6, 8\}$. In the shallow net the paths of active neurons are

$$\text{path}(1, 9) = \big\{\{1, 4, 9\}, \{1, 5, 9\}, \{1, 6, 9\}, \{1, 8, 9\}\big\},$$
$$\text{path}(2, 9) = \big\{\{2, 4, 9\}, \{2, 5, 9\}, \{2, 6, 9\}, \{2, 8, 9\}\big\},$$

while in the deep net, we have

$$\text{path}(1, 9) = \big\{\{1, 4, 5, 8, 9\}, \{1, 4, 6, 8, 9\}\big\},$$
$$\text{path}(2, 9) = \big\{\{2, 4, 5, 8, 9\}, \{2, 4, 7, 8, 9\}\big\}.$$

In order to compare the representational power of the two nets, we can simply compare their possible configurations of active neurons. However, in the shallow net, we only have six "different" configurations, which correspond to the number of hidden neurons. The reason is that OHL nets exhibit a huge number of equivalent configurations. In both cases, the overall number of active configurations is

$$\sum_{i=1}^{6} \binom{6}{i} = \sum_{i=0}^{6} \binom{6}{i} - 1 = 2^6 - 1 = 63.$$

All groups with the same number of hidden units yield the same output, which is determined by summing up over the hidden layer. The case of deep net is more involved. The net requires at least three active hidden units, otherwise the output is detached from the input. Hence we are in front of a combinatorial structure such that, depending on the inputs, we have 3, 4, 5, or 6 active hidden neurons. As already seen, intralayer permutations do not enrich the computational capabilities, but layer permutation leads to different nets. More precisely, each of the six permutations of the three layers corresponds with different configuration. Depending on this number the corresponding patterns of active neuron in the same layer are shown in Table 5.2.

For example when there are 4 active neurons, they can be distributed as indicated above, that is, 2 active neurons in one layer and 1 active neurons in the remaining two layers. Similarly, for 5 and 6 active units. Overall, there are 8 different configurations.

This example is instructive and can nicely be generalized. It is clear that as the depth D of layered networks increases, the number of different configurations explodes.

Table 5.2 Distribution of active patterns.

Active Units	Patterns	Configuration/Pattern
3	$\{1, 1, 1\}$	1
4	$\{2, 1, 1\}, \{1, 2, 1\}, \{1, 1, 2\}$	3
5	$\{2, 2, 1\}, \{2, 1, 2\}, \{1, 2, 2\}$	3
6	$\{2, 2, 2\}$	1
overall	−	8

EXERCISES

1. [*18*]   Let us consider a feedforward neural network with two inputs and one output that is charged for classifying the domain below.

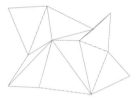

Suppose we want to use a depth-3 network that operate according to the constructing scheme proposed in Section 5.3. How many hidden units can be minimally used? Draw the network.

5.4 Convolutional networks

Convolutional networks are neural architectures that are mostly used in computer vision. An in-depth understanding of their meaning, however, goes well beyond their marriage with vision, and involves general and fundamental principles on the *contextual information* associated with a focused point, which leads to the extraction of *invariant features*. Hence this is a fundamental topic in machine learning, which arises when involving spatiotemporal information and, especially, perception. Most of the topics covered so far have been unified by the simplified assumption that an intelligent agent is asked to take decisions on the basis of a compact representation of the environment, that is expressed by a vector $x \in X \subset \mathbb{R}^d$. In Section 1.3, we have given a classic example of the way patterns of the real world can be represented in the memory of a computer. In that case images of handwritten characters can nicely be given a natural vectorial representation that comes from de-sampling high-resolution images. Interestingly, it has been shown that high resolution might not be required for the purpose of character recognition, so as simplified pattern representations can work effectively. While the MNIST learning task has a strong educational content, it might hide important issues on character segmentation. In most real-world problems one needs to carry out segmentation before recognition, a problem which could be as difficult as that of pattern recognition! As already discussed in Section 1.4.1, in typical problems of vision, we are trapped into the chicken-and-egg dilemma on whether classification of objects or their segmentation takes place first. The solution of most interesting cognitive problems in perception is trapped in this dilemma, which seems to be tightly related to the structure of the information that is captured by humans. In natural vision, while the information acquired by the retina somewhat resembles the processing taking place in cameras, the saccadic movements of the eyes act with the purpose of focusing attention, at any time, on a specific point. Focus of attention in vision seems to be just a nice way of breaking the chicken-and-egg dilemma: If we focus on a specific point, we can get rid of information flooding, since we can spot visual concepts that emerge from the point of focus. Clearly, this is tricky, since we don't know what context of the pixel must be taken into account for the decision. This computational scheme is not restricted to vision. Humans mostly experience a spatiotemporal interaction in their perceptual acquisition of environmental information. The process of speech and language understanding is also interwound with temporal and spatial coordinates, a property that is shared by all five senses, including taste, smell, and touch.

5.4.1 Kernels, convolutions, and receptive fields

The previous discussion on the role of the spatiotemporal context and of the focus of attention suggests that we need to discover a way of providing a compact expression of contextual information.

One of the most challenging issues in vision is human ability to jump easily from pixelwise to more abstract visual concept recognition that involve portions of a video.

This is Pink Panther

There is a blanket

hat   This is Pink Panther

sheriff star

FIGURE 5.15

Focus of attention and extracted concepts. (Left) The focus on the point leads to the emergence of the concept of `Pink Panther`. There are many different "objects" in the picture, including a `blanket`. (Right) The role played by the context: The largest blob evokes the presence of the `Pink Panther`, whereas the other two smaller blobs (smaller contexts) refer to the `hat` and to the `sheriff star`.

When focusing on a given pixel in a picture, humans can easily make a list of "consistent objects" that reflect the visual information around that pixel. Interestingly, that process takes place by automatically adapting the "virtual window" to be used for the decision.

This results in the typical detection of objects with dimension which is growing as that virtual window gets larger and larger. More structured objects detected at a given pixel are clearly described by more categories than simple primitive objects, but, for humans, the resulting *pixelwise* process is surprisingly well-posed from a pure cognitive point of view. Fig. 5.15 provides motivations for an appropriate representation of the visual information. On the left a visual agent is expected to attach a class category to the picture in the indicated pixel. The pixel clearly belongs to `Pink Panther`, but that label might not be necessarily the only one; when considering a wider context, one could consistently return the class `Pink Panther room`. On the other hand, depending on the communication of the class label, a visual agent can be instructed in remarkably different way. When pointing to a specific point, there is *strong supervision*, whereas the linguistic statement "`there is a blanket`" is only reporting the presence of a blanket somewhere in the picture. This is referred to as *weak supervision*, namely a kind of supervision with less informative value, which consequently requires more cognitive skills to acquire the related visual concept. On the right the role of context is emphasized. Smaller blobs are identifying the `sheriff star` and `hat`, respectively, whereas the largest blob is identifying the `Pink Panther`. Clearly, as the context becomes larger and larger, the associated concepts increase their degree of abstraction.

Now let $\Omega \subset \mathbb{R}^2$ be the *retina*, where each pixel is identified by $z = (z_1, z_2)$. Let $v(z) \in \mathbb{R}^{\vdash m}$ be the brightness of pixel z, where $\vdash m = 1$ for black–white pictures and $\vdash m = 3$ for color pictures. Let us introduce a set of *kernel-based filters* defined by the symmetric vectorial function $g : \Omega^2 \to \mathbb{R}^m$, $(u, z) \to g(u, z) \in \mathbb{R}^{m, \vdash m}$. Now we can take

$$y(z) := g(z, \cdot) * v = \int_\Omega g(z, u)v(u)du \qquad (1)$$

as a compact representation of the contextual information associated with z. The symmetry of g means that $g(u, z) = g(z, u)$. Each component $g_{\alpha, \beta}(u, z)$, $\alpha = 1, \ldots, m$, $\beta = 1, \ldots, \vdash m$ of g contributes to determining the output according to matrix multiplication $g(z, u)v(u)$. A remarkable case is that in which $g(z, u) = h(z - u)$. As it will be shown later, the choice of peaked filters leads to computing values that are independent of the pixel position. Whenever $g(z, u) = h(z - u)$, the computation defined by Eq. (1) returns the map $y(\cdot)$ that is the *convolution* of filter $g(\cdot)$ with the video signal $v(\cdot)$. It's worth mentioning that the notion of convolution is typically given for $\vdash m = m = 1$. This extension over vectorial signals yields a similar, yet hybrid, marginalization on $u \in \Omega$, defined by the operator $\int_\Omega$, and of β, defined by the operator $\sum_\beta$. Another interesting case arises if we provide a distributional interpretation of $g_{\alpha, \beta}$; this allows us to consider

$$\forall \alpha = 1, \ldots, m, \quad \forall \beta = 1, \ldots, \vdash m, \quad g_{\alpha, \beta}(z, u) = \delta(z - u)$$

so as, from (1), we get $y(z) = v(z)$. Basically, in this case, the extraction of contextual information results in the degeneration of convolution, and the output simply corresponds with the brightness $v(z)$ of the pixel.[10] The brightness v is characterized by $\vdash m$ features (e.g., the RGB components), while the convolutional map returns an output y represented by m features. While for any pixel located by z, the maps $v(\cdot)$ and $y(\cdot)$ share the principle of returning a certain number of associated features, one might argue that map $y(\cdot)$ represents contextual information, whereas map $v(\cdot)$ only expresses lighting properties of the single pixel, regardless of its neighbors. In other words, $v(\cdot)$ returns local (pixel-based) features, while $y(\cdot)$ returns global (context-based) features. However, this is not a very solid argument! Map $y(\cdot)$ produced by Eq. (1) is just another view of the image. When considering the physical nature of the video signal, it turns out that each picture (frame) on the retina, expressed by the brightness v, can always be regarded as the outcome of the convolution with a certain filter. One could always regard v as the output of filter $\bar{h}$ with input $\bar{v}$, that is, $v = \hat{h} * \bar{v}$. In so doing, the previous convolutional computation yields

$$y = h * v = h * (\bar{h} * \bar{v}) = (h * \bar{h}) * \bar{v} = \tilde{h} * \bar{v},$$

where the output y comes from the new brightness $\bar{v}$ processed by the filter $\tilde{h} := h * \bar{h}$. As a consequence, Eq. (1) is just a natural representation that arises at some level of processing of the light reflected by an abject. The brightness $v(z)$ on pixel z can itself be reasonably regarded as the output of another filter, so we can conceive the output of Eq. (1) as a cascade-based processing with multiple convolutions.

---

[10] This degeneration property is very useful for motion invariant features.

The convolution returns a feature vector $y(z) \in \mathbb{R}^m$ that depends on the pixel z, on which we are focusing attention. In a sense, $y(z)$ can be regarded as a way of providing a compact representation of the pair (v, z), where here we consider v as a representation of the picture on the retina Ω. Hence any decision process on (v, z) turns out to be reduced to the processing of $y(z)$. The relation $y(z) \equiv (v, z)$ is clearly dependent on the choice of the filters that are expected to model the context. In addition, the convolutional representation of the context provides an amount of information that clearly depend on m. The incorporation of larger and larger contexts requires representations with increasing values of m.

As shown in Exercise 1, under mild conditions, the convolution is not only associative — a property that has been used in the above equation — but also commutative, which turns out to be very useful from a computational point of view. The numerical computation of convolution can be gained simply by considering the associated definition in the discrete domain (see also Exercise 2).

Interestingly, we can easily see that, under an appropriate choice of g, when m becomes large, $y(z)$ offers an arbitrary good approximation of the information $(v, z$ in the retina. We can gain this interesting result simply by approximating the convolution in the discrete setting. Of course, this is important in itself for computational purposes. Suppose we divide the retina Ω by

$$z = (i_z, j_z)\Delta_z, \qquad i_z = 0, \ldots, n_i - 1, \quad j_z = 0, \ldots, n_j - 1.$$

Suppose that the brightness $v^\sharp$ is defined by the features $0, 1, \ldots, \vdash m - 1$. Then a discrete approximation of Eq. (1) turns out to be

$$\forall \alpha, i_z, j_z, \quad y^\sharp(\alpha, i_z, j_z) = \sum_{\beta=0}^{\vdash m-1} \sum_{i_u=0}^{n_i-1} \sum_{j_u=0}^{n_j-1} g^\sharp(\alpha, \beta, i_z, j_z, i_u, j_u) v^\sharp(\beta, i_u, j_u).$$

Here we have introduced the tensors $g^\sharp \in \mathbb{R}^{m \times \vdash m, n_i, n_j}$, $v^\sharp \in \mathbb{R}^{\vdash m \times n_i \times n_j}$, $y^\sharp \in \mathbb{R}^{m \times n_i \times n_j}$, so the above equations can concisely be written as

$$y^\sharp = g^\sharp v^\sharp. \tag{2}$$

The 3D tensor $v^\sharp$ is mapped to a consistent 3D tensor $y^\sharp$ by means of the 6D tensor g, and the above multiplication keeps the coordinates α, i_z, j_z, while marginalizing the β, i_u, j_u. In case of convolution, $g^\sharp(\alpha, \beta, i_z, j_z, i_u, j_u)$ is replaced with the 4D tensor $h^\sharp(\alpha, \beta, i_z - i_u, j_z - j_u)$, so

$$\forall \alpha, i_z, j_z, \quad y^\sharp(\alpha, i_z, j_z) = \sum_{\beta=1}^{\vdash m} \sum_{i_u=0}^{n_i-1} \sum_{j_u=0}^{n_j-1} h^\sharp(\alpha, \beta, i_z - i_u, j_z - j_u) v^\sharp(\beta, i_u, j_u).$$

$$\tag{3}$$

Interestingly, these discrete versions of kernel/convolution based computations turn out to be more natural than the associated Eq. (1), where, as already pointed out, there is a hybrid marginalization of u and β. In the discrete setting of computation the contributions from $v^\sharp(\beta, i_u, j_u)$ come from the homogeneous (integer) indexes (β, i_u, j_u), so that there is no hybrid accumulation of $\int_\Omega$ and $\sum_\beta$ anymore.

When processing the brightness by a cascade of filters, one might reasonably wonder whether we are progressively loosing information and, under which conditions, the convolution can be inverted. The extreme case, $m = 1$, is not of interest for this kind of representations, but is frequently used for image filtering (see Exercise 5). In this case, for invertibility to take place, $g^\sharp$ must be of full rank.

Now let Ω_h, Ω_s, and Ω_p be subsets of the retina which represent the context of the hat, star, and Pink Panther concepts, respectively. If we go back to Fig. 5.15 and consider extracting information on the hat, we expect that for some of the visual features $\int_\Omega g_\kappa(z, u)v(u)du \simeq \int_{\Omega_h} g_\kappa(z, u)v(u)du$, that is, they emerge only from the "hat context." To sum up, visual recognition tasks based on the (v, z), like the one in Fig. 5.15–(left), can be still faced by processing a vector $y(z)$ that nicely summarizes all the information needed to decide when focusing on z. This is good news, since we can reuse all the apparatus that has been described so far to make decision on the visual task defined by (v, z). However, as we will see, things are more involved, since the role of the filters is of crucial importance, and we need to address the problem of their appropriate selection. As we move towards visual interpretation tasks that involve the whole retina, or even entire portions of video, typical visual tasks that are solved by humans become extremely difficult. Interestingly, this difficulty arises even in the formulation of the tasks. There is in fact an interwound story of vision and language in challenging visual tasks that is briefly discussed in Chapter 7. In order to get a context in the processing of video streams, we can simply reuse the definition of convolution given by Eq. (1) when considering that the video signal is now represented by $v(t, z)$, where the retina domain Ω now becomes $\Gamma = \Omega \times \mathcal{T}$, with $\mathcal{T} = [t_o, t_1]$ being the temporal domain of the video. Hence if we define $\zeta := (t, z)$ and $\mu := (\tau, u)$ then we have

$$y(\zeta) := g(\zeta, \cdot) * v(\cdot) = \int_\Gamma g(\zeta, \mu)v(\mu)d\mu. \tag{4}$$

This time, vector $y(\zeta)$ represents the information described by the pair $\mathcal{I} := (v, \zeta)$ Again, the relevance of the association $\mathcal{I} \equiv y(\zeta)$ essentially depends on the filter function g. The following spatiotemporal factorization of g yields a simple, yet natural structure

$$g(t, z, \tau, u) = g_t(t, \tau) \odot g_z(z, u)$$

where $\odot$ is the Hadamard product. This leads us to a cascading computation of the temporal and spatial convolution, that is,

$$y(t, z) := \int_\Gamma g_t(t, \tau) \odot g_z(z, u)v(\tau, u)d\tau du = \int_\mathcal{T} g_t(t, \tau) \odot \int_\Omega g_z(z, u)v(\tau, u) \, du d\tau$$

so if we pose $V(\tau) = \int_\Omega g_z(z, u)v(\tau, u)du$ we get $y(t, z) = \int_\mathcal{T} g_t(t, \tau) \odot V(\tau)d\tau$. A further simplification arises when the convolutional filters g_t and g_z are radial functions, that is, whenever $g_t(t, \tau) = h_t(t - \tau)$ and $g_z(z, u) = h_z(\|z - u\|)$.

In speech and language understanding we are in front of a very similar problem, the only difference being that in this case we only process information along the temporal dimension. In the case of a speech signal, the filter $g_t(t, \tau)$ can be used to capture the information that is distributed in the neighborhood of t, by simply collecting $y(t) = \int_{\mathcal{T}} g(t, \tau)v(\tau)d\tau$. This is clearly a special case of the spatiotemporal convolution analyzed in vision, and we can attach a similar meaning to the convolutional filter. It is worth mentioning that when we need real-time processing of the voice signal then we are in front of a *causal filter*, which is characterized by the property

$$\forall t, \tau, \quad \text{with} \quad \tau > t, g(t, \tau) = 0$$

Let $\mathcal{T} = \mathcal{T}_- \cup \mathcal{T}_+$. Whenever the convolution returns a value $y(t)$ that only depends on the past of the current time t, we can restrict the integration to $y(t) = \int_{\mathcal{T}} g(t, \tau)v(\tau)d\tau = \int_{\mathcal{T}_-} g(t, \tau)v(\tau)d\tau$. Semantic concepts can only be extracted when considering a context that is large enough to stimulate human interpretation.

This discussion suggests that the convolutional feature extraction discussed in spatiotemporal problems, like vision and speech, can always be based on Eq. (4). Now we make three fundamental hypotheses that, as we will see in the remainder of this section, have a fundamental impact on computational issues:

(*i*)   Kernel $g(\cdot, \cdot)$ satisfies the property $g(\zeta, \mu) = h(\zeta - \mu)$;
(*ii*)   Function $h(\cdot)$ can be approximated by a kernel expansion over Γ;
(*iii*)   Brightness v and filter h both vanish on the border of their domain.

Now we analyze the consequences of these assumptions. We begin noting that the discrete formulation of convolution on the retina, given by Eq. (3), can be straightforwardly extended to a sampling of Γ, where $t = i_t \Delta_t, i_t = 0, \ldots, n_t - 1$. In so doing we have $\forall \alpha, i_t, i_z, j_z$

$$y^{\sharp}(\alpha, i_t, i_z, j_z) = \sum_{\beta=1}^{\vdash m} \sum_{i_\tau=0}^{i_{n_t}-1} \sum_{i_u=0}^{n_i-1} \sum_{j_u=0}^{n_j-1} h^{\sharp}(\alpha, \beta, i_t - i_\tau, i_z - i_u, j_z - j_u)v^{\sharp}(\beta, i_\tau, i_u, j_u).$$

(5)

Now, given any grid $\Gamma^{\sharp}$ on Γ, suppose we choose a set of points $\mathcal{R} := \{\zeta_i : i = 1, \ldots, r\}$ to be used for kernel expansion. If we restrict to retina-based convolutions, $\mathcal{R}$ is a set of r points chosen in Ω. We have

$$h(\zeta) = \sum_{\zeta_i \in \mathcal{R}} \omega_i k(\zeta - \zeta_i).$$

(6)

Here $\omega_i \in \mathbb{R}^m$ is the vector, whose components characterize the specific convolutional filter. As a consequence, we have

$$y(\zeta) = \int_\Gamma h(\zeta - \gamma)v(\gamma)\,d\gamma$$

$$= \int_\Gamma \sum_{\zeta_i \in \mathcal{R}} \omega_i k(\zeta - \gamma - \zeta_i)v(\gamma)\,d\gamma$$

$$= \sum_{\zeta_i \in \mathcal{R}} \omega_i \int_\Gamma k(\zeta - \gamma - \zeta_i)v(\gamma)\,d\gamma \qquad (7)$$

$$= \sum_{\zeta_i \in \mathcal{R}} \omega_i \xi_i(\zeta)$$

where[11] $\xi_i(\zeta) := \int_\Gamma k(\zeta - \alpha - \alpha_i)v(\alpha)\,d\alpha$. According to hypothesis (iii), convolutional filters vanish on the border of their domain. Hence, we can simplify the computation by setting $\Omega = \mathbb{R}^2$ or $\Gamma = \mathbb{R}^3$ in vision and $\mathcal{T} = \mathbb{R}$ in temporal tasks like speech and language understanding. As shown in Exercise 1, in this case, since both arguments of the convolution are *absolutely summable*, the convolution is commutative, so Eq. (7) can also be derived after commuting h with v. In that case, we have

$$y(\zeta) = \int_\Gamma h(\gamma)v(\zeta - \gamma)\,d\gamma$$

$$= \int_\Gamma \sum_{\zeta_i \in \mathcal{R}} \omega_i h(\gamma - \zeta_i)v(\zeta - \gamma)\,d\gamma$$

$$= \sum_{\zeta_i \in \mathcal{R}} \omega_i \int_\Gamma h(\gamma - \zeta_i)v(\zeta - \gamma)\,d\gamma \qquad (8)$$

$$= \sum_{\zeta_i \in \mathcal{R}} \omega_i \xi_i(\zeta).$$

We can promptly see that this expression of $y(\zeta)$, where

$$\xi_i(\zeta) = \langle v(\zeta - \cdot), h(\cdot - \zeta_i)\rangle, \qquad (9)$$

allows us to decrease significantly the computational burden with respect to Eq. (7), since the computation of $y(\zeta)$ does not require recomputing the output of the filters $h(\zeta - \gamma - \zeta_i)$ for any given ζ, since we can precompute and store the values of $h(\gamma - \zeta_i)$. Eq. (8) suggests that $y(\zeta)$ can be computed according to a neural computational scheme with inputs $\xi_i(\zeta)$ and connection weights ω_i.

Another strong reduction of the computation burden arises if we make the additional assumption that we are dealing with peaked filters capable of taking into account only information in a "small neighborhood" of ζ. This corresponds with the biological evidence that only a specific *receptive field* close to the chosen point ζ

---

[11] From now on, for the sake of simplicity, we will drop the dependence on ζ, that is, $\xi_i(\zeta) \to \xi_i$.

reacts in the computation of the activation $y(z)$. As it will be clear in the following, this assumption turns out to be very effective also for favoring motion invariance of the extracted features, an issue that likely has had an important role also in biological evolution. There's more! It will be seen that, while motion invariance favors receptive fields, the local computational restriction does require its combination with deep architectures to gain high level concepts involving large contexts. While this issue will be covered in the next section, we begin noting that the kernel expansion over $\mathcal{R}$, which has been chosen only on the basis of Ω (or Γ), can conveniently be made dependent on ζ. Hence we can expand h in the associated receptive field $\mathcal{R}_\zeta$ so that

$$y(\zeta) = \sum_{\zeta_i \in \mathcal{R}_\zeta} \omega_i \xi_i(\zeta). \tag{10}$$

The discrete sets that characterize the receptive fields $\mathcal{R}_\zeta$ are typically chosen according to simple geometrical structures. A possible choice is

$$\mathcal{R}_\zeta = \{\zeta_i : \zeta_i = \zeta + (\Delta_z, 0)i_z + (0, \Delta_z)j_z\},$$

where $i_z \in \lfloor -r_{i_z}, r_{i_z} \rfloor$ and $j_z \in [-r_{j_z}, r_{j_z}]$. In general, we can think of convolutional filters constructed over different kernel representations, so as to return different views of the contextual information (e.g., small and large contexts). As we can see, the different components of the filter are expected to capture different levels of details and context. In the extreme case, in which $k(\alpha) = \delta(\zeta)$, from Eq. (9), the distributional interpretation leads us to conclude that $\xi_i(\zeta) = v(\zeta - \zeta_i)$. Hence we get

$$y(\zeta) = \sum_{\zeta_i \in \mathcal{R}_\zeta} \omega_i v(\zeta - \zeta_i). \tag{11}$$

Notice that the distributional interpretation leads to collecting the contextual information only from pixels of the receptive field $\mathcal{R}_\zeta$, since the receptive input degenerates to the brightness. Computational issues on the convolution in the classic case of distributional kernel, that is, $h = \delta$, are discussed in Exercise 6.

Finally, notice that all the analysis carried out here for the convolution also applies for the more general case of kernels, the only difference being that the expansion clearly requires replacing $k(\zeta - \gamma)$ with $k(\zeta, \gamma)$.

EXERCISES

1. [16] Prove that convolution is commutative, that is, $u * v = v * u$.

2. [18] Let us consider convolutional filters for which $g(z, u) = h(z - u)$ and, furthermore, the case in which $h(\alpha) \in \mathbb{R}$. The notion of convolution given by Eq. (1), in case $\Omega = \mathbb{R}^2$, can be given the following associated definition:

$$\forall (m, n) \in \mathbb{N}^2 \quad y_{m,n} = \sum_{h=1}^{\infty} \sum_{\kappa=1}^{\infty} h_{m-h, n-\kappa} v_{h,\kappa}$$

What if $\Omega \subset \mathbb{R}^2$? Propose an extension of the above definition in the case $\mathcal{D}_\Omega = [1, \ldots, n_i] \times [1, \ldots, n_j]$. Prove that the proposed extended definition is just a numerical approximation of (1). *Hint:* Be careful about the definition on the border of $\mathcal{D}_\Omega$.

3. [M18] Prove that for $z \in \mathbb{R}$ we have

$$\frac{d}{dz}(u * v) = \frac{du}{dx} * v = u * \frac{dv}{dz}.$$

Hint: This an easy consequence of commutativity, and consequently the convolution of two functions has the better of the differentiability properties of the two individual functions.

4. [18] Let us consider the functions defined on $[-1, 1]$ as $u(z) = z$ and $v(z) = 1 - z$. Does the commutativity property $u * v = v * u$ hold on $[-1, 1]$?

5. [25] Suppose that $\forall z \in \Omega$ you are given $y(z) = g(z, \cdot) * v(\cdot)$. Given $y(z)$, can we reconstruct $v(z)$? This problem is referred to as *deconvolution*.

6. [18] Let us consider the case of convolutional filters where $h = \delta$. Write an algorithm for computing the convolution over a given retina. What happens at the borders of the retina?

5.4.2 Incorporating invariance

In the previous section, we have given an interpretation of spatiotemporal information in terms of the convolutional operators. For example, for images, we have seen that they can always be regarded as 3D tensors, with growing feature dimension m as the contexts become larger and larger. It has been pointed out that convolution is a natural operator to represent in a compact way local information in spatiotemporal environments. In vision, instead of processing the information $\mathcal{I} := (v, \zeta)$, which corresponds with the act of extracting information from v when focusing on ζ, one can replace $\mathcal{I}$ with the single vector $y(\zeta) = (h * v)(\zeta)$. When appropriate filters are chosen, we can foresee a computation aimed at returning features that are invariant under appropriate transformation of the input.

By and large, convolution has gained importance and popularity in the machine learning community because of its *translational equivariance*. Let us consider the case of images, but the analysis can easily be extended to video. Suppose that a small portion of the retina $\bar{\Omega}_1 \subset \Omega$ is repeated in two different locations, so one is obtained from the other by a certain translation. Basically, given any $z_2 \in \bar{\Omega}_2 \subset \Omega$ there exists $z_1 \in \bar{\Omega}_1 \subset \Omega$ such that $z_2 = z_1 + \rho$, with $\rho \in \Omega$ being the vector that defines the translation. Furthermore, suppose that $\gamma = (\gamma_{z_1}, \gamma_{z_2})$ and define the set

$$\mathcal{C} := \{\gamma \in \mathbb{R}^2 : (|\gamma_{z_1}| < \epsilon_{z_1}) \wedge (|\gamma_{z_2}| < \epsilon_{z_2})\},$$

for certain given thresholds ϵ_{z_1} and ϵ_{z_2}. The set $\mathcal{C}$ turns out to be the continuous counterpart of the receptive field $\mathcal{R}_\zeta$. It is useful to express $\bar{\Omega}_1 = \{z_1 : (z_1 = \bar{\zeta}_1 + \gamma) \wedge (\gamma \in \mathcal{C})\}$ and $\bar{\Omega}_2 = \{z_2 : (z_2 = \bar{\zeta}_2 + \gamma) \wedge (\gamma \in \mathcal{C})\}$. Now suppose we choose the filter $h(\cdot)$ according to

$$h(\gamma) = [\gamma \in \mathcal{C}]\tilde{h}(\gamma), \tag{1}$$

where $\tilde{h}(\cdot)$ is a generic filter. This assumption means that the filter $h(\cdot)$ only reacts for values in the box $\mathcal{C}$. Now, due to the assumption of vanishing on the border, we have

$$y(z_2 \in \bar{\Omega}_2) = \int_\Omega h(z_2 - \gamma)v(\gamma)\,d\gamma = \int_\Omega h(\gamma)v(z_2 - \gamma)d\gamma$$
$$= \int_\Omega [\gamma \in \mathcal{C}]\tilde{h}(\gamma)v(z_2 - \gamma)\,d\gamma.$$

Since the two portions $\bar{\Omega}_1$ and $\bar{\Omega}_2$ share the same picture, if ζ_2 is obtained from the translation of z_1 by $z_2 = \rho + \zeta_1$ then $\zeta_2 - \gamma$ is obtained from the translation of $\zeta_1 - \gamma$ and, consequently, we have

$$[\gamma \in \mathcal{C}]v(z_2 - \gamma) = [\gamma \in \mathcal{C}]v(z_1 - \gamma).$$

Hence

$$y(z_2 \in \bar{\Omega}_2) = \int_\Omega [\gamma \in \mathcal{C}]\tilde{h}(\gamma)v(z_1 - \gamma)d\gamma = \int_\Omega h(\gamma)v(z_1 - \gamma)d\gamma = y(z_1 \in \bar{\Omega}_1).$$

To sum up, translational invariance is gained because of the convolutional expression of $g(z, \gamma) = h(z - \gamma)$ and because of the receptive field assumption.

An additional assumption is that $y(t, z) = \int_\Omega h(\|z - \gamma\|)v(t, \gamma)d\gamma$ where the translation invariance is expressed by the additional property of radial dependency. Something similar holds for speech and language: The extracted features don't depend on the absolute position in the text, but on the distance between the inputs in the sequence.

The analysis on translational invariance suggests that the choice of expanding the h on the receptive field $\mathcal{R}_\zeta$ turns out to be useful when we assume that h is a peaked filter, with the receptive field hypothesis stated by Eq. (1). We can quickly derive translational invariance from Eq. (10) by sharing the weights ω_i in the receptive field. The receptive inputs computed, like in the previous case, at $z_2 = z_1 + \rho$ yield $\xi_i(z_2) = \xi_i(z_1)$. We notice in passing that the receptive field assumption dramatically simplifies the computation of $\xi(z)$ defined by Eq. (9), since we have

$$\xi_i(z) = \int_\Gamma h(u - z_i)v(z - u)du = \int_\Gamma h(\gamma)v(z - \gamma - z_i)d\gamma$$
$$= \int_\mathcal{C} \tilde{h}(\gamma)v(z - \gamma - z_i)d\gamma.$$

In vision, scale, rotation, and elastic invariance are other fundamental requirements for the features to be effective during the recognition process. In speech, scale invariance is also meaningful since we can deliver the same message at a remarkably different speed but, clearly, there is no rotation invariance. Let us focus on vision, which offers the most difficult and interesting perceptual invariance.

Again, we restrict attention to images, so that we involve the retina Ω instead of the visual domain Γ. Suppose we are given a picture and that we focus attention on z, since we want to extract convolutional filters related to $\mathcal{I} = (v, z)$. We want to extract features that are invariant under scale transformations and see how the receptive field $\xi(z)$ is affected by a local scaling of the picture in z. This time, suppose that $h(\gamma) = \tilde{h}(\gamma)[\|\gamma\| < r]$, that is, we assume that there is a circular receptive field of radius r reacting to the visual stimulus. Furthermore, let us consider the scaling map which transforms $u \rightarrow z + \alpha(u - z)$, where $\alpha \in \mathbb{R}$ is the scale factor. Now we compare the receptive input coming from a picture centered in z with another one, still centered in z such that $v_1(u) = v_2(z + \alpha(u - z))$. If we compute the receptive input associated with v_1, we get

$$\xi_i^1(z) = \int_{\Gamma} h_1(u - z_i^{(1)}) v_1(z - u) du = \int_{\mathcal{C}} \tilde{h}_1(u - z_i^{(1)}) v_1(z - u) du$$

$$= \int_{\mathcal{C}} \tilde{h}_1(u - z_i^{(1)}) v_2(z - \alpha u) du = \int_{\mathcal{C}_\alpha} \frac{1}{\alpha} \tilde{h}_1\left(\frac{\mu}{\alpha} - z_i^{(1)}\right) v_2(z - \mu) d\mu,$$

where the last equality arises by the change of variables $\mu = \alpha u$. From these equalities, if we set

$$\tilde{h}_2(\mu - z_i^{(2)}) := \frac{1}{\alpha} \tilde{h}_1\left(\frac{\mu}{\alpha} - z_i^{(1)}\right) \tag{2}$$

then

$$\xi_i^1(z) = \int_{\mathcal{C}_\alpha} \tilde{h}_2(\mu - z_i^{(2)}) v_2(z - \mu) d\mu = \xi_i^{(2)}(z).$$

Again, if we share the weights ω_i, the receptive input and, consequently, the convolutional filters are invariant under scale transformation. However, this time, we only need to properly change the kernel according to Eq. (2). It requires remapping the points of the receptive fields $z_i^{(1)}$ according to the scale factor α. In case of Gaussian kernels, we have

$$\frac{1}{\sqrt{2\pi}\sigma_1\alpha} \exp\left(-\frac{\|\mu - \alpha z_i^{(1)}\|^2}{2\sigma_1^2\alpha^2}\right) = \frac{1}{\sqrt{2\pi}\sigma_2} \exp\left(-\frac{\|\mu - z_i^{(2)}\|^2}{2\sigma_2^2}\right).$$

The equality requires setting $z_i^{(2)} = \alpha z_i^{(1)}$ and $\sigma_2 = \alpha\sigma_1$. In a sense, the analogy with translation invariance is limited to the principle of weight sharing, but in case of scale invariance one also needs to properly choose an appropriate value of the variance that depends on the scale. Of course, this is not gained if we use one single neuron for detecting a corresponding feature. In order to conquer scale invariance, one can think of using more filters for each single feature with different σ, and let the learning of the ω_i parameters detect which filter better resonates at a certain scale.

While we have been witnessing a significant progress in the field of computer vision, the search for very expressive features is still an open problem. The extraction of an invariant feature seems to be a major problem. While translational invariance is built-in in convolutional networks with receptive fields, scale invariance must properly be developed by learning. The same holds for rotations and elastic deformations. Is that all? Not really. Humans likely capture other invariant properties that might be hard to be formally stated. It looks like something is wrong. As already pointed out, as a matter of fact, we are likely facing a problem more difficult than nature does! What are we missing? We have been mostly focusing attention on images than on video streams. This seems to be the natural outcome of well-established pattern recognition methods working on images, which have given rise to current emphasis on collecting big labeled image databases, with the purpose of devising and testing challenging machine learning algorithms. While this framework is the one in which most of nowadays state-of-the-art object recognition approaches have been developing, there are strong arguments to start exploring the more natural visual interaction that humans experiment in their own environment. Visual processes are immersed in the temporal dimension, an issue that has been mostly neglected. Now translation and scale invariance, which has been previously discussed, gives in fact examples of invariance that can be gained whenever we develop the ability to detect features that are invariant under motion. If my thumb moves closer and closer to my eyes then any feature that is motion invariant will also be scale invariant. The finger will become bigger and bigger as it approaches my face, but it's still my thumb. Clearly, translation, rotation and complex deformation invariance also derives from motion invariance. Human life always experiences motion, so the gained visual invariance might only arise from motion! Animals with foveal eyes also move the focus of attention when looking at fixed objects, which means that they continually experience motion. Translation, rotation, scale, and other invariance might be somewhat artificial since, unlike motion invariance, it cannot rely on continuous teaching from nature. How does information on motion invariance arise? Suppose we are moving a finger and focus attention on the nail. Now its features, which come from its shape and color, are not expected to change as the nail is moving! The nail is just the nail, regardless of its movement. Hence any feature vector $y(z)$ deriving from the corresponding convolutional filters is expected not to change along the motion trajectory defined by $z(t)$, over the interval $[t_0, t_1]$. Without loss of generality, we restrict to the case of a scalar feature, and assume that the convolutional filter changes over time, so that $h(\alpha)$ is replaced with $h(t, \alpha)$. Hence motion invariance can be compactly expressed by imposing[12] $dy(z)/dt = (d/dt) \int_\Omega h(t, \alpha) v(t, z(t) - \alpha) d\alpha = 0$. This yields

$$\forall (t, z) \in \Omega, \quad \int_\Omega \left(v(t, z - \gamma) \partial_t h + h \left(\partial_t v + \dot{z} \cdot \nabla_z v \right) \right) d\gamma = 0, \tag{3}$$

which enforces restriction on the space of candidate convolutional filters. This is the *motion invariance constraint*. Notice that if $h(t, z - \alpha) = \delta(z - \alpha)$ then the above equation turns out to be the classic *brightness invariance* condition used in computer vision to estimate the optical flow. In this special case we have

$$\forall (t, z) \in \Omega, \quad Dv = \frac{dv}{dt} = \partial_t v + \dot{z} \cdot \nabla_z v = 0. \tag{4}$$

---

[12] A similar idea can be used to express other invariances by appropriate constraints (see Exercise 1 for scale invariance).

Moreover, we notice in passing that the motion invariance condition is a linear bilateral constraint. It is associated with the linear map

$$\mu \colon \mathcal{F} \to \mathbb{R}, \quad \mu(h) = \int_{\Omega} (v(t, z - \gamma)\partial_t h + h\,(\partial_t v + \dot{z} \cdot \nabla_z v))\,d\gamma$$

so as motion invariance consists of determining $\mathcal{N}\mu$. Clearly, for convolutional features to posses motion invariance, the condition

$$\mathcal{N}\mu \neq \emptyset \tag{5}$$

must hold. A remark on map μ is in order to grasp the very nature of the problem. For a filter h which doesn't change with time, that is, $\partial_t h = 0$, motion invariance prescribes $dv = \partial_t v + \dot{z} \cdot \nabla_z v = 0$. This only happens provided that all points of the retina are translating with the same velocity $\dot{z}$. In general, $\partial_t v(t, \alpha) + \dot{z} \cdot \nabla_z v(t, \alpha) \neq 0$ since α doesn't translate with z. As a consequence, we can promptly see that the filter typically needs to change over time, that is, $\partial_t h \neq 0$. Convolutional neural networks that are currently used in the literature don't respect this property: They are characterized by their own weights that, after learning has come to an end, are used invariantly in the experiments, that is, there is no change at test time! On the other hand, if we go back to the requirement of scale invariance, we can recognize the same need to adopt time-variant filters. Now we give intriguing insights on the fulfillment of condition (5). We identify four interesting conditions, which, on the one hand, comply with the structure of the visual signals and, on the other hand, provide an information-based support to biological evidence on computational processes of vision.

(*i*)   *Finiteness of visual information*: The condition $\mathcal{N}\mu \neq \emptyset$ is favored by low bandwidth of the brightness. This is quite reasonable: An agent likely experiences more difficulty when trying to gain motion invariance in a very informative video. Amongst others, tracking is also more difficult.

(*ii*)   *Blurring in newborns*: Suppose that the input $v(\cdot, \cdot)$ is filtered by a spatiotemporal low pass filter. This corresponds with blurring the video, a process which is known in newborns. Clearly, strong blurring favors the condition $\mathcal{N}\mu \neq \emptyset$ for the above explained reasons. Hence it looks like there are very good reasons for protecting newborns from visual information flooding!

(*iii*)   *The evolutionary solution of receptive fields*: The evolutionary solution discovered by nature to use receptive fields is another way of forcing the condition (5) — we can easily see that the associated algebraic linear system coming from discretization is strongly sparsified.

(*iv*)   *Emergence of focus of attention*: Notice that map μ can only be given with an approximate representation, which is due to the approximation that is necessarily introduced in the solution of the optical flow problem. Since the brightness invariance constraint is just one of the many constraints that can be associated with different filters, its joint satisfaction makes it possible to improve the overall feature extraction, including the computation of the velocity. Finally, motion invariance can also be imposed in case of images, since the saccadic movements always yield movement.

We can enforce motion invariance by the isoperimetric constraint (3). An interesting simplification, which comes with a dramatic complexity cut, arises when relying on kernel representations of the convolutional filter, according to Eq. (6). In that case we get

$$\int_\Omega \left(v(t, z - \gamma) \frac{\partial}{\partial t} \sum_{z_i \in \mathcal{R}} \omega_i k(\gamma - z_i) + (\partial_t v + \dot{z} \cdot \nabla_z v) \sum_{z_i \in \mathcal{R}} \omega_i k(\gamma - z_i) \right) d\gamma = 0,$$

which yields

$$\left(\sum_{z_i \in \mathcal{R}} \int_\Omega (v(t, z - \gamma) k(\gamma - z_i) d\gamma \right) \dot{\omega}_i + \left(\int_\Omega (\partial_t v + \dot{z} \cdot \nabla_z v) \sum_{z_i \in \mathcal{R}} k(\gamma - z_i) d\gamma \right) \omega_i = 0.$$

Now let us define

$$a_i(t, z) := \int_\Omega v(t, z - \alpha) k(\gamma - z_i) \, d\gamma,$$

$$b_i(t, z, \dot{z}) := \int_\Omega (\partial_t v + \dot{z} \cdot \nabla_z v) \, k(\gamma - z_i) d\gamma.$$

Then we get

$$\sum_{\alpha_i \in \mathcal{R}} a_i(t, z) \dot{\omega}_i + \sum_{\alpha_i \in \mathcal{R}} b_i(t, z, \dot{z}) \omega_i = 0. \tag{6}$$

Notice that, as already seen, the receptive field assumption only leads to replacing $\mathcal{R}$ with $\mathcal{R}_z$, that is, a set of points that are arranged according to a predefined geometrical structure in the neighborhood of z. Furthermore, the equation relies on the assumption that the velocity field $\dot{z}$ is given, though one can regard $\dot{z}$ as an unknown of brightness invariance — distributional degeneration of the filters. The adoption of the kernel expansion of the convolutional filter dramatically simplifies the problem, since the temporal structure is now captured by a linear differential equation. Focusing on a precise point at any time which, for instance, is carried out in foveal eyes, yielding another dramatic complexity cut! In that case we only need to enforce the constraint (6) on a single point of focus at any time. The actual choice of that point is obtained by appropriate saccadic movements. While the convolutional features can be determined by an overall supervised learning process, pairing motion invariance with unsupervised learning is definitely the most natural solution! Exercise 2 proposes using maximum mutual information to extract the features under motion invariance constraint expressed by Eq. (6).

EXERCISES

1. [*18*]   Propose an algorithm to incorporate scale invariance using the consequences of the condition

$$\frac{d}{dt} \int_\Omega h(t, \gamma(z - \alpha)) v(t, \gamma(z - \alpha)) d\alpha = 0.$$

2. [*HM50*]   Use maximum mutual information (MMI) to extract the features under motion invariance constraint expressed by Eq. (6). *Hint:* The MMI constraint, joined with motion invariance, yields an integro-differential equation. Use focus of attention based on virtual eye movement to convert it to a differential equation, and get a local computational model. Formulate the problem according to the principle of Least Cognitive Action (see Section 6.5.1.).

5.4.3 Deep convolutional networks

The convolutional features that have been considered so far can be computed according to a linear neuron scheme defined by Eq. 5.4.1–(10). If we also use the neuron nonlinearity then the computation over the retina produces the features defined by the tensor

$$y_\sigma^\sharp = \sigma(y^\sharp).$$

The nonlinearity of $\sigma(\cdot)$ performs a very useful clustering process.

A fundamental issue that has been previously addressed concerns the structure of the filter to be adopted. The motion invariance analysis has made it clear that receptive field based filters are definitely preferable. On the other hand, while these filters are better suited to gain invariances, they are limited in terms of the context from which they extract information. This naturally leads to promoting deep convolutional networks, which are a natural outcome of invariance principles and representational power. We can easily see that as we go higher and higher in the layered deep net, the receptive field based convolutional features depend on increasingly larger virtual windows. In particular, any time we add a convolutional block, the dimension of the virtual window covered by the receptive field increases the size of the window.

At each level of the network the 3D tensor that represents the processed image (video) is processed by filters according to the kernel representation expressed by Eq. 5.4.1–(6). The choice of the kernel $k(z - z_i)$ must take into account the increasing dimension of the context as we go higher and higher in the network levels. While in the layers close to the inputs those kernels are likely to be Gaussians with small σ, when we consider layers close to the output layer, where high level concepts are involved, the value of σ must be increased accordingly. As a matter of fact, the 3D tensor $y^\sharp$, which expresses different views of the given image, exhibits the property of progressively increasing its feature dimension as we go towards the output. The reason is simple: Layers close to the output must represent information that can only be gained by large contexts. However, this remarkable unbalancing can naturally be compensated by an appropriate tensor *pooling* along the retina dimensions (i_z, j_z). The representation in layers close to the output is highly redundant. We can reduce the degree of redundancy by filtering in many different ways. A common solution is to use *max-pooling*, where each output is constructed by returning the max over a sliding window. Let $W_{z^\sharp}(\cdot)$ be the *pooling block* defined as

$$W_{z^\sharp}(u^\sharp) = H\big(p - |(z^\sharp - u^\sharp)|_1\big)$$

where $|\cdot|_1$ is the L_1 norm, $H(\cdot)$ is the Heaviside function, and $p \in \mathbb{N}$ is an integer which defines the width $2p + 1$ of the window. Formally,

$$y_\sigma^\sharp(i_z, j_z, \beta) = \max_{u^\sharp}\big\{W_{z^\sharp}(\cdot) \oslash y_\sigma^\sharp(\cdot, \cdot, \beta)\big\}, \quad \beta = 1, \ldots, m$$

where the pooling operator $\oslash$ corresponds with 2D discrete convolution $*$. (In the following, to emphasize this property, we overload the symbol $*$ to denote both continuous and discrete convolution.) Overall, the input $y^\sharp$ follows a three-step cascading process based on convolution with h, nonlinear neural map $\sigma(\cdot)$, and pooling, that can be regarded as the cascade of convolutional filter $W(\cdot)$ with max operator. The cascading of these blocks is referred to as a *convolutional block*

$$\max_{u^\sharp} \left\{ W_{z^\sharp}(\cdot) \oslash y_\sigma^\sharp(\cdot, \cdot, \beta) \right\} \qquad \text{pooling}$$
$$\uparrow$$
$$y_\sigma^\sharp = \sigma(y^\sharp) \qquad\qquad \text{nonlinear map} \qquad\qquad (1)$$
$$\uparrow$$
$$y^\sharp = h^\sharp * v^\sharp \qquad\qquad \text{convolution}$$
$$\uparrow$$
$$v^\sharp \qquad\qquad\qquad \text{input}$$

The pooling process can be thought of as a form of regularization which also increases the invariance of convolutional filters. The computation that takes place in convolutional networks with multiple blocks can nicely be grasped when gaining the notion of *convolutional feature maps*. These maps are created when the input v is forwarded to the upper layers of a convolution block according to Eq. (1). Hence the input $v(t, z) \in \mathbb{R}^{\vdash m}$ is mapped to m different convolutional feature maps, compactly[13] denoted by the tensor $y_\sigma^\sharp$. Finally, this tensor is properly de-sampled by a pooling operator, which yields the de-sampled tensor $y_{\sigma,p}^\sharp$. Since there is one of these tensors for any convolutional block, they are located also by the block of the deep network. Each of the m feature maps expresses a specific property of the image, which gains levels of abstraction as we move higher and higher in the network hierarchy. This tensor notation makes it possible to express deep convolutional nets in a very straightforward way. As an example, let us focus on the MNIST benchmark. For instance, we can start from a 28×28 gray level picture representing handwritten chars and construct a deep convolutional network with the architecture defined by which consists of two convolutional layers followed by three fully connected layers. In the convolutional layer, each neuron is based on a 5×5 receptive field with distributional kernels, that is, $h = \delta$. At each layer the pooling performs a downsampling by returning one out of four values.

---

[13] In order to better focus on images, we drop time in the representation of the tensor.

5.5 Learning in feedforward networks

In this section, we show that the feedforward architecture is a fundamental ingredient to make learning algorithms efficient. This holds regardless of the learning protocol, since the efficiency nicely emerges thanks to the partial ordering on the vertexes of the graph.

5.5.1 Supervised learning

We start discussing supervised learning that is translated as the optimization of the error function. We use gradient descent heuristics, which is in fact the simplest numerical method for performing the optimization. Higher-order methods are also mentioned, but it is claimed that in most cases of relevant applicative interest, they are likely to be useless, since we typically deal with high-dimensional input spaces, which makes those methods very computationally expensive. Basic numerical techniques are also reviewed.

5.5.2 Backpropagation

Backpropagation is likely to be the most popular word in machine learning. Yet, it is quite often the source of a surprising misunderstanding. More than a learning algorithm, it is an efficient gradient computation algorithm which, as it will be seen later on in this section, is in fact optimal! Learning algorithms typically require computing the gradient of the loss for any example v, that is, ∇e, where $e(w, v, y) = V(v, y, f(w, v))$. In order to grasp the idea, it is important to realize that the derivatives of a function can either be computed numerically or symbolically. For instance, if we want to compute $\sigma'(a)$, where $\sigma(a) = 1/(1 + e^{-a})$, the symbolic derivation immediately leads us to notice that

$$\sigma'(a) = \sigma(a)(1 - \sigma(a)). \tag{1}$$

Alternatively, one can use numerical schemes that are typically based on clever approximations; for instance, we have

$$\sigma^{(1)}(a) = \frac{\sigma(a + h) - \sigma(a - h)}{2h} - \frac{h^2}{6}\sigma^{(3)}(\tilde{a}), \tag{2}$$

where $\tilde{a} \in (a - h, a + h)$. Of course, for "small" h we have $\sigma^{(1)}(a) \simeq (\sigma(a + h) - \sigma(a - h))/2h$, which gives rise to a good numerical scheme to compute $\sigma^{(1)}(a)$. In Exercise 3 we propose discussing why this numerical approximation of the derivative is preferable[14] with respect to the asymmetric approximation $\sigma^{(1)}(a) \simeq$

---

[14] Notice that we can use also numerical schemes with precision better than the one achieved by (2). For example, if we keep five instead of three samples, we get the $O(h^4)$ approximation

$$\sigma^{(1)}(a) = \frac{-\sigma(a + 2h) + 8\sigma(a + h) - 8\sigma(a - h) + \sigma(a - 2h)}{12h} + \frac{h^4}{30}\sigma^{(5)}(\tilde{a}).$$

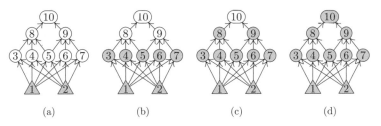

FIGURE 5.16

Data flow computation: The input is applied at the first layer (a). It is then propagated forward to the second (b), third (c), and fourth (d) layer, which contains the output.

$(\sigma(a + h) - \sigma(a))/h$. However, no matter what numerical algorithm is used, it cannot reach the perfect expression achieved by the symbolic computation (1), once we know $\sigma(a)$. Interestingly, this difference between symbolic expression and numerical computation of the gradient has an impact not only on the precision. When dealing with high dimensional problems, there is in fact a fundamental impact on computational complexity.

Suppose use a numerical computation of the gradient, where any of its components $\partial e/\partial w_{ij}$ is computed using the idea sketched in Eq. (2). Hence

$$\frac{\partial e}{\partial w_{ij}} \leftarrow \frac{e(w_{ij} + h, v, y) - e(w_{ij} + h, v, y)}{2h}. \tag{3}$$

Let $\mathcal{N} = (V, A)$ be a feedforward network. According to the above equation, the computation of $\partial e/\partial w_{ij}$ requires three floating-points operations. However, since we are interested in an asymptotical analysis, we can promptly see that we are reduced to determining the complexity of computing $e(w, v, y)$ that, in turn, is reduced to establishing the complexity of the computation of $F(w, v)$.

Fig. 5.16 shows how such a computation takes place in a feedforward network with two hidden layers. First in (a), the input is applied to the inputs (gray level, units 1, 2). Then it is propagated forward to the second (b), third (c), and fourth layer (d) (output). Any of the three forward computations, which construct the outputs of the two hidden layers (b, c) and of the output (d), require (asymptotically) as many floating-point operations as the number of connections with the previous layer. For example, on layer (b), we need to compute $x_i = \sigma(w_{i1}x_1 + w_{i2}x_2 + b_i)$ for every $i = 3, \ldots, 7$. When considering a neural network as simple as this, modeling the cost of x_i is not a trivial issue. In particular, it is important to know what kind of threshold function $\sigma(\cdot)$ we are considering and, in addition, how we compute it on the given platform. As an extreme case, we can dramatically optimize the computation of σ if we use its tabular-based approximation. Clearly, with large numbers of inputs and neurons one can regard the cost of σ as $O(1)$. Hence, the computation of outputs of any layers requires a number of floating-point operations that corresponds with the number of weights (including the bias) that are connected with the previous layer.

Overall, the complexity grows proportionally to the number of weights. Of course, this holds true also in case of a generic DAG, since the data flow computation requires finding $x_i = \sigma(\sum_j w_{ij} x_j)$, where $j \in \text{pa}(i)$ is any parent of i in the DAG.

Algorithm F (*Forward Propagation*). Given a neural network $\mathcal{N} = (\mathcal{G}, w)$ based on the DAG $\mathcal{G}$ and on the weights w, a vector m, that will be used as a *weight modifier*, and a vector of inputs v, for all $i \in V \setminus I$ the algorithm computes the state of vertex i and stores its value into the vector x_i. We assume that we have already defined TOPSORT(S, s) that takes a set S equipped with the an ordering $\prec$ and copies the elements of this set into the topologically sorted array s, so that for each i and j with $i < j$ we have $s_i \prec s_j$. In what follows this algorithm will be invoked by FORWARD$(\mathcal{G}, w, m, v, x)$.

F1. [Initialize.] For all $i \in I$ set $x_i \leftarrow v_i$ and initialize an integer variable $k \leftarrow 1$.

F2. [Topsort.] Invoke TOPSORT on $V \setminus I$, so that now the vector s contains the topological sorting of the nodes of the net. Set the variable l to the dimension of the vector s.

F3. [Finished yet?] If $k \leq l$ go on to step F4, otherwise the algorithm stops.

F4. [Compute the state x.] If $m = (1, 1, \ldots, 1)$ set $x_{s_k} \leftarrow \sigma\left(\sum_{j \in \text{pa}(s_k)} w_{s_k j} x_j\right)$ otherwise set $x_{s_k} \leftarrow m_{s_k} \sum_{j \in \text{pa}(s_k)} w_{s_k j} x_j$. Increase k by one and go back to step F3. ∎

This is sketched in Algorithm F, which relies on the topological sort of the neurons identified by the vertices $V \setminus I$ in $\mathcal{N} = (V, A)$. For example, in Fig. 5.16, $V \setminus I = \{3, 4, 5, 6, 7, 8, 9, 10\}$ and, amongst the possible topological sorts, we can clearly choose $s = (3, 4, 5, 6, 7, 8, 9, 10)'$. Exercise 5 raises a discussion on the number of possible topological sorts of a given DAG. While this is a trivial issue in multilayer nets, in a general digraph, there are clearly a lot of different ways of sorting the vertices, but one of them can be found in linear time — this is in fact the cost of TOPSORT. (A brief discussion is in Section 5.8.) The subsequent loop for the forward step takes $O(|A|)$, since we need to accumulate the value of the activations for all the arcs. This dominates in the algorithm, so that the computation of $F(w, x_\kappa)$ and, consequently, of $\partial e / \partial w_{ij}$ is clearly optimal, that is, $\Theta(|A|)$, since it is also $\Omega(|A|)$. Let $m = |A|$ be the number of arcs, which corresponds to the number of weights. Hence the numerical computation of all the m components of the gradient requires $O(m^2)$. Feedforward neural networks are sometimes applied in problems where m is on the order of millions! The numerical computation of the gradient in those cases would require teraflops. This is a remarkable computational burden when considering that this is only for the computation of the gradient associated with a single pattern! As we will see, Backpropagation is a clever algorithm to dramatically cut this bound to $O(m)$.

In order to come up with a solution to compute the gradient smarter than based on Eq. (3), one should realize that the same forward step is repeated for all the weights m times, and we do not capitalize from previous computations. Let us attack the problem by analytically expressing the gradient with symbolic manipulations. We start noticing that

$$\frac{\partial e}{\partial w} = \frac{\partial V}{\partial f} \cdot \frac{\partial f}{\partial w} = \sum_{o \in O} \frac{\partial V}{\partial f_o} \frac{\partial f_o}{\partial w}. \tag{4}$$

Whenever we are given a symbolic expression for $V(y, F(w, v))$, the first term in Eq. (4) can also be given a corresponding symbolic expression. For example, in case $V(y, f) = (y - f)^2/2$, we have $\nabla_f V = y - f$ and, therefore, its computation requires a forward step to determine $F(w, v)$. (Interesting cases are those in which the loss is not always differentiable in its domain.) The symbolic expression of $\partial f/\partial w$ can be gained if we exploit the DAG structure of feedforward nets. Consider the derivative of $F_o(w, v)$ with respect to the (i, j)th weight w_{ij}, and call this quantity g_{ij}^o; by using the chain rule, we get

$$g_{ij}^o = \frac{\partial x_o}{\partial w_{ij}} = \frac{\partial x_o}{\partial a_i} \frac{\partial a_i}{\partial w_{ij}} = \frac{\partial x_o}{\partial a_i} \frac{\partial}{\partial w_{ij}} \sum_{h \in \mathrm{pa}(i)} w_{ih} x_h = \delta_i^o x_j, \tag{5}$$

where we have defined $\delta_i^o := \partial x_o/\partial a_i$. This definition, which is motivated by the computation of g_{ij}^o, can be generalized when considering the transfer of the activation a_i onto the unit j. That is, we can replace δ_i^o with δ_i^j by assuming that the role of $o \in O$ is moved to $j \in H$. Clearly, $\delta_i^j = 0$ whenever $i \succ j$. We can immediately determine the gradient with respect to the bias, since[15] $\partial x_o/\partial b_i = \delta_i^o$. The term δ_i^o is referred to as the *delta error*.

Let $m \in O$ be the index of an output neuron. Then, by definition, the delta error is different from zero only when $m = o$, and in that case we have

$$\delta_o^o = \sigma'(a_o). \tag{6}$$

For asymmetric sigmoidal functions, from Eq. (1), we get $\delta_o^o = x_o(1 - x_o)$. In case of symmetric sigmoidal functions $\sigma(a) = \tanh(a)$, similarly, we have

$$\delta_o^o = \frac{1}{2}(1 + x_o)(1 - x_o),$$

and related symbolic expressions can be found for other LTU units that directly involve the value of x_o. Basically, once the forward step has been completed and x_o is known, we can compute δ_o^o directly. If $i \in H$ is the index of any hidden unit then δ_i^o cannot be directly expressed like for the case of output units. However, by using the chain rule we have

$$\delta_i^o = \frac{\partial x_o}{\partial a_i} = \sum_{h \in \mathrm{ch}(i)} \frac{\partial x_o}{\partial a_h} \frac{\partial a_h}{\partial x_i} \frac{\partial x_i}{\partial a_i} = \sigma'(a_i) \sum_{h \in \mathrm{ch}(i)} w_{hi} \delta_h^o. \tag{7}$$

---

[15] For the sake of simplicity, in the following discussion we will incorporate the bias as an ordinary weight, by assuming the x has been enriched, as usual, by $\hat{x} = (x', 1)'$.

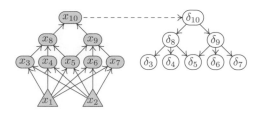

FIGURE 5.17

The backward step propagates recursively the delta error beginning from the output through its children. For example, $\delta_5 = \sigma'(a_5)(w_{85}\delta_8 + w_{95}\delta_9)$. Since there is only one output, we don't bother to write down the index o.

Eqs. (6) and (7) allow us to determine δ_i^o by propagating backward the values δ_o^o throughout the hidden units $i \in H$. This is shown in Fig. 5.17, where we can see the recursive propagation based on the children of the output. Since δ_i^o is needed in Eq. (5) for the gradient computation, we can immediately see that there is no propagation throughout the inputs.

Now suppose that instead of the derivative of each output x_o we want to calculate the derivative of the loss V with respect to the generic weight w_{ij}. We immediately realize that we can follow the steps outlined above since we can exploit the chain rule

$$\frac{\partial V}{\partial w_{ij}} = \frac{\partial V}{\partial a_i} \frac{\partial a_i}{\partial w_{ij}} = \delta_i x_j,$$

where this time δ_i is simply $\partial V/\partial a_i$. As before, after the forward phase, we can immediately evaluate the δ_i for $i \in O$ once we know the symbolic expression of V; for example, in case of quadratic loss $V(y, f) = (y - f)^2/2$, then $\delta_o = (\sigma(a_o) - y_o)\sigma'(a_o)$. Of course, we can recursively evaluate all the other δ_i using an analogue of Eq. (7),

$$\delta_i = \sum_{h \in \mathrm{ch}(i)} \frac{\partial V}{\partial a_h} \frac{\partial a_h}{\partial x_i} \frac{\partial x_i}{\partial a_h} = \sigma'(a_i) \sum_{h \in \mathrm{ch}(i)} w_{hi}\delta_h.$$

We will now show how these ideas can be used to write an algorithm that computes the derivatives either of the output or of the loss function for a general DAG with respect to the weights.

Algorithm B (*Backward propagation*). Given a network $\mathcal{N} = (\mathcal{G}, w)$ based on the DAG $\mathcal{G}$, all the states x_i of the vertices of $\mathcal{G}$, a parameter q, and the symbolic expression of a loss function $V(y, f)$, depending on whether q is positive or not, it returns the derivatives g_{ij}^q, if $q > 0$ and $q \in O$; otherwise for any $q \le 0$, it returns the derivatives of the loss $\partial V/\partial w_{ij}$. In what follows the algorithm is invoked as BACKWARD($\mathcal{G}, w, x, q, V$), where V is the name of the loss.

B1. [Loss or output?] If $q \le 0$ go to step B2, otherwise jump to step B3.

B2. [Initialize the loss.] For all $o \in O$ set $v_o \leftarrow \partial V/\partial a_o$ and go to step B4.

B3. [Initialize x_q.] For each $o \in O$ if $o \neq q$ set $v_o \leftarrow 0$, else if $o = q$ make the assignment $v_o \leftarrow \sigma'(\sigma^{-1}(x_o))$.

B4. [Compute backwards.] For each $k \in V \setminus I$ set $m_k \leftarrow \sigma'(\sigma^{-1}(x_k))$, then invoke FORWARD($(\mathcal{G} \setminus \mathcal{I})', w', m, v, \delta$).

B5. [Output the gradient.] For each $i \in V \setminus I$ and then for each $j \in \text{pa}(i)$ set $g_{ij} \leftarrow \delta_i x_j$ and output g_{ij}. Terminate the algorithm. ∎

In the latter algorithm, we are assuming — specifically in step B2 — that we are able to handle symbolic differentiation; this topic will be discussed in more details in Section 5.5.3. It's interesting to notice that the assignments $v_o \leftarrow \sigma'(\sigma^{-1}(x_o))$ in step B3 and $m_k \leftarrow \sigma'(\sigma^{-1}(x_k))$ are almost immediate once we have chosen a specific σ function; for example if $\sigma = \tanh$ we have that $\sigma'(\sigma^{-1}(x_k)) = 1/2(1 + x_k)(1 - x_k)$. A last comment on this algorithm is in order: In step B4 FORWARD is invoked on the graph $(\mathcal{G} \setminus I)'$; we have used this notation to indicate the graph obtained by $\mathcal{G}$ by pruning the input nodes (together with the arcs attached to those nodes) and reversing the direction of the arrows in the remaining graph. This way of performing the calculations is what characterize the algorithm as the "backward propagation".

Now that we have defined Algorithms F and B, we are ready to introduce the backpropagation algorithm.

Algorithm FB (*Backpropagation*).   Given a network $\mathcal{N} = (\mathcal{G}, w)$ based on the DAG $\mathcal{G}$, a vector of inputs v, and a loss function V, the algorithm returns the gradient of the loss with respect to w.

FB1. [Forward.] Invoke FORWARD($\mathcal{G}, w, (1, 1, \ldots, 1), v, x$).

FB2. [Backward.] Invoke BACKWARD($\mathcal{G}, w, x, -1, V$) and terminate the algorithm. ∎

The algorithm requires that the forward step has already been carried out, whose effect is that of determining all the x_κ. Once the output values for all the neurons are given, we start computing the δs by means of the backward step. At this point the gradients are obtained using Eq. (5). Notice also that in Algorithm F, we have introduced the modifier vector m so that we could use the same algorithm to compute the backward step; this is necessary since in order to compute the i-th delta error we need to multiply the i-th activation by $\sigma'(a_i)$. From the analysis of the algorithm we can easily draw the conclusion that it has complexity $O(m^2)$, just like for the forward step of Algorithm F. The cost arises from the computation of the delta error as well as form the computation of the gradients. As clearly shown also in Fig. 5.17, the backpropagation of the error is restricted to the hidden units, but the gradient computation involves all the weights, which indicates that it is dominant and involves all the weights of the neural network.

In case of layered structures, which is common in many applications, the forward/backward steps get a very simple structure that is sketched in Fig. 5.16 and Fig. 5.17, respectively. We can also express the forward/backward equations by referring to the indexes of the layers using the tensor formalism. We have

$$\hat{X}_q = \sigma(\hat{X}_{q-1}\hat{W}_q), \quad q = 0, \ldots, Q. \tag{8}$$

This expression clearly shows the composition of the map that takes place on a layered architecture with Q layers, which returns

$$\hat{X}_Q = \sigma(\ldots\sigma(\sigma(\hat{X}_0\hat{W}_1)\hat{W}_2)\ldots\hat{W}_Q).$$

For $Q = 3$, we have $\hat{X}_3 = \sigma(\sigma(\sigma(\hat{X}_0\hat{W}_1)\hat{W}_2)\hat{W}_3)$, which indicates a nice symmetry: The input $\hat{X}_0$ is right-multiplied by the weight matrices and left-processed by the σ. Likewise the backward step returns both the delta error and the gradient according to

$$\Delta_{q-1} = \sigma' \odot (\Delta_q W_\ell), \tag{9}$$

$$G_q = \hat{X}'_{q-1}\Delta_q, \tag{10}$$

where $\sigma' \in \mathbb{R}^{L \times q-1}$ is the matrix with coordinates $\sigma'(a_{i,\kappa})$, $\odot$ is the Hadamard product, and $\Delta_q := (\delta_1, \ldots, \delta_{n(q)}) \in \mathbb{R}^{q,n(q)}$.

Now we use similar arguments rooted in the graphical structure of feedforward nets to express the Hessian matrix, which turns out to be useful to investigate the nature of the critical points of the error function. Let $v_\kappa := V(x_\kappa, y_\kappa, f(x_\kappa))$ be, then a generic coordinate of the Hessian matrix can be expressed as

$$
\begin{aligned}
h_{ij,lm} = \frac{\partial^2 v_\kappa}{\partial w_{ij}\partial w_{lm}} &= \frac{\partial}{\partial w_{ij}}\sum_{o \in O}\frac{\partial v_\kappa}{\partial x_{\kappa o}}\frac{\partial x_{\kappa o}}{\partial w_{lm}} \\
&= \sum_{o \in O}\frac{\partial^2 v_\kappa}{\partial w_{ij}\partial x_{\kappa o}}\frac{\partial x_{\kappa o}}{\partial w_{lm}} + \sum_{o \in O}\frac{\partial v_\kappa}{\partial x_{\kappa o}}\frac{\partial^2 x_{\kappa o}}{\partial w_{ij}\partial w_{lm}} \\
&= \sum_{o \in O}\sum_{q \in O}\frac{\partial^2 v_\kappa}{\partial x_{\kappa o}\partial x_{\kappa q}}\frac{\partial x_{\kappa q}}{\partial w_{ij}}\frac{\partial x_{\kappa o}}{\partial w_{lm}} + \sum_{o \in O}\frac{\partial v_\kappa}{\partial x_{\kappa o}}\frac{\partial^2 x_{\kappa o}}{\partial w_{ij}\partial w_{lm}} \\
&= \sum_{o \in O}\sum_{q \in O}\frac{\partial^2 v_\kappa}{\partial x_{\kappa o}\partial x_{\kappa q}}\delta^q_{\kappa i}\delta^o_{\kappa l}x_{\kappa j}x_{\kappa m} + \sum_{o \in O}\frac{\partial v_\kappa}{\partial x_{\kappa o}}\hbar_{ij,lm}
\end{aligned}
\tag{11}
$$

where

$$\hbar_{ij,lm} := \frac{\partial^2 x_{\kappa,o}}{\partial w_{ij}\partial w_{lm}}. \tag{12}$$

When using backpropagation rule again, we get

$$
\begin{aligned}
\hbar_{ij,lm} = \frac{\partial}{\partial w_{ij}}\left(\delta^o_{\kappa l}x_{\kappa m}\right) &= x_{\kappa m}\frac{\partial \delta^o_{\kappa l}}{\partial w_{ij}} + \delta^o_{\kappa l}\frac{\partial x_{\kappa m}}{\partial w_{ij}} \\
&= x_{\kappa m}\frac{\partial}{\partial w_{ij}}\frac{\partial x_{\kappa o}}{\partial a_{\kappa l}} + [i \prec m]\delta^o_{\kappa l}\delta^m_{\kappa i}x_{\kappa j} \\
&= x_{\kappa m}\frac{\partial^2 x_{\kappa o}}{\partial a_{\kappa l}\partial a_{\kappa i}}\frac{\partial a_{\kappa i}}{\partial w_{ij}} + [i \prec m]\delta^o_{\kappa l}\delta^m_{\kappa i}x_{\kappa j}
\end{aligned}
\tag{13}
$$

$$= \frac{\partial^2 x_{\kappa o}}{\partial a_{\kappa l} \partial a_{\kappa i}} x_{\kappa m} x_{\kappa j} + [i \prec m] \delta^o_{\kappa l} \delta^m_{\kappa i} x_{\kappa j}$$

$$= \delta^{o2}_{\kappa li} x_{\kappa m} x_{\kappa j} + [i \prec m] \delta^o_{\kappa l} \delta^m_{\kappa i} x_{\kappa j}$$

where

$$\delta^{o2}_{\kappa li} := \frac{\partial^2 x_{\kappa o}}{\partial a_{\kappa l} \partial a_{\kappa i}} \qquad (14)$$

is referred to as the *square delta error*. Now we show that, just like the delta error, it can be optimally computed in feedforward nets. We start noticing that

$$\delta^{o2}_{\kappa li} \neq 0 \quad \text{if and only if} \quad i \text{—} l \in A,$$

meaning that either $i \longrightarrow l$ or $l \longrightarrow i$ is an arch of the DAG $\mathcal{G}$, or said in another way i and l must be connected vertexes, with a directed path from one to the other. For example, neurons on the same layer do not satisfy this property and, consequently, $\delta^{o2}_{\kappa li}$. Like for the delta error, we distinguish the case of $i, l \in O$ and $i, l \in H$. If $i, l \in O$, we get $\delta^{o2}_{\kappa li} = 0$ if $i \neq l$. If $i = l = o$, we have

$$\delta^{o2}_{\kappa oo} = \frac{\partial}{\partial a_{\kappa o}} \frac{\partial x_{\kappa o}}{\partial a_{\kappa o}} = \frac{\partial}{\partial a_{\kappa o}} \sigma'(a_{\kappa o}) = \sigma''(a_{\kappa o}) \qquad (15)$$

For the generic term $\delta^{o2}_{\kappa li}$, we have

$$\begin{aligned}
\delta^{o2}_{\kappa li} = \frac{\partial}{\partial a_{\kappa l}} \delta^o_{\kappa i} &= \frac{\partial}{\partial a_{\kappa l}} \left(\sigma'(a_{\kappa i}) \sum_{j \in \mathrm{ch}(i)} w_{ji} \delta^o_{\kappa j} \right) \\
&= \frac{\partial}{\partial a_{\kappa l}} \frac{d\sigma(a_{\kappa i})}{d a_{\kappa i}} \sum_{j \in \mathrm{ch}(i)} w_{ji} \delta^o_{\kappa j} + \sigma'(a_{\kappa i}) \sum_{j \in \mathrm{ch}(i)} w_{ji} \frac{\partial \delta^o_{\kappa j}}{\partial a_{\kappa l}} \\
&= \frac{d}{d a_{\kappa i}} \frac{\partial x_{\kappa i}}{\partial a_{\kappa l}} \sum_{j \in \mathrm{ch}(i)} w_{ji} \delta^o_{\kappa j} + \sigma'(a_{\kappa i}) \sum_{j \in \mathrm{ch}(i)} w_{ji} \delta^{o2}_{\kappa jl} \\
&= \frac{d\delta^i_{\kappa l}}{d a_{\kappa i}} \sum_{j \in \mathrm{ch}(i)} w_{ji} \delta^o_{\kappa j} + \sigma'(a_{\kappa i}) \sum_{j \in \mathrm{ch}(i)} w_{ji} \delta^{o2}_{\kappa jl}.
\end{aligned} \qquad (16)$$

Now we are ready to define an algorithm for the computation of the Hessian. The generic term $h_{ij,lm}$ is computed by Eq. (11) that relies on the chain of variables $\hbar_{ij,lm}$ and $\delta^{o2}_{\kappa li}$ according to the following expression tree[16]:

---

[16] The value $h_{ij,lm}$ is computed by using the postorder visit of the tree.

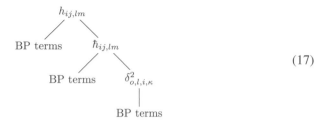

$$(17)$$

This expression tree indicates how to compute $h_{ij,lm}$ beginning from the leaves of the tree, where the vertexes also indicate the equations to be used. We can easily see that this computational scheme, referred to as *Hessian BP*, allows us to compute the Hessian with complexity $\Theta(m^2)$, where m is the number of weights. Exercise 9 proposes the detailed formulation of Hessian BP. Exercise 10 proposes the adoption of the Hessian BP scheme to the simple case of a single neuron (see Eq. (5)).

EXERCISES

1. [*HM48*]   Consider a feedforward neural network whose connections are modeled by weights depending on the applied input. The generic weight of the connection can be modeled by a function $w_{i,j} : \mathbb{R} \to \mathbb{R}$ so that the activation of neuron i is dictated by

$$x_i = \sum_{j=1}^{d} x_j w_{i,j}(x_j) + b_i,$$

where $j \prec i$ and $w_{i,j}(x_j)$ is the corresponding weight of the connection. Here, the output function of the neuron is linear, but all the connections are functionally dependent on the input. This neuron is used to compose feedforward architectures according to graphic connections based on DAGs. Formulate supervised learning in the framework of regularization proposed in Section 4.4. Prove that

$$w_{i,j}(x_j) = \sum_{\kappa=1}^{\ell} \lambda_{i,j,\kappa} g(x_j - x_{j,\kappa}),$$

so that the activation of the neuron looks like

$$x_i = \sum_{j=1}^{d} \sum_{\kappa=1}^{\ell} x_j g(x_j - x_{j,\kappa}) \lambda_{i,j,\kappa} + b_i$$

This can be given a simple interpretation: Any input x_j, which is fed through connection (i, j) to neuron i, is filtered out by the training set so as to return the equivalent input $x_j^{\kappa} := x_j g(x_j - x_{j,\kappa})$. In so doing, we end up with a classic ridge linear neuron model. Propose a corresponding learning algorithm with regularization based on the new weight $\lambda_{j,\kappa}$.

2. [*M39*]   Given the neural network as defined in Exercise 1, prove that we can construct an equivalent network where the nonlinearity is moved to the vertexes, that is,

$$x_i = \gamma \left(\sum_{j=1}^{d} w_{i,j} x_j \right).$$

Prove that also the inverse construction is possible, and reformulate the learning algorithm of Exercise 1 for this new feedforward network.

3. [*17*] Using the Lagrange remainder in Taylor expansion, give a proof of Eq. (2). Then consider the asymmetric approximation and prove that it is significantly worse than the symmetric one.

4. [*17*] Given the neural network of Fig. 5.16, count the number of different topological sorts.

5. [*21*] How many slices of pizza can we obtain by n straight cuts with a pizza knife? More formally, how many regions of the plane can be created by n cuts?

6. [*22*] Let us consider the feedforward neural network in the figure on the side, where $w = 4$, $w_u = 2$, $b = -2$, and the units are based on the sigmoidal function $\sigma(a) = 1/(1 + e^{-a})$. This network is the *time-unfolding* of $x_{t+1} = \sigma(wx_t + b + u_t)$, where $u_0 = 1$ and $u_t = 0$, for $t > 0$. What happens when $t \to \infty$? Determine $\lim_{t \to \infty} x_t$ and $\lim_{t \to \infty} \nabla e(w, b)$.

7. [*18*] Suppose you are given a multilayered network with one output such that all its weights (including the bias terms) are null. Compute the gradient for the cases of hyperbolic tangent, logistic sigmoid, and rectifier units.

8. [*21*] Let us consider a feedforward network whose output neurons are computed by smx. What are the backprop equations in this case?

9. [*M17*] Based on the Hessian BP computational scheme defined by (17) construct a correspondent algorithm for the Hessian computation.

10. [*M21*] Use Hessian BP to reproduce Eq. (5).

5.5.3 **Symbolic and automatic differentiation**

Learning algorithms rely on the computation mostly of the gradient and of the Hessian of appropriate objective functions. Differentiation can be done manually, but there are nice tools to perform *symbolic differentiation*. As discussed in the previous section, we can use numerical differentiation, but we can do better by properly exploiting the structure of the function to be optimized. Backpropagation neither performs numerical nor symbolic differentiation. Unlike numerical analysis, it returns a precise expression for the gradient and, unlike symbolic differentiation, it computes the gradient on a given point with optimal complexity, but it doesn't return the symbolic expression. Since, nowadays, learning schemes go well beyond supervised learning with feedforward nets, one may be interested in understanding the generality of the discussed backpropagation computational scheme.

In order to shed light on the essence of this computational scheme, let us consider the following example. Suppose we want to compute the gradient of the function

$$y_o = f(x_1, x_2) = (1 + x_2) \ln x_1 + \cos x_2.$$

We can provide the following *expression DAG*:

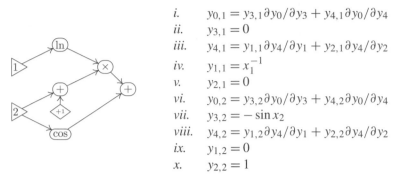

i. $y_{0,1} = y_{3,1} \partial y_0/\partial y_3 + y_{4,1} \partial y_0/\partial y_4$

ii. $y_{3,1} = 0$

iii. $y_{4,1} = y_{1,1} \partial y_4/\partial y_1 + y_{2,1} \partial y_4/\partial y_2$

iv. $y_{1,1} = x_1^{-1}$

v. $y_{2,1} = 0$

vi. $y_{0,2} = y_{3,2} \partial y_0/\partial y_3 + y_{4,2} \partial y_0/\partial y_4$

vii. $y_{3,2} = -\sin x_2$

viii. $y_{4,2} = y_{1,2} \partial y_4/\partial y_1 + y_{2,2} \partial y_4/\partial y_2$

ix. $y_{1,2} = 0$

x. $y_{2,2} = 1$

which suggests a recursive structure for the computation of ∇f. The end of the recursion is reached for the vertexes of the graph whose parents are the variables x_1 and x_2. We can easily see that the computation of ∇f can be carried out by a forward step. We distinguish two different differentiations: $y_{i,j} = \partial y_i/\partial x_j$, with $j = 1, 2$, and $\partial y_i/\partial y_\kappa$. The objective of the computation is to determine $y_{i,j}$ for $i = 0$. We can determine $y_{1,j}$, $y_{2,j}$, and $y_{3,j}$ directly, whereas we need a forward propagation for computing $y_{4,j}$ and $y_{0,j}$. The difference with $\partial y_i/\partial y_\kappa$ is that they can be symbolically determined immediately from the expression DAG. We have

$$\partial y_0/\partial y_3 = 1, \quad \partial y_0/\partial y_4 = 1, \quad \partial y_4/\partial y_1 = y_2, \quad \partial y_4/\partial y_2 = y_1.$$

Now, if we refer to the DAG expression and to the side equations, then the computation of $\nabla f = (y_{0,1}, y_{0,2})'$ takes place according to a data flow scheme which follows the sorting:

$$y_{0,1} \to \{iv, v, ii\}, iii, i, \quad y_{0,2} \to \{vii, ix, x\}, viii, vi,$$

where the numbers in curly braces indicate that there are no sorting constraints, so they can be associated with parallel computations. Exercise 1 proposes the computation of the network sensibility by automatic differentiation using the described forward step. Notice that the computation of the network sensibility is closely related to the computation of the gradient with respect to the weights connected to the inputs. In case of feedforward networks, because of the bilinear structure in which weights and inputs are involved, the equations are very similar (see Exercise 2). It is easy to realize that the described automatic differentiation has a general structure, which is dictated by the DAG expression. (Sometimes, it is called a computational graph.)

The computation of ∇f can be carried out also by using a generalization of backpropagation. In particular, we have

$$y_{0,1} = \frac{\partial y_1}{\partial x_1}\frac{\partial y_0}{\partial y_1} + \frac{\partial y_2}{\partial x_1}\frac{\partial y_0}{\partial y_2} + \frac{\partial y_3}{\partial x_1}\frac{\partial y_0}{\partial y_3} = \frac{1}{x_1}\frac{\partial y_0}{\partial y_1},$$

$$y_{0,2} = \frac{\partial y_1}{\partial x_2}\frac{\partial y_0}{\partial y_1} + \frac{\partial y_2}{\partial x_2}\frac{\partial y_0}{\partial y_2} + \frac{\partial y_3}{\partial x_2}\frac{\partial y_0}{\partial y_3} = \frac{\partial y_0}{\partial y_2} - \frac{\partial y_0}{\partial y_3}\sin x_2.$$

The auxiliary variables $\partial y_0/\partial y_1$, $\partial y_0/y_2$, and $\partial y_0/\partial y_3$ correspond with backprop-agation *delta error* and can be determined by a backward step. (In the automatic differential literature, this is also referred to as *reverse accumulation*.)

Like for the forward step computation, this scheme is general and can be applied as soon as the DAG expression is given. It is not difficult to realize that the forward step technique is more efficient than the backward step for functions $f : \mathbb{R}^d \to \mathbb{R}^m$, where $m \gg d$, whereas the backward step technique is more efficient in case $d \gg m$ (see Exercise 3).

EXERCISES

1. [*21*]   Use the automatic differentiation by forward step to determine the gradient of the output with respect to the input (network sensibility).

2. [*21*]   Discuss the differences between the network sensibility and the gradient equations with respect to the weights connected to the inputs.

3. [*16*]   Prove that the forward step technique is more efficient than the backward step for functions $f : \mathbb{R}^d \to \mathbb{R}^m$, where $m \gg d$, whereas the backward step technique is more efficient in case $d \gg m$.

4. [*31*]   Let us consider a single neuron with logistic sigmoid as a nonlinear transfer function, and the XOR training set $\mathcal{L} = \{(a, 0), (b, 1), (c, 0), (d, 1)\}$, where $a = (0, 0)$, $b = (1, 0)$, $c = (1, 1)$, and $d = (0, 1)$. Determine the stationary points and discuss their nature.

5. [*26*]   Suppose we are given the training set $\mathcal{L} = \{((0, 0)', \underline{y}), ((1, 0)', \overline{y})\}$ and a single neuron with logistic sigmoid. Discuss the solution of the loading problem. What about generalization to new examples? What is the relationship between the choice of the nonasymptotic targets and the generalization? Determine which values of $\underline{y}$ and $\overline{y}$ lead to the same solution that one would determine by maximizing the maximum margin.

5.5.4 **Regularization issues**

In this section we discuss regularization issues beginning from the analysis carried for kernel machines in Section 4.4.4. The corresponding view of learning is that which drives most of the methods discussed in this book: Minimize a parsimony index under the satisfaction of the environmental constraints. Basically, we want to unify the analysis carried out in kernel machines with regularization in feedforward networks. The first remark is that they provide a clever way of discovering the feature map and return the output

$$f(x) = \hat{w} \cdot \widehat{\phi}(x) = w \cdot \phi(x) + b.$$

The essence of the neural computation is in the developed feature map, which is characterized by the hidden structure of the network, while the parameters in vector $\hat{w}$ are only those of output connections. The parsimony index is based on the idea of enforcing smoothness in the solution, which leads to considering the norm $\|f\|^2 = \langle Pf, Pf \rangle$, where $P = \sum_{i=0}^{m} \alpha_i D^i$. In Section 4.4.4, it has been shown that $\|f\|^2 = \lambda \cdot G\lambda$ can be expressed as a quadratic function in the dual space parameters λ, where G is the Gram matrix of the Green function of the regularization operator

$L = P^*P$. In case of feedforward networks, which consists of a representation of f in the primal space, we need to do something similar. Hence we can express $\|f\|^2$ by using regularization operators just like for kernel machines. We have

$$\|f\|^2 = \langle P\hat{w} \cdot \widehat{\phi}, P\hat{w} \cdot \widehat{\phi} \rangle$$

$$= \left\langle \sum_{r=1}^{m} \alpha_r D^r \left(\sum_{h=0}^{D} w_h \phi_h + b \right), \sum_{s=1}^{m} \alpha_s D^s \left(\sum_{k=0}^{D} w_k \phi_\kappa + b \right) \right\rangle. \tag{1}$$

Now, let $\alpha_0 = 0$. This means that $\|f\|^2$ doesn't care about the magnitude of f, but only focuses on its smoothness. A discussion on the case in which $\alpha_0 \neq 0$ is initiated in Exercise 1. We have

$$\|f\|^2 = \left\langle \sum_{r=1}^{m} \sum_{h=1}^{D} \alpha_r w_h D^r \phi_h, \sum_{s=1}^{m} \sum_{k=1}^{D} \alpha_s w_k D^s \phi_k \right\rangle$$

$$= \sum_{r=1}^{m} \sum_{s=1}^{m} \sum_{h=1}^{D} \sum_{k=1}^{D} \alpha_r w_h \alpha_s w_k \langle D^r \phi_h, D^s \phi_k \rangle$$

$$= \sum_{r=1}^{m} \sum_{s=1}^{m} w_h w_k \sum_{h=1}^{D} \sum_{k=1}^{D} \alpha_r \alpha_s \langle D^r \phi_h, D^s \phi_k \rangle \tag{2}$$

$$= \sum_{h=1}^{D} \sum_{k=1}^{D} w_h w_k \left\langle \sum_{r=1}^{m} \alpha_r D^r \phi_h, \sum_{s=1}^{m} \alpha_s D^s \phi_k \right\rangle$$

$$= \sum_{h=1}^{D} \sum_{k=1}^{D} w_h w_k \langle P\phi_h, P\phi_k \rangle = \sum_{h=1}^{D} \sum_{k=1}^{D} w_h w_k \langle L\phi_h, \phi_k \rangle.$$

This holds regardless of the choice of ϕ and P. Now, given the feature map ϕ, suppose there exists P such that for the differential operator $L = P^*P$, for all $j = 1, \ldots, D$, the following condition holds true:

$$L\phi_j(x) = \gamma_j \phi_j(x). \tag{3}$$

In so doing, we are assuming that the features ϕ_j are the eigenfunctions of the regularization operator L. As a consequence,

$$\langle L\phi_h, \phi_k \rangle = \gamma_h \langle \phi_h, \phi_k \rangle = \langle \phi_h, L\phi_k \rangle = \gamma_k \langle \phi_h, \phi_k \rangle,$$

which holds true only if the eigenfunctions are orthogonal, that is, $\langle \phi_h, \phi_k \rangle = \delta_{h,k}$. Finally, we have

$$\|f\|^2 = w'\Gamma w, \tag{4}$$

where $\Gamma = \text{diag}(\gamma_i)$. Interestingly, also in the general case in which we choose a feature map that doesn't satisfy Eq. (3), the value of $\|f\|^2$ can still be computed by

Eq. (4). Of course, in this case, we need to choose $\gamma_{h,k} := \langle L\phi_h, \phi_k \rangle$. If we interpret the regularization at layer level, the application of this idea leads to extend Eq. (4) to all the weights of a feedforward network.

Once we know how to express $\|f\|^2$ in the primal space, we can immediately formulate supervised learning as the following optimization problem:

$$\hat{w}^* \in \arg\min_{\hat{w}} \left(E(\hat{w}) + \mu w' \Gamma w \right).$$

Here we overload the symbol w to denote a vector with all the neural network weights. Unfortunately, it's hard to provide an estimate of Γ.

The simplest heuristic solution is simply to choose $\Gamma = \text{Id}$. When using gradient descent, we get

$$w_{i,j}^{r+1} = w_{i,j}^r - \frac{\eta\mu}{2} w_{i,j}^r - \eta \nabla E |r = \beta w_{i,j}^r - \eta \nabla E|_r \tag{5}$$

where η is the learning rate, while $\beta := 1 - \eta\mu/2$. This learning scheme driven by Eq. (5) is referred to as *weight decay*. Here the learning rate is chosen small enough to guarantee that $0 < \beta < 1$, so the processing dictated by Eq. (5) consists of generating a gradient descent trajectory that is properly filtered out by the term $\beta w_{i,j}$. A related learning process is based on the *momentum term*, which acts like weight decay, the only difference being that it involves the weight variation. Eq. (5) becomes

$$w_{i,j}^{r+1} = w_{i,j}^r + \beta(w_{i,j}^r - w_{i,j}^{r-1}) - \eta \nabla E|_r. \tag{6}$$

This resembles weight decay as stated by Eq. (5). If we pose $\Delta w_{i,j}^{r+1} := w_{i,j}^{r+1} - w_{i,j}^r$ then the weight updating stated by Eq. (6) becomes

$$\Delta w_{i,j}^{r+1} = +\beta \Delta w_{i,j}^r - \eta \nabla E|_r.$$

The momentum term $\beta \Delta w_{i,j}^r$ closely resembles weight decay of Eq. (5). Like for weight decay, the gradient term is properly low-pass filtered, but in this case, we end up with a second-order difference equation. A nice interpretation of the momentum term will be given in Section 6.5 when formulating the learning process as the outcome of computational laws that involve the temporal dimension.

EXERCISES

1. [20] Remove the assumption $\alpha_0 = 0$ by using Eq. (1) for the expression of $\|f\|$.

5.6 Advances in deep learning

It is widely known that two of the key factors that strongly contributed to the popularity of deep learning are the larger availability of affordable hardware with advanced parallel computing capabilities (Graphics Processing Units, GPUs) and the massive diffusion of larger scale datasets. On one hard, this motivated researcher to inspect and evaluate neural architectures composed of several stacked computational layers, as the ones discussed in the previous sections. On the other hand, the scientific community started to explore novel ways of propagating the signal through the network, or to assemble advanced architectures composed of modules that either cooperate or compete one against each other. In this section we focus on several neural-network-related approaches that play an important role not only in developing more powerful applications but also in gaining a deeper understanding about important properties of neural architectures, of their training procedure, and of their intrinsic limits.

5.6.1 Residual networks

Neural networks composed of several stacked layers with non-linear activation functions might lead to issues in gradient computations. These issues could be due to the numerical explosion of the gradients, when back-propagating the signal down to the lower layers, or, more commonly, to the vanishing of the gradient, that can become extremely weak when computed with respect to the parameters belonging to the bottom layers of the network. Let us indicate with $F^l(w^l, o^l)$ the function computed by the l-th layer of a feed-forward neural network with multiple layers. The weights and biases of the l-th layer are compactly represented by w^l, while o^l is the vector with the layer input, usually the output of the $(l-1)$-th layer. Formally, the usual computational scheme is the following one,

$$o^l = F^l\left(w^l, o^{l-1}\right) = \sigma\left(f^l\left(w^l, o^{l-1}\right)\right), \tag{1}$$

that repeats for all valid l's. This straightforward scheme models a learnable non-linear function F^l which transforms o^{l-1} into o^l, where σ is a non-linearity applied to the usually linear activation score, here indicated as the outcome of $f^l\left(w^l, o^{l-1}\right)$. The example of Fig. 5.18 (left) shows an instance of a 3-layer network, using the just introduced notation.

A residual mapping introduces a related but structurally different computational scheme, in which f^l is responsible of computing the offset vector to be added to o^{l-1} in order to get o^l, which is the outcome of the so-called *residual connection*. Formally, a *residual block* is based on function $\hat{F}^l$, that we introduce to differentiate it from F^l, defined as

$$o^l = \hat{F}^l\left(w^l, o^{l-1}\right) = \sigma\left(o^{l-1} + f^l\left(w^l, o^{l-1}\right)\right), \tag{2}$$

where we considered the simplest form of such a block and we explicitly included σ in the computations. Notice that f^l could also be composed of multiple compu-

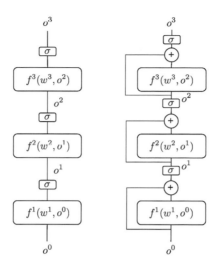

FIGURE 5.18

Classic Multilayer Feed-Forward network (left) and Residual Network (right). The external input is indicated with o^0, while o^3 is the network output ($L = 3$ layers).

tational layers interleaved with non-linearities. This novel way of transforming the data that traverses though the net has an important property, that can be immediately grasped once we realize that o^{l-1} explicitly appears twice in the formula that computes o^l. In Eq. (2) we see that the first instance of o^{l-1} is not influenced at all by the weights of the layer, implementing a simple form of skip connection, where the layer input signal also "skips" the computations in f^l and it is added to the output of f^l itself. This means that the layer can more easily compute an identity mapping, by simply setting w^l to zero (assuming $\sigma(0) = 0$), and the learning dynamics are completely focused in predicting how o^{l-1} should be "perturbed" in order to compute the novel representation o^l. This is different from vanilla layers of Eq. (1), where setting w^l to zero yields a zero output (still assuming $\sigma(0) = 0$ and zero bias). As a result, residual blocks make it easier to keep the signal propagated even in case of small or null weights. Of course, if the size of the output of f^l has a different number of components with respect to the output of layer $l-1$, then o^{l-1} must be further projected to match the dimensionality of the right term in the sum of Eq. (2). Stacking residual blocks yields the so-called *residual network*, as shown in Fig. 5.18 (right).

It is interesting to consider the case in which σ is the identify function, $\sigma(x) = x$, yielding complete identity-mapping-based models. This is the case in which it is easier to show the benefits of residual blocks in back-propagating gradients through the network. Let us consider a network composed of L residual layers of the type formalized in Eq. (2), with $l = 1, \ldots, L$ and with σ set to the identity function (o^0 is the external input). We have

$$o^L = o^{L-1} + f^L\left(w^L, o^{L-1}\right)$$
$$o^{L-1} = o^{L-2} + f^{L-1}\left(w^{L-1}, o^{L-2}\right)$$

$$\ldots$$

$$o^1 = o^0 + f^1\left(w^1, o^0\right).$$

Then, by recursively replacing from top to bottom each instance of o^{j-1} not included in f^j with its definition from the line below, we get

$$o^L = o^0 + \sum_{j=1}^{L} f^j\left(w^j, o^{j-1}\right),$$

or, more specifically, for each $l = 0, \ldots, L-1$,

$$o^L = o^l + \sum_{j=l+1}^{L} f^j\left(w^j, o^{j-1}\right),$$

where it is easy to see that the contributions of the different f^j's are summed up. This is significantly different from the classic case of Eq. (1), in which the output of each f^j is only provided as input to f^{j+1} (discarding the role of non-linearities to simplify the description). As a result, whenever we compute, for example, the derivative of the network output o^L with respect to some internal weights w^{l+1}, that might belong to the bottom layers of the network, we get

$$\frac{\partial o_L}{\partial w^{l+1}} = \frac{\partial \sum_{j=l+1}^{L} f^j(w^j, o^{j-1})}{\partial w^{l+1}} = \frac{\partial f^{l+1}(w^{l+1}, o^l)}{\partial w^{l+1}} + \frac{\sum_{j=l+2}^{L} \partial f^j(w^j, o^{j-1})}{\partial w^{l+1}}.$$

We basically have two distinct paths that propagate the gradients with respect to w^{l+1}. While the dependencies on f^j's with $j > l+1$ are modeled by the second term of the summation above, the derivative of f^{l+1} with respect to "its own weights" w^{l+1} explicitly appears in the first term of the summation, without any dependencies on the other f^j's with $j > l+1$, thus helping gradients to directly propagate to the considered layer.

5.6.2 Adversarial machine learning

The research field of *adversarial machine learning* investigates the vulnerability of neural networks and, more generally, of machine learning-based models, to carefully-altered input samples aimed to mislead predictions of a given model. Let us consider a predictor that has been trained from a finite set of supervised pairs, for example a

PREDICTION $\rightarrow$ "CAT"   HIGHLIGHTED DIFFERENCE   PREDICTION $\rightarrow$ "LAPTOP"

FIGURE 5.19

The picture of a cat is correctly classified by a neural network (left). Then, a few pixels are altered (middle–dots are artificially highlighted for better visibility), and the network predicts the picture as belonging to the "laptop" class (right). The altered picture looks almost unchanged to a human.

neural network. The network learns to be coherent with the provided data, thus mimicking the input-output pairs of the training set. However, the network will develop its own criterion to reach such a goal, that might be different from what one might expect. The network could learn that pictures of cows are classified as such because there is a large lawn of grass in the background, completely ignoring the cows themselves! If the training set is composed of pictures of cows in which the background is always like that (and the same background is never shown in pictures of other categories), the solution found by the network perfectly fits the training data. What if, at test time, we provide an image of a cow on a sandy beach? The network will end up in not being able to correctly classify it. This simple example clearly suggests that there is an inherent issue in the model that exposes it to potential vulnerabilities, due to the biases the network inherits from the training data. There are regions of the input space that are not covered (or covered with low probability) by the distribution of the training examples, and if a test example belongs to such regions then the final prediction of the network is not easy to define in advance. As a matter of fact, an attacker can exploit this issue to favor prediction errors. It turns out that neural networks can learn to be very sensitive to variations in some regions of the input space. It has been shown that changing the values of a few pixels in an image, in a way that is not perceptible for a human, can make the network change its decision. A correctly classified image of a cat can be classified as the one of a laptop after having slightly changed the intensity of some pixels, as sketched in the toy example of Fig. 5.19. These comments open to the need of studying how an attacker could manipulate a machine learning model in order to make it misclassify test examples. Of course, improving the knowledge on how a model could be corrupted or on how data could be altered to make the network behave in wrong way is very important in order to propose appropriate defenses or to make the network more robust.

The two most common families of attacks against machine learning models are *evasion* attacks and *poisoning* attacks. The former consists of those attacks that directly manipulate the test data in order to let the machine take a wrong decision, as we already anticipated in the toy example of Fig. 5.19. Let us consider a multi-class problem, and let us assume that we are given a test example x that is correctly classified as belonging to the k-th class by a neural network with weights w. We use the notation $f(w, \cdot)$ to indicate the function computed by the network, that returns a vector of confidence scores, one for each class. If we indicate with $f_k(w, x)$ the confidence on the k-th class, the most straightforward goal of the attacker is to minimize $f_k(w, x)$ by altering x, thus, for example, solving the following problem by gradient descent,

$$\min_{x'} f_k(w, x'),$$

with x' at the first step of the descent initialized to the value of x, i.e., $x'_0 = x$. The generic update step, indexed by t, is

$$x'_{t+1} = x'_t - \alpha \left. \frac{\partial f_k(w, s)}{\partial s} \right|_{s=x'_t},$$

being $\alpha > 0$. This can also be described in terms of the perturbation that is added to x in order to yield x', so $\min_\delta f_k(w, x + \delta)$, where $x' = x + \delta$. However, this is usually not enough. First of all, unless we are dealing with a binary classification problem with competing classes, the attacker will also try to make the classifier develop a very high confidence on another class, different from k. Moreover, it is pretty common to assume that x is not altered in a too "evident" manner (it is also eventually constrained to fit some specific bounds). The first open problem is how to model the latter condition, i.e., how to measure if x is modified in a too strong manner or not. A naive approach consists in defining a distance-like function $d(x', x)$ that returns high values when x' is "far" from x, and to ensure that $d(x', x) \leq \tau$, for some $\tau > 0$. Function d is frequently assumed to be $||x' - x||_p$, where $|| \cdot ||_p$ is an L_p norm. However, there might be directions in the input space on which it is possible to strongly alter x without making this change "evident" to a human or to other automatic systems, while, vice-versa, some directions should be left untouched. If we think about an image, there might be pixels that are located in messy areas, so that they can be altered without being noticed by the observer, while adding some red pixels in a large-white wall could be easily spotted. For this reason, $d(\cdot, \cdot)$ could also be a form of perceptual similarity. Alternatively, one might restrict the number of dimensions on which the optimization is applied, for example by altering a single pixel of an image. Formally, the attacker can alter x, transforming it into $\tilde{x}$, solving

$$\tilde{x} \in \arg\max_{x'} \left(\max_{i \neq k} f_i(w, x') - f_k(w, x') \right)$$

$$\text{s.t.} \quad d(x', x) \leq \tau$$

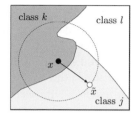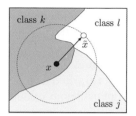

FIGURE 5.20

An evasion attack in which an example x of class k is transformed into an adversarial example $\tilde{x}$, predicted as belonging to another class. Different levels of gray are about space regions in which the network predicts different classes. The dotted circle is about the constraint on the maximum distance from x. In the first picture (left) the target class is not defined, and x is moved to the closest class. In the second picture (right) the target class is l.

while constraining $d(x', x) \leq \tau$. The generated $\tilde{x}$ will be close to x accordingly to the notion of distance encoded by $d(\cdot, \cdot)$, and it minimizes the confidence on the k-th class. The term $\max_{i \neq k} f_i(w, x')$ selects the class (different from k) on which the classifier is more confident, to facilitate the optimization of the constrained problem. Basically, a wrong class on which the classifier currently has the largest confidence is a good candidate for making it even more confident (the "closest" class), as shown in Fig. 5.20 (left). Of course, an attacker might also be interested in misclassifying x while having in mind a specific target class. For example, he could aim to specifically alter x, of class "car", to get $\tilde{x}$ classified as an "animal." In this case, the previous problem becomes,

$$\tilde{x} \in \arg\max_{x'} \left(f_l(w, x') - \max_{z \neq l} f_z(w, x') \right),$$

where l is the target class ("animal"), whose confidence $f_l(w, x')$ is maximized, while the term $-\max_{z \neq l} f_z(w, x')$ is a way to reduce the probability of other classification outcomes (including the correct one–"car"), as shown in Fig. 5.20 (right). Both the problems are usually solved by gradient-based optimization algorithms, starting with $x' = x$ and progressively updating x' while still fulfilling the distance constraint, for example by projected gradient descent. Of course, the optimizer could also get stuck in local minima or other stationary points.

The second family of attacks we consider, i.e., *poisoning attacks*, is the one in which the attacker can inject some malicious examples into the training data, so that the resulting network will be biased by them and it will misclassify some or several test examples. In principle, the simplest form of poisoning consists in adding a fraction of wrongly labeled examples to the training set, thus usually making the classifier misclassify data that is close to the selected wrongly labeled examples. When the goal is to generically make the classifier maximize the number of prediction errors on test data, adding wrongly labeled training example is still a possibility but, of course, there might be malicious examples located in space regions that, overall, will have a stronger negative effect on the quality of the trained classifier with respect to others. For this reason, the attacker specifically alters a set points so that, when they are added to the training set, they end up in making the classifier strongly increase the test errors. Formally, given an original training set $\mathcal{T}$, the attacker aims to build the labeled set $\tilde{\mathcal{P}}$ of malicious examples given by

$$\tilde{\mathcal{P}} = \arg\max_{\mathcal{P}} \mathcal{L}(\mathcal{V}, f, w^*)$$

$$\text{s.t.} \quad w^* = \arg\min_{w'} \mathcal{L}(\mathcal{T} \cup \mathcal{P}, f, w')$$

where $\mathcal{L}$ measures the loss of the network when making predictions on the set represented by its first argument, and $\mathcal{V}$ is a set of untainted points excluded from the training procedure and used to evaluate the generalization skills of the model. The term $\mathcal{L}(\mathcal{V}, f, w^*)$ basically tells how good the network is in generalizing to never-seen-before examples, that is what the attacker tries to compromise (larger values of $\mathcal{L}(\mathcal{V}, f, w^*)$ imply worse generalization skills). The weights of the network (w^*) are the outcome of training the net using $\mathcal{T} \cup \mathcal{P}$ as training data, as modeled by the constraint of the problem above. Poisoning attacks can also be performed with the goal of making the network misclassify examples of specific classes, and not in a generic manner as we described so far. In both the cases, poisoning procedures are more costly than evasion attacks, since the network must be retrained every time that $\mathcal{P}$ is updated, thus, in practice, several approximations of the described procedure are commonly exploited.

An important variable in the attack procedure involves the amount of knowledge that the attacker has access to. While it is evident that, in poisoning attacks, the attacker needs to have the capability of adding malicious training data to the target classifier, the specific setting in which adversarial attacks are crafted differs in function of the attacker's knowledge. As a matter of fact, if the attacker has full access to the neural network that he is planning to attack, to the training data, to type of input representation, to the learning algorithm, etc., then we talk about *white-box attacks*. In this case, the attacker can make queries and compute gradients exploiting the target model in the exact conditions in which it is used by the provider of the service that relies on such a model. This is a very extreme setting, that represents a natural framework for worst-case evaluations of the security of a learning algorithm. Of course, it seems more realistic to assume that the attacker is only able to access the predictions of the network and that he does not have any other information (network structure, values of the weights, etc.), so that we talk about *black-box attacks*. The number of queries the attacker can submit to $f(w, \cdot)$ might be limited within specific bounds introduced by the provider of the service that hosts the network, or there might exist more practical limits related to the computational times. For example, one might be tempted to compute gradients by numerical differentiation, that would require a large number of queries and that is likely to quickly become unfeasible on a service that operates over a slow network or with long response times. There are several other intermediate configurations in-between black and white box, such as when the attacker operates in a relaxed black-box setting, knowing also the network structure, and/or the input representation, and/or the learning algorithm, so that we talk about *gray-box attacks* (notice that the values of the weights are still not known, as well as the original training data). In black-box scenarios and, more frequently, in gray-box ones, the attacker can rely on a surrogate model that is similar to the target $f(w, \cdot)$, and that he uses to craft adversarial examples. As a matter of fact, it has been shown that adversarial attacks can be easily *transferred* from one model to another. If two models are similar enough, then an attack which is effective on the former will be likely effective also on the latter. The surrogate model is created by the attacker, for example training his own neural net, and it is what the attacker will use to compute gradients and solve the optimization problem to generate the malicious $\tilde{x}$ or $\tilde{\mathcal{P}}$. In this case, the attacker has to find a reasonable way to train the surrogate model, for example collecting a private dataset of images whose labels are obtained by querying the target $f(w, \cdot)$ (still fighting with the aforementioned limits to the number of queries), or by collecting another dataset in the same domain in which the network is operating.

Being able to craft powerful attacks to machine learning models is not the only focus of the research community that studies adversarial machine learning. In fact, a lot of attention is also devoted to improving the security of a target model by means of different strategies, i.e., procedures that are aimed at making the model more robust to a variety of attacks. On one hand, the machine learning model must be deeply kept under control, to mitigate the effects of attacks that could have happened in the past. This usually involves semi-automated procedures to verify the correctness of the classifier decision, with the help of human experts. On the other hand, it is very important to include defenses to protect from attacks that could take place in the future. Among a variety of possibilities, we mention the so-called adversarial training, in which adversarial examples are generated at each step of the training procedure, and used to make the network more robust. The training procedure of the neural models consists in minimizing a loss on the training data $\mathcal{T}$ and on a set $\mathcal{A}$ composed of pairs of adversarial examples and the labels of the original training examples from which they were generated,

$$\min_{w'} \mathcal{L}(\mathcal{T} \cup \mathcal{A}, f, w').$$

The set $\mathcal{A}$, is recreated at each epoch of the training procedure, following the aforementioned evasion procedure for a network whose weights are the ones at the considered epoch. In other words, evasion attacks are created as long as learning proceeds, and the network is trained to make correct predictions in those space regions (represented by the data in $\mathcal{A}$) in which it is more error-prone at the current epoch.

Another popular family of the defenses consists of those approaches that detect evasion attacks and reject the malicious samples. They are based on the idea that adversarial examples appear in regions of the space not populated by training data. As a consequence, a possible heuristic to reject potential malicious examples consists in evaluating if they fall in areas that are out of the estimated distribution of the training data. This immediately triggers the question on how to estimate such distribution in an effective manner and on how to define a criterion to measure if a point should be rejected or not. A feasible road is the one of indirectly doing it. Consider the case of multi-label classification, and let us assume that we have access to knowledge on the relationships among classes. For example, we could know that class "dog" and "cat" are mutually exclusive, and they both imply class "animal" and not class "vehicle." The classifier will be very confident in fulfilling these relationships on the training data, i.e., when $f_{dog}(w, x)$ is large on the input x (being $f_{dog}(w, x)$ the confidence on the class dog), then also $f_{animal}(w, x)$ will be large, while $f_{vehicle}(w, x)$ will be very small. In the space regions where training data is not distributed, the network does not guarantee to be strongly coherent with this knowledge, thus it might happen that a large f_{dog} is paired with a not-so-small $f_{vehicle}$. It is indeed possible to evaluate the fulfillment of the known relationships, building a rejection criterion to discard examples that are not coherent with the domain knowledge.

We conclude this description mentioning that, in principle, whenever a defense gets proposed to make a network robust to a certain attack, a novel type of attack that circumvents it can be designed. This triggers a loop that is only broken whenever defense strategies are not only specifically designed to handle known attacks, but that are also general enough to mitigate potential future attacks. Of course, this is far from being trivial.

5.6.3 Generative adversarial networks

The main principles behind the generation of adversarial examples are exploited to formulate the learning criterion of the so-called Generative Adversarial Networks (GANs). Imagine to have the use of a machine that is able to generate data from scratch ("fake" data), and imagine that such data is so realistic that is extremely hard or even impossible to distinguish it from real samples ("real" data), i.e., samples that were specifically collected in the considered domain, and not artificially generated. Let us consider the domain of pictures of human faces as running example. The real samples are pictures that were acquired by a camera, taking photographs of existing people, while fake samples are pictures artificially generated by a machine. Notice that such fake samples are associated to faces of people that do not exist. GANs are special neural models designed to generate hardly-distinguishable fake data, trained exploiting a strategy that pretty much resembles ideas related to adversarial machine learning. In particular, a GAN is composed of two networks that, in a sense, compete one against the other. The first network, the *discriminator D*, is a binary classifier which is responsible of determining if its input belongs to the class of real data or to the class of fake data. In our running example, D processes an input picture of a face and it predicts whether it is real or fake. Of course, D is expected to classify as real those pictures that were collected using a camera, and to classify as fake what was artificially generated by a machine. The second network, the *generator G*, is a neural network that, given a random stimulus, outputs a fake sample, such as a fake picture of a human face. Changing the random input makes the network generate a different, but still fake, picture. Training a GAN is the outcome of an iterative process in which D learns to better discriminate real data and data generated from the current G, and, in turn, G learns to generate data on which D makes mistakes, thus fake examples that are misclassified as real.

Let us indicate with $D(w_D, x)$ the function computed by the discriminator for an input x, overloading the already introduced notation. Let us assume D to return a value in $[0, 1]$, where 1 is the score associated to inputs that are predicted as real with the largest confidence and 0 is associated to data strongly predicted as being fake. We used w_D to refer to the weights of the network. Similarly, we indicate with $G(w_G, z)$ the function of the generator, being z a vector of random numbers. Given a collection of n_r real samples, x_i, $i = 1, \ldots, n_r$, and given n_f vectors of random numbers z_j, $j = 1, \ldots, n_f$, the learning criterion that is minimized in order to train a GAN is

$$\min_{w_G} \max_{w_D} \left(\sum_{i=1}^{n_r} \log D(w_D, x_i) + \sum_{j=1}^{n_f} \log\left(1 - D(w_D, G(w_G, z_j))\right) \right)$$

where, for simplicity, we assumed to aggregate losses on each input using the sum. The learning criterion basically consists of a minimax game. The discriminator maximizes such a criterion whenever it outputs 1 for all the real samples and 0 for all the ones generated by G. The generator minimizes the criterion when it is able to generate samples on which the discriminator outputs 1 (real), considering all the n_f fake samples that are generated exploiting the n_f random stimuli. To make this point easier to follow, we can restrict the previous learning problem to what involves the discriminator D,

$$\max_{w_D} \left(\sum_{i=1}^{n_r} \log D(w_D, x_i) + \sum_{j=1}^{n_f} \log\left(1 - D(w_D, G(w_G, z_j))\right) \right),$$

while from the point of view of G, the problem is

$$\min_{w_G} \sum_{j=1}^{n_f} \log\left(1 - D(w_D, G(w_G, z_j))\right).$$

The first problem is basically the learning objective of a binary classification task, where the classifier is D and we have real samples and fake samples to distinguish. Differently, when minimizing the second problem we are basically making G able to generate *adversarial examples* for D, i.e., artificially generated examples that are misclassified (as real) by the discriminator. The two learning problems can be optimized in an alternate manner, making a (or some) gradient-based update(s) of w_D (first problem) and a gradient-based update of w_S (second problem), and repeating for multiple iterations. The data used in each iteration can be a randomly selected

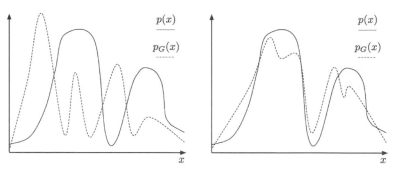

FIGURE 5.21

The distribution (p) of real data (solid) and the one (p_G) of the fake/generated data (dashed), comparing them during the early stage of the GAN learning procedure (left) and when closer to the end of training (right).

sub-portion of the original one, thus applying an instance of stochastic gradient descent for each of the two problems.

It is interesting to better investigate what happens during the training procedure of a GAN. Let us indicate with $p(x)$ the distribution of the real data in the input space, such as the distribution of real human faces in our previous example. This distribution is static in the considered problem, that is, it does not change while training proceeds. The set of real samples that is used in the learning objective is sampled from $p(x)$. Let us indicate with $p_G(x)$ the distribution of the fake data generated by G which, differently from $p(x)$, undergoes several changes during the training procedure. As a matter of fact, the data generated by G potentially changes every time that w_G is updated. During the first training epochs, with randomly initialized w_G, it is very likely that $p_G(x)$ will be pretty different from $p(x)$. As long as training proceeds, G improves, and it yields fake data that progressively gets harder to distinguish from real data. This means that the two distributions, p and p_G, are becoming more similar and the discriminator D progressively faces more difficulties in distinguishing samples coming from p and those coming from p_G. In Fig. 5.21, we show a plot of the two distributions at the early stages of learning and when closer to the end. Of course, the degree of similarity between them depends on several factors, among which we mention the structure of the neural architectures of D and G. The way D and G are implemented is actually arbitrary, but the representational capabilities of the two networks must be good enough to capture detailed information from the distribution of the real and fake data in the input domain, in order to benefit from the competition between discriminator and generator. In the case of data with meaningful spatial components, such as in images, D and/or G can be eventually implemented using convolutional architectures. Finally, we mention that GANs benefit from the availability of large sets of real data, that allows the model to better capture the properties of $p(x)$.

5.6.4 Transformers

A Transformer is a neural network architecture built around the concept of *attention*. In order to describe how a Transformer works, we have to introduce three basic and strongly related topics, with increasing degree of complexity. The first one is the generic notion of (*i*) *attention* in neural networks, that we will briefly present by providing a simple description of its role in well-known tasks. Then, such a notion will be generalized to introduce the second topic, that is about the so-called (*ii*) *self-attention*. We will provide a detailed description of self-attention before moving to the third topic, where multiple self-attention schemes are instantiated, yielding (*iii*) *multi-head attention*. As a matter of fact, Transformers are neural networks that massively use the idea of multi-head attention over multiple computational layers. As a result, they handle the input data by performing operations that are easily parallelizable.

From a very qualitative standpoint, the term *attention* in the deep learning literature has to do with a procedure in which different sub-portions of the input are appropriately weighed to create a new representation that is then used for further processing. The intriguing property of an attention model is that the weight assigned to each sub-portion is predicted by a function that includes learnable components, thus the network learns to predict what is more relevant in a certain configuration. Let us assume that the input of our network is divided into n sub-portions, whose contents are represented by vectors $x_1, \ldots, x_n$, with d components each, also referred to as *embeddings*. For example, x_i could be a feature vector that represents a portion of an image, or a vector representing a word in an n-word sentence. We will indicate with X the $n \times d$ matrix whose rows are the embeddings x_i's. An attention model learns to predict a (column) vector of probabilistically normalized scores $\alpha = (\alpha_{(1)}, \ldots, \alpha_{(n)})$, $\sum_{i=1}^{n} \alpha_{(i)} = 1$, $\alpha_{(i)} \geq 0$, and it computes the representation z that, in the simplest case, is a linear combination of the x_i's weighed by α,

$$ z = \phi(\alpha, X) = X^T \alpha = \sum_{i=1}^{n} \alpha_{(i)} x_i, $$

where the superscript T indicates the transpose operator. The new representation z is basically obtained by giving more emphasis to the sub-parts of the input associated to larger $\alpha_{(i)}$'s, and this representation is what is used as a key element to make a prediction in the task at hand. For example, it could be used to classify the input image, or the input text. The learnable function that has the major role in predicting α is what we will refer to as f_{att} in the following. Interestingly, this component is latent, and it does not directly receive any supervision signals. In other words, the network autonomously learn what to focus on.

Let us skip, for a while, the description about how α is exactly computed, and let us introduce two intuitive cases in which attention might be precious. Let us consider a squared image which is provided to a neural network with the goal of generating a textual description of its contents, and let us assume that the image is partitioned into n non-overlapping squared patches, where the i-th patch is represented by vector x_i.

Imagine the picture of a living room, with a sofa in the middle and a person sitting on it, talking on the phone. The neural network is expected to generate the sentence "a person sitting on the sofa and taking on the phone" in a progressive manner, word-by-word. For each word of the output description, one or more sub-parts of the image are indeed more important than others. Even if we avoid describing the details of the language generation process, it can be intuitively understood that, when the network is in principle of generating the word "person" (resp. "phone"), the salient sub-part of the image is the one with the person (resp. phone), thus we would like the network to learn to focus on it, assigning a large $\alpha_{(\cdot)}$ to the corresponding patch, and giving less importance to the rest, thus generating a representation z in which the focused patch provides a larger contribute, and then using z to generate the next word(s). When generating the word "sitting" (resp. "taking"), the network should focus on the person and also on what the person is sitting on, i.e., the sofa (resp. phone), so that the model should compute a compact representation z that gives relevance to both the image parts. Notice that building such a representation could involve also sub-portions of the input that are far away one from each other. Another example is the one of language translation, in which, for example, the neural network processes words $x_1, \ldots, x_n$ of the input sentence and it is expected to output the words of the corresponding sentence in the target language. In the case of a destination language whose syntax and grammar differ from the ones of the source language, when generating a word of the translation different parts (words) of the input sentence could be more important than others, even if located in significantly different positions, an intuition that an attention model implements by appropriately predicting the $\alpha_{(i)}$'s of the input words and generating representation z.

We can go beyond the idea of transforming the whole input into a single vector z to be used for prediction purposes, and introduce the more generic notion of *self-attention*. Given a set of n elements, we can build an attention-based representation for each of them, thus repeating multiple times the aforementioned procedure. Reusing the previously introduced notation, let us assume that, for each of the n sub-portions of the input, we have the use of a vector of attention scores, $\alpha_i = (\alpha_{i,(1)}, \ldots, \alpha_{i,(n)})$, $i = 1, \ldots, n$. We can compute a new representation z_i for each sub-portion, by linearly combining $x_1, \ldots, x_n$ by means of α_i, i.e., $z_i = \phi(\alpha_i, X)$. Basically, we are transforming matrix X into matrix Z, whose rows are $z_1, \ldots, z_n$. We focus on the case in which the x_i's are learnable representations, as it happens in Transformers, and we can now introduce the way each attention score vector α_i is computed, thanks to the attention function f_{att},

$$f_{att}(x_i, X) = \frac{X x_i}{\sqrt{d}} = \left(\langle x_i, x_1 \rangle / \sqrt{d}, \ldots, \langle x_i, x_n \rangle / \sqrt{d} \right),$$

that returns a vector of n not-normalized attention coefficients. Such a function can be implemented in several different ways (for example, using a multi-layer perceptron), and here we instantiated it using the scaled dot product $\langle \cdot, \cdot \rangle / \sqrt{d}$, that is what is exploited in Transformers. Then,

$$\alpha_i = \mathrm{smx}\left(f_{att}(x_i, X)\right)$$
$$= \mathrm{smx}\left(\left(\langle x_i, x_1 \rangle / \sqrt{d}, \ldots, \langle x_i, x_n \rangle / \sqrt{d}\right)\right)$$
$$= \left(\frac{\exp(\langle x_i, x_1 \rangle / \sqrt{d})}{\sum_{j=1}^{n} \exp(\langle x_i, x_j \rangle / \sqrt{d})}, \ldots, \frac{\exp(\langle x_i, x_n \rangle / \sqrt{d})}{\sum_{j=1}^{n} \exp(\langle x_i, x_j \rangle / \sqrt{d})}\right).$$

The softmax function is only used to enforce a probabilistic normalization of the raw attention scores (soft-attention). It is easy to enforce the attention coefficients to be ≈ 0 in some input portions, if needed, by replacing with $-\infty$ the components of $f_{att}(x_i, X)$ from which attention must be excluded. When this happens, we talk about *masked self-attention*.

We can now compactly indicate with `attention(x)` the function that maps X into Z, i.e.,

$$Z = \mathrm{attention}(X)$$
$$= (\phi(\alpha_1, X); \ldots; \phi(\alpha_n, X))$$
$$= (\phi(\mathrm{smx}(f_{att}(x_1, X)), X); \ldots; \phi(\mathrm{smx}(f_{att}(x_n, X)), X)),$$

where the notation $(a_1; \ldots; a_n)$ indicates a matrix whose rows store the contents of the vectors listed in square brackets. For reasons that will become clear shortly, it is useful to exploit different names to refer to some of the arguments that appear in the equations above. In particular, we will talk about *query* when referring to the first argument of f_{att}, we will talk about *keys* when referring to the rows of the matrix in the second argument of f_{att}, and we will name *values* the rows of the last argument of ϕ. The last equations can then be rewritten using a different notation, introducing q_i for the i-th query, K for the matrix of keys, and V for the matrix of values,

$$Z = \mathrm{attention}(Q, K, V)$$
$$= (\phi(\alpha_1, V); \ldots; \phi(\alpha_n, V))$$
$$= (\phi(\mathrm{smx}(f_{att}(q_1, K)), V); \ldots; \phi(\mathrm{smx}(f_{att}(q_n, K)), V)),$$

where Q is the matrix that collects all the queries (row-wise) and where we redefined `attention` to make Q, K, V explicit. In other words, up to this point we only focused in the case of `attention(X, X, X)`. This notation helps understand the intuitive principle behind self-attention, that pretty much resembles a hash-map. A query q_i is compared with all the keys in K by function $f_{att}(q_1, K)$. The outcome of this comparison is a set of not-normalized attention coefficients that, in a sense, tell how strongly q_i is similar/related to each key in K. The softmax-normalized coefficients are used to compute a weighted combination of the values in V, yielding z_i. In the most extreme case, when only one key is given an attention score of 1 (and all the others are scored 0), z_i is equivalent the value associated to such a key.

What we described so far allowed us to transform X into Z by means of a single self-attention module. We can go beyond this idea, moving to *multi-head attention*, where Z is the outcome of having executed $\overline{h}$ instances of self-attention, also referred to as *attention heads*, fusing their results. We indicate with Z^h the outcome of the h-th head, which is a self-attention module that exploits its own queries, keys, and values. In the case of Transformers, the head-specific queries, keys, and values are obtained by linearly projecting the rows of Q, of K, of V, by means of independent matrices W^{Q_h}, W^{K_h}, W^{V_h}, respectively, each of them of size $d \times d_h$, being d_h the number of dimensions of the target spaces (in general, each of the three matrices can have its own specific number of columns). The outputs of the heads, $\{Z^h, \ h = 1, \ldots, \overline{h}\}$, are then fused and projected, in order to generate the final output Z of size $n \times d$, as in the case of a single self-attention module. Formally,

$$Z = \texttt{multihead\_attention}(Q, K, V)$$
$$= \texttt{cat}(Z^1, \ldots, Z^{\overline{h}}) W^Z$$

where $\texttt{cat}$ yields a matrix in which each row is the concatenation of the corresponding rows of the matrices in the argument list, and

$$Z^h = \texttt{attention}(Q W^{Q_h}, K W^{K_h}, V W^{V_h}), \quad h = 1, \ldots, \overline{h}.$$

The projection matrix W^Z is a matrix of size $\overline{h} d_h \times d$, restoring the dimensionality d of the rows of Z.

We are now ready to describe the architecture of Transformers that is sketched in Fig. 5.22. We analyze it in the context of machine translation, i.e., in problems where an input sentence in a source language is translated into another sentence belonging to a target language. The input of the Transformer is a sentence, let's say of length n, that, for simplicity, we assume to be tokenized at word level. As a result, the sentence is a sequence of n words. Each word is assigned to its own (eventually learnable) embedding, so that $(x_1, \ldots, x_n)$ is the sequence of embeddings of the words in the input sentence, stored (row-wise) into matrix X. The output of the Transformer is the sentence translated into the target language, let's say a c-word sentence composed of words $(y_1, \ldots, y_c)$, where each y_j is the symbol associated to a word in the target language (not the embedding). This is achieved by first encoding the input sentence, and using the encoded data as input of a decoder that as a key role in generating the output sequence. However, the output sequence is not decoded in a single shot, while it is produced in an iterative manner, one word at a time. The decoder acts in an autoregressive regressive way, exploiting the embeddings of the words generated in the previous steps as additional inputs when generating the next word. Fig. 5.22 shows that a Transformer is composed of two main parts, referred to as *encoder* and *decoder*. We anticipate that, in practice, several applications of Transformers in computer vision and natural language processing use either only the structure of the encoder or only the decoder, adapting them to different tasks.

The *encoder* is responsible of processing the embeddings in X, $n \times d$, and returning their novel representations, that are the outcome of having applied self-attention

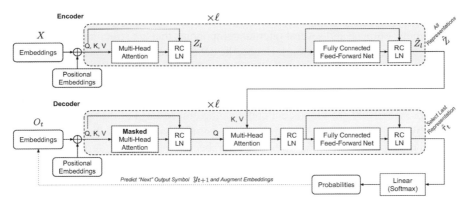

FIGURE 5.22

Architecture of a Transformer in a machine translation problem. The embeddings of the tokens of the input sentence are collected in matrix X and, at step t, the next predicted word of the output sequence is y_{t+1}. An encoder (top) and a decoder (bottom), composed of ℓ layers each (that in the text are referred to as ℓ encoding blocks and ℓ decoding blocks, respectively) are used to transform different embeddings into new representations, by means of multi-head attention, and other components (RC LN stands for Residual Connections followed by Layer Normalization). The last representation yielded by the decoder is used to compute the probability of the next word. The embeddings provided as input to the decoder, O_t, are progressively populated while predicting the output sentence.

(and other transformations) multiple times (Fig. 5.22, top). In particular, the encoder is based on ℓ stacked modules ($\ell = 6$), each of them with the same structure of the other ones, that, in this description we will refer to as *encoding blocks* (sometimes also simply referred to as layers). An encoding block consists of a multi-head attention module followed by a feed-forward neural network. Such a network is composed of 2 fully connected layers (ReLU activation after the first layer), with d input neurons and d output neurons, usually with a larger number of neurons in the hidden layer. Each encoding block receives as input the output of the block below, and the learnable parameters of a block are not shared with the ones of the others. In the first encoding block, the multi-head attention transforms X into new representations collected in matrix Z_1, while the following feed-forward network further projects/transforms each row of matrix Z_1, yielding matrix $\hat{Z}_1$ as output. The l-th block, with $l > 1$, transforms $\hat{Z}_{l-1}$ into $\hat{Z}_l$. Both the self-attention module and the feed-forward net exploit residual connections (i.e., the output of each of them is summed up to the input, before using it for further operations), followed by an output normalization procedure (also known as "layer normalization," where each output representation is shifted in order to have zero mean and it is divided by its standard deviation, sometimes also including d additional learnable scaling factors and d additive biases). The output of the decoder is $\hat{Z}_\ell$, that, for simplicity, we simply indicate as $\hat{Z}$. Notice that each encoding block can encode its input embeddings in parallel.

The *decoder* is based on ℓ stacked modules that we refer to as *decoding blocks*, with inputs, outputs, and several internal components that are structured as in the encoding blocks (Fig. 5.22, bottom). Hence, the decoder transforms a set of embeddings into another set of representations by means of multiple multi-head self-attention modules. The embeddings that are provided to the decoder are collected row-wise in matrix O_t, where t is a step index. The decoder is called multiple times, and the size of O_t progressively grows, since it contains the embeddings of the words generated so far by the Transformer. At step $t = 1$, O_1 is only composed of the embedding of a special "start" token, since no words have been generated yet; at step $t > 1$, O_t collects also the embeddings of the words generated during the previous steps. The first component which constitutes a decoding block is a masked multi-head attention. Masking is needed for practical reasons, to prevent queries from attending to keys/values belonging to subsequent positions. Thus, at time t, the components of $f_{att}(q_j, K)$ with index $> j$ are masked out, for each $j \leq t$. Then, the decoding block features a (non-masked) multi-head attention that receives $\hat{Z}$ as key matrix and as value matrix (see Fig. 5.22), while the queries come from the output of the masked attention. This basically bridges the encoder and the decoder, allowing queries to attend over the representations of the sentence in the source language. The output of the non-masked attention is then provided to a fully connected feed-forward neural network, as we described for the encoding blocks. Residual connections and layer normalization are present right after each of the described components.

For what we described so far, the output of the decoder at time t is a set of t representations, each of them with dimension d. The last of them, which we indicate with $\hat{r}_t$, is projected into a space with a number of dimensions equal to the size of the output vocabulary, i.e., $|\mathcal{V}^{\mathcal{Y}}|$. This is implemented with a linear transformation modeled by matrix W^Y of size $|\mathcal{V}^{\mathcal{Y}}| \times$ d, followed by a softmax normalization to yield a probability distribution. The word with the largest probability is selected,

$$y_t = \arg\max \left(\text{smx} \left(W^Y \hat{r}_t \right) \right),$$

which is the first word of the translated sentence (in principle, softmax is not needed in the just described operation, and it is only computed when a probability distribution is needed). This generation process is repeated for c steps, if the length of the target sentence is known, or until a special "stop" symbol is predicted.

It turns out that the described self-attention mechanism is not affected by the order in which keys and values are stored. For simplicity, let us focus on single-head attention. Given a query q_i, if we change the order of the rows of K and we apply the same ordering to the rows of V, then the generated z_i will not change. Similarly, when changing the order of the rows in the query matrix Q, the resulting Z will be simply re-ordered accordingly. This is not a desirable feature in the case in which the order of the input elements has a precise meaning (think about the words in a sentence). For this reason, Transformers also exploit a positional encoding to alter the embeddings provided as inputs to the encoder and decoder, in order to make the model sensitive to changes in the way these inputs are ordered. Different approaches are possible, such as adding an eventually learnable position dependent signal to each embedding of the input sequence and, similarly to those of the output sequence.

Given a loss function that measures the mismatch between the predicted next-word and the target one in the training data, Transformers can be trained by gradient descent. At training time, it is convenient to exploit the previously described parallel computational scheme of the encoder in the decoder as well. This is done by filling up O_1 not only with the "start" embedding, but also with the c embeddings of the target sentence (independently on what the Transformer predicts), and predicting the next word for each output representation. We recall that the learnable parameters of a Transformer are the embeddings of the words in the input and target languages (provided to the encoder and decoder), the multi-head projection matrices, the weights in the feed-forward networks, the eventually learnable positional encodings, and the additional scalings/biases in the layer normalization operations. Of course, the matrix W^Y of the final projection must be learned as well.

The machine translation case we focused so far represents a setting in which both the Transformer encoder and decoder are exploited. Let us consider the case of language modeling, in which the goal is to estimate the probability of a given sequence of words. Without going into further details, this task is commonly faced by letting the machine learn to predict the probability of the next word of a sentence, given the precedings words (if we keep working at a word level). A variant of this task is the one in which a word is masked in a given sentence, and the machine must learn to predict it given the surrounding words, also known as masked language modeling. Transformers can indeed be used to face these tasks, even if instead of exploiting the whole encoder-decoder model, either the principles behind the encoder or the decoder are considered, with some adaptations.

More specifically, in the case of classic language modeling, the Transformer decoder can be directly used (removing the previously described non-masked multi-head attention of the decoding blocks). The embeddings belonging to the words of a portion of text are provided as input to the decoder, and the Transformer learns to predict the next word. As we already described, the decoder exploits a masked self-attention procedure that avoids introducing dependencies on the succeeding words, that well copes with the requirements of the considered task (i.e., next-word prediction given the previous ones). Differently, in the case of masked language modeling, only the Transformer encoder is considered. The input text (e.g., a sentence) is provided to the multi-layer encoder, that yields $\hat{Z}$, masking a few words from the input, that are replaced by a special token "[MASK]." A classifier is exploited with the goal of predicting the hidden words from the output representations of the masked tokens. Both these models can be trained from a given text corpus without any external supervision.

Of course, also several classic supervised tasks can be faced with Transformers (text classification, question answering, sentence paraphrasing, etc.). Let us consider the case of text classification as a simple example. We are given a portion of text (e.g., a document) and the Transformer must be trained to predict the category to which it belongs. In the case of a decoder-based model, as the one we used in classic language modeling, the embeddings of the words of the input text can be provided to the Transformer decoder, and the output prediction will be about the class label, instead of being about the next word. In the case of an encoder-based model, as the one of masked language modeling, the embeddings of the whole input text are provided to the encoder, augmented by another special token artificially added at the beginning, named "[CLS]." The output representation of "[CLS]" is handled by a classifier to predict the class of the document. Large training sets are usually needed to let Transformer learn robust representations and yield high quality predictions. In several real-world tasks the number of supervised data is indeed pretty limited. For this reason, transfer learning is commonly exploited, pretraining the Transformer in a language modeling task (masked or not), where it is easy to collect large text corpora.

Finally, we mention that Transformers are not only limited to text, since what we presented so far still holds when we replace words with other symbols, or word embeddings with other given or pre-computed representations. For example, when we consider an image classification task, the input image can be divided into non-overlapping patches, and each patch can be flattened and linearly projected to dimensionality d. As a result, X is the matrix of input embeddings, in which each row is a projected patch of size d. We are now in the exact conditions of the already described text classification with a Transformer encoder. Hence, the embedding of a special "[CLS]"-like token is added at the beginning of the projected-patch-sequence, and the first output representation is what is provided to a final classifier. We remark that patches can also be created after having processed the input image with a convolutional neural net, thus they can be portions of convolutional feature maps.

5.7 **Complexity issues**

The experiments carried out by feedforward neural networks are typically driven by *trial and error*, which has been the subject of many discussions on the lack of scientific foundations. In this section, we compose a puzzle with some theoretical results that help understanding the complex optimization process driven by backpropagation. In addition, to respond to a truly scientific curiosity, we address complexity issues that shed light on real-world experiments.

5.7.1 **On the problem of local minima**

The formulation of learning as an optimization problem leads us to analyze the chances of returning the expected optimal solution. In this section we cover the problem of local minima of the error function that is used for learning with the primary purpose of understanding its connection with the given learning problem and the corresponding complexity issues. The study of linear and kernel machines carried out in Chapters 3 and 4 gives for granted that the solution of the related optimization problem leads to the optimal solution.

In these cases, we are dealing with quadratic functions and quadratic programming, for which numerical algorithms are expected to return the optimum. However, this is not necessarily the case for neural networks! Local minima can arise for different reasons. In particular, they can arise whenever the training set is nonrealizable, in the sense that there is no weight vector for which the error function is zero.

The adoption of some pairs of loss and neural transfer functions may result in a very critical problem, though the explosion of local minima is directly associated with the lack of realization capabilities. For example, if the loss and transfer functions are bounded and the domain is unbounded, then there exist areas of saturation where the error function is essentially flat. In addition, the loss function is additive, which means that the different effects cumulate. However, the presence of a local minimum in a certain limited area of the domain is not changed by other examples, whenever we deal with saturated neurons. Additional insights on local minima for nonrealizable loading problems are given in Section 5.8. However, it is worth mentioning that the lack of realizability is something that we typically avoid in practice.

Realizable loading problems are those on which we focus attention, even though it is hard to get a formal guarantee on the verification of the property. The discussion in Sections 5.2 and 5.3 on function realization has given many insights to understand the issue in both the cases of Boolean and real-valued functions. The thumb rule that summarizes most of the given realization techniques is that large enough architectures exhibit universal approximation.

As it will be pointed out in Section 5.7.3, the presence of suboptimal local minima is not the only source of problems during the optimization. However, unlikely what is sometimes claimed, suboptimal configurations corresponding to local minima do exist and are inherently companions of supervised learning in feedforward neural networks. If $\min_w E(w) = 0$ we can understand the emerge of critical configurations,[17] by considering the corresponding delta error Δ that, as it will be shown, plays a crucial role also in their classification. We can distinguish critical configurations depending on whether or not $\Delta = 0$. The presence of suboptimal configurations is clearly characterized by the condition $\Delta \neq 0$ that can emerge in different ways.

We start showing that, unlike linear and kernel machines, there are in fact classification problems in which the formulation of learning in feedforward neural networks yields a function with stationary points that are nonseparating. This corresponds with a class of configurations for which $\Delta \neq 0$. How can this be possible? The analysis of Section 3.2.3 on "failing to separate" and the discussion in Section 3.2.2 shed light on this undesirable property. Basically, if we choose the quadratic error function and targets $\underline{y}$ and $\overline{y}$ that are not asymptotical values (e.g., $\underline{y} = 0.1$ and $\overline{y} = 0.9$), then we cannot jointly exploit the nice sign properties stated by Eqs. 3.2.2.2–(3) and 3.2.2.2–(4). An in-depth discussion on this problem is proposed by Exercise 1. Interestingly, these spurious configurations can be global minima of the error function, but the corresponding optimum is not desirable. We can easily realize that this problem can easily emerge in practice, thus raising issues on the appropriateness of the formulation of learning as optimization. However, like in Section 3.2.3, we can prove that these spurious configurations turn out to be the outcome of the wrong translation of learning into the associated optimization problem, more than the sign of inherent complexity! In order to fully catch this concept, suppose that we are still using nonasymptotical targets but, instead of the quadratic function, we use a penalty of a threshold LMS type, i.e., a function that, like the following one, is zero for values "beyond" the desired target values:

$$E(w) = \sum_{\kappa \in P} (\overline{y} - f(x_\kappa)_+ + \sum_{\kappa \in N} (f(x_\kappa) - \underline{y})_+,$$

where P and N are the set of positive and negative examples, respectively. We can promptly see that for this penalty the sign property stated by the delta sign Eq. (3) still holds, which leads to extending the result stated in Section 3.2.2 for single logistic sigmoidal neurons acting on linearly separable examples. Likewise, when restricting to monotone nonlinear neural units σ, the sign property holds for the hinge function. In the case of a single output, we have

$$\frac{\partial e_\kappa}{\partial a_\kappa} = -\sigma'(a_\kappa) y_\kappa$$

whose sign is defined by y_κ, with $\sigma'(a_\kappa) > 0$ and e_κ is the value of the loss on the κ-th example. Spurious formulations correspond with the violation of the sign property, and even though they can correspond to global minima, they might not satisfactorily separate the examples of the training set. While this analysis on the violation of the

---

[17] Critical configurations are those for which the gradient is null.

non-linearly separable   non-linearly separable

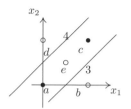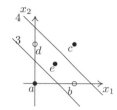

FIGURE 5.23

The XOR5 learning task. The configuration in the middle corresponds with perfect separation (global minimum). The rightmost configuration is associated with a suboptimal local minimum.

delta sign equation is restricted to penalty functions used to model supervised learning, it applies also to other penalties that, as will be shown in Chapter 6, are used to handle constraint-based learning environments.

This discussion on spurious configurations, however, does not allow us to conclude that the problem of suboptimal solutions is avoided whenever we use penalty functions that respect the sign property. There are in fact other undesirable configurations that correspond with local minima that are not global! While we can get rid of spurious configurations by appropriate adoption of the penalty function, some configurations exhibit a sort of structural problem of complexity.

The following example clearly shows the presence of similar configurations. We consider the three neuron network already used for the XOR predicate, but we add the additional labeled example[18] $e = ((0.5, 0.5)', 0)$. Let us compute the gradient corresponding to the class of configurations in which $w_{31} = w_{32} = \alpha$ and $w_{41} = w_{42} = \beta$, regardless of the other values of the weights. We have

$$\frac{\partial E}{\partial w_{31}} = \delta_{b3} + \delta_{c3} + \frac{1}{2}\delta_{e3}, \qquad \frac{\partial E}{\partial w_{32}} = \delta_{c3} + \delta_{d3} + \frac{1}{2}\delta_{e3}.$$

From the backpropagation rule we have $\delta_{b3} = \sigma'(a_{b3})w_{53}\delta_{b5}$ and $\delta_{d3} = \sigma'(a_{d3})w_{53}\delta_{d5}$. Now, because of the symmetry, $a_b = a_d$ and therefore $\delta_{b5} = \delta_{d5}$. Hence we get $\delta_{b3} = \delta_{d3}$, which implies that $\partial E/\partial w_{31} = \partial E/\partial w_{32}$. As a consequence, any update in the subspace defined by the coordinates (w_{31}, w_{32}) has direction given by $(1, 1)$, that is, gradient updating only leads to translations of separating line 3. Clearly, the same analysis holds for line 4. To sum up, when starting from the configuration defined by $w_{31} = w_{32} = \alpha$ and $w_{41} = w_{42} = \beta$, any gradient descent algorithm can only produce translations of the lines 3 and 4, thus making it impossible to perform any rotation to break this configuration, for reaching the optimal solution! Clearly, the translation to the configuration drawn in the figure leads to a local minimum which is different from

---

[18] This learning task is referred to as $XOR5$.

the global one corresponding with null error. We can see more about this configuration. For the gradient vectors $(\partial E/\partial w_{31}, \partial E/\partial w_{32})'$ and $(\partial E/\partial w_{41}, \partial E/\partial w_{42})'$ to be null, we clearly need that lines 3 and 4 achieve the configuration shown in the figure, where patterns a and c are correctly separated. Now if we impose $\partial E/\partial w_{53} = 0$ and $\partial E/\partial w_{54} = 0$ and consider that, because of the symmetry, b, d, and e get the same value, we get

$$\delta_{b5}x_{b5} + \delta_{d5}x_{d5} + \delta_{e5}x_{e5} = 0 \Rightarrow$$
$$\delta_{b5} + \delta_{d5} + \delta_{e5} = (x_{b5} - \overline{y}) + (x_{d5} - \overline{y}) + (x_{e5} - \underline{y}) = 0,$$

from which we get

$$x_{b5} = x_{d5} = x_{e5} = \frac{2\overline{y} + \underline{y}}{3}.$$

Instead of the global minimum $E = 0$, this configuration corresponds with $E = (1/3)(\overline{y} - \underline{y})^2$. It is a local minimum, since as already discussed, there is no way of breaking the symmetry. One could determine the Hessian to see whether it is positive definite in this configuration. We could also be curious to see what happens when line 3 and/or 4 rotate by a small angle ϕ. We claim that infinitesimal rotations lead to changes of the activations for patterns b, d, and e that are infinitesimals of higher order than the changes for patterns a, c (see Exercise 2). As a result, any such rotation clearly results in a worst separation of a, c, with a correspondent increment of the error.

We conjecture that we can construct plenty of similar configurations, unified by the idea of *unbreakable symmetries*, like the above discussed configuration (see Exercise 3).

The XOR5 and related learning tasks correspond with stationary configurations in which not only $\Delta \neq 0$, but, in addition, this is inherently connected to the structure of the problem. In these cases, unlike what is sometimes claimed, the corresponding suboptimal configuration is not necessarily acceptable! If $\underline{y} = 0$ and $\overline{y} = 1$, then in the $XOR5$ example, the output corresponding with the e pattern is badly predicted as $2/3$, which clearly results in a classification error! In real-world situations, it is quite common to observe experiments, where different runs, initialized randomly, lead to slightly different errors. This likely depends on the solution space which, as discussed in Section 5.3.3, contains a huge number of different solutions, whereas sometimes the different solutions are interpreted as local minima that are very close to the global one! This remark drives attention towards the actual existence of suboptimal configurations, whereas the belief that local minima only slightly differ from global minima does not rely on any concrete evidence.

A good picture of the general structure of the error surface arises when thinking of peaks and valleys in the middle of a mountainous environment. It is not clear whether we can get to the sea level just by following the steepest descent strategy. While there are basins of attraction towards the sea, there are also others where one gets stuck when following the steepest descent! The above discussion on structural local minima indicates that something similar happens for the error surfaces of the error function of a feedforward neural network.

Now it's time to discuss how to discover paths to the sea. When do gradient descent learning algorithms end up with an optimal solution? We know that this is possible but, in general, this happens only provided that we are lucky enough to initialize the algorithm in the basin of attraction of a global minimum. While there is a huge literature on numerical optimization, one should be very well aware of the very local nature of optimization algorithms. This is not restricted to the gradient; higher-order methods make no exception. Hence, the structure of the error surface, which is inherited by the loading problem, offers a sort of inherent complexity measure. As a consequence, while some loading problems exhibit an error surface where gradient descent likely leads to an optimal solution, for others it is unlikely to reach it in reasonable time.

In Section 3.2.2, it has been shown that a case in which the loading problem is local minima free is the one in which the training set is linearly separable. Since the presence of suboptimal local minima arises because of the presence of hidden units, one might wonder whether the property of a local minima free error function is kept also in the case of multilayered networks. Lucky enough, the nice property derived for a single neuron is not destroyed by multilayered nets. We give a proof of this property for OHL nets with one output only, which is used for classification of two classes. Exercise 4 proposes an extension to the case of multiple outputs, Exercise 5 discusses the case of deep nets, while Exercise 6 involves any feedforward architecture.

In the case of an OHL with a single hidden layer and a single output, the gradient equation 5.5.2–(10) becomes[19] $G_1 = \hat{X}_0' \Delta_1$. Now we can use the same idea adopted in Section 3.2.2 for single neuron nets. Since the patterns are linearly separable, there exists a separating hyperplane defined by α such that Eq. 3.2.2.2–(4) holds true. If we left-multiply by α' both sides of the chain rule that expresses G_1, we get

$$\alpha' G_1 = (\alpha' \hat{X}_0') \Delta_1 \tag{1}$$

Now, from backpropagation Eq. (9), due to the assumption of dealing with a single output net, we have

$$\Delta_1 = \sigma' \odot (\Delta_2 W_2) \tag{2}$$

Now it takes a while to realize that, basically, the property for Δ_1 holds true just like for the output. This only requires a check on the sign of the elements of W_2.

First, let us assume that $w_{oi} \neq 0$. If it is positive, since $\sigma'(a_{\kappa i}) > 0$, the sign property for $\delta_{\kappa 0}$ clearly holds also for Δ_1. If it is negative, Δ_1 gets the flipped sign, but this holds for all the coordinates of the matrix. Hence, when using the same arguments as those of Section 3.2.2, we get

$$(\alpha' \hat{X}_0') \Delta_1 = 0 \Rightarrow \Delta_1 = 0.$$

---

[19] Here, for the sake of simplicity, we omitted the index o of the output units, since we are dealing with single output nets.

Since $w_{oi} \neq 0$, we can promptly see that $\Delta_1 = 0 \Rightarrow \Delta_2 = 0$. Now, from the analysis carried out in Section 3.2.2 for the case of logistic sigmoid, we already know that having $\Delta_2 = 0$ means that the only stationary point of the error function is a global minimum. Exercise 6 discusses the implications of the condition $\Delta_2 = 0$ in the case of rectifier nonlinear transfer function.

What if $w_{oi} = 0$? Unfortunately, we cannot drive any conclusion using the previous arguments, since there is no backpropagation of the delta error to the hidden units. From Eq. (2) we can immediately see that $\Delta_1 = 0$ and, consequently, $G_1 = 0$. While nothing can be said about G_2, the condition $G_1 = 0$ is enough to create a trap for the learning process. We notice in passing that if also the weights, which connect the first hidden layer with the inputs, are null ($W_1 = 0$) and the nonlinear transfer function is the hyperbolic tangent then we also get $X_1' \Delta_2 = 0$, since in this case, $X_1 = 0$. Hence for all output weights w_{oj} we have $\partial E / \partial w_{oj} = 0$. As discussed in Exercise 7, the gradient with respect to the bias gets stuck in a configuration where the bias depends on the difference between the number of positive and negative examples. Fortunately, this is not a local minim. If we move slightly in a neighborhood then gradient heuristics slowly gets away from this trap (see Exercise 8 for an in-depth understanding of this issue). This also leads us to the conclusion that the error surface, corresponding to OHL nets with one output only used for classification, is local minima free if the training set is linearly separable.

The discussion on linearly separable patterns can be extended to the case of quadratic separable patterns by replacing the ridge neurons with radial basis function neurons. Some insights on the analysis are given in Section 5.8.

While most of the discussion carried out so far sheds light on the deep connection between the loading problem and the structure of the error surface, it does not cover most common experimental conditions, where the networks that are used are pretty big! This is essentially the outcome of clever experimental setup driven by a nice mixture of intuition and trial and error. How can, as a matter of fact, a big net circumvent local minima? We will provide arguments to support this thesis in the case of OHL with many hidden units and of big deep nets. There is more. While the analysis of these two cases resembles close analogies, there is also an interesting difference, since we will consider rectifier units in the case of deep nets and logistic sigmoids for OHL nets.

We begin to discuss big nets with OHL architecture with one output units used for classification, where the big dimension is gained because of the choice of the number of hidden units. As soon the input code has been defined, the given learning task does in fact characterize the input. In particular, given the training set with ℓ examples, suppose we choose an OHL net which satisfies the condition

$$|H| \geq \ell - 1. \tag{3}$$

The nullification of the gradient on the last layer yields $\hat{X}_1' \delta_2 = 0$. Since $\hat{X}_1 \in \mathbb{R}^{\ell \times |H|+1}$, matrix $\hat{X}_1'$ is generally full rank,[20] which means that only $\delta_2 = 0$ is possible, which, in turns, implies that the error function is local minima free. This suggests the importance of condition (3), though it is not sufficient to guarantee that $\hat{X}_1$ is of full rank. Exercise 9 discusses the details of the formal proof to address the configurations connected with rank deficiencies of $\hat{X}_1$.

Now we explore deep networks where the large dimension is gained also by the number of layers of the net. In order to enlighten the effect of the dimension, in this case we assume that the neurons use the rectifier as a nonlinear transfer function. We conjecture that, again, "large nets" get around local minima. As shown in Section 5.1.2, the behavior is deep network with rectifier units can be understood by considering the clustering of those inputs which activate the same neurons. In do doing, in any phase of gradient descent, we aggregate the data by partitioning the inputs, so as any partition yields a convex penalty function. During the learning process, the partition changes dynamically under the constraint that the weights that are shared during the commutation do not change. As the dimension of the weight space increases, these constraints are less and less important, so as any single penalty is optimized like in a free domain. Since they are convex, we end up with a global minima (see Exercise 10 on this conjecture).

EXERCISES

1. [*M31*] Let us consider the following loading problem:

$$\mathcal{L} = \{((-1,0)', 0.1), ((1,0)', 0.9), ((0,1)', 0.9), ((0,-5)', 0.9)\}.$$

This is clearly a linearly separable problem. Suppose we are using an LTU network with a logistic sigmoidal neuron and quadratic error function. In addition, suppose that the decision is based on the following thresholding criterion: If $x < 0.1$ then the class is $-$, whereas the class is $+$ if $x > 0.9$, and there is no decision in the remaining range of x. Prove that there exists a stationary point that does not yield a separating solution.[21]

2. [*M28*] Look at Fig. 5.23–(c) and prove that an infinitesimal rotation of the separation lines yields changes of $a_{5,b}, a_{5,d}$, and $a_{5,e}$ that are infinitesimals of higher order with respect to $a_{5,a}$ and $a_{5,c}$.

3. [*HM45*] Based on the $XOR5$ learning task, let us consider any associated learning task such that A and C are fixed and B, D, E can move under the constraint of keeping the separation structure of XOR5. Is any such configuration a suboptimal local minimum?

4. [*M19*] Extend the result given in Section 5.6.1 on local minima error surfaces under the hypothesis of OHL nets and linearly separable examples to the case of multiple output.

5. [*M39*] Extend the result given in Section 5.6.1 on local minima error surfaces under the hypothesis of OHL nets and linearly separable examples to the case of deep nets, that is, with multiple hidden layers.

---

[20] It is full rank with probability one.
[21] This loading problem has been proposed in M. Brady, R. Raghavan, J. Slawny, *IEEE Trans. Circuits Syst.* **36** (1989), 665–674.

6. [*M34*]   Based on the analyses of Section 5.6.1, complete the proof that OHL nets with single output for linearly-separable examples yield local minima free error functions by proving that the point with all null weights is not a local minimum. In Section 3.2.2, when discussing linearly-separable examples, it has been proven that for the case logistic sigmoidal units, the condition $\Delta = 0$ characterizes different stationary points, but the only one which attracts when performing gradient descent is the global minimum. That is, the error surface is local minima free. What happens in the case of the rectifier?

7. [*M21*]   In Section 5.6.1 the case of all null weights for neurons with the hyperbolic tangent has been discussed. It has been proven that both G_1 and G_2 are null for OHL networks. However, the gradient with respect to the bias term was not considered. What is its contribution? Can we extend the stationarity of the gradient to the case of logistic sigmoid?

8. [*M21*]   Let us consider an OHL net in the configuration where all its weights (including the biases) are null. Prove that this stationary point *is not* a local minimum.

9. [*HM46*]   Let us consider condition (3). The proof that the error function is local minima free given in Section 5.6.1 assumes that X_1 is a full rank matrix. While this is generally true whenever condition (3) is met, there are configurations with rank deficiency, that is, weights w such that rank $X_1(w) < \min\{\ell, 1 + |H|\}$. Prove that also these configurations are not local minima. This problem has been addressed in T. Poston, C. Lee, Y. Choie, Y. Kwon, *Int. Joint Conf. on Neural Netw.* (1991), 173–176, X.H. Yu, Neural Netw. **3** (1992) 1019–1020, X.H. Yu, G.A. Chen, *IEEE Trans. Neural Netw.* **6** (1995) 1300–1303. X.H. Yu and G.A. Chen, *IEEE Trans. on Neural Netw.* **6** (1995), 1300–1303. Some comments published in L.G.C. Hamey, *IEEE Trans. Neural Netw.* **5** (1994) 844–844 have raised doubts on the proofs.

10. [*HM48*]   Beginning from the arguments given in Section 5.6.1, give formal conditions under which the learning in a deep network with rectifier units is local minima free.

5.7.2 Facing saturation

Continuous gradient descent only gets stuck in local minima. However, the gradient heuristics vanishes also in plateaux, where the error function is nearly constant. To make things easy, let us consider a single sigmoidal neuron with a single input and null bias used for classification, along with a quadratic penalty. In this case we have $E(w) = \sum_{\kappa=1}^{\ell} (d_\kappa - \sigma(wx_\kappa))^2$. We can promptly see that $\lim_{w \to \pm\infty} \partial E(w)/\partial w = 0$, which simply indicates the problem emerging from neural saturation! This is clearly related to choosing sigmoidal transfer units. In this case, the activation $a = \hat{w}'\hat{x}$ can easily get values that correspond with neural saturation. This yields configurations for which $\sigma'(a) \simeq 0$, which prevents from learning! Local variations of w do not affect $\sigma'(a)$, which indicates the vanishing of the gradient heuristics. Neural saturation is typically a problem which arises at the beginning of learning, when we need to choose the initial configuration for driving the gradient descent. The analysis on local minima of Section 5.7.1 suggests not to choose weights that are too small, to avoid a stationary point of the gradient. Even though it is not a local minimum, escaping from this configuration is computationally expensive. Likewise, we need to avoid the neural saturation that arises when using sigmoidal transfer functions. Clearly, this depends on both the input x and the weight w. A common suggestion to face saturation is that of normalizing the inputs, so $x \to x/(\max_i |x_i|)$. In so doing, each coordinate is upper-bound by 1. Once the inputs have been normalized, we need an appropriate choice of $\hat{w}$ that must face the dimensionality of the input. Suppose we want to keep the values of a below an appropriate value B to prevent from saturation, that is, $|\hat{w}'\hat{x}| \leq B$. Of course, when using Schwartz inequality $|\hat{w}'\hat{x}| \leq \|\hat{w}\| \cdot \|\hat{x}\| \leq B$. Hence $\|\hat{w}\| \leq B/\|\hat{x}\|$. Let f_{in} and f_{out} be the fan-in and fan-out of the neuron. The most restrictive condition arises when all inputs are equal, which leads to $\|\hat{x}\| = \sqrt{1+f_{in}}$. Now we need to choose $\hat{w}$ such that $\|\hat{w}\| \leq B/\sqrt{1+f_{in}}$. Likewise, when considering weights of the same value w (including the bias), we get $\|\hat{w}\| = w\sqrt{1+f_{in}}$, and therefore saturation is prevented provided that

$$w \leq \frac{B}{1+f_{in}}.$$

However, while this choice prevents from saturation, it is not adequate as an initialization, since it is a symmetrical configuration, whose breaking is computationally expensive. (The analysis on the gradient corresponding to this configuration is proposed in Exercise 1.) In most practical problems, there is no heuristics that can help with the weight initialization, and therefore a random initialization is quite common. In particular, we make the following assumptions:

(*i*)   The weights belong to $[-w, w]$ and are regarded as random variable with uniform distribution and zero mean (including the bias term);

(*ii*)   The inputs x_i, $i = 1, \ldots, d$ are uncorrelated;

(*iii*) The inputs are zero mean random variables, and are normalized to have the same variance;

(*iv*)   The inputs and the weights are uncorrelated random variables.

We consider a neuron of the first hidden layer, whose activation can be associated with the random variable $A_i = \sum_{j=1}^{d} W_{ij} X_j + B_i$. Because of (*iv*), we immediately see that

$$\mathrm{E}\, A_i = \sum_{j=1}^{d} \mathrm{E}(W_{ij} X_j) + \mathrm{E}\, B_i = \sum_{j=1}^{d} \mathrm{E}\, W_{ij}\, \mathrm{E}\, X_j + \mathrm{E}\, B_i = 0.$$

In order to avoid saturated configurations, we can inspect the root mean square deviation σ_{A_i}, since its bounded value allows us to draw conclusion on the probability that $A \in [-a, +a]$. From Chebyshev's inequality, we get

$$\Pr(|A_i| > \beta \sigma_{A_i}) \leq \frac{1}{\beta^2}. \tag{1}$$

If we choose $\beta = 5$ and $\sigma_{A_i} = 1/5$, this means that with probability at least of 96% we have $A_i \in [-1, +1]$, which is good to face the saturation of sigmoidal units.[22]

---

[22] Clearly, you can play with these numbers, but this clearly indicates that the variance $\sigma_{A_i}^2$ has a fundamental role for understanding the saturation after random initialization.

Exercise 2 proposes a bound that is sharper than (1) and is based on the remark that the probability distribution of A_i can be very well approximated by means of a Gaussian.

Now, in order to avoid saturation, we use the principle of clamping the variance of the output of neuron i and the variance of the inputs X_j, that is, $\sigma^2_{X_j} = \sigma^2_{A_i}$. Let us express $\sigma^2_{A_i}$ as follows:

$$
\begin{aligned}
\sigma^2_{A_i} = \mathrm{E}\, A_i^2 &= \mathrm{E}\left(\sum_{j=1}^{d} W_{ij} X_j + B_i^2\right)^2 \\
&= \mathrm{E}\left(\sum_{\alpha=1}^{d}\sum_{\beta=1}^{d} W_{i\alpha} X_\alpha W_{i\beta} X_\beta\right) + \mathrm{E}\left(\sum_{j=1}^{d} W_{ij} X_\alpha B_i\right) \\
&= \sum_{\alpha=1}^{d}\sum_{\beta=1}^{d} \mathrm{E}\left(W_{i\alpha} X_\alpha W_{i\beta} X_\beta\right) + \sum_{j=1}^{d} \mathrm{E}\left(W_{ij} X_\alpha B_i\right) \\
&= \sum_{\alpha=1}^{d}\sum_{\beta=1}^{d} \mathrm{E}(W_{i\alpha} W_{i\beta})\, \mathrm{E}(X_\alpha X_\beta) + \sum_{j=1}^{d} \mathrm{E}(W_{ij} B_i)\, \mathrm{E}\, X_\alpha \\
&= \sum_{\alpha=1}^{d} \mathrm{E}\, W_{i\alpha}^2\, \mathrm{E}\, X_\alpha^2 \\
&\simeq d\sigma^2_{W_{i,\alpha}} \sigma^2_{X_\alpha},
\end{aligned}
\tag{2}
$$

where the last inequality holds for large d. If we impose $\sigma^2_{A_i} = \sigma^2_{X_\alpha}$ then

$$
\sigma_{W_{i\alpha}} = 1/\sqrt{d}.
\tag{3}
$$

Let us consider the case of uniform and Gaussian distribution. If $W_{i\alpha}$ is a uniformly distributed random variable, we know (see Exercise 3) that from $W_{ij} \in [-w, +w]$ we have $\sigma^2_{W_{i\alpha}} = (2/3)w^3$ and, therefore,

$$
w = \sqrt[3]{\frac{3}{2}\sigma^2_{W_{i\alpha}}} \propto \frac{1}{\sqrt[3]{d}}.
\tag{4}
$$

If $W_{i\alpha}$ is a Gaussian random variable, then we can assume that the choice $w = \gamma \sigma_{W_{i\alpha}}$ with $\gamma > 4$ leads to a consistent generation. Hence in this case, we get[23]

$$
w = \gamma \sigma_{W_{i\alpha}} \propto \frac{1}{\sqrt{d}}.
\tag{5}
$$

---

[23] From Chebyshev's inequality (1), we can promptly realize that this holds for any probability distribution. Notice that condition (5) is clearly sharper than Eq. (4), since it allows us initializations in larger intervals.

Notice that this analysis prevents saturation of hidden units connected with the inputs. When considering hidden and output units that receive connections from other hidden units, the problem is more involved, but we end up with the same conclusion (see Exercise 4), that is, $\sigma^2_{W_i} = 1/f_{in}$. A more careful analysis on backpropagation allows us to realize that neural saturation might arise also during the backward step! There is in fact a perfect mirroring with respect to the saturation that arises during forward steps. Hence we can immediately conclude that we need to respect the condition $\sigma^2_{W_i} = 1/f_{out}$, which now involves the children of i, instead of its parents. Hence a good choice is expected to respect both condition, that is,

$$\sigma^2_{W_i} = \min\{1/f_{in}, 1/f_{out}\}.$$

While sigmoidal transfer functions nicely fit the needs of representing Boolean-like functions, the discussed problems of neural saturation, along with the proposals for its treatment, suggest a more careful analysis to understand their source. Let us go back to the simple case of a single neuron, but this time, suppose we look for a different penalty. In the case of the quadratic error the saturation arises since the growth of the activation doesn't create any benefit after a certain threshold, as there is nothing which faces neural saturation. Now suppose we use the relative entropy penalty function

$$E = -\sum_{\kappa=1}^{\ell}\left(y_\kappa \log x_{\kappa o} + (1 - y_\kappa)\log(1 - x_{\kappa o})\right)$$

in the case of neural network with a single output units o. We start noticing that the neural saturation, this time related to using logistic or hyperbolic tangent neural transfer function, doesn't prevent learning. Suppose $x_{\kappa o} \to 0$ or $x_{\kappa o} \to 1$ and that we are using logistic sigmoid. The case $x_{\kappa o} \to 0$ corresponds with $a_{\kappa o} \to -\infty$. When using the quadratic error, we get stuck in configurations where $y_\kappa = 1$, whereas by using the relative entropy, we have

$$-\log(\text{logistic}(a_{\kappa i})) = -\log\frac{1}{1 + e^{-a_{\kappa i}}} \simeq -\log\frac{1}{e^{-a_{\kappa i}}} = a_{\kappa i},$$

which indicates that the gradient heuristic is still strong, since

$$\frac{\partial}{\partial w_{oi}}(-\log(\text{logistic}(a_{\kappa o}))) \simeq \frac{\partial a_{\kappa o}}{\partial w_{oi}} = x_{\kappa i}.$$

The same conclusions can be drawn in the case $x_{\kappa o} \to 1$ (see Exercise 5). Is the marriage of the logistic sigmoid with the relative entropy successful also in case of deep nets? From backpropagation Eqs. 5.5.2–(5), 5.5.2–(6), and 5.5.2–(7) we see that in case of saturated neurons the only possibility to face the vanishing of the gradient is to move $\delta_{\kappa i}$. Interestingly, this happens for the output neurons, which behave just like in the case of a single neuron. However, hidden units are also penalized by neural saturation because of the term $\sigma'(a_{\kappa i})$ of backward Eq. (7).

While this discussion shows that the problems connected with the saturation of sigmoidal units can be faced by an appropriate initialization of the weights and by an opportune selection of the penalty function, we shouldn't neglect their effect, which can really prevent from learning if no appropriate precautions are taken. The adoption of rectifier neurons presents a two-state behavior (active/nonactive) that, however, does not suffer from saturation. However, to some extent, saturated configurations of sigmoidal units can be related to nonactive configurations in rectifier units. Furthermore, the discussion in the next section indicates that the choice of a monotone transfer function for ridge neurons doesn't affect the complexity of learning. The differences that are typically reported are related to the specific numerical algorithm.

EXERCISES

1. [*29*]   Suppose we are given an OHL net in a configuration in which all its weights are equal. Discuss the evolution of the learning process when beginning from this configuration by expressing the correspondent value of the gradient.

2. [*M21*]   Let us consider Chebyshev's inequality (1). First, discuss the reason why A_i can be regarded as a random variable with normal distribution. Second, provide a sharper bound that relies on the normal distribution.

3. [*M16*]   Given a random variable W with uniform distribution in $[-w, +w]$, calculate its variance.

4. [*24*]   Let us consider the analysis carried out for defining the random weight initialization, which is summarized by Eq. (3). Generalize this property to the case of a generic neuron, either output or hidden unit, that receives inputs from other hidden units. Prove that Eq. (3) can simply be replaced with $\sigma_{W_i} = 1/f_{in}$.

5. [*23*]   Complete the proof on the asymptotic behavior of the cross-entropy penalty in the case $x_{KO} \to 1$.

5.7.3 Complexity and numerical issues

The discussion in this section indicates that the complexity of the loading problem is related to the presence of suboptimal configurations (local minima that are not global) and to different forms of ill-conditioning. This is a fundamental distinction since these two sources of complexity may result in very different computational efforts. The presence of ill-conditioning in the error function can arise in different cases. As shown in this section, spurious formulations of learning can lead to failures in classifications, but in these cases one can typically come up with different choices of the penalty to face the problem. Another context in which ill-conditioning arises is the lack of independent learning examples. The case of a linear machine offers an example of such a problem, which is clearly expressed by Eq. 3.4.3–(9). It is not only the case in which $\det(\hat{X}'\hat{X}) = 0$ that creates problems in the naive formulation of linear machines; whenever the condition number $\text{cond}(\hat{X}'\hat{X})$ is small, we are still in trouble! As shown in Section 3.1.3, this type of ill-conditioning is very well handled by regularization. Clearly, the injection of regularization may somewhat mask the lack of examples to characterize the concepts to be learned. However, it represents a very effective way of facing ill-conditioning. When LTUs are involved, the error function associated to feedforward networks is alternatively composed of plateaux and ravines. This gives rise to a numerical kind of complexity that is due to the need of properly adapting the learning rate to follow hardly predictable sequences of plateaux and ravines. It is argued that classic formulations of computational complexity in the continuum setting lead to the identification of *circuit complexity* and *numerical complexity* that are captured by a proper notion of *condition number*.

When the inherent complexity of the loading problem is due to the presence of suboptimal local minima, we are in a very different situation. The most remarkable difference is that, in this case, there is no effective methodology to parallel the role of regularization for ill-conditioning! This crucial issue is often neglected, and the distinction between these two different sources of complexity is not carefully taken into account most of the times. When the error surface reveals a multimodal structure, we need to involve global optimization methods that might be seriously plagued by the presence of suboptimal local minima.

5.8 Scholia

Section 5.1 In real-world applications, feedforward architectures are mostly multi-layered networks, where the role of hidden layers has been the subject of theoretical and experimental debate. Some visual techniques [H. Bischof, A. Pinz and W.G. Kropatsch, *Proc. 11th IAPR Int. Conf. on Pattern Recognit.* **2** (1992), 581–585,], like Hinton's diagrams, have significantly supported the intuition for investigations on architectural issues. The choice of the nonlinearity in the neuron is of remarkable importance. For many years, feedforward architectures have been based on sigmoidal units with logistic sigmoid or hyperbolic tangent. A few years ago, some people began realizing that the rectifier function is a very good alternative to sigmoidal-like functions [see X. Glorot, A. Bordes, Y. Bengio, *Proc. of the JMLR* **15** (2011), 315–323 and A.L. Maas, A.Y. Hannun, A.Y. Ng, *ICML Workshop on Deep Learning for Audio, Speech and Language Processing* (2013)] In particular, it gives rise to a remarkably different learning process, since the optimization trajectory is basically driven under a risk function involving piecewise nonlinearity. The refraining comments of the 1990s on deadly saturation have been overcome in big networks, where the probability of finding, for any input, a nonsaturated path from the input to the output is very high.

Section 5.2 The study of methods for realizing a Boolean function is extremely useful to capture many interesting computational aspects of feedforward neural networks. Whenever we are interested in classification tasks, the corresponding computational processes share many common issues with Boolean circuits. The analysis carried out in this chapter on these topics has been largely inspired by the work reported in *Discrete Neural Networks* (Prentice Hall, 1995) by K.Y. Siu, V. Roychowdhury and T. Kailath. An in-depth understanding of representational aspects is nicely interwound with circuit complexity. The classical work of Claude Shannon [*Bell Syst. Tech. J.* **28** (1949), 59–98] early pointed out that all but a vanishingly small fraction of Boolean functions with d variables require an exponential number $\Omega(2^d)$ of AND–OR gates for their computation. The notion of circuit complexity early emerged to understand the limitations connected with the architectural choices. The computational analysis typically involves the *fan-in* and *fan-out* of the gates, their specific structure, along with the tradeoffs between size and depth of the overall architecture. As pointed out in Section 5.2, LTUs are typically more effective than AND–OR gates, and the depth plays an important role in the computation. The analysis of bounded vs unbounded fan-in/fan-out leads to important results on the realization of Boolean functions. It can be proven that all circuits whose gates have bounded fan-in, in order to compute any non-degenerate function, must have size $\Omega(d)$ and depth $\Omega(\log(d))$. Hence we need to relax to unbounded fan-in to get better complexity bounds.

There are a number of techniques to attack the design of Boolean functions using a given type of gate. Interesting results are given by using analytic techniques, rational approximation, and communication complexity arguments [see K.Y. Siu, V. Roychowdhury and T. Kailath *Discrete Neural Networks* (Prentice Hall, 1995)]. The analysis illustrated in Section 5.2.4 on telescopic techniques yields interesting realization with quite sharp complexity bounds.

Telescopic techniques all share a unifying principle that arises in telescopic series, where the partial sum eventually exhibits only a fixed number of terms after cancellation. For example, the series $\sum_{n=1}^{\infty} 1/n(n+1)$ has a telescopic structure, since it can be computed simply looking far away for the remaining noncanceled term. We have

$$\sum_{n=1}^{\infty} \frac{1}{n(n+1)} = \lim_{m \to +\infty} \sum_{n=1}^{m} \left(\frac{1}{n} - \frac{1}{n+1} \right)$$
$$= \lim_{m \to +\infty} \left[\left(1 - \frac{1}{2} \right) + \left(\frac{1}{2} - \frac{1}{3} \right) + \cdots + \frac{1}{m} - \frac{1}{m+1} \right]$$
$$= 1 - \lim_{m \to +\infty} \frac{1}{m+1} = 1.$$

Telescopic techniques play an important role in the design of Boolean circuits, but they must be carefully treated so as to avoid pitfalls and paradoxes (see Exercise 3). Their adoption leads to the dramatic complexity cut, from $O(d)$ to $O(\sqrt{d})$, in the realization of parity — see Section 5.2.4, as a nice example of their power.

The realization and the corresponding circuit analysis of Boolean functions has been the subject of in-depth investigations. The notion of symmetric function presented in Section 5.2.4 has been naturally extended to that of *generalized symmetric functions*, where

$$f(x_1, \ldots, x_d) = \psi \left(\sum_{i=1}^{d} w_i x_i \right),$$

that is, the dependence on $\sum_{i=1}^{d} x_i$ that characterizes symmetric functions is extended to the case of a weighted sum. We can easily see that in case of integer weights, this representational extension possesses the features of symmetric functions. For example, $f(x_1, x_2, x_3) = \psi(x_1 + 3x_2 + 2x_3)$ can be rewritten as $f(x_1, x_2, x_3) = \overline{f}(x_1, x_1, x_1, x_2, x_3, x_3)$. That is, the generalized symmetric function f is transformed into the symmetric function $\overline{f}$, so one can use the realization techniques presented for symmetric functions also in this more general case.

Sums, products, ratios, and powers can be realized very efficiently by LTU units. Once again, these units turn out to be remarkably more efficient than classic AND–OR gates. Depth-3 circuits are known for computing the sum with d bits by using "block-save" techniques and results on symmetric functions, while depth-4 realizations have been developed for multiplication, exponentiation, and division [K.Y. Siu, V. Roychowdhury and T. Kailath *Discrete Neural Networks* (Prentice Hall, 1995)].

The realization techniques for symmetric functions presented in Section 5.2.4 suggest that, while for AND–OR gates the configuration detection takes place at min-term/max-term level, the adoption of LTU units makes it possible to detect $\sum_{i=1}^{d} x_i \in \mathbb{N}$. This clearly allows us to come up with realizations with integer numbers. The case of generalized symmetric functions is an example of similar realizations. Interestingly, it can be proven that in case of Boolean functions with d inputs, real weights can always be replaced with integers of $O(d \log(d))$ weights [K.Y. Siu, V. Roychowdhury and T. Kailath *Discrete Neural Networks* (Prentice Hall, 1995)]. In-depth analyses have also been devoted to the realization with "small weights." As shown in Section 5.2.4, there are functions for which the increment of depth leads to circumventing the exponential explosion of the weights.

Section 5.3 When at the end of the 1980s the interest on neural networks exploded, scientists with strong background in mathematics were soon attracted by the problem of understanding the computational capabilities of feedforward architectures. While for Boolean functions their expression in canonical forms immediately suggested the universal representation by means of one hidden layer, the same property for real-valued functions originated an interesting novel field of investigation. In general, the problem can be attacked in the framework of function approximation. In the case of classification, however, the appropriate segmentation of the input space so as to isolate positive from negative examples can naturally be formulated in the framework of computational geometry.

The seminal paper by Richard Lippmann [*IEEE ASSP Mag.* **4** (1987) 4–22.], which is partly reviewed in Section 5.3.1, offered nice insights on the geometrical structure of approximation mostly for classification tasks. In this case, there are some intriguing connections with the realization of Boolean functions, the difference being that a function defined on a real-valued domain requires a partition into convex sets of the set in which it is active (positive set). The cover of the positive set reminds us of the classic problem of a set cover.

The set cover problem is a classical problem in computer science, one of Karp's 21 NP-complete problems shown to be NP-complete in 1972 [see M.R. Garey, D.S. Johnson, *Computers and Intractability: A Guide to the Theory of NP-Completeness* (Freeman, 1979)]. Given $\mathcal{A} = \{1, 2, \ldots, m\} \subset \mathbb{N}$ (the universe) and a collection $\mathcal{F} = \{\mathcal{A}_i : i = 1, \ldots, n\}$ of sets such that $\mathcal{A} = \bigcup_{i=1}^{n} \mathcal{A}_i$, set cover is the problem of identifying the smallest subcollection of $\mathcal{F}$, whose union equals the universe. In the set cover decision problem, the question is whether there is a set cover of a least a given size s. The decision version of set cover is NP-complete, while the optimization/search version of a set cover is NP-hard. If each set is assigned a cost, it becomes a weighted set cover problem. Set cover has a geometric version, which perfectly fits the learning problem that a feedforward neural network solves in classification. Given the pair $(\mathcal{A}, \mathcal{F})$, where $\mathcal{A}$ is the universe and $\mathcal{F}$ is a family of *range sets*, a *geometric set cover* problem consists of determining the minimum-size subset $\mathcal{S} \subset \mathcal{F}$ of ranges such that every point in the universe $\mathcal{A}$ is covered by some range in $\mathcal{S}$. While the problem is still intractable, there are approximation algorithms that, due to the geometric nature of the problem, are quite efficient and accurate. Notice that cover problems are closely related to packing problems, but learning does require to cover the positive set with convex domains more than packing them inside.

Early studies based on calculus and related topics, like those shown in Section 5.3.2, were used to conclude that the universal representation is gained by a network with one single hidden layer. A fundamental distinction on approximation issues concerns existential vs constructive approaches. In the first case one is only interested in proving that there exists a realization, without bothering with any constructive process. Overall, the subject has been massively investigated, and a number of surveys have been published, which nicely review the state-of-the-art [e.g., M. Sanguineti, *Open Appl. Math. J.* **2** (2008), 31–58 or F. Scarselli, A.C. Tsoi, *Neural Netw.* **11** (1998), 15–37]. In particular, Marcello Sanguineti discusses most relevant approximation approaches based on Fourier analysis, the Hahn–Banach Theorem, the Radon Transform, and the Stone–Weierstrass Theorem.

Existential approaches offer a nice way of exploring the approximation capabilities. Let us consider the class of functions $\mathbb{N}$ generated by a single hidden layer feedforward network

$$\mathcal{N}_g^3 = \{f \in \mathcal{F} : f(x) = \sum_{i=1}^n w_i g(w, b, x) + w_0\}.$$

The approximation of continuous functions in $C(\mathbb{R}^d)$ can be reformulated by studying whether $\mathcal{N}_g^3$ is dense in $C(\mathbb{R}^d)$. An elegant approach along this direction is reported in the seminal paper by George Cybenko *Math. Control Signals Syst.* **3** (1989), 303–314, which is based on the Hahn–Banach theorem. Related studies flourished at that time, which covered technical aspects on approximation in different functional spaces [see, e.g., K. Hornik, M. Stinchcombe, H. White, *Neural Netw.* **2** (1989), 359–366 or V. Kurkova, *Neural Netw.* **5** (1992), 501–506]. Kurt Hornik also pointed out a curious property, which involves the bias. He proved [*Neural Netw.* **6** (1993), 1069–1072] that if $\sigma(\cdot)$ is an analytic function then there exists a bias value b such that $\mathcal{N}_\sigma(w, \{b\})$ is dense in $L^p(\mu)$, and in $C(K)$, where $K \in \mathbb{R}^d$ is a compact set. In-depth investigations on density properties came to the conclusion that the universal approximation capabilities can only be achieved by using nonpolynomial neurons [see also M. Leshno, V.Y. Lin, A. Pinkus, S. Schocken, *Neural Netw.* **6** (1993), 861–867]. Moshe Leshno et al. proved that feedforward nets with one layer of locally bounded piecewise continuous activation functions can approximate any continuous function to any degree of accuracy if and only if the activation function is not a polynomial. The necessary condition of nonpolynomial functions can be understood when extending the analysis carried out in Section 5.1.2 for the case of polynomial neurons, whereas the sufficient condition requires a more involved analysis. Gori and Scarselli in *IEEE Trans. Neural Netw.* **9** (November 1998), 1086–1098, provided a somewhat opposite view of approximation by exploring the expressiveness of feedforward networks with one hidden layer composed of a given number of units. While the studies on the approximation of a given function conclude that any degree of approximation can be achieved with "enough" hidden units, they analyzed the class of functions generated by a network with a limited number of units. Basically, they proved that there are functions, which can be approximated up to any degree of accuracy, without having to increase the number of the hidden nodes. It was proven that rational functions, products of polynomials, and exponentials are expressible by networks with sigmoid activation function.

The discussion in Section 5.3.3 on symmetries is based on multilayered nets. What if we deal with any DAG? In this case, the layer symmetry is broken, and the number of equivalent solutions drops dramatically. Marco Gori, M. Maggini, L. Sarti in *IEEE Trans. Pattern Anal. Mach. Intell.* **27** (2005), 1100–1111, used the idea behind the popular PageRank algorithm to address graph isomorphisms. Interestingly, it turns out that the algorithm generally yields distinct values on the nodes whenever there are no symmetries. Now the computational scheme of PageRank is very much related to forward propagation in case of DAGs. This suggests that graphs without symmetries give rise to a forward propagation with a very rich expressiveness. This is a nice issue that might enrich the discussion on shallow vs deep nets.

In Section 5.3.2 we have seen constructive realizations. The first two of them basically extend the scheme adopted in Section 5.3.1 for classification tasks, whereas the third realization scheme is based on Fourier expansion [see B. Irie, S. Miyake, *Int. Joint Conf. on Neural Netw.* (1988), 641–648 and K. Funahashi, *Neural Netw.* **2** (1989), 183–192]. Constructive realizations clearly make us wonder how we can conquer a desired degree of precision. The results based on density in functional spaces suggest that the precision can be arbitrarily high, but there is no upper bound on the required number of hidden units. Unfortunately, this holds also for most realization methods; the work of Andrew R. Barron [*IEEE Trans. Inf. Theory* **39** (1993), 930–945] is worth mentioning in this respect.

Feedforward nets have been massively used in real-world problems, so it might be hard to discover a domain where nobody has tried their application. The availability of many software libraries has been facilitating this process, which is not rarely driven by an intriguing mixture of magic and mysteries. Sometimes it looks like there are excessive expectations from these "magic boxes." In the early 1990, just after one of the authors completed his Ph.D., with Angelo Frosini and Paolo Priami, he was stimulated by the possibility of using these neural nets technologies in banknote verification. They designed a neural-based banknote acceptor, which achieved state-of-the-art performance by using low cost opto-electronic sensors. For that project he learned a lot of things, but he mostly realized how important is to go beyond, and magic and the mysteries of machines just regarded as black boxes. Our banknote acceptor detected the length, so as to make a first hypothesis on the class. Then, for each class, a feedforward net was used to check whether the incoming banknote was a counterfeit. The *rejection criterion* was simply based on threshold checking: Whenever the output was below the threshold, the machine rejected the banknote. The network had learned from a collection of different banknotes and counterfeits using batch-mode learning. Early results were extremely promising; the machine could discriminate sophisticated counterfeits, which made us pretty excited. However, the satisfaction for those results, only lasted a few days: A lottery-like ticket, properly adjusted, so as to fit the dimensions of a banknote, passed through the machine, which returned FF, the hexadecimal code to express highest acceptance rate! The explanation of this very critical behavior comes out when considering the type of separation surfaces created by MLP nets. In order to favor the intuition, suppose we deal with bi-dimensional patterns, so that true banknotes are represented by •, whereas counterfeits are represented by ▲ and ■:

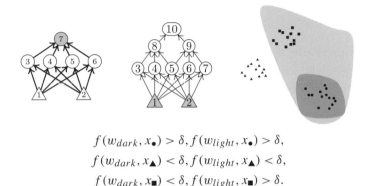

$$f(w_{dark}, x_\bullet) > \delta, f(w_{light}, x_\bullet) > \delta,$$
$$f(w_{dark}, x_\blacktriangle) < \delta, f(w_{light}, x_\blacktriangle) < \delta,$$
$$f(w_{dark}, x_\blacksquare) < \delta, f(w_{light}, x_\blacksquare) > \delta.$$

Once a certain threshold δ has been selected, the verification process is simply driven by a comparison with the output of the MLP. Now we can promptly see that the two neural networks in the figure lead to a very different behavior. They share the same architecture, but different weights. The corresponding separation surfaces are represented on the right: Thick connections are associated with the dark domain and, likewise, thin connections correspond with the lighter domain. While the first domain is bounded, and nicely envelopes the true banknote patterns, the second one is unbounded and also includes the boxes (■) that are counterfeits. Now both networks may work correctly for years, since the counterfeits, represented as triangles (▲), are perfectly separated also by the right-side network. However, this holds until somebody feeds the machine with a lottery-like ticket (box patterns)!

The study of the separation surfaces offers a clear interpretation of successes and failures of feedforward classifiers in problems of pattern recognition and verification [see, e.g., M. Gori, F. Scarselli, *IEEE Trans. Pattern Anal. Mach. Intell.* **20** (1998), 1121–1132 and M. Bianchini, P. Frasconi, M. Gori, *IEEE Trans. Neural Netw.* **6** (1995), 512–515].

Years later, a number of people realized that the spectacular results of deep convolutional nets can lead to apparently baffling results on ImageNet! We could select an image — a "panda" — and change it almost imperceptibly to human interpretation, but the neural network returns a wrong classification. There's more: The way the errors arise doesn't have an apparent easy interpretation. However, when referring to the above discussion on separation surfaces, this is not so surprising. In any case, once a classifier has been trained, an oracle acting as an adversary, can take any pattern x and generate $\tilde{x}$, which is expected to be very close to x, with the property of maximizing the loss difference. In particular, $\tilde{x}$ can be generated by

$$\tilde{x} \in \arg\max_{x \in X} \left((\tilde{x} - x)' \nabla_x V \right),$$
$$\|\tilde{x} - x\|_\infty \le \epsilon,$$

with ϵ being a "small" threshold that is used to model the imperceptible movement in a neighborhood of x and $\nabla_x V$ the gradient of the loss $V(x, y(x), f(x))$ computed in x. We can promptly see that $\tilde{x} = x + \epsilon \nabla_x V / \|\nabla_x V\|_\infty$ yields the maximum change. A good choice is also

$$\tilde{x} = x + \epsilon \, \text{sign} \left(\nabla_x V \right).$$

This way of generating an adversarial attack is referred to as the "fast gradient sign method." Interestingly, this attack requires an efficient gradient computation that can be carried out by Backprop [see I.J. Goodfellow, J. Shlens, C. Szegedy, arXiv preprint arXiv:1412.6572 (2014)]. This clever way of discovering an attack can be joined with a corresponding adversarial training algorithm that relies on the principle of adding, for any given x, its associated adversarial example $\tilde{x}$ in the training set. In so doing, the loss $V(x, y(x), f(x))$ is replaced with

$$\alpha V(x, y(x), f(x)) + (1 - \alpha) V(\tilde{x}, y(x), f(\tilde{x})),$$

where we assume that the generated pattern $\tilde{x}$ has the same category as x, that is, $y(\tilde{x}) = y(x)$, and $0 < \alpha < 1$ (in arXiv:1412.6572, $\alpha = 0.5$ is chosen). In a sense, learning with the introduction of adversarial examples produces a clever augmentation scheme that gives rise to a sort of regularization. The topic of adversarial learning has received considerable attention is the literature [see, e.g., B. Biggio, I. Corona, D. Maiorca, B. Nelson, N. Šrndíc, P. Laskov, G. Giacinto, F. Roli, *Joint European Conf. on Mach. Learn. and Knowl. Discovery in Databases* (2013), 387–402, D. Lowd, C. Meek, *Proc. of the Eleventh ACM SIGKDD Int. Conf. on Knowl. Discovery in Data Mining* (2005), 641–647 and I.J. Goodfellow, arXiv:1701.00160 (2017)]. However, the underlying problem has been put forward also in different contexts, especially when one is interested in designing methodologies for adaptive classifiers [see, e.g., C. Alippi, M. Roveri, *IEEE Trans. Neural Netw.* **19** (2008), 1145–1153 and *IEEE Trans. Neural Netw.* **19** (2008) 2053–2064]. The novelty of the idea of adversarial learning doesn't remove the inherent limitations behind many machine learning approaches. In the case of vision, the confusion of a "panda" with a gibbon is a serious problem that is not far away from the mentioned failures in banknote acceptors. It looks like these dramatic differences in perceptual capabilities arise because of the lack of incorporations of fundamental invariances in the neural architectures. The discussion in Section 5.4 advocates a substantial reformulation of convolutional nets in which we incorporate motion invariance.

The analysis on Boolean functions suggests that the benefit from circuital complexity that arises from exploiting deep nets is likely extendable to real-valued functions. While this is quite obvious, for about two decades the scientific community has been driven by the reassuring property of universal approximation that can be gained when using only one hidden layer. Beginning from the end of the 1980s, this universal property, together with the belief that learning with deep nets early deteriorates because of the gradient vanishing on deep paths, was the fuel of most design choices in the impressive number of different applications of feedforward neural networks.

In addition, the massive application to many different tasks and early studies on the structure of the error surfaces reinforced this simple choice. It was early clear that when the number of hidden units increases, in addition to a progressively better degree of approximation, also the optimization of the error function becomes simpler! The circulation of these statements reinforced in the scientific community the belief that concrete experiments must rely on OHL nets. Interestingly, for years, no one raised criticisms on shallow nets! When in mid-1990s kernel machines appeared in the community of machine learning, shallow nets with one hidden layer were once again another sign of the natural structure of computational models of learning. There was more. As shown in Section 4.4.4, shallow nets in the form of kernel machines are the optimal outcome of the classic formulation based on regularization principles.

Suppose you are given a piece of paper about 0.01 cm high and that you start folding it.[24] Each time we fold the paper, the thickness doubles. By the time you get to 27 foldings, we get $10^{-4} \times 2^{27} \simeq 12,8$ km, that is more than Mount Everest. Interestingly, it only takes 42 foldings of a paper to get from the Earth to the Moon, and only about 94 foldings of a paper to make something the size of the entire visible Universe! Clearly, while this folding cannot be concretely implemented, the conceptual scheme is technically sound. Another way of capturing the flavor of the outcome of this progressive folding is that it produces an exponential number of small faces. Interestingly, this is very much related to the way feedforward nets with rectifier neurons partition the input space. To grasp the issue, Montúfar et al. [arXiv:1402.1869v2] introduced the notion of *folding map*. For example, the function $g \colon \mathbb{R}^2 \to \mathbb{R}^2$

$$g(x_1, x_2) = (|x_1|, |x_2|)'$$

can be regarded as the outcome of the computation of a layer of a feedforward network. Basically, it returns an output which is defined by the first quadrant, since all other points are determined by horizontally and vertically folding the domain. Deep structures are rich and offer intriguing representational properties. Yoshua Bengio and Yann LeCun early realized the fundamental role of deep structured to provide appropriate representations for concrete AI problems, and published a seminal book chapter in *Large Scale Kernel Machines, NIPS'12* (MIT Press, 2007). Other interesting papers that recently shed light on the reasons of the success of deep architectures are H. Mhaskar, Q. Liao, T.A. Poggio, arXiv:1603.00988 (2016) H.W. Lin, M. Tegmark, arXiv:1608.08225 (2016) and H. Mhaskar, T.A. Poggio, arXiv:1608.03287, 2016.

Section 5.4 Convolution nets were early introduced at the end of the 1980s by Yann LeCun and associate in a number of papers [see, e.g., Y. LeCun, B. Boser, J.S. Denker, D. Henderson, R.E. Howard, W. Hubbard, L.D. Jackel, *Neural Comput.* **1** (1989), 541–551 and Y. Le Cun, B. Boser, J.S. Denker, R.E. Howard, W. Habbard, L.D. Jackel, D. Henderson, *Advances in Neural Information Processing Systems* **2**

---

[24] http://scienceblogs.com/startswithabang/2009/08/31/paper-folding-to-the-moon.

(1990), 396–404 and later on in Y. LeCun, Y. Bengio, *The Handbook of Brain Theory and Neural Networks* (MIT Press 1995, pages 255–258 edited by M.A. Arbib]. The basic ideas that characterize current convolutional architectures were already included in those papers, but for its strong impact we had to wait for the second wave of connectionist models. There was in fact a very strong impact in computer vision for models involved in both classification and segmentation [see A. Krizhevsky, I. Sutskever, G.E. Hinton, *Proc. of the 25th Int. Conf. on Neural Inf. Process. Syst.* (2012), 1097–1105, M.D. Zeiler, R. Fergus, arXiv:1311.2901 (2013), K. Simonyan, A. Zisserman, arXiv:1409.1556 (2014), E. Shelhamer, J. Long, T. Darrell, *IEEE Trans. Pattern Anal. Mach. Intell.* **39** (2017), 640–651. P. Sermanet, D. Eigen, X. Zhang, M. Mathieu, R. Fergus, Y. Le-Cun, arXiv:1312.6229 (2013)]. Remarkable results on document recognition that exploit convolutional architectures are found in Y. LeCun, L. Bottou, Y. Bengio, P. Haffner, *Proc. of the IEEE* **86** (1998), 2278–2324. A somewhat different convolutional scheme has been used in M. Gori, M. Lippi, M. Maggini, S. Melacci, *Comput. Vis. Image Underst.* **146** (2016), 9–26. for semantic labeling. Convolutional nets have been successfully applied also to text classification [X. Zhang, J. Zhao, Y. LeCun, *Proc. of the 28th Int. Conf. on Neural Inf. Process. Syst.* (2015), 649–657] and text rank [A. Severyn, A. Moschitti, *Proc. of the 38th Int. ACM SIGIR Conf. on Res. and Develop. in Inf. Retrieval* (2015), 373–382].

Convolutional nets exhibit the fundamental feature of incorporating spatial invariance. However, since the dawn of computer vision, people realized that other invariance properties like rotation and scale are also very important to conquer visual skills. Another relevant mechanism for enforcing invariance is the tangent distance, where the surfaces of possible transforms of a pattern is approximated by its tangent plane [P.Y. Simard, Yann LeCun, J.S. Denker, B. Victorri, *Lecture Notes in Computer Science* **1524** (Springer, 1996), 239–274]. The studies on *scale-invariant feature transform (SIFT)* by David Lowe [D.G. Lowe, *Int. J. Comput. Vis.* **60** (2004), 91–110] have dramatically influenced the research in computer vision. Beginning from visionary work on hierarchical models of the cortex by Hubel and Wiesel [D.H. Hubel, T.N. Wiesel, *J. Physiol.* **160** (1962), 106–154.], Maximilian Riesenhuber and Tomaso Poggio, in a seminal paper on models of object recognition in the cortex [Nat. Neurosci. **2** (1999), 1019–1025], discussed the issue of invariance also in the framework of neuroscience. They proposed a model based on a MAX-like operation that may have an important role in cortical function. Another interesting biologically inspired model with focus on feature invariance can be found in [T. Serre, L. Wolf, T. Poggio, *Proc. of the 2005 IEEE Comput. Soc. Conf. on Comput. Vision and Pattern Recognit* (2005), 994–1000.], where the authors stressed the property of using the proposed feature-based representation for learning with a few examples. An interesting approach to invariance rooted on computational foundations is given in F. Anselmi, L. Rosasco, T.A. Poggio, arXiv:1503.05938 (2015) and T. Poggio, F. Anselmi, *Visual Cortex and Deep Networks: Learning Invariant Representations* (MIT Press, 2016). The authors focus on transformations that are at the same time selective, where two patterns share the representation only if one is a transformation of the other. The analysis carried out in Section 5.4 proposes focusing on the extraction of features that are invariant under movement, which assumes that the agent lives in its own environment, thus conquering position, rotation, and scale invariance by concrete interactions. Interestingly, motion invariance seems to be candidate to naturally capture invariances, like those associated to ironed shirts and shirts thrown into the laundry basket! This is an open research issue that raises fundamental issues on appropriate cognitive architectures for computer vision [A. Chella, M. Frixione, S. Gaglio, *Artif. Intell.* **89** (1997), 73–111.].

The remarkable experimental results that have been obtained by convolutional networks, where the propagation of visual information is locally restricted by receptive fields, suggests that one should seriously consider that the computation needn't take place necessarily over all the retina with the same degree of resolution. Interestingly, human eye movements, with saccadic quick scans alternated to fixations [K. Rayner, *Psychol. Bull.* **124** (1998), 372–422.], can focus on most informative parts of the visual frames to collect a global picture.[25] This nicely resonates with the emphasis on motion invariance that is advocated in Section 5.4. In addition, there are abstract visual skills that are unlikely to be attacked by pixelwise computation. Humans provide visual interpretations that go beyond the truly visual pattern (see, e.g., Kanizsa's illusions [*Sci. Am.* **234** (1976), 48–52]). This might be facilitated by the focus of attention, which somehow locates the object to be processed. As the focus is on a certain pixel, the corresponding object can be given an abstract geometrical interpretation by its shape expressed in term of its contour. While pixel-based processes are based on all the visual information of the retina associated with a given pixel, shape-based recognition emerges when recognizing objects on the basis of their contour, once we focus attention on a point of the object. In the process of feature extraction, one shouldn't neglect that some objects receive a linguistic description that is not primarily based on their shape, but mostly on their function. For instance, a bowl is very well identified by its role in containing liquids more than by its shape. Pixelwise processes don't seem to be adequate to draw conclusions on actions, whose understanding does require involving portions of a video. Moreover, the notion of *object affordance* indicates the strict connection with that of an action. Humans carry out many object recognition processes on the basis of actions in which they are involved, so as objects are detected because of their role in the scene. In other words, the affordance involves the *functional role* of objects, which is used for the emergence of abstract categories.

Section 5.5 The Backprop algorithm can fully be understood as we set the computation within the framework of data flow. This requires familiarity with topological sort, consisting of consistently ordering the vertices and the edges of a DAG in a way to respect the partial ordering on its vertexes. Topological sort is also known as a linear extension. We can promptly see that in the extreme case the number of possible sorts

---

[25] It's worth mentioning that the adoption of LSTM nets has been recently proven to succeed in the problem of saliency estimation [M. Cornia, L. Baraldi, G. Serra, R. Cucchiara, arXiv:1611.09571 (2016)].

explodes with the dimension of the graph. This surely happens when $\mathcal{A} = \emptyset$, since a graph with no arcs between the vertexes can be sorted in $|V|!$ different ways. A linear extension of a given DAG can easily be found by $O(|V|)$ algorithms that are optimal since the problem is clearly $\Omega(|V|)$. A possible algorithm performs a depth-first search of the DAG. This is very important for the data flow computational scheme of feedforward neural networks, since the existence of such an algorithm indicates that we can always perform the forward step using an optimal algorithm. As suggested by the extreme case of graphs without arcs, computation of all the linear extensions can be prohibitively expensive. The number of linear extensions can grow exponentially in the size of the graph, and counting the number of linear extensions is NP-hard.

In 1986, a seminal trilogy of books was published that, better than anything else, contributed to the spread of principles on *Parallel Distributed Processing* (PDP) D.E. Rumelhart, J.L. McClelland and PDP Research Group Parallel Distributed Processing (MIT press, 1987) volumes 1, 2 and 3 (see also the compact view given in D.E. Rumelhart, G.E. Hinton, R.J. Williams, *Nature* **323** (1986), 533–536). By and large, backpropagation was one of the most distinguishing achievements in the rising wave of connectionism at the end of the 1980s. The PDP group contributed in a crucial way to the spread of interest in the algorithm, though its roots can be found in different topics. Traces of backpropagation had already appeared in the context of optimal control theory [A.E. Bryson, W.F. Denham *J. Appl. Mech.* **29** (1962) 247–257 and *Applied Optimal Control* (Blaisdell, 1969) by A.E. Bryson, Y.C. Ho]. Hecht-Nielsen [*Int. Joint Conf. on Neural Netw.* (1989) 593–605] cited the work of Bryson and Ho and of Paul Werbos [*Beyond Regression: New Tools for Prediction and Analysis in the Behavioral Sciences*, Ph.D. thesis, Harvard University, 1974.] as the two earliest sources of the backpropagation idea. In the neural network community, the method has been reinvented a number of times [e.g., D.B. Parker, *Tech. Report TR–47, Center for Comput. Res. in Econ. and Manage. Sci.*, (MIT, 1985) and D.B. Parker, *IEEE Int. Conf. on Neural Netw.* **2** (1987), 593–600], until it was eventually brought to fame by the PDP group. Yan Le Cun also contributed independently to the spread the basic ideas behind Backprop in his PhD thesis [*Une procédure d'apprentissage pour réseau a seuil asymmetrique* in *Proc. of Cognitiva* **85** (1985), 599–604]. Additional evidence was given a few years later in collaboration with Francoise Fogelman-Souliánd Patrick Gallinari [F. Fogelman-Soulié, P. Gallinari, Y. LeCun, S. Thiria, *Proc. of the First Int. Conf. on Neural Netw.* (1987), Y. le Cun, *The 1988 Connectionist Models Summer School* (Morgan Kauffman, 1988) 21–28, edited by D. Touretzky, G. Hinton, T. Sejnowski]. The impact on many related disciplines was remarkable in many fields, including pattern recognition [see, e.g., C.M. Bishop, *Neural Networks for Pattern Recognition* (Oxford University Press, 1995), B.D. Ripley, N.L. Hjort, *Pattern Recognition and Neural Networks* Cambridge University Press, 1995) 1st edition, S. Marinai, M. Gori, G. Soda, *IEEE Trans. Pattern Anal. Mach. Intell.* **27** (2005), 23–35], bioinformatics [see, e.g., P. Baldi, S. Brunak, *Bioinformatics: The Machine Learning Approach* (MIT Press, 2001.) 2nd edition, P. Baldi, S. Brunak, Y. Chauvin, C.A.F. Andersen, H. Nielsen, Bioinformatics **16** (2000), 412–424], automated control and identification [K.S. Narendra, K. Parthasarathy, *IEEE Trans. Neural Netw.*

1 (1990), 4–27, E.B. Kosmatopoulos, M.M. Polycarpou, M.A. Christodoulou, P.A. Ioannou, *IEEE Trans. Neural Netw.* **6** (1995), 422–431.]

As late as 2004, about 20 years after the circulation of backpropagation, the excitement for the novel paradigm of neural nets that was experienced beginning from mid-1980s has mostly vanished, and kernel machines were mostly replacing, and slightly improving, experimental results in real-word applications. Yet, this didn't really result as a breakthrough in the rest of the AI academic research. The same year, Geoffrey Hinton, with a small amount of funding from the Canadian Institute for Advanced Research (CIFAR), together with Yan Le Cun and Yoshua Bengio, founded the *Neural Computation and Adaptive Perception program*, with a few hand-picked researchers dedicated to creating computing systems that mimic organic intelligence. Hinton believed that creating such a group would spur innovation in AI and maybe even change the way the rest of the world treated this kind of work.[26] No matter whether deep learning really mimics organic intelligence, he was right! Instead of relying on OHL nets, the group launched deep architectures, namely feedforward nets with many layers, in real-word applications. In 2011, Andrew Ng, a researcher at Stanford, launched a deep learning project at Google that seems to use currently this technology to recognize voice commands on Android phones and tag images on the GooglePlus social network. Baidu opened AI labs in China and Silicon Valley, and Microsoft incorporated deep learning techniques into its own voice recognition research. Facebook also opened an AI research group, headed by Yan Le Cun. Meanwhile, the academic interest was exploding. A fundamental contribution came from Yoshua Bengio. As early as 2005, he circulated a technical report (the report was published in an extensive form years later [Y. Bengio, *Found. Trends Mach. Learn.* **2** (2009), 1–127]) that went well beyond the peaceful interlude of OHL nets! He focused on representational issues, looking from a corner that had been mostly neglected: The expressiveness of the developed representation. A wide coverage of the fundamental idea of deep learning also appeared in a book chapter *Scaling learning algorithms towards AI* in *Large Scale Kernel Machines, NIPS'12* (MIT Press, 2007) by Bengio and LeCun. Other two seminal papers [G.E. Hinton, S. Osindero, Y. Teh, *Neural Comput.* **18** (2006), 1527–1554 and Y. Bengio, P. Lamblin, D. Popovici, H. Larochelle, *Proc. of the 19th Int. Conf. on Neural Inf. Process. Sys.* (MIT Press, 2006), 153–160] provided clear evidence on how to efficiently train deep belief networks. At first glance, nothing seemed to be really new, since depth/size tradeoffs are ordinary stuff in switching logic, and it was quite obvious that generalization to new examples was what really mattered. Yet, for about 20 years, the OHL credo had not been really challenged. The excitement on deep learning exploded, thus leading to an unbelievable resurgence of the interest in neural nets. Most importantly, deep learning approaches early became the unified approach for facing hot and fundamental AI topics, ranging from speech and language understanding to vision. Why is the impact in the AI community and related field so strong? According to [Y. Bengio, *Found.*

---

[26] http://www.wired.com/2014/01/geoffrey-hinton-deep-learning.

Trends Mach. Learn. **2** (2009), 1–127], representational issues play a crucial role. It was clearly a major step to replace the shallow architectures of kernel machines. The deep structures succeeded in many real-world problems because of their compositional structures. Moreover, the easy access to GPU along with the access to huge training sets made it possible to crown a dream cultivated since the end of the 1980s! Fundamental novelties on the exploitation of unsupervised and semisupervised learning, along with crucial "tricks," like the adoption of the rectifier nonlinearity, and novel regularization methods, like dropout [N. Srivastava, G. Hinton, A. Krizhevsky, I. Sutskever, R. Salakhutdinov, *J. Mach. Learn. Res.* **15** (2014), 1929–1958], did the rest.

When looking carefully at the foundations of backpropagation, it becomes early clear that it has an intertwined history with reverse mode automatic differentiation. Ideas underlying automatic differentiation date back to the 1950s [see J.F. Nolan, *Analytical Differentiation on a Digital Computer*, Ph.D. thesis, Massachusetts Institute of Technology, 1953 and L.M. Beda, L.N. Korolev, N.V. Sukkikh, T.S. Frolova, *Tech. report, Inst. for Precise Mechanics and Computation Techn.* (1959)]. When casting reverse mode automatic differentiation in the continuous-time setting, we end up with the Pontryagin Maximum principle, which is well-known in the control theory community. Speelpenning [*Compiling Fast Partial Derivatives of Functions Given by Algorithms* Ph.D. thesis, University of Illinois at Urbana-Champaign, 1980] provided the first linguistic formalization and concrete implementation of reverse mode automatic differentiation. An excellent survey on the automatic differentiation in machine learning is the paper by A.G. Baydin, B.A. Pearlmutter, A. Andreyevich Radul arXiv:1502.05767 (2015).

Section 5.6 One of the simplest and most effective advances in the context of deep learning is due to the introduction of residual blocks [K. He, X. Zhang, S. Ren, J. Sun. Deep Residual Learning for Image Recognition. *Proc. of the IEEE Conf. on Computer Vision and Pattern Recognition* (2016), 770–778], that are the basic tools exploited to to setup residual networks. Two of the most widely known residual networks are ResNet-152 and ResNet-50 (152 layers and 50 layers, respectively — 18, 34, 101 layer versions exist too), used in image classification and implementing residual blocks based on f^l's that perform convolutions. The ResNet model is the one that won the popular ImageNet Large Scale Visual Recognition Challenge (ILSVRC) 2015, it still represents a state-of-the art model for image classification purposes, and it is exploited as basic feature extractor (backbone net) in several other networks for tackling computer vision problems. Exploiting the same notation of Section 5.6.1, in ResNet-152 and ResNet-50 the function f^l consists of three convolutional layers interleaved with ReLU activations, and σ of Eq. (2) is a ReLU function as well. The three convolutional layers that implement f^l specifically consist of a layer with receptive fields with a spatial coverage of 1×1 pixels, another layer in which such a coverage is 3×3, and a final 1×1 convolutional layer, respectively. In some parts of the network, o^{l-1} is also transformed by an additional 1×1 convolutional layer in order to match the number of feature maps returned by f^l and to perform the sum inside the residual block. For the specific sizes of the convolutions and full details on the network structure, refer to the work of He et al. that originally introduced these architectures [K. He, X. Zhang, S. Ren, J. Sun. Deep Residual Learning for Image Recognition. *Proc. of the IEEE Conf. on Computer Vision and Pattern Recognition* (2016), 770–778]. In Section 5.6.1, we also discussed the case of identity-mapping-based models, where $\sigma(x) = x$, that were proposed in the paper of K. He, X. Zhang, S. Ren, J. Sun, on Identity Mappings in Deep Residual Networks [*Proc. of the European Conf. on Computer Vision* (2016), 630–645], where further details can be found.

The activity in the area of adversarial machine learning traces back to 2004, even if it has become more popular in the machine learning community nearly a decade later, due to specific studies in the context of deep networks for computer vision [see, e.g., B. Biggio, F. Roli. Wild patterns: Ten years after the rise of adversarial machine learning. *Pattern Recognition* **84** (2018), 317-331]. The sensitivity of neural networks to small-specific perturbations of the input (Fig. 5.19) was highlighted in image classification, and remarked in papers focusing on learning representations with deep models, such as [C. Szegedy, W. Zaremba, I. Sutskever, J. Bruna, D. Erhan, I. Goodfellow, R. Fergus, Intriguing properties of neural networks. *Proc. of the Int. Conf. on Learning Representations* (2014)]. Neural networks are actually black-boxes, that return predictions based on a set of "criteria" that the network autonomously develops while learning from the given training data. Of course, the way such data is distributed plays a major role in the learning process. While we do not have direct control on what the network will actually learn, it sounds natural to assume that whenever a test pattern is "far" from the observed distribution of the training data, then it might be a malicious example or simply something that is so-different from the training data that it is hard to return predictions with a large confidence. The authors of [S. Melacci, G. Ciravegna, A. Sotgiu, A. Demontis, B. Biggio, M. Gori, F. Roli. Domain Knowledge Alleviates Adversarial Attacks in Multi-Label Classifiers. *IEEE Trans. on Pattern Analysis and Machine Intelligence* (2021)] implemented this principle to build a defence procedure that indirectly evaluates if an input pattern looks like an out-of-distribution example. This is done by exploiting domain knowledge in multi-label classification problems.

Generative Adversarial Networks (GANs) were proposed in the popular paper of Ian Goodfellow [I. Goodfellow, J. Pouget-Abadie, M. Mirza, B. Xu, D. Warde-Farley, S. Ozair, A. Courville, Y. Bengio. Generative Adversarial Nets. *Advances in Neural Information Processing Systems* **27** (2014)], that was a source of inspiration for several variants of GANs. As a matter of fact, the original GAN model has been extended in a large number of different ways, introducing novel functionalities or revisiting the learning criteria to better cope with high-resolution images. We mention the case of those GANs that introduce a cyclic dependence between generations on two different spaces [J.Y. Zhu, T. Park, P. Isola, A.A. Efros. Unpaired image-to-image translation using cycle-consistent adversarial networks. *Proc. of the IEEE Int. Conf. on Computer Vision* (2017), 2223–2232]. This idea is at the basis of image-to-image translations, in which, for example, the image of a horse is transformed in the one of a zebra, preserving several features of the source picture, or when transforming a human face from male to female and vice versa.

The concept of attention in deep neural networks was popularized by several works in machine translation and automatic image captioning. For the former task and the way attention is used, we mention the work of [D. Bahdanau, K. Cho, Y. Bengio. Neural Machine Translation by Jointly Learning to Align and Translate. *Proc. of the Int. Conf. on Learning Representations* (2015)]. In the case of the latter task, the reader can refer to [K. Xu, J. Ba, R. Kiros, K. Cho, A. Courville, R. Salakhudinov, R. Zemel, Y. Bengio. Neural Image Caption Generation with Visual Attention. *Proc. of the Int. Conf. on Machine Learning* **37** (2015) 2048-2057] for more details. By further developing the ideas behind attention models, Transformers were proposed in [A. Vaswani, N. Shazeer, N. Parmar, J. Uszkoreit, L. Jones, A.N. Gomez, L. Kaiser, I. Polosukhin. Attention is All you Need. *Advances in Neural Information Processing Systems* **30** (2017)]. Transformers were originally studied in processing text corpora, in particular in the context of machine translation, where the input of the model is a sentence in a source language and the expected output is the sentence translated into the target language. However, they did not only become popular due to the outstanding results they achieved, but also due to the way they handle the input data, that is based on an easily parallelizable procedure. Their popularity strongly increased due to the outstanding results achieved by Transformer-based language models, such as GPT and BERT. In the case of GPT, including the popular variant GPT-3 [T. Brown, et al. Language models are few-shot learners. *Advances in Neural Information Processing Systems* **33** (2022) 1877-1901] the Transformer decoder is exploited in same way we described in Section 5.6.4, thus we have an instance of classic language modeling in which the machine learns to predict the next word. Trained models turn out to have an outstanding capability of generating text (almost) from scratch. The case of BERT is pretty different, since it is built on the Transformer encoder [J. Devlin, M.W. Chang, K. Lee, K. Toutanova. Bert: Pre-training of deep bidirectional transformers for language understanding. *Proc. of NAACL-HLT* (2019) 4171-4186.], and it is then trained by masking words in the input sequence. GPT-3, BERT, and other language models include a huge number of learnable parameters, that, for example, is around 175 billions in the case of GPT-3. These models are trained on several GPUs, using very large collections of text. In the case of vision we mention Vision Transformers (ViTs) [A. Dosovitskiy, et al. An image is worth 16x16 words: Transformers for image recognition at scale. *Proc. of Int. Conf. on Learning Representations* (2021)]. These Transformers are based on the Transformer encoder on top of a patch-based representation of the input image. Moreover, the layer normalization procedure is anticipated before each of the elements inside the encoding blocks, instead of being applied after having considered the residual connections.

Section 5.7 The adoption of the definition of the derivative for finite difference approximation commits both cardinal sins of numerical analysis: "Thou shalt not add small numbers to big numbers," and "thou shalt not subtract numbers which are approximately equal" [A.G. Baydin, B.A. Pearlmutter, A. Andreyevich Radul arXiv:1502.05767 (2015)]. The optimal bound $\Theta(m)$ to compute the gradient by backpropagation has been playing a crucial role in concrete applications. While in theoretical computer science the complexity reduction from $O(m^2)$ to $O(m)$ is typically regarded as a negligible issue, since both cases are associated with the class of polynomial algorithms, the need to work with millions of weights makes a significant difference! The importance of cutting polynomial bounds is rarely an important issue, but there are a few cases which share this requirement. A noticeable example is the `PageRank` algorithm of the scoring of Web pages, for which the same optimal bound $\Theta(|\mathcal{A}|)$ given for backpropagation can be exhibited [M. Bianchini, M. Gori, F. Scarselli, *ACM Trans. Internet Technol.* **5** (2005), 92–128]. As a matter of fact, nowadays we can concretely train neural nets with millions of weights. A classic numerical algorithm that doesn't exploit the power of the DAG structure, in case of about 33 million weights, would be doomed to compute a number of floating-point operations on the order of one trillion!

The efficiency in the computation of the gradient is clearly one of the secrets of the success of learning with feedforward nets, especially in case of huge deep structures. However, there are a number of other important issues that must be neglected. First, especially when choosing some nonlinear functions, like the hyperbolic tangent and the logistic sigmoid, we must carefully avoid the premature saturation problem. Strategies for the initialization have been proposed in a number of papers [see, e.g., Y. LeCun, L. Bottou, G.B. Orr, K. Müller, *Efficient backprop*, in *Neural Networks: Tricks of the Trade. Lecture Notes in Computer Science, vol 7700* (Springer, 2012) and Y. Lee, S. Oh, M.W. Kim, *Neural Netw.* **6** (1993), 719–728].

Since the dawn of the connectionism, the potential presence of suboptimal local minima has always been considered as one of the major problems. This is not surprising. In a sense, the gradient heuristics is a way of exploring the solution space. The corresponding search problem can inherently be difficult! There are in fact intriguing connections with classic algorithms in problem solving, where the well know suboptimality deriving from hill-climbing is faced by algorithms like `A*` [N.J. Nilsson, *Principles of Artificial Intelligence*, (Tioga, 1980)], and `IDA*` [R.E. Korf, *Artif. Intell.* **27** (1985), 97–109]. The suboptimality of the heuristics is always regarded as a potential warning, which clearly propagates to machine learning.

M.L. Brady et al. early began noticing that it could also be the case that "backpropagation fails where perceptrons succeed," which appeared to be a major theoretical problem [M. Brady, R. Raghavan, J. Slawny, *IEEE Trans. Circuits Syst.* **36** (1989), 665–674]. Nearly at the same time Pierre Baldi and Kurt Hornik published a nice paper [*Neural Netw.* **2** (1989), 53–58], in which they prove that it is in fact possible to "learn from examples without local minima." The main statement was somewhat on the opposite side with respect to the previous result. Interestingly, they were formally consistent, the only difference being that Baldi & Hornik paper stated results under the assumption of linear computational units. On the other hand, it was also early clear that the negative result published in [M. Brady, R. Raghavan, J. Slawny, *IEEE Trans. Circuits Syst.* **36** (1989), 665–674] was derived under a sort of spurious configuration. The issue was put forward in a couple of elegant papers by E.D. Sontag and H.J. Sussman who also proposed using LMS-threshold loss functions to get rid of the problem raised in [M. Brady, R. Raghavan, J. Slawny, *IEEE Trans. Circuits Syst.* **36** (1989), 665–674, E.D. Sontag, H.J. Sussman, *Complex Syst.* **3** (1989), 91–106 and E.D. Sontag, H.J. Sussman, *Int. Joint Conf. on Neural Netw.* (IEEE Press, 1989), 639–642]. It became clear the local minima identified in [M. Brady, R. Raghavan, J. Slawny, *IEEE Trans. Circuits Syst.* **36** (1989), 665–674] were spurious, since they depend on the "wrong" joint choice of the loss function and the training set. However, the presence of structural local minima was pointed early out by one of the authors in his PhD thesis [M. Gori, *Apprendimento con Supervisione in Reti Neuronali*, PhD thesis, Università degli Studi di Bologna, 1990] — in Italian — and, later on, in [M. Gori, A. Tesi, *Parallel Architectures and Neural Netw.* (1990) and J.M. McInerny, K.G. Haines, S. Biafore, R. Hecht-Nielsen, *Int. Joint Conf. on Neural Netw.* (1989) 627–627]. Related discussions on the topic can be found in L.F.A. Wessels, E. Barnad, *IEEE Trans. Neural Netw.* **3** (1992), 899–905, K. Matsuoka, J. Yi, *Int. Joint Conf. on Neural Netw.* (1991), 1117–1122, J. Barhen, J.W. Burdick, B.C. Cetin, *IEEE Int. Conf. on Neural Netw.* (1993), 836–842, J. Chao, W. Ratanasuwan, S. Tsujii, *Int. Joint Conf. on Neural Netw.* (1991), 1079–1083, D. Gorse, A. Shepherd, J.G. Taylor, *Avoiding Local Minima by a Classical Range Expansion Algorithm* in *ICANN'94*, L.G.C. Hamey, *IEEE Trans. Neural Netw.* **5** (1994), 844–844 and K. Fukumizu, S. Amari, *Neural Netw.* **13** (2000), 317–327. In addition to the analysis on local minima free conditions given in P. Baldi, K. Hornik, *Neural Netw.* **2** (1989), 53–58, it was pointed out that multilayer neural networks with hidden layers don't generate suboptimal local minima in the case of linearly separable patterns [see M. Gori, A. Tesi, *IEEE Trans. Pattern Anal. Mach. Intell.* **14** (1992) 76–86]. A detailed analysis on this condition is given in P. Frasconi, M. Gori, A. Tesi, *IEEE Int. Conf. on Neural Netw.* (1993), 1818–1822. A related analysis was carried out in M. Bianchini, P. Frasconi, M. Gori, *IEEE Trans. Neural Netw.* **6** (1995), 749–756 for radial basis functions. The role of hidden units in the presence of suboptimal local minima in case of OHL nets was studied by Poston and Yu in T. Poston, C. Lee, Y. Choie, Y. Kwon, *Int. Joint Conf. on Neural Netw.* (1991), 173–176, X.H. Yu, *IEEE Trans. Neural Netw.* **3** (1992) 1019–1020 and X.H. Yu and G.A. Chen, *IEEE Trans. on Neural Netw.* **6** (1995), 1300–1303. They claimed that local minima disappear when the number of hidden units is greater or equal to the number of training sets. The result they discovered didn't have such a great impact, since in real-world experiments people had been using architectures with much less hidden units. However, recent experiments have been carried out also with really huge architectures. Interestingly, many people have developed the feeling that the issue of local minima is somewhat negligible in practice. They're basically right! As a matter of fact, very big neural nets don't suffer from this problem. Moreover, it's worth mentioning that if we add a quadratic regularization term, which is associated with the classic weight decay updating learning rule, it is unlikely that adding this convex term to the risk function will generate suboptimal minima. Notice that this statement formally holds in case the risk is convex, while in this case we have a local minima free error function. A couple of independent contributions, Y.N. Dauphin, R. Pascanu, C. Gulcehre, K. Cho, S. Ganguli, Y.

Bengio, *Proc. of the 27th Int. Conf. on Neural Inf. Process. Sys.* (2014), 2933–2941 and A. Choromanska, M. Henaff, M. Mathieu, G.B. Arous, Y. LeCun, *Proc. of the Eighteenth Int, Conf. on Artif. Intell. and Stat.* (2015) 192–204, provided additional arguments that as the size of the net increases, the probability of bad local minima becomes very small also in case of deep networks. Most critical points along the training path turn out to be saddle points and the local minima will congregate at a low value of the cost, not much higher than the global minimum. The case of online learning is clearly different, since the actual learning process doesn't really follow the true gradient like in batch mode. A companion of Rosenblatt's PC (perceptron convergence) theorem for feedforward networks stating that pattern mode Backprop converges to an optimal solution for linearly separable patterns was proven in M. Gori, M. Maggini, *IEEE Trans. Neural Netw.* **7** (1996), 251–253. Surveys on the overall topic of optimal convergence of learning algorithms can be found in M. Bianchini, M. Gori, *Neurocomputing* **13** (1996), 313–346, P. Frasconi, M. Gori, A. Tesi, *Progress in Neural Networks* (Ablex Publishing, 1993), 205–242, M. Bianchini, P. Frasconi, M. Gori, M. Maggini, *Neural Netw. Sys., Techn. and Appl.* (1998) 1–51.

Learning with constraints

*In strange contrast to the hardly tolerable constraint
and nameless invisible domineerings of the captain's table,
was the entire care-free license and ease,
the almost frantic democracy
of those inferior fellows the harpooneers*
Herman Melville, *Moby Dick* (1851)

*We cannot arrest the Professor,
because he has done no crime,
nor can we place him under constraint,
for he cannot be proved to be mad.
No action is as yet possible.*
Sherlock Holmes, in *The Adventure of the Creeping Man*, (1923)

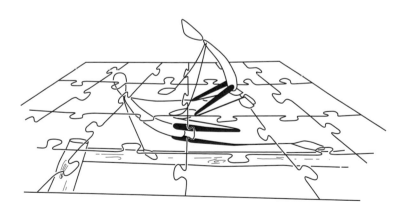

THIS CHAPTER provides a unified view of learning and inference in structured environments that are formally expressed as constraints that involve both data and tasks. A preliminary discussion has been put forward in Section 1.1.5, where we began proposing an abstract interpretation of the ordinary notion of constraint that characterizes human-based learning, reasoning, and decision processes. Here we make an effort to formalize those processes and explore the corresponding computational aspects. A first fundamental remark for the formulation of a sound theory is that most interesting real-world problems correspond with learning environments that are heavily structured, a feature that has been mostly neglected in the previous chapters on linear and kernel machines, as well as on deep networks. So far we have been mostly concerned with machine learning models where the agent takes a decision on patterns represented by $x \in \mathbb{R}^d$, whereas we have mostly neglected the issue of constructing appropriate representations from the environmental information $e \in E$. The discussion in Section 1.1.5 has already stimulated the need of processing information organized as lists, trees, and graphs. Interestingly, in this chapter, it is shown that computational models, like recurrent neural networks and graph neural networks can also be regarded as a way for expressing appropriate constraints on environmental data by means of diffusion processes. In these cases the distinguishing feature of the computational model is that the focus is on uniform diffusion processes, whereas one can think of constraints that involve both data and tasks in a more general way. Basically, different vertexes of a graph that model the environment can be involved in different relations, thus giving rise to a different treatment. As a result, this yields richer computational mechanisms that involve the meaning attached to the different relations.

The discussion in Section 1.1.5 has put forward a number of rich and expressive constraints that operate on the tasks. A constraint can yield restrictions on tasks that process the information in the nodes. As an example, the supervision on one pattern (node of the graph) is a pointwise constraint that involves the meaning of the associated task. Like other constraints, it propagates information through the graphical structure used as a data model of the environment. We use the notion of *semantic-based constraint* to describe restrictions that operate directly on the output returned by the functions that model the environmental tasks. The underlying idea is that those restrictions somehow express the knowledge on the environment and convey the meaning of the tasks involved. Throughout this book, learning machines working with constraints are referred to as *constraint machines*. It's worth mentioning that while some tasks are modeled by functions which return semantics directly from inference on the perceptual space, others operate on higher conceptual levels. In order to grasp this additional feature of constraint-based environments, we use the notion of *individual* as an extension of the traditional term *pattern* that is more popular in machine learning. We need to introduce this additional concept, since there are interesting cases where we want to draw conclusions on the basis of relational structures only. An *individual* is generally regarded as a pair (identity-label, pattern), where the feature-based part — the pattern — might be missing. For example, consider the text classification problem in Section 1.1.5 and suppose that the agent is operating on the books of a library, where one is also interested in associating the subjects with the corresponding location in the building. A good strategy might be that of allocating the books in appropriate areas according to their topics. Hence one could benefit from introducing a function $f_{sa}: Y^2 \to \{T, F\}$, which establishes whether two books must be located in the same area. For example, $\forall x_1, x_2, [f_{sa}(f_{na}(x_1), f_{ml}(x_2))] = 1$ states that documents on numerical analysis and machine learning must be located in the same place. This constraint has a relational flavor and, moreover, operates at a higher level with respect to the functions involved in text classification. When interpreting x_1 and x_2 as individuals, instead of patterns, we are clearly enriching the range of inferential processes. Basically, x_1, x_2 can contain semantic information on the books as well as the code that can be used also for their allocation. Like in this example, the coherence constraint shown in Section 1.1.5 indicates that, in general, the enforcement jointly takes place on information that is spread in the graph — remember the sign agreement concerning an environmental input $e \in E$ that is represented by mean of two different "pattern views."

According to the unified learning paradigm used throughout this book, once the interactions of the agent with the environment have been properly modeled in terms of a collection of constraints, one can think of *learning as the process which minimizes a parsimony index, while satisfying the given environmental constraints*. This very much reminds us of the approach that gives rise to kernel machines, where the constraints reduce to the presentation of a list of supervised examples — the training set. Once the minimization process has been carried out, just like in neural networks and kernel machines, the resulting (learned) function is used for the inference on any individual of the environment, which is characterized by a given feature vector $x \in \mathbb{R}^d$. The distinguishing property of such inferential processes is that they are very efficient, since they rely on the direct computation of the learned function — think of handwritten char recognition by neural networks. It will be shown that once we adopt the constrained-based representation of the environment, other inferential processes naturally arise that are closely related to what is typically modeled by logic-based descriptions. Basically, whenever we consider relational environments, *the uniform principle of constraint satisfaction can also be used for activating inferential processes on individuals that posses no features*. Surprisingly enough, there's no need to distinguish between individuals represented by features with respect to individual that are only characterized by their identifier: In both cases we can use the same computational framework.

The systematic representation given of the environment enables the adoption of machine learning concepts that have been already discussed when dealing with deep learning and kernel machines. However, something really new arises that strongly motivates the adoption of life-long learning/inference computational schemes. Complex environments are likely associated with a huge number of constraints, with different degree of abstraction. Should we uniformly learn all together or should we define appropriate focus of attention schemes to filter out what is too complex at a certain stage of agent development? How can an agent pose "smart questions" to quickly acquire a concept? Throughout the chapter we give insights and methods to capitalize from the distinction of specific constraints — think of supervised pairs of a certain function — with respect to high level constraints that cross the border of logic.

6.1 Constraint machines

In this section, we provide an in-depth discussion on most important environmental interactions of the agent, which can also acquire abstract descriptions on the available prior knowledge. Based on the parsimony principle, which is translated according to the mathematical formalisms of Chapter 4 on kernel machines, a variational analysis is presented, which discloses the functional structure of an agent while satisfying the constraints. This gives rise to *support constraint machines*, a computational model that closely resembles and generalizes support vector machines.

6.1.1 Walking through learning and inference

We show that in many interesting cases constraints come up with a remarkably different structure that plays a fundamental role in the functional representation of the agent. Before offering a unified view of semantic-based constraints, we consider a few examples to appreciate the different mathematical structure that, as it will be shown, has quite an important impact on the corresponding learning algorithms. Let's start with the translation of the principle of coherent decision. This is very well exemplified by properly extending one of the learning tasks described in Chapter 3, namely the one which deals with the prediction of the average weight of adults given their height. It will be used as a running example for drawing some conclusion on the structure of the involved constraints. We enrich the learning environment in such a way that also the age is involved. We can introduce the following learning tasks:

$$f_{\omega h} : W \to H, \quad h \to \omega(h),$$
$$f_{ah} : W \to A, \quad h \to a(h),$$
$$f_{\omega a} : A \to W, \quad a \to \omega(a),$$

where $f_{\omega h}$ estimates the weight from the height, f_{ah} estimates the age from the height, and $f_{\omega a}$ estimates the weight from the age. Suppose we are simply using linear functions for prediction on the basis of a collection of supervised examples. Then we can learn the weights of the three functions by independent LMS. However, in so doing, we are clearly missing the fundamental information that the predictions are intertwined with each other, since the following constraint holds true:

$$f_{\omega h}(h) = f_{\omega a} \circ f_{ah}(h).$$

This functional equation is imposing the circulation of coherence. Since the functions are linear, this constraint can be converted to $w_{\omega h}h + b_{\omega h} = w_{\omega a}w_{ah}h + (w_{\omega a}b_{ah} + b_{\omega a})$. The equivalence $\forall h \in \mathbb{R}_+$ yields

$$
\begin{aligned}
w_{\omega a}w_{ah} - w_{\omega h} &= 0, \\
w_{\omega a}b_{ah} + b_{\omega a} - b_{\omega h} &= 0.
\end{aligned}
\tag{1}
$$

Notice that we have ended-up with constraints that hold regardless of h, a, and ω, that is, they are independent of the training set. Basically, the circulation of coherence requires satisfying a set of equations that only depend on the weights of the machines. This suggests a learning process where f_{ah} $f_{\omega a}$, and $f_{\omega h}$ are determined by tracking the supervisions, as well as by satisfying the above pair of coherence constraints (1). The precise formulation is discussed in Exercise 1. Notice that in this learning problem, the coherence and pointwise constraints coming from the supervised data are fundamentally different in their nature. While we typically tolerate errors with respect to the supervised pairs of the training set, it makes sense to interpret the consistency condition (1) as a truly hard constraint. The difference seems to be rooted in the different structure of the constraints. While supervised learning directly involves the pairs of the training set, the consistency constraints make a general statement on the environment. In particular, as already stated, we can see from Eq. (1) that there is no direct involvement of the input and output data, but only of the model parameters. This is not restricted to this example! Whenever consistency leads to a general condition, which is independent of the specific instances of data, its translation by a corresponding hard constraint makes sense.

A broader investigation on this prediction problem leads to discovering other nice aspects connected with its qualitative interpretation in ordinary life, which is typically based on common sense reasoning. As an example, it is quite obvious to notice that tall and heavy persons are not children. This could be written as[1]

$$
\sigma(h - H)\sigma(f_{\omega h}(h) - W)(1 - \sigma(f_{ah}(h)) - A) = 0,
\tag{2}
$$

where σ is the logistic function. Here, we need a fuzzy definition of tall, heavy, and children. (Interestingly, we will see the rest of the chapter, these parameters can also be learned.) We could choose $H = 180$ cm, $W = 100$ kg, and $A = 10$ years old. Why is Eq. (2) an appropriate formalism for expressing the above knowledge granule? We can promptly check that the concepts of tall and heavy people are represented by $\sigma(h - H) \simeq 1$ and $\sigma(\omega - W) \simeq 1$, respectively, while the concept of non-child is represented by $\sigma(a - A) \simeq 1$. Eq. (2) arises when considering the estimates $\omega = f_{\omega h}(h)$ and $a = f_{ah}(h)$. Under the assumption of linear prediction, the above constraint can be converted into the weight space by imposing that $\forall h \in \mathbb{R}^+$ the following equation

---

[1] In order to better express the related concepts one can replace $\sigma(h - H)$, $\sigma(\omega - W)$ and $\sigma(a - h)$ with $\sigma(\alpha_h(h - H))$, $\sigma(\alpha_\omega(\omega - W))$ and $\sigma(\alpha_a(a - h))$ by appropriate choices of α_h, α_ω and α_a.

holds true:

$$\sigma(h - H)\sigma(w_{\omega h}h + b_{\omega h} - W)(1 - \sigma(w_{ah}h + b_{ah} - A)) = 0. \qquad (3)$$

Like Eq. (1), this is a constraint in the parameters of the model. However, notice that it also directly involves the environmental variable h. Like for the pointwise constraints coming from supervised learning, which also involve the environmental variables, partial violation of this constraint is inherently connected with its deep meaning.

To sum up, constraint (1) and (3) are quite different. Once we enforce the quantifier on h, the structure of constraints (1) becomes independent of h itself! The same does not hold for (3), simply because the nonlinearity does not allow us such a direct reduction. How can we incorporate constraints where the universal quantifier does not easily lead to data-independent equations like (1)? We can basically use the same idea of supervised learning to measure the satisfaction of the constraint by an appropriate risk function — penalty in the terminology of optimization theory. Let $p(h)$ be the probability distribution of the height. Then we can associate the constraint with the minimization of the associated penalty function

$$V = \int_H p(h)\sigma(h - H)\sigma(w_{\omega h}h + b_{\omega h} - W)(1 - \sigma(w_{ah}h + b_{ah} - A)) \, dh$$

$$\propto \sum_{h \in H^\sharp} \sigma(h - H)\sigma(w_{\omega h}h + b_{\omega h} - W)(1 - \sigma(w_{ah}h + b_{ah} - A)).$$

Here, the sum is an approximate translation of the continuous penalty over the finite set $H^\sharp$. Notice that since the integrand is positive, whenever $V(h) = 0$ we restore the hard satisfaction of constraints (3). However, as already pointed out, this a typical method for a soft-enforcing of the constraints.

We can naturally extend this simple computational model by considering learning tasks with two independent variables. For instance, we can involve the new learning task $\omega = f_{\omega|(h,a)}(h, a)$ that can be the subject of constraints of the already discussed types (see Exercise 2).

This example highlights other nice aspects on semantic-based constraints. Notice that the sampling $H^\sharp$ of H makes it possible to acquire the probability density directly from the given data. However, suppose one wants to estimate p on the basis of a collection of samples $\{(h_\kappa, y_\kappa) : \kappa = 1, \ell\}$, where y_κ is the probability density in h_κ. The problem of learning p is that of fitting the above data, while respecting the probabilistic normalization. It can be conveniently framed as the following minimization problem:

$$p^* \in \arg\min_{p \in F} \left(\langle (p(h) - y_\kappa)^2, \delta(h - h_\kappa) \rangle + \mu \langle Pp, Pp \rangle \right), \qquad (4)$$

with

$$\int_H p(h) \, dh = 1, \qquad (5)$$

where $\langle(p(h)-y_\kappa)^2, \delta(h-h_\kappa)\rangle = \int_{h\in H}(p(h_\kappa)-y_\kappa)^2\delta(h-h_\kappa)\,dh$ and P is a differential operator, so that the term $\langle Pp, Pp\rangle$ implements the parsimony principle. Notice that, like the coherence constraint (1), this probabilistic normalization condition can be regarded as one to be hardly ever satisfied! However, unlike (1), which must be enforced for any point of the space, this probabilistic constraint exhibits an inherent global structure. As we will see in Section 6.1.3, their impact in learning is remarkably different. In addition, while constraint (5) acts in a variational optimization problem, constraint (1) requires working at finite dimension.

Now we want to better capture the essence of the functional expression of knowledge granules, a topic which will be deeply investigated in Section 6.2. The common sense reasoning stated by Eq. (2) is centered around the asymmetric structure of the premises and the conclusions in logic statements. For example, assume that $f_i\colon X \to [0,1]$, $i=1,2$ and bear in mind the association with Boolean-like decisions, so $f_i(x) \simeq 0$ and $f_i(x) \simeq 1$ mean that the corresponding concept is false and true, respectively. Any constraint of the form

$$\forall x \in X, \qquad f_1(x)(1-f_2(x)) = 0 \tag{6}$$

shares the underlying inferential principle behind Eq. (2). If $f_1(x)=1$ then the satisfaction of the constraint implies $f_2(x)=1$. However, the converse doesn't hold true, since if $f_2(x)=1$ then any value of $f_1(x)$ is admitted. We can immediately see that this is in fact a property of the images of the functions, so we can generalize to functions operating on different domains. Clearly, in case $f_1\colon X_1 \to \mathbb{R}$ and $f_2\colon X_2 \to \mathbb{R}$, with $X_1 \neq X_2$, the constraint $f_1(x_1)(1-f_2(x_2)) = 0$ enjoys the same properties. Now suppose that also $f_2(x_2)(1-f_1(x_1)) = 0$ holds true. Of course, if we sum them up we get

$$f_1(x_1) + f_2(x_2) - 2f_1(x_1)f_2(x_2) = 0, \tag{7}$$

which has gained a symmetric structure. We can see that it is satisfied only if $f_1(x_1) = f_2(x_2) = 0$ or $f_1(x_1) = f_2(x_2) = 1$ (see Exercise 3). Basically, when we interpret the values of the functions in the Boolean domain, this constraint translates the concept of logic equivalence. It's quite easy to realize that this is not the only way of expressing equivalence and, likewise, the implication stated by Eq. (6) can be expressed in different ways — although not all of them properly express the deep meaning of decision-like equality (see Exercise 4).

Now, let us jump to another example, which is substantially different in its deep nature. We consider the extreme case of constraints based on symbols, namely the case in which there is no feature-based representation of the environment. So far the learning tasks have been regarded as real-valued functions over a certain vector space X, whereas here we consider the case of domains, where no features are given. For example, the prediction of the person weight is a regression problem which makes sense because of the information coming from the height and age of the person, and it is quite clear that, once we miss a similar feature-based representation, we get in trouble concerning prediction. However, this is not the case in all intelligent tasks which present a strong logical structure. Let us consider the problem of accommodating n queens on a given chessboard in nonattacking positions. This can be regarded as a problem with constraints on the task, since the decision must be coherent with the permitted moves of the queens. Interestingly, apart from the rules of the game, there is no feature-based information available from the environment. While in the previous case the inference can be done by knowing the height of a person, in this case the inferential process acts with a dummy input representation. Whenever we are given a queen allocation on the chessboard, we can draw conclusions on whether or not it respects the rule of the game, but we must construct the configuration during the inference, whereas in the previous examples the inference emerges from the given feature-based representation of the input.

Now $\forall i, j \in \mathbb{N}_n := \{1, 2, \ldots, n\}$, let $q_{i,j} \in [0, 1]$ be real numbers, which are expected to return the decision on whether or not the corresponding position (i, j) is occupied by a queen. Basically, we would like our intelligent agent to return $q_{i,j} = 1$ in those locations (i, j) where the queens can be accommodated, whereas $q_{i,j} = 0$ in attacking positions. We accept fuzzy decisions, thought we would very much like our agent return a crisp decision. This means that the ideal solution is one where $q_{i,j} \in \{0, 1\}$. Now the nonattacking constraint conditions for rows and columns can be written as

$$
\begin{aligned}
(i) \quad & \forall i \in \mathbb{N}_n : \quad \sum_{j \in \mathbb{N}_n} q_{i,j} = 1, \\
(ii) \quad & \forall j \in \mathbb{N}_n : \quad \sum_{i \in \mathbb{N}_n} q_{i,j} = 1.
\end{aligned}
\tag{8}
$$

The conditions for the diagonals are kind of tricky! We must impose nonattacking conditions on both the forward and backward diagonals, and we must bear in mind that all conditions on positions on the same diagonals are equivalent. Hence only one of them must be expressed as the representative of the diagonal. All other conditions are redundant, so it makes sense to get rid of them in the constraint-based formulation. Let's start with forward diagonals: If we locate a queen on position $(1, j)$, $j = 1, \ldots, n$, we can check upper diagonal configurations, whereas if we locate on $(i, 1)$, $i = 1, \ldots, n$, we can check lower diagonal configurations. Notice that it suffices to check nonattacking configurations from these positions, since all others turn out to be redundant. Hence we impose:

$$
\begin{aligned}
(i) \quad & \forall j = 1, \ldots, n : \quad q_{1,j} + \sum_{k=1}^{n-j} q_{1+k, j+k} \leq 1, \\
(ii) \quad & \forall i = 2, \ldots, n : \quad q_{i,1} + \sum_{k=1}^{n-i} q_{i+k, 1+k} \leq 1.
\end{aligned}
\tag{9}
$$

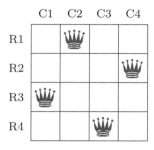

FIGURE 6.1

The two solution of the 4-queens problem: One *fundamental solution* only! There are 25 variables and 24 constraints. The left-side solution can be obtained by a π rotation of the right-side solution.

Notice that the diagonal is included among the forward diagonals. Likewise for the backward diagonals:

$$(i) \quad \forall j = 1, \ldots, n : q_{1,j} + \sum_{k=1}^{n-j} q_{1+k, j-k} \leq 1,$$

$$(ii) \quad \forall j = 2 \ldots, n : q_{i,n} + \sum_{k=1}^{n-i} q_{i+k, n-k} \leq 1. \tag{10}$$

We have $6n$ constraints with n^2 variables. As n increases, the constraints can be satisfied by an increasing number of configurations. While the row and column nonattacking conditions are translated by bilateral constraints (equalities), the corresponding conditions for the diagonal are expressed by unilateral constraints (inequalities). The reason for this difference can promptly be gained when looking at the configurations of Fig. 6.1. Any integer solution $q_{i,j}$ that jointly satisfies the constraints (8), (9), and (10) allocates queens in nonattacking positions. There's more! The satisfaction of constraints (8) guarantees that we allocate exactly n queens in the chessboard, which is what we are expected to do. Hence it suffices to use the constant objective function $V(q) = 1$ and satisfy the discussed constraints to discover a generic solution. Now suppose you prefer to allocate queens in specific positions defined by the set $Q := \{\bar{q}_{i,j} : (i, j) \in P \subset \mathbb{N}_n^2\}$. Hence in this case we can look for q^* such that

$$q^* = \arg\min_{q \in Q} -\frac{1}{2} \sum_{(i,j) \in P} (q_{i,j} - \bar{q}_{i,j})^2. \tag{11}$$

We are in front of 0–1 integer programming problems, which are known to be NP-complete. Whenever we deal with similar inferential problems, we must seriously consider the warning of computational intractability. The formal nature of the problem — detachment of $q_{i,j}$ from the environment — often dooms the inferential processes to exponential explosion of the computational complexity. As previously noticed, we need to construct a chess configuration, whose consistency can be efficiently checked. This scheme is quite common; its complexity explosion is often — not always — connected with the difficulty of generating solutions. The n-queens problem doesn't reflect the common situation of similar combinatorial problems, since in this case we can find good approximations also by relaxing to real-valued interpretation of $q_{i,j}$. Interestingly, although n-queens has a combinatorial structure, the problem of finding one configuration can be solved polynomially, while finding them all is exponential (see Section 6.6).

The n-queens problem gives us the opportunity to see inference processes from a different corner with respect to classic machine learning approaches. So far, regardless of the kind of model, the goal of learning has been shown to be that of determining the unknown functions, e.g., by discovering appropriate parameters under a certain functional representation. Once f has been learned, the inference on x is regarded as the computation of $f(x)$. This holds true also in the learning from the constraints scheme discussed herein. A distinguishing feature of this kind of inference scheme is that the computation of $f(x)$ is typically very efficient — think of neural networks and kernel machines. However, in cases like that of n-queens, we are missing the features, and consequently we miss the beauty of these efficient inferential schemes for computing $f(x)$. As it will be clear in Section 6.2, this is exactly what happens in formal logic, which often leads to inferential problems that are computationally intractable! In any case, also for n-queens, unlike what happens in feedforward nets and kernel machines, the inferential process of discovering the nonattacking configurations can be framed as constraint satisfaction that one can solve by optimization. Curiously, there's no learning from the dummy representation, but the optimization comes out during inference. This is the opposite of classic learning machines where the optimization is needed for learning.

Now let's have look at a different face of inference that enjoys both the computational efforts of learning and online propagation of evidence required in case of dummy variables, i.e., no features. The environmental graph $\mathcal{G} = \{V, A\}$ describes individuals that are characterized by the *perceptual map* $\mu \colon \bar{V} \subset V \to X$, $x = \mu(v)$, which simply links the vertexes to the associated features, if any. The graph describes the factual knowledge by also expressing the relationship among the individuals. In particular, suppose we enrich the running example at the beginning of this section by introducing an additional task, the purpose of which is that of establishing whether a person is a professional basketball player. It's reasonable to restrict the search for this class of persons to athletes which possess appropriate features: They are expected to be pretty high $h > H_B$, and their weight and age must lie in a certain interval, that is, $W_l \le \omega \le W_h$, and $A_l \le a \le A_h$, respectively. Now we make a different assumption that is due to the factual knowledge on the problem. Suppose that professional basketball players are friends of other professional basketball players, so one could state that athletes who are friends of at least one professional basketball player are professional basketball players. The associated function $f_{pbp} \colon W \times A \times H \to [0, 1]$ yields decisions that depend on the friendship graph, as well as on the features. Interestingly, in some cases, we might know the status of professional basketball player of a person — associated with vertex v — even though we might completely miss his features. This is represented by function $q_{pbp} : V \times \to [0, 1]$ that must be clearly coherent with f_{pbp}, that is,

$$\forall v| \quad \mu(v) = (w, a, h)' : f_{pbp}(w, a, h) = q_{pbp}(v). \tag{12}$$

The knowledge about the professional basket player can be promptly expressed by following the same arguments used at the beginning of the section. The notion of athlete can be represented by function f_{at} defined as

$$f_{at}(w, a, h) = \sigma(A_h - f_{ah}(h))\sigma(f_{ah}(h) - A_l)$$
$$\sigma(W_h - f_{\omega h}(h))\sigma(f_{\omega h}(h) - W_l)\sigma(h - H_B). \tag{13}$$

This equation simply translates the conjunction of the single properties that an athlete is expected to possess, that is, $f_{at}(w, a, h) \simeq 1$ only if all the factors of Eq. (13) are nearly one. The friendship of $u, v \in V$ is expressed by the corresponding graph $\mathcal{G} = \{V, A\}$, so if $(u, v) \in A$ then $f_{fr}(u, v) = [(u, v) \in A]$. Graph $\mathcal{G}$ is just a way of representing the factual knowledge in the environment. It's quite obvious that other graphs, where the relations specified by the arcs A are different, can express other facts, e.g., marriage or citizenship relations. Let $(\bar{w}, \bar{a}, \bar{h})' = \mu(u)$ and $(w, a, h)' = \mu(v)$. Then the previous linguistic description of the knowledge about professional basket players can be translated by

$$f_{at}(w, a, h)f_{fr}(u, v)f_{pbp}(\bar{w}, \bar{a}, \bar{h})(1 - f_{pbp}(w, a, h)) = 0,$$
$$f_{at}(w, a, h)f_{fr}(u, v)f_{pbp}(\bar{w}, \bar{a}, \bar{h})(1 - q_{pbp}(v)) = 0. \tag{14}$$

Of course, both equations present the same underlying logic structure that has already been seen for the constraint expressed by Eq. (6), where the premise f_1 is the conjunction of three propositions. It's worth mentioning that, unlike Eq. (2), where the constraint only involves feature-based functions that depend on the height, here the propagation of evidence also relies on a truly Boolean variable that establishes the friendship relation between u and v. The friendship graph, which reports a description of facts in this environment, enables a new form of inferential process that takes place by a diffusion of evidence through the persons. Notice that an explicit supervision on some nodes is in fact propagated through the graph by a diffusion mechanism that is different from the inference described so far, where the concepts are gained by satisfying constraints that only involve a single individual. It's quite obvious that the more specific factual information we introduce, the more we enrich and restrict higher level inferential processes. Needless to say, as we express the *pattern identity* by the environmental graph, we can enrich the computation of $f_{\omega h}$, $f_{\omega a}$, and f_{ah}. Hence we have composed a knowledge base involving persons, athletes, and professional basketball players, which is associated with the constraints (2), (1) (in general, in its functional form $f_{\omega h} = f_{\omega a} \circ f_{ah}$), (12), (13), and (14). The knowledge base contains constraints that involve properties of single individuals (patterns), factual relations, and mixtures of them. We can do pretty interesting things with these constraints, like answering questions that involve jointly discrete and continuous variables. For instance, if we know that a certain person is a professional basketball player, what can we say about his weight, height, and age? Interestingly, also the features can became variables to be determined by constraint satisfaction.

All the inferential processes described so far are based on the idea that the abstract notion of implication can be translated by Eq. (6). In Section 6.2, a discussion on the theory of t-norms will lead us to argue that this is not necessarily the best choice. Regardless of technical details, the discussion based on t-norm still relies on the principle that logical evidence is propagated on the basis of the value of the single variables. We can go one step further, by extending the propagation of evidence that takes place unidirectionally by means of the constraint defined by Eq. (6). Such a transfer of evidence is only a function of the value of $f_1(x_1)$, whereas one might be interested in richer propagation schemes that yield a sort of *conditional transfer of evidence*. This is expressed by the constraint

$$f_1(x_1)f_{1,2}(x_1, x_2) = f_2(x_2). \tag{15}$$

We can promptly see that we can restore the meaning of constraint (6). This requires a slight restriction on functions f_1 and f_2, where we assume that the codomain is the open set $(0, 1)$. If we choose

$$f_{1,2}(x_1, x_2) := (1 - f_2(x_2)) + \frac{f_2(x_2)}{f_1(x_1)},$$

we can promptly see that we are back to Eq. (6). While the constraint $f_1(x_1)(1 - f_2(x_2)) = 0$ is only a property of the values of f_1 and f_2, the constraint defined by Eq. (15) is more general. The first case conveys the classic notion of $f_1 \to f_2$, while Eq. (15) can play a richer role by conditioning the implication to the value of $f_{1,2}(x_1, x_2)$. For instance, suppose we have $f_1(x_1) := H(x_1) - H(3 - x_1)$ and $f_{1,2}(x_1, x_2) := H(2 - x_2) - H(3 - x_1)$. Then we have $f_2(x_2) := H(2 - x_2) - H(3 - x_2)$. In this case, function $f_{1,2}$ is deeply involved in establishing the relation between the information attached with the vertexes 1 and 2. The evidence stated by $f_1(x_1) := H(x_1) - H(3 - x_1)$ doesn't circulate when $x_1 \in (1, 2)$, that is, we are in front of a conditional implication. Unlike constraints like that expressed by Eq. (6), in this case, there is no function α that can aggregate the single operands of the implications so as $f_{1,2}(x_1, x_2) = \alpha(f_1(x_1), f_2(x_2))$.

This reminds us of the propagation that takes place with probability. We can borrow the graphical structures used in probabilistic graphical models to provide a more general propagation of dependencies. If one considers the simple graph $\mathcal{G} = \{(v_1, v_3), (v_2, v_3)\}$ then one can propagate evidence according to

$$f_3(x_3) = \psi(f_1(x_1)f_{1,2}(x_1, x_3), f_2(x_2)f_{1,2}(x_2, x_3)).$$

Here, ψ is the *aggregation function* that collects evidence from the parents of vertex v_3. A special case is that in which $\psi(\alpha, \beta) = \alpha\beta$ and $f_{1,2}(x_1, x_3) = f_{1,2}(x_1, x_3) = 1$, which yields $f_3(x_3) = f_1(x_1)f_2(x_2)$.

EXERCISES

1. [*28*] Consider the learning task discussed in Section 6.1.1 concerning the prediction of the weight of adult people. Suppose that the three functions are learned from examples constructed from the same persons, so that we are given the training sets $\{(h_\kappa, a_\kappa^o), \kappa = 1, \ell\}$, $\{(h_\kappa, \omega_\kappa^o), \kappa = 1, \ell\}$, and $\{(a_\kappa, \omega_\kappa^o), \kappa = 1, \ell\}$, where we use ^{o} to denote the target value for the corresponding function. Suppose that the problem is over-determined, so we needn't involve regularization. Find the solution by assuming soft-satisfaction of the pointwise constraints associated with supervision and hard-satisfaction of the coherence constraints expressed by Eq. (1).

2. [*18*] Extend the formulation of learning given in Section 6.1.1 with the task $\omega = f_{\omega|(h,a)}(h, a)$.

3. [*M16*] Prove that Eq. (7) only admits the solutions $f_1(x_1) = f_2(x_2) = 0$ or $f_1(x_1) = f_2(x_2) = 1$.

4. [*17*] Consider the constraint of Eq. (7) and then replace $f_1(x_1)$ with $f_1(x_1)^2$ and $f_2(x_2)$ with $f_2(x_2)^2$, so as we get the homogeneous quadratic form $f_1(x_1)^2 - 2f_1(x_1)f_2(x) + f_2^2(x_2) = (f_1(x_1) - f_2(x_2))^2 = 0$, which yields $f_1(x_1) = f_2(x_2)$. Can you notice any difference between the constraint stated by Eq. (7) and $f_1(x_1) = f_2(x_2)$?

5. [*20*] Prove that for $n = 4$ the only solutions are given in Fig. 6.1, and that there is only one *fundamental solution*.

6. [*17*] Give an example in which the enforcing of some type coherence constraints is not opportune. *Hint:* Consider the running example of Section 6.1.1 and suppose that the weight prediction must be carried out also in case of pregnancy, which is known by a Boolean variable.

6.1.2 A unified view of constrained environments

Learning and inference are traditionally kept separate in human sciences, where learning seems to be properly conceived for supporting inference. However, it has been shown that, at least in the Bayesian view, learning and inference are strictly related. We can in fact think of learning as an inferential process for estimating the parameters that are regarded as random variables. The discussion in the previous section suggests that something similar holds true also when adopting the general constraint-based description of the interactions of the agent with the environment. This is quite interesting, since this analogy is gained outside the probabilistic mathematical framework, while, on the other hand, the unification of learning and inference doesn't arise in other statistical approaches: Learning and inference are remarkably different in the frequentist approach.

Now we make an effort to generalize and clarify the deep connections between learning and inference, while capturing some neat differences in the associated computational processes. Apart from the uniform mathematical framework, we can gain intriguing connections if we try to bridge constraint satisfaction in tasks like handwritten character recognition and n-queens. Why is inference so different in these two cases? What is striking is the dummy input of n-queens, which seems to lead to a learning task without features. However, suppose we allocate a queen or, more generally, we are in certain configuration $Q_m \in \mathbb{N}^2$, with $m < n$ allocated queens, and we want to continue by sequentially allocating queens. The partial configuration with m nonattacking queens is not a dummy input anymore! The successive allocation steps will necessarily require taking decisions on the basis of Q_m, which can be thought of as a feature-based representation of an individual. This means that an agent could be instructed to learn how to allocate the $(m+1)$th queen on the basis of Q_m. If a supervisor is so kind to offer collections of pairs $(Q_m, q_{i,j}(m+1))$ then one can regard inference as a sequence of forward steps driven by the learned function $f(Q_m)$. In other words, the constraint satisfaction that is based on integer programming is converted into a sequence of forward operations that very much resemble what happens when we recognize handwritten chars.

In order to gain an in-depth understanding of inference and learning in constraint-based environments, however, it's convenient to keep the distinction between tasks based on the perceptual space of features and tasks which are based on a dummy input space. Given the environmental graph $\mathcal{G}$ along with the perceptual space X, the agent returns the function over the individual. Roughly speaking, we are interested in agents which can make decisions on the basis of the information associated with a given vertex of the environmental graph, but in some cases we have missing features. This can be contemplated by enriching X with the `nil` symbol, so $X_o := X \cup \{\text{nil}\}$. For any vertex $v \in V$ we can consider the pair $(x^{\uparrow}, x)$, where $x^{\uparrow}$ is the identifier of $x \in X \subset \mathbb{R}^d$. Hence the environment is populated by individuals of $J = V \times X_o$, where the features are present only for the set $\bar{V} \subset V$. As already seen, the features can be linked to the identifier by the perceptual map $\mu: \bar{V} \to X$, $v \mapsto x = \mu(v)$ that can be enriched to $\mu: V \to X$, $v \mapsto x = \mu(v)$ by assuming that $\forall v \in (V \setminus \bar{V})$ we have $x = \mu(v) := \text{nil}$. It turns out that the environmental graph characterizes *factual knowledge* by expressing relations among some elements of the environment. Instead of regarding individuals as elements of J, we think of a more compact representation where they are characterized by either the feature vector $x \in X$, whenever it is available, or by the vertex $v \in V$, in case of missing features. Basically, the individual corresponds with the feature vector x, when the proposition $\mathsf{p_x} = (v \in \bar{V}) \vee (x \in X \setminus \bar{X}_o)$ is true. Hence, *individuals* are formally defined by

$$\chi := [\mathsf{p_x}]x + [\neg \mathsf{p_x}]v \in I = X \cup V.$$

Here, the operator "+" is the ordinary sum in $\mathbb{R}^d$ enriched with the sum over operators, where one of them is $0 \in \mathbb{R}^m$, with $m \neq d$. The extension consists of assuming that the result takes the dimension of the nonnull argument. Hence, instead of expressing properties that involve abstract concepts of the single pattern x, here we start thinking of factual knowledge on the overall environment that allows us to draw conclusions on X. In this framework an agent is represented by function $f: I \to \mathbb{R}^n$. Both learning and inference consist of enforcing constraints, but the difference is that while in learning processes the agent is expected to perform the satisfaction over a collection of individuals (training set), in inferential processes it is expected to satisfy the constraints for the given individual $\check{\chi} \in I$. Hence

$$
\begin{aligned}
\forall \chi \in \bar{I}, \qquad & \psi(v, f(\chi)) = 0, \\
\check{\chi} \in I \setminus \bar{I}, \qquad & \psi(\check{v}, \check{f}(\check{\chi})) = 0.
\end{aligned}
\tag{1}
$$

As pointed out at the beginning of the chapter, learning machines working with constraints are referred to as *constraint machines*. They can be represented by a function depending on a set of learning parameters at finite dimensions, but can also be regarded as elements of a functional space (see the interpretation given of kernel machines in Section 4.4).

Now let us restrict our analysis on constraints to the case in which individuals coincide with patterns. On the one hand, this simplifies the general description of constraints, while on the other hand, this restriction only slightly reduces the generality of the given definitions and analyses. As shown in the previous section, however, this restriction mostly misses optimization-based inference. In the previous discussion on constraints we have encountered both *bilateral* (equality) and *unilateral* (inequality) constraints. For example, constraints 6.1.1–(1) that are used to enforce coherence are bilateral, whereas the pointwise constraints that are used in kernel machines are unilateral. As it will be put forward in Section 6.1.3, this difference has a remarkable impact on the structure of learning machines that are supposed to satisfy constraints. An obvious difference concerns their effect of the measure of the admissible space. In a given space, the measure of the admissible set of bilateral constraints is less than that of unilateral constraints. Roughly speaking, bilateral constraints provide more information on the learning tasks, thus reducing more significantly the dimension of the search space. Constraint (1) is a nice example of the general principle of compositional coherence, which arises whenever different tasks yield a computation by following different paths. As already pointed out in Section 1.1, other coherent schemes arise by comparing the decision of different machines — committee machines.

The environmental constraints can either be relational or content-based, while the constraints can exhibit a wide range of structures. Let us analyze the structure of content-based constraints. Eq. (1) turns out to be the translation of the functional coherence constraint $f_{\omega h} = f_{\omega a} \circ f_{ah}$, once we make an assumption on the structure of the learning tasks. This makes it possible to impose this functional constraint at finite dimensions. A generalization of this case can be captured by considering constraints characterized by

$$
\mathcal{W}_x = \{w \in \mathbb{R}^m : \psi(x, w) = 0\}.
\tag{2}
$$

The compositional coherence constraint of Eq. (1) belongs to the class defined by this equation, but it is independent of $x \in X$. Hence it enforces constraints on the parameters of the model, but there is no direct dependence on the perceptual space. While Eq. (2) defines a wide class of constraints, it relies on the assumption that the learning agent is given a parametric structure in advance. A more general view can be gained if we don't make any specific parametric assumption on the model of the agent — in the previous case, we were assuming linearity. Hence let F be a space of functions $f: X \to \mathbb{R}^n$, X_κ open subsets of X, $\psi_\kappa: X_\kappa \times \mathbb{R}^n \to \mathbb{R}$ and $\check{\psi}_\kappa: X_\kappa \times \mathbb{R}^n \to \mathbb{R}$ continuous functions. The functional equations

$$\forall x \in X_\kappa \subseteq X \quad \psi_\kappa(x, f(x)) = 0, \quad \kappa = 1, \ldots, \ell_H,$$

$$\forall x \in X_\kappa \subseteq X \quad \check{\psi}_\kappa(x, f(x)) \geq 0, \quad \kappa = 1, \ldots, \check{\ell}_H$$

are referred to as bilateral and unilateral *holonomic constraints*, respectively. The asset allocation problem formulated in Section 1.1.5 by Eqs. 1.1.5–(5) is an example of bilateral holonomic constraint. Exercise 1 stimulates a discussion on the role of the explicit dependence on x in $\psi_\kappa(x, f(x))$. Whenever holonomic constraints are restricted to samples $X^\sharp$ of X they are referred to as *pointwise constraints*. Most classic supervised and unsupervised learning used in machine learning belong to this class. A more general class of constraints is the one defined by

$$\forall x \in X_\kappa \subseteq X \quad \psi_\kappa(x, f(x), Qf(x)) = 0, \quad \kappa = 1, \ldots, \ell_{NH}, \tag{3}$$

$$\forall x \in X_\kappa \subseteq X \quad \check{\psi}_\kappa(x, f(x), Qf(x)) \geq 0, \quad \kappa = 1, \ldots, \check{\ell}_{NH}, \tag{4}$$

where Q is a differential operator. The constraint which translates the concept of brightness invariance discussed in Section 1.1 [see Eq. 1.1–(8)] is an example of a nonholonomic constraint. Brightness appears in the functional dependence also with its gradient, that is, $Q = \nabla$.

While the above constraints are enforced locally, we can think of imposing global satisfaction over the functional space as follows:

$$\Psi_\kappa(f) = 0, \quad \kappa = 1, \ldots, \ell_I,$$

$$\check{\Psi}_\kappa(f) \geq 0, \quad \kappa = 1, \ldots, \check{\ell}_I,$$

where $\Psi_\kappa: F \to \mathbb{R}$ and $\check{\Psi}_\kappa: F \to \mathbb{R}$ are continuous functionals. They are referred to as bilateral and unilateral *isoperimetric constraints*. Constraint 6.1.1–(5) is an example of this class. The probabilistic normalization is a global constraint which can be checked only when the probability is given on all the points of its domain. For notational simplicity, when dealing with constraints of the same type, the notation "ℓ_H", "ℓ_I", "$\check{\ell}_H$", and "$\check{\ell}_I$" will be replaced simply by "ℓ." We consider both the case in which perfect constraint satisfaction is needed and the case where constraint violations are allowed. The former situation corresponds to a *hard* (hr) interpretation of the constraints, whereas the latter to their *soft* (sf) interpretation. For the sake of simplicity and with a little abuse of terminology, we refer to these cases as *hard constraints* and *soft constraints*, respectively. Soft-constraints are more common in many machine learning tasks, e.g., think of supervised learning. Sometimes strong fulfillment may be desirable (e.g., coherence constraints), which will be shown to be more computationally challenging. Based on the notion of constraint, we can accommodate into the same framework stimuli of very different kinds, like those shown

Table 6.1 Examples of constraints from different environments.

	Description	Math Representation	Classification
(i)	supervised pairs	$y_i \cdot f(x_i) - 1 \geq 0$	(in,pw,sf,un)
(ii)	prob. normalization	$\forall x \in X \quad f_1(x) + f_2(x) + f_3(x) = 1$	(in,ho,hr,bi)
		$\forall x \in X \quad f_i(x) \geq 0$	(in,ho,hr,un)
(iii)	prob. normalization	$\forall x \in X \quad f_i(x) = \frac{e^{a_i}}{\sum_j e^{a_j}}$	(fi,ho,hr,bi-un)
(iv)	density normaliz.	$\int_X f(x)\,dx = 1,$	(in,is,hr,bi)
		$\forall x \in X \quad f(x) \geq 0$	(in,ho,hr,un)
(v)	multiview coherence decision agreement	$\forall x = (x_1, x_2) \in X$ $f_1(x_1) \cdot f_2(x_2) \geq 0$	(in,ho,hr,un)
(vi)	compositional coherence	$f_{\omega h} = f_{\omega a} \circ f_{ah}$	(in,co,hr,bi)
		$w_{\omega a} w_{ah} - w_{\omega h} = 0$	(fi,co,hr,bi)
		$w_{ah} b_{ah} + b_{\omega a} - b_{\omega h} = 0$	
(vii)	asset allocation	$\forall x \in X$	(in,ho,hr,bi)
	portfolio in USD	$f_c^d(x) + f_b^d(x) + f_s^d(x) = t_d(x);$	
	portfolio in Euro	$f_c^e(x) + f_b^e(x) + f_s^e(x) = t_e(x);$	
	overall portfolio	$t_d(x) + c \cdot t_e(x) = T$	
(viii)	optical flow	$\partial v_t + \dot{x} \cdot \nabla v = 0$	(in,nh,sf,bi)
(ix)	diabetes	$[(m \geq 30) \wedge (p \geq 126)] \Rightarrow d$	(in,ho,sf,un)
(x)	doc classification	$\forall x \quad f_{na}(x) \wedge f_{nn}(x) \Rightarrow f_{ml}(x)$	(fi,-,sf,-)
(xi)	N-queens (Exercise 1)	see Eqs. 6.1.1–(8), 6.1.1–(9) and 6.1.1–(10)	(fi,-,hr,bi)
(xii)	manifold reg.	$\forall x_1, x_2 \in X \quad [\rho(x_1, x_2) \leq \delta] \Rightarrow$ $[f(x(x_1)) f(x_2) \geq 1]$	(in,-,hr,-)

The first six are problem-independent, whereas the others are descriptions of granules of knowledge in specific domains. The classification of the constraints in the last column concerns local vs global properties, bilateral vs unilateral constraints, and finite vs infinite dimension. The classification hard- vs soft-constraint is referred to the way in which the constraints are usually dealt with in practice.

in Table 6.1. The labels (fi,in,ho,nh,is,bi,un,pw,hr,sf) are used to classify the different constraints. The classification considers three categorical variables that represent the specific local/global nature of the constraint. In this respect, the labels pw,ho,nh, and is stand for *pointwise, holonomic, nonholonomic,* and *isoperimetric,* respectively. The labels *bi,un* stand for *bilateral/unilateral,* while *hr,sf* stand for the *hard/soft.* Finally, satisfaction can take place on finite- or infinite-dimensional spaces, which is denoted by labels fi and in, respectively. It's worth mentioning that sometimes a given problem can be formulated by different types of constraint. A noticeable case in which this happens is the classical learning from supervised examples, which corresponds to constraints that can be regarded either as pointwise, holonomic, or isoperimetric (see Exercise 2).

Table 6.1 gives examples of constraints that are informally described in column 2. This description is kept concise, but it could be much more detailed. Column 3 contains a translation of constraints into real-valued functions that are the formal environmental description of the environment. Most of the examples have already been discussed in Section 1.1. Example (i) is the classic case in which the pair (x_κ, y_κ) is used for supervised learning in classification, where x_κ is the κth supervised example $y_\kappa \in \{-1, 1\}$ its label. If f is the function that the agent is expected to compute then the corresponding real-valued representation of the constraint, which is reported in column 3, is just the translation of the classical "robust" sign agreement between the target and the function to be learned. Examples (ii) and (iv) are classic probabilistic normalizations. Notice that the adoption of softmax (iii) makes it possible to frame the problem in finite dimension, whereas the solution is searched in an infinite-dimensional space. Example (v) imposes coherence between the decisions taken on x_1 and x_2, for the visual recognition of object x, where x_1 and x_2 are two different views of the same object x. In (vi) another example of coherence constraint is given, which involves composition of the learning tasks. While this is formulated in an infinite dimensional space, once we make the linear assumption on the tasks then we end up with a constraint which acts on the finite-dimensional space of parameters (see Eq. 6.1.1–(1)). Example (vii) describes the constraints needed to impose consistency in portfolio asset allocation when investing money (USD and Euro) in bonds and stocks. Here we remind that f_c^d, f_b^d, and f_s^d denote the allocations in cash, bond, and stock in USD on the basis of the financial feature vector x, while f_c^e, f_b^e, and f_s^e are the corresponding allocations in Euro. The constraints simply express the consistency imposed by the overall amount of available money, denoted by T (in USD), with c being the Euro/USD conversion factor. Example $(viii)$ comes from computer vision and concerns the classical problem of determining the optical flow. It is a nonholonomic constraint of the form given by Eq. (3), since the brightness is involved also with its gradient. Example (ix) is a holonomic constraint on a learning task concerning the diagnosis of diabetes. It reminds us of supervised learning, but instead of being imposed on a finite collection of data, it is enforced on boxes. Example (x) is a logic constraint that involves Boolean variables. Interestingly, the predicates operate on the perceptual space of documents. Likewise, (xi) and (xii) operate on Boolean variables. While the N-queen problem is a truly discrete constraint satisfaction problem, constraint (xii) expresses a condition that is typically used in *manifold regularization*. Here, if x_1 and x_2 are patterns such that $\rho(x_1, x_2) = \|x_1 - x_2\| < \delta$ (close to each other) then we impose the same classification (see Section 6.6).

Now we turn our attention to soft-satisfaction by focusing on holonomic constraints. The softness of pointwise and isoperimetric constraints can be directly understood once we have the notion of holonomic constraints. Whereas isoperimetric constraints yield directly a measure of their violation (given, for instance, by $|\Psi(f)|$ and $(-\check{\Psi}(f))_+$), for holonomic constraints we can express a global degree of mismatch in terms of some given (possibly generalized, e.g. expressed in terms of Dirac deltas) data probability density p (more generally, a weighted average of their degree of violation). For example, for a bilateral holonomic constraint ψ associated with an open perceptual space X and $q \in \mathbb{N}_+$, one can express the global degree of mismatch as

$$E_\psi = \int_X |\psi(x, f(x))|^q p(x)\, dx \tag{5}$$

and, for a continuous unilateral holonomic constraint $\check{\psi}$, as

$$E_\psi = \int_X (-\check{\psi}(x, f(x)))_+^q p(x)\, dx. \tag{6}$$

However, when different kinds of constraint are involved, the quantities defined in Eqs. (5) and (6) might not satisfactorily represent the interactions of the agent with the environment. Basically, the constraints come with their own specificity and the agent might be willing to express a belief on them. Hence, whereas (5) and (6) represent inherent degrees of mismatch, which depend on the probability density, we need to introduce the *belief* of a soft constraint. Given the ith constraint of a collection $\mathcal{C}$ of soft constraints, its *belief* is defined as follows:

- For isoperimetric constraint by a nonnegative constant β;
- For holonomic constraint by either a function from X to $\mathbb{R}^+$ or a linear combination of Dirac deltas with positive coefficients;
- For pointwise constraint by a vector of $|X|$ nonnegative constants.

For the pointwise case, the belief can be reduced to that of the holonomic case, when the latter is expressed by a linear combination of Dirac deltas with positive coefficients. So, in the following definition, we deal only with isoperimetric and holonomic constraints. Given the ith constraint of a collection $\mathcal{C}$ of soft constraints (possibly of different kinds), its *qth-order degree of mismatch* is defined as follows:

- For isoperimetric bilateral by $E_\Psi^q(f) := |\Psi(f)|^q \beta$;
- For isoperimetric unilateral by $E_{\check{\Psi}}^q(f) := |(-\check{\Psi}(f))_+|^q \beta$;
- For holonomic bilateral by $E_\psi(f) := \int_X |\psi(x, f(x))|^q \beta(x)p(x)\, dx$;
- For holonomic unilateral by $E_{\check{\psi}}^q(f) := \int_X |(-\check{\psi}(x, f(x)))_+|^q \beta(x)p(x)\, dx$.

The qth-order degree of mismatch of $\mathcal{C}$, denoted by $E_\mathcal{C}^{(q)}(f)$, is the sum of the degrees of mismatch of each constraint in $\mathcal{C}$.

Of course, $E_\mathcal{C}^{(q)}(f) = 0$ iff $\mathcal{C}$ is a collection of constraints that are strictly satisfied by f. Notice that, in the holonomic case, it is reasonable to expect $\beta_\kappa(x) \equiv c_\kappa$ for every $\kappa \in \mathbb{N}_m$, which expresses a *uniform belief* on the holonomic constraints. It might be the case that $c_\kappa = c$ for every $i \in \mathbb{N}_m$, when one has no reason to express different beliefs on different constraints. In other cases, the choice of the belief is not obvious, since it may actually involve local properties of the constraints. Exercise 5 gives an insight into the joint role of the probability density and of the belief of the constraints.

Constraint-based environments might not be univocally represented by a set of constraints. The simplest example comes from supervised learning. Given the training set as a collection of supervised pairs, different risk functions are equivalent in terms of perfect satisfaction, since a typical requirement for a loss function is $V(x, y, f(x)) = 0$ whenever $y = f(x)$. Hence all constraints $\psi(f) = \sum_{\kappa=1}^{\ell} V(x_\kappa, y_\kappa, f(x_\kappa)) = 0$, which are induced by any loss v, are equivalent in terms of perfect satisfaction. We can promptly see that this is not a special property of this pointwise constraint. For example, the coherence constraints 6.1.1–(1) can obviously be replaced with $\alpha(w_{\omega a} w_{ah} - w_{\omega h}) = 0$ and $\beta(w_{ah} b_{ah} + b_{\omega a} - b_{\omega h}) = 0$ where $\alpha \neq 0$ and $\beta \neq 0$. Since $w_{ah} \neq 0$, we can also replace the first constraint with $w_{\omega a} - w_{\omega h}/w_{ah} = 0$. We can do more! Any nonnegative function $\alpha \colon X \to \mathbb{R}$ can be used as a factor $\alpha(w_{\omega a}, w_{ah})$ without affecting the space of satisfiability of the original constraint. This is clearly a general property of constraints that can be represented by $\psi(x, f(x)) = 0$. Clearly, whenever $\alpha(x) > 0$, we can replace this constraint with $\psi_\alpha(x) := \alpha(x)\psi(x, f(x)) = 0$. This equivalence deserves attention, since it is at the basis of the logic structure of constraints. Hence it makes sense to define ψ_1 and ψ_2 as equivalent constraints, and write $\psi_1 \sim \psi_2$, provided that there exists a nonnegative function α such that $\psi_2(x, f(x)) = \alpha(x)\psi_1(x, f(x))$. This means that in general we deal with a class of equivalent constraints characterized by any representative element $\bar{\psi}$, and by the quotient set $F/\sim$. The presence of this equivalence relationship indicates that constraints are entities characterized by the quotient set, and that they express regularities that are deeply rooted in the structure of the tasks defined by f. As pointed out in Section 6.1.4, there is in fact a sort of logic structure.

EXERCISES

1. [15] Consider the asset allocation problem defined in Section 1.1.5 by Eq. 1.1.5–(5). Write down the function ψ to show that the associated constraint belongs to the class of bilateral holonomic constraints.

2. [13] Prove that supervised learning can be interpreted as pointwise, holonomic, and isoperimetric constraint.

3. [20] Prove that pointwise constraints can be interpreted as appropriate discretizations of constraints of holonomic type.

4. [20] Any holonomic bilateral constraint $\psi_i(x, f(x)) = 0$ can be expressed in terms of the pair of unilateral constraints $\{\psi_i(x, f(x)) \geq 0, -\psi_i(x, f(x)) \geq 0\}$. Prove that unilateral constraints can be expressed in terms of an appropriate choice of bilateral constraints.

5. [20] Let us consider the following holonomic constraints, along with their beliefs:

$$\forall x \in X \quad \psi_1(f_1(x), f_2(x)) := f_1(x)(1 - f_2(x)) = 0; \quad \beta_1(x) = \frac{1}{2};$$

$$\psi_2(f_1(x), f_2(x)) = f_1(x) - y_1 = 0; \quad \beta_2(x) = \frac{1}{4}\delta(x - \bar{x});$$

$$\psi_3(f_1(x), f_2(x)) = f_2(x) - y_2 = 0; \quad \beta_3(x) = \frac{1}{4}\delta(x - \bar{x}).$$

Determine the overall 2nd degree of mismatch.

6. [24]   Let us consider the constraints

$$\psi_1(x, f(x)) = f_1(x) + f_2(x) = 0,$$
$$\psi_2(x, f(x)) = f_1(x) + 2f_2(x) = 0,$$
$$\psi_3(x, f(x)) = 2f_1(x) + 3f_2(x) = 0,$$

which we want to softly enforce by using the isoperimetric criterion

$$\int_X \left(\psi_1^2(x, f(x)) + \psi_2^2(x, f(x)) + \psi_3^2(x, f(x)) \right) dx = 0.$$

Then consider the solutions f^* for the pair $\{\psi_1, \psi_2\}$ and for the set $\{\psi_1, \psi_2, \psi_3\}$. What happens for high values of λ (multiplier associated with the previous isoperimetric constraint)? How is the reaction shared among the above terms of the integral?

6.1.3 Functional representation of learning tasks

In this section we address a foundational topic involving the deep structure of learning agents. We ask ourselves what is the "most natural structure" of agents dealing with (content-based) constrained environment, where individuals are reduced to patterns, that is, $I = X$. As already pointed out, environments with this kind of individuals are strongly based on learning, which yields efficient inference — once the optimal solution is discovered. We follow a path that has already been traced in Section 4.4 concerning supervised learning. Instead of learning in a finite-dimensional set of parameters, we formulate a variational problem with the purpose of discovering the optimal solution in a functional space. In supervised learning, we used the parsimony principle under pointwise constraints derived from the collection of supervised pairs. We can use a more general view of learning that involves content-based constraints of different types that is inspired by the same idea: *Discover the most parsimonious solution under the satisfaction of the given constraints.* Basically, we use the same mathematical apparatus of kernel machines to express the parsimony of a given solution. However, there's something new when dealing with constraints which depend on their structure. The kernel-based expansion of Eq. 4.4.4–(8) in Chapter 4 comes out from the special type of supervised learning constraints. Other content-based constraints admit different functional solutions. In order to grasp the idea, let us consider the class of holonomic constraints. Furthermore, suppose we restrict attention to a single unilateral constraint $\check{\psi}(x, f(x)) \geq 0$ that we want to softly enforce. The constraint that is imposed for any supervised pair is a special case that arises when adopting a distributional interpretation of the holonomic constraint, that is, when posing $(-\check{\psi}(x, f(x)))_+ p(x) = V(x_\kappa, y_\kappa, f(x_\kappa)) \delta(x - x_\kappa)$. This can promptly be seen, since the soft-interpretation of the constraint can be given by means of the association of its degree of mismatch, that is, by the penalty

$$E_{\check{\psi}}^q = \int_X (-\check{\psi}(x, f(x)))_+^q \, p(x) \, dx = V(x_\kappa, y_\kappa, f(x_\kappa)),$$

which comes from choosing $\beta(x) = 1$ and $q = 1$. To sum up, if we want to learn by softly enforcing $\check\psi(x, f(x)) \geq 0$, we can simply minimize the functional

$$E(f) = V(f) + \frac{1}{2}\mu\langle Pf, Pf\rangle \tag{1}$$

where $\tilde\psi(x, f(x)) := (-\check\psi(x, f(x)))_+^q$ and $V(f) := \int_X \tilde\psi(x, f(x))p(x)\,dx$. Function $\tilde\psi$ is the loss corresponding with the constraint $\check\psi(x, f(x)) \geq 0$. Notice that since the argument f in this constraint takes on vectorial values, we need to define precisely the norm $\langle Pf, Pf\rangle$. We follow the same principles behind kernel machines, so as we define

$$\|f\|^2 = \langle Pf, Pf\rangle = \sum_{i=1}^{n}\langle Pf_i, Pf_i\rangle, \tag{2}$$

which accumulates the norm of the single functions, thus disregarding dependencies amongst tasks. This is a reasonable parsimony criterion which relies on the principle that smoothness is guaranteed by the single terms $\langle Pf_i, Pf_i\rangle$, while the dependencies amongst the tasks are expressed by the constraints. The application of the parsimony principle for learning under the unilateral holonomic constraint $\check\psi(x, f(x)) \geq 0$ is then converted into the problem of finding $f^* \in \arg\min_{f\in F} E(f)$. We parallel the variational analysis of Section 4.4, so as we consider the variation $f_j \to f_j + \epsilon h_j$. Of course, this time we need to consider the variations with respect to each single task. Again, we choose $\epsilon > 0$, while the variation h still needs to satisfy the boundary condition stated by Eq. 4.4.4–(6) that are put forward in Section 4.4. For the parsimony term we have

$$\delta_j\langle Pf, Pf\rangle = 2\epsilon\langle Lf_j, h_j\rangle. \tag{3}$$

Now for the penalty term we have

$$\delta_j V(x, f(x)) = \int_X \tilde\psi(x, f(x) + \epsilon e_j h_j(x))p(x)\,dx - \int_X \tilde\psi(x, f(x))p(x)\,dx$$

$$= \epsilon\int_X p(x)h_j(x)\partial_{f_j}\tilde\psi(x, f(x))\,dx = \epsilon\,\langle p\,\partial_{f_j}\tilde\psi, h\rangle,$$

where $e_j \in \mathbb{R}^n$ is defined by $e_{j,i} := \delta_{i,j}$. Hence, $\delta_j E(f) = 0$ yields

$$\langle \mu Lf_j^* + p\,\partial_{f_j}\tilde\psi, h_j\rangle = 0.$$

Finally, from the fundamental lemma of variational calculus we get

$$Lf^* + \frac{p}{\mu}\nabla_f\tilde\psi = 0.$$

Here, we overload symbol L to denote the same operation over all f_j. Now if g is the Green function of L then

$$f^* = g * \omega_{\tilde\psi}, \tag{4}$$

$$\omega_{\tilde{\psi}}(x) = -\frac{1}{\mu} p(x) \nabla_f \tilde{\psi}(x, f^*(x)).\tag{5}$$

This returns kernel expansions whenever the unilateral holonomic constraint $\tilde{\psi}$ is quantized according to a distribution, which corresponds to assuming that

$$\tilde{\psi}(x, f(x)) p(x) = \tilde{\psi}(x_\kappa, f(x_\kappa)) \delta(x - x_\kappa)$$

and

$$
\begin{aligned}
f^*(x) &= -\frac{1}{\mu} g * \left(\nabla_f \tilde{\psi}(x_\kappa, f(x_\kappa)) \delta(x - x_\kappa) \right) \\
&= -\frac{1}{\mu} \nabla_f \tilde{\psi}(x_\kappa, f(x_\kappa)) g(x - x_\kappa) \\
&= \lambda_\kappa g(x - x_\kappa),
\end{aligned}
\tag{6}
$$

where

$$\lambda_\kappa := -\frac{1}{\mu} \nabla_f \tilde{\psi}(x_\kappa, f(x_\kappa)).\tag{7}$$

Finally, because of linear superposition we get the classic kernel expansion over the training set. Eqs. (4) and (5) offer a natural generalization of kernel machines and suggest emphasizing the role of $\omega_{\tilde{\psi}}$, which is referred to as the *constraint reaction* of $\tilde{\psi}$. In supervised learning, this interpretation leads us to formulate the presence of a reaction to pattern x_κ that is expressed by $\omega_{\tilde{\psi}} \propto \delta(x - x_\kappa)$. Clearly, when dealing with general unilateral holonomic constraints, things are more involved, but the principle is the same! The reaction of $\tilde{\psi}$ depends on the probability distribution and on $\nabla_f \tilde{\psi}$. As one could expect, the reaction $\omega_{\tilde{\psi}}$ takes on high values $\|\omega_{\tilde{\psi}}\|$ in high density regions of the perceptual space X. Interestingly, the reaction vanishes when $p(x) \to 0$. Hence, in high-dimensional spaces, where data are distributed on manifolds, the constraint reaction is generally null, which suggests the development of methods for speeding up the computation of the convolution, which is intractable at high dimensions. Moreover, the reaction depends on the *cognitive field* $c(x) = \nabla_f \tilde{\psi}(x, f^*(x))$, so large values of the reaction correspond with large values of the field. Like in supervised learning, if $\tilde{\psi}$ is a pointwise constraint then we still enjoy the simplicity of kernel expansion, since we have exactly the same mathematical structure behind. Basically, instead of dealing with a set of supervised examples, the availability of an unsupervised example still induces the probability distribution $p(x) \propto \delta(x - x_\kappa)$ for each pattern, which formally leads to Eq. (6). In this case, the only difference concerns parameter λ_κ that, however, can be learned as already seen for kernel machines. (More details are in Section 6.4.) It's worth mentioning that when $X \to X^\sharp$, and the constraints are pointwise, we needn't determine the probability distribution p, which comes out from the given data. The spectral interpretation of Eq. (4) leads to state that the Fourier transform of the solution is

$$\hat{f}^*(\xi) = \hat{g}(\xi) \cdot \hat{\omega}_{\tilde{\psi}}(\xi).\tag{8}$$

This provides a traditional filtering interpretation: The Green function acts as a filter on the spectrum of the constraint reaction. When abandoning pointwise constraints, things are more involved. Eq. (4) does in fact involve the unknown f^* on both sides, since $\forall x \in X$ the reaction is $\omega_{\bar{\psi}}(x, f^*(x))$. Hence, $\forall x \in X$ we need to solve $f^*(x) = g * \omega_{\bar{\psi}}(x)$, which is in general a complex functional equation. An in-depth discussion on this functional structure and on corresponding algorithmic solution is given in Section 6.4.

What happens for other constraints? Interestingly, the representational structure of Eq. (4) still holds true. While the technical analysis requires facing some issues, the basic idea is still the same, and leads to blessing the solution based on the convolution of the Green function of L with the constraint reaction. Again, the Green function is simply the outcome of the parsimony principle; it is in fact the technical interpretation of the smoothness condition arising from the choice of differential operator P. The reaction is dependent on the structure of the corresponding constraint. If we still consider soft-enforcement, the same holonomic problem considered so far in case of bilateral constraint, namely $\psi(x, f(x)) = 0$, can be approached by the same analysis. Here, the only difference concerns the way the constraint is mapped to a corresponding penalty function. For example, if we choose

$$V(f) = \int_X |\psi(x, f(x))|^p \, p(x) \, dx,$$

then we end up with the same conclusions concerning the representation of the optimal solution f^*. Clearly, the same holds true for isoperimetric problems, where a single constraint corresponds with the global satisfaction of an associated holonomic constraint.

When the constraints must be hard-enforced, we cannot rely on the same idea of an associated penalty. We can think at most of a family of approximating problems where the penalty comes with an increasing weight. Let's assume that the bilateral holonomic constraint $\psi(x, f(x)) = 0$ must be hard satisfied over the perceptual space X. The Lagrangian in this case is

$$\mathcal{L} = \langle Pf, Pf \rangle + \int_X \lambda(x)\psi(x, f(x)) \, dx. \tag{9}$$

Intuitively, this corresponds with imposing the satisfaction over an infinite number of constraints $\forall x \, \psi(x, f(x)) = 0$. We have a corresponding Lagrangian multiplier $\lambda(x)$ for each point $x \in X$ on which the constraints are enforced. We can promptly see that Eq. (9) has exactly the same mathematical structure as Eq. (1). As a consequence, any stationary point satisfies

$$Lf^*(x) + \lambda(x)\nabla_f \psi(x, f^*(x)) = 0, \tag{10}$$

that is,

$$f^* = g * \omega_\psi, \tag{11}$$

$$\omega_\psi(x) = -\lambda(x)\nabla_f \psi(x, f^*(x)). \tag{12}$$

The formal equivalence of Eq. (4) and Eq. (5) arises when matching $p(\cdot)$ with $\lambda(\cdot)$. However, this formal equivalence hides remarkable differences! In case of soft-constraints, it's reasonable to assume that $p(\cdot)$ is a given density, which reflects the nature of the problem at hand. In case it is not given, one can add the normalization condition $\int_X p(x)dx = 1$ and $\forall p(x) \geq 0$. This contributes to discovering also the probability distribution. Suppose that we are also given a collection of points that can be regarded as a supervised/unsupervised training set $X^\sharp$. Clearly, pairs $(x_\kappa, p(x_\kappa))$ yield the supervised constraint expressed by the corresponding risk function, just like in Eq. 6.1.1–(4). Overall, when opening also to the discovering of the probability distribution, we need to solve a more challenging problem, that is discussed in Exercises 1 and 2. In case of hard constraints we also need to impose $\partial_{\lambda(x)}L = 0$ that restores $\psi(x, f(x)) = 0$.

The difference between p and λ naturally arises when going back to the discussion on different functional representations of the same functional space, and to the notion of equivalence of constraints. In the case of soft-constraints, the mechanism used to convert the given constraint into a penalty function is simple, but it introduces a degree of freedom which might have a nonnegligible impact in the solution. As already seen, the holonomic constraint $\forall x \in X \; \psi(x, f(x)) = 0$ is equivalent to $\psi_\alpha(x, f(x)) := \alpha(x)\psi(x, f(x))$, where α is a nonnegative function. From Eq. (4) we can promptly see that the constraint reaction becomes $\omega_{\tilde{\psi}}(x) = \alpha(x)p(x)\nabla_f\tilde{\psi}(x, f(x))$. This indicates that the equivalent constraint $\psi_\alpha(x, f(x))$ can be thought of as one which is based on the new probability distribution

$$p_\alpha(x) = \frac{p(x)\alpha(x)}{\int_X p(x)\alpha(x)dx}.$$

Hence, while $\psi \equiv \psi_\alpha$, their corresponding solutions might be remarkably different! This is very well known in case of supervised learning where the different loss functions that are used can lead to appreciably different results. Something different happens for hard constraints, where the solution is clearly independent of α. In that case, when replacing ψ with ψ_α we can see that if $\nabla\psi(x, f^*(x)) \neq 0$ then $\lambda_\alpha(x) = \lambda(x)/\alpha(x)$ (see Exercise 3).

The results stated so far consider the presence of single constraints. Only the case of supervised learning has been considered with the joint presence of multiple pointwise constraints. In that case, things are pretty easy and we can use the superposition principle straightforwardly. Of course, the same holds for any constraint, but in case of soft-constraint we need some additional thoughts. Two holonomic constraints ψ_1 and ψ_2 are generally combined to get

$$\int_X \left(p(x)\beta_1(x)\psi_1(x, f(x)) + p(x)\beta_2(x)\psi_2(x, f(x)) \right) dx.$$

Things can become more involved because of the presence of constraints that are somewhat dependent on each other. This is nicely illustrated in Exercise 6. Whenever one constraint can be formally derived from a given collection, learning is not affected. This issue is covered with some details in the following section.

A final remark is in order on the actual possibility of discovering f^*. What has been presented is based on the Euler–Lagrange equations, which only return a stationary point of $E(f)$. It's quite obvious that the actual discovering of the global minimum depends on the class of constraints that characterize the learning environment. In case of supervised learning, convex losses functions yield convex risks, since the sum of convex loss over different points is still convex and the regularization term is convex, too.

EXERCISES

1. [*M30*]   Consider the problem of learning from holonomic unilateral constraint where also $p(\cdot)$ is unknown. Suppose we are given a training set with supervised pairs for $p(\cdot)$. Determine the functional solution of learning from constraint.

2. [*HM47*]   Consider the problem of learning from holonomic unilateral constraint where also $p(\cdot)$ is unknown. Determine the solution by adding the probabilistic conditions $\forall x \in X$ $p(x) \geq 0$; $\int_X p(x)\, dx = 1$. Discuss both the cases in which these constraints are hard- and soft-enforced, respectively.

3. [*20*]   Prove that if $\psi_\alpha(x, f(x)) := \alpha(x)\psi(x, f(x))$ then $\lambda_\alpha(x) = \lambda(x)/\alpha(x)$.

4. [*15*]   Give the expression of the cognitive field for supervised learning under quadratic loss. Under which condition we get $c(x) = 0$? What if we consider the hinge loss?

6.1.4 Reasoning with constraints

Now we want to highlight the common logic structure behind real-valued constraints in terms of inferential mechanisms that are also driven by the parsimony principle. There are in fact common properties in the class of constraints that are typically the subject of investigation in logic. Real-valued constraints can be given an abstract picture along with the notions of constraint equivalence and deduction. Before getting into more technical details, we start by giving an insight into main concepts and results by using three remarkably different examples.

FOL constraints. As sketched in Table 6.1–x, document classification can be modeled by FOL constraints that express the relationships among the categories. Now let us consider the related artificial example sketched in Fig. 6.2, where we are interested in highlighting the basic idea behind the inference process that involves both a collection of logic statements and supervised examples. We inherit the fundamental assumption coming from tasks, like document classification, that the logic variables which express the category are in fact functions of the perceptual input $x \in \mathbb{R}^2$. More precisely, the categories can be regarded as four unary predicates, and we expect our agent to carry out a decision that is based on both the logic statements and a collection of supervised pairs. The connections between the input space and the document class is very well expressed in Fig. 6.2. In principle, these predicates can be learned from examples, but we know that this becomes harder and harder as the dimension

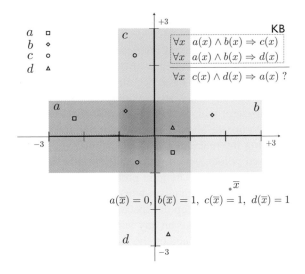

FIGURE 6.2

An example of inference process where the predicates $a(\cdot)$, $b(\cdot)$, $c(\cdot)$, and $d(\cdot)$ depend directly on the perceptual information given by the points $x \in \mathbb{R}^2$.

of the input space increases. Now suppose that an oracle offers the following ground truth:

$$
\begin{aligned}
a(x) = \mathsf{T} &\Leftrightarrow x \in [-3, 1] \times [-1, +1], \\
b(x) = \mathsf{T} &\Leftrightarrow x \in [-1, 3] \times [-1, +1], \\
c(x) = \mathsf{T} &\Leftrightarrow x \in [-1, +1] \times [-1, +3], \\
d(x) = \mathsf{T} &\Leftrightarrow x \in [-1, 1] \times [-3, +1].
\end{aligned}
\tag{1}
$$

The agent is expected to guess this ground truth on the basis of the environmental knowledge

$$
\mathsf{KB} := \{\forall x : a(x) \wedge b(x) \Rightarrow c(x), \quad a(x) \wedge b(x) \Rightarrow d(x)\},
\tag{2}
$$

and of a collection of supervised pairs that are shown in Fig. 6.2. Now let's consider the unary predicates just as formal logic variables by dropping the dependence on x. We want to see whether the argument $\mathsf{KB} \vdash c \wedge d \Rightarrow a$ is valid.

From a quick check, we can promptly see that the argument is not valid, since the choice $(0, 1, 1, 1)$ for a, b, c, d violates the conclusion, while the premises are true. As shown in Fig. 6.2, if we also restore the unary predicates with the dependence on the perceptual space, clearly the argument is violated when choosing $x \in X$ such that $(a(\overline{x}), b(\overline{x}), c(\overline{x}), d(\overline{x})) = (0, 1, 1, 1)$. However, if the data are distributed in such a way that the predicates depend on the environment as indicated in Eq. (1), then we have $\mathsf{KB} \models \forall x \quad c(x) \wedge d(x) \Rightarrow a(x)$. Unlike formal inference, where the turnstile symbol $\vdash$ denotes formal derivation, this new inference, which takes place under the restrictions imposed by the environment, is denoted by $\models$. Notice that the deduction now holds because of the hypothesis of restricting the space of possible predicates, which is due to the functional dependence of the Boolean variables on the perceptual space. A concrete step towards inference mechanisms that breaks the traditional symbolic deduction schemes arises when we are given supervised pairs of the unary predicates that are consistent with the definition of Eq. (1). Basically, an intelligent agent is supposed to acquire the above KB along with training points like those shown in Fig. 6.2. The resulting inference mechanism is significantly enriched by the presence of those supervised points, which restrict the hypothesis space from which we search for parsimonious solutions. The joint presence of logic statements and supervised pairs enforces a learning mechanism where the logic inference is restricted to the environment defined by the perceptual space. In particular, while in formal inference the literals are expected to take on any value on the Boolean hypercube, here the literals can only take values defined by the maps a, b, c, and d. The point $\bar{x}$ which violates the previous argument is not coherent with the additional constraints that are enforced by supervised learning. The more examples compose the training set the more we restrict the range of the literals, which clearly favors the argument truth. In a sense, it is like working with the traditional logic formalisms, with the additional information that some occurrences of literals are not used for checking the derivation! Hence supervised examples somewhat restrict the range of the literals, thus changing significantly the inferential processes.

Now suppose that the above KB is updated so as

$$\text{KB} := \{\forall x : a(x) \wedge b(x) \Rightarrow c(x), \quad a(x) \wedge b(x) \Rightarrow d(x), \quad c(x) \Rightarrow d(x)\}.$$

We can promptly see that this is a *reducible set of constraints*, since one of the premises can be derived from the premises themselves, that is, if we get rid of the premise $c(x) \Rightarrow d(x)$, then

$$\forall x : \{a(x) \wedge b(x) \Rightarrow c(x), \quad a(x) \wedge b(x) \Rightarrow d(x), \quad c(x) \wedge d(x) \Rightarrow a(x)\}$$

is now irreducible, that is, no premise can be derived from the others. Interestingly, this is not true in the environment with the restriction stated by Eq. (1), which is shown in Fig. 6.2. Again this corresponds with the restriction to the class of functions used to check the inference. Under this restriction, $c(x) \wedge d(x) \Rightarrow a(x)$ can be derived from the others premises.

Linear constraints. The inferential mechanism described for FOL constraints can nicely be exported to a large class of constraints. We can immediately gain an insight on the generality of that notion when considering the case of real-valued linear constraints. For example, let us assume that $\forall x \in \mathbb{R}^2$ the knowledge base KB consists of

$$3f_1(x) + 2f_2(x) - f_3(x) - 1 = 0,$$
$$f_1(x) - 2f_3(x) - 2 = 0.$$

Like for logic statements, we can define formal inference and inference in the environment. Concerning formal inference, if we drop x, we can easily check that

$$\text{KB} \vdash 4f_1 + 2f_2 - 3f_3 - 3 = 0,$$
$$\text{KB} \nvdash f_2^3 + f_2 f_3^2 + f_1 = 0.$$

In the first case, $\forall f_1, f_2, f_3$ which makes the premises true then $4f_1 + 2f_2 - 3f_3 = 3$ is true, since this just comes from adding the premises. On the other hand, the second argument is false. It suffices to notice that while $f_1 = 2$, $f_2 = -5/2$, $f_3 = 0$ satisfy the premises, the conclusion is violated. Now something similar to the restriction on the range of the previous Boolean functions a, b, c, d holds in this case. Suppose that the environment on which the agent lives leads to restricting the class of admissible functions f_1, f_2, and f_3. Let's consider the class of functions

$$\begin{aligned} f_1(x) &= 0, \\ f_2(x) &= \cos x, \\ f_3(x) &= -\sin x. \end{aligned} \tag{3}$$

Like in the previous example, the range of $f = (f_1, f_2, f_3)'$ is restricted because of the dependency on the environmental variable x. We can easily see that the premises are only met when we also restrict the domain of those function to the countable set defined by $S = \{x \in \mathbb{R} : x = (\pi/2) + 2\kappa\pi, \ \kappa \in \mathbb{N}\}$, and that on these points also the conclusion is verified. Notice that, unlike in the previous example, there is a remarkable drop of measure in the perceptual space $X = \mathbb{R}$, which is restricted to the countable set $S \subset X$. In the previous case, the box-like restrictions of $X = \mathbb{R}^2$ didn't result in a drop of measure. Generally speaking, in high-order dimensional spaces it makes sense to assume that there's a significant drop of measure, so that data are only distributed on manifolds.

Supervised pairs. The same principle of inference that has been used for FOL predicates and linear constraints can be adopted for the constraints imposed on functions by a classic training set of supervised pairs.

However, in this case, the pointwise singularity of the constraints, which can nicely be read when expressing the generic supervision on pair (x_κ, y_κ) as $V(x_\kappa, y_\kappa f(x))\delta(x - x_\kappa) = 0$ by means of the Dirac distribution, needs a more careful analysis. Unlike other constraints which act on all elements of nonzero measure subsets of the perceptual space X, here the constraints only impose a condition on f on a finite collection on points. Hence the adoption of the formal inference defined by $\vdash$ doesn't really provide any meaningful derivation! Basically all $f \in F$, apart from those which don't meet the condition $y_\kappa = f(x_\kappa)$, must be considered for checking the conclusion, which leads to the degeneration of $\vdash$. If we immerse the inference into the environment then things become more interesting. As the environment brings any sort of regularity on f then the sketched ideas on inference can be generalized to the class of pointwise constraints. Notice that, also in the first example, we can

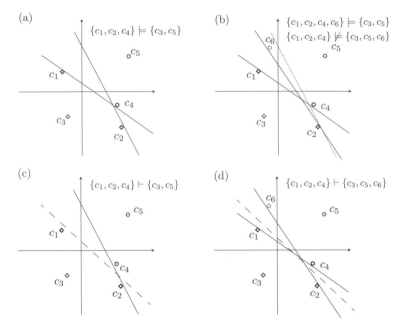

FIGURE 6.3

(a) Any line which separates c_1 and c_2 from c_4 also separates c_3 and c_5. (b) When adding c_6, the same conclusion holds true, though the class of separating lines reduces significantly. In (c) and (d) the dashed separation line indicates the unique robust line separation of kernel machines.

clearly see the relevance of the choice of the admissible class of functions, since the inferential mechanisms are different when considering the underlying probability distribution that is reflected in the definition of the predicates (1). On the other hand, the postulation of hypothesis on the space of admissible functions is in fact typical of any machine learning approach. The idea is sketched in Fig. 6.3, where, for the sake of simplicity, we restrict the functions to linear transformations of the feature space. It is worth mentioning that the inferential principle can be applied when using other classes of functions. From Fig. 6.3–(a), we can see that

$$\{c_1, c_2, c_4\} \models c_3, c_5. \tag{4}$$

Here, c_i, $i = 1, \ldots, 5$ denote the pointwise constraints connected with the corresponding point in $\mathbb{R}^2$, that is, $c_i \sim (y_i - f(x))^2 \delta(x - x_i) = 0$ (quadratic loss), where $f(x) = \hat{w} \cdot x$. As we can see, all possible lines that separate the premises $\{c_1, c_2, c_4\}$ also separate the conclusions c_3 and c_5. Clearly, this comes out from the reasoning in the perceptual environment $X = \mathbb{R}^2$. However, from Fig. 6.3–(b) we can see that we cannot draw conclusion c_6, that is, $\{c_1, c_2, c_4\} \not\models c_6$. Basically, the derivation doesn't hold if we choose the LTU associated with the dotted line. However, if we add c_6 to the premises we have

$$\{c_1, c_2, c_4, c_6\} \models c_3, c_5. \tag{5}$$

Like for the previous examples, it turns out that $\{c_1, c_2, c_4\}$ and $\{c_1, c_2, c_4, c_6\}$ are two irreducible sets of formal support constraints for the case of Fig. 6.3–(a) and Fig. 6.3–(b), respectively.

    The common logic structure that arises from these three significantly different examples reveals the generality of reasoning with constraints. Like in the first two examples, we may require hard enforcement of the constraints, or, like in the last example, only enforcement of soft-constraints. The connection with the perceptual space suggests translating the inference from the space of functions F (e.g., literals in logic) to the perceptual space X. In so doing, the computational barrier of inferential schemes in F is moved to X, where we still need to check whether joint satisfiability of the premises leads to the satisfiability of the conclusion. However, when working on X we can think of the joint satisfiability of the premises as learning from constraints, that is, to a *parsimonious satisfaction*. In the last example, this corresponds with discovering of the linear function which maximizes the class separation margin. This can be seen in Fig. 6.3–(c) and Fig. 6.3–(d), which shows that a *parsimonious decision* is supported by a few constraints. While this is well-known and leads to kernel machines, we notice that the same principle holds for the other two examples, so as we can always carry out a parsimonious inference, which will be denoted by the operator $\models^*$.

    The difference between $\models$ and $\models^*$ clearly emerges when looking at Fig. 6.3–(c) and Fig. 6.3–(d), since we have

$$c_1, c_2, c_4 \not\models c_6,$$
$$c_1, c_2, c_4 \models^* c_6.$$

While the deduction $\models$ inherits the classic enumerative check which characterizes logic inference, $\models^*$ is in fact the decision that comes from a learning agent! However, such a parsimonious decision takes exactly the same face for general constraints. Hence the interplay of learning and inference, which is well-known in supervised learning, does hold also for a large class of constraints. Like for Support Vector Machines, the *support constraints* are those which are in charge of decisions and are enabled by the smoothness hypothesis on the class of functions. They are characterized by the corresponding Lagrangian multiplier, so for support constraints $\lambda_i \neq 0$, whereas for *straw constraints* we have $\lambda_i = 0$. Like for kernel machines, in case of straw constraints the reaction ω_i is null and, therefore, they do not play any role in the functional representations discussed in the previous section. The (parsimonious) learning from constraints creates a partition into the set of constraints $\mathcal{C}_\psi := \{\psi_1, \ldots, \psi_\ell\}$ into the sets of support $\hat{\mathcal{C}}_\psi$ and straw $\tilde{\mathcal{C}}_\psi$ constraints. We can immediately see that, since the functional representation of the solution is independent of straw constraints,

$$\mathcal{C}_\psi \models^* \tilde{\mathcal{C}}_\psi.$$

An in-depth view on the logic structure of constraints and on the corresponding inference process which generalizes the intuitive description given by these examples is given in Section 6.2.

EXERCISES

1. [*20*] Express the N-queen problem by First-Order-Logic constraints.

2. [*15*] Let $\epsilon > 0$ and consider the constraints

$$\check{\psi}_1(f_1, f_2, f_3) = f_1 f_2(f_3 - 1) - \epsilon \geq 0,$$

$$\check{\psi}_2(f_1, f_2, f_3) = \frac{f_1 f_2 f_3^2 (f_3 - 1)}{f_1^2 + f_2^2} - \epsilon \frac{f_3^2}{f_1^2 + f_2^2} \geq 0,$$

$$\check{\psi}_3(f_1, f_2, f_3) = \frac{f_3^2 (f_3 - 1)}{f_1 f_2} - \epsilon \left(\frac{f_3}{f_1 f_2} \right)^2 \geq 0.$$

Prove that these constraints are equivalent.

3. [*15*] Consider the bilateral holonomic constraints

$$\psi_1(x, f_1, f_2, f_3, f_4) = f_1^2 - f_2^2 - f_3 = 0,$$
$$\psi_2(x, f_1, f_2, f_3, f_4) = f_1 - f_2 - f_4 - 1 = 0.$$

Prove that $\{\psi_1, \psi_2\} \models \overline{\psi}$, where $\overline{\psi}(x, f_1, f_2, f_3, f_4) = f_1 + f_2 + f_1 f_4 + f_2 f_4 - f_3$.

6.2 Logic constraints in the environment

The formulation of learning with constraints given in the previous chapter relies on a very general notion of constraint, which also embraces relationships between real-valued and Boolean variables. Because of its fundamental importance, the case of logic constraints, around which the fundamental mechanisms of symbolic reasoning revolve, is conveniently set apart. For Boolean-valued variables to satisfy a set of constraints, one typically needs to explore combinatorial structures, which smell of computational intractability. The math behind constraint satisfaction involving real-valued variables has a truly different form. More than searching methods in a combinatorial space, one moves by continuous-argument processes in the parameter space. In both cases, we are driven by proper heuristics, which is in fact the dominant component of learning.

This section covers the translation of logic constraints into real-valued constraints so as to get congruent representations that can be used in the unified formulation of learning from constraints. Like in classic supervised learning, where the choice of the different loss function yields remarkably different results, in the case of logic constraints there are in fact different choices that are related to the notion of triangular norm.

6.2.1 Formal logic and complexity of reasoning

While machine learning and automated reasoning are definitely intertwined, their treatment is often surprisingly kept separate in terms of basic methods! Roughly speaking, the roots of this separation are in the different math on which we construct theories. While machine learning relies mostly on continuous math, automated reasoning is primarily built up on logic. Of course, there are plenty of practical problems in which one can conceive excellent engineering solutions just by constructing an appropriate hybrid architecture, in which separate modules face different tasks and then communicate with each other to achieve the goal. But that's not always the case.

Let us consider the classical AI animal recognition problem based on the rules summarized in Table 6.2. This is a nice prototype of those problems where an intelligent agent takes decision solely on the basis of a logic-based inferential process. For example, suppose we know that an animal has hair and hoofs, it has a long neck, tawny color, and dark spots. From R1, we know that it is a mammal, while from R7 we deduce that it is an ungulate. Finally, from R12 we draw the conclusion that the animal is a giraffe. This inferential process is activated by the knowledge of hair, hoofs, longneck, tawny, and darkspots, that is, by some triggering literals that we are given to take the decision.

Suppose that the previously described KB is fully available in its triggering literals, so that they are fully specified by symbols, without connection with the environment. Basically, we do not possess any picture of the animals, but only linguistic features. In this case, the animal recognition problem can nicely be formulated as inference in KB. For example, once the following animal features hair, hoofs, longneck, longlegs, tawny, darkspots are true, the quest for which animal is identified by these linguistic

Table 6.2 A simplified animal identification problem due to Patrick Winston.

	Natural Expression of the Rule	Proposition
R1	If the animal has hair then it is a mammal.	`hair` $\Rightarrow$`mammal`
R2	If the animal gives milk then it is a mammal.	`milk` $\Rightarrow$`mammal`
R3	If the animal has feathers then it is a bird.	`feathers` $\Rightarrow$`bird`
R4	If the animal flies and it lays eggs then it is a bird.	`flies` $\wedge$`layseggs` $\Rightarrow$`bird`
R5	If the animal is a mammal and it eats meat then it is a carnivore.	`mammal` $\wedge$`meat` $\Rightarrow$`carnivore`
R6	If the animal is a mammal and it has pointed teeth and it has claws and its eyes point forward then it is a carnivore.	`mammal` $\wedge$`pointedteeth` $\wedge$`claws` $\wedge$`forwardeye` $\Rightarrow$`carnivore`
R7	If the animal is a mammal and it has hoofs then it is an ungulate.	`mammal` $\wedge$`hoofs` $\Rightarrow$`ungulate`
R8	If the animal is a mammal and it chews cud then it is an ungulate.	`mammal` $\wedge$`cud` $\Rightarrow$`ungulate`
R9	If the animal is a mammal and it chews cud then it is even-toed.	`mammal` $\wedge$`cud`$\Rightarrow$`eventoed`
R10	If the animal is a carnivore and it has tawny color and it has dark spots then it is a cheetah.	`carnivore`$\wedge$`tawny` $\wedge$`darkspot` $\Rightarrow$`cheetah`
R11	If the animal is a carnivore and it has tawny color and it has black stripes then it is a tiger.	`carnivore` $\wedge$`tawny` $\wedge$`blackstripes` $\Rightarrow$`tiger`
R12	If the animal is an ungulate and it has long legs and it has a long neck and it has tawny color and it has dark spots then it is a giraffe.	`ungulate` $\wedge$`longlegs` $\wedge$`longneck` $\wedge$`tawny` $\wedge$`darkspots` $\Rightarrow$`giraffe`
R13	If the animal is an ungulate and it has white color and it has black stripes then it is a zebra.	`ungulate` $\wedge$`white` $\wedge$`blackstripes` $\Rightarrow$`zebra`
R14	If the animal is a bird and it does not fly and it has long legs and it has a long neck and it is black and white then it is an ostrich.	`bird` $\wedge\neg$`fly` $\wedge$`longlegs` $\wedge$`longneck` $\wedge$`black`$\Rightarrow$`ostrich`
R15	If the animal is a bird and it does not fly and it swims and it is black and white then it is a penguin.	`bird` $\wedge\neg$`fly` $\wedge$`swims` $\wedge$`blackwhite` $\Rightarrow$`penguin`
R16	If the animal is a bird and it is a good flyer then it is an albatross.	`bird` $\wedge$`goodflier` $\Rightarrow$`albatross`

feature can be written in terms of truth of the following logic argument:

$$\{\texttt{hair, hoofs, longneck, longlegs, tawny, darkspots}\} \models \texttt{giraffe}. \qquad (1)$$

In this framework, the given features can be regarded as assumptions that we number as $A1 = $`hair`, $A2 = $`hoofs`, $A3 = $`longneck`, $A4 = $`longlegs`, $A5 = $`tawny`, and $A6 = $`darkspots`. The inferential process to conclude whether the argument is true is based on the following steps:

1. mammal $(A1, R1 \Rightarrow E)$,
2. ungulate $(\text{mammal}, A2 \Rightarrow E)$,
3. giraffe $(\text{ungulate} \wedge A3 \wedge A4 \wedge A5 \wedge A5, R12 \Rightarrow E)$.

Here, $\Rightarrow E$ denotes the classic *modus ponens* inference rule. Notice that the above argument is equivalent to establishing the truth of the proposition

hair $\wedge$ hoofs $\wedge$ longneck $\wedge$ longlegs $\wedge$ tawny $\wedge$ darkspots $\Rightarrow$ giraffe.

This formula, like all the formulas of the KB, is a *Horn clause*, which is special interesting logic fragment for which there exist polynomial algorithms for the automated proof.

The KB of Table 6.2 gives the meaning of terms by specifying the required properties, that is, the necessary and sufficient conditions for belonging to the set being defined. This is referred to as *intensional knowledge* and represents an abstract definition of the categories. On the other hand, the explicit list of items belonging to a given category is referred to as *extensional knowledge*. For example, suppose we consider a set of animals and we name them all. This is simply represented by a list of unary predicates; for example,

cheetah(Abby)	tiger(Randy)	giraffe(Lala)
zebra(Fajita)	ostrich(Jeniveve)	penguin(Pinky)
albatross(Azul)	tiger(Tacoma)	zebra(Skumpy)

is the extensional knowledge associated with the intensional knowledge of Table 6.2. Suppose we extend the KB of Table 6.2 with

$\forall x \forall y \forall z$ tiger(x) $\wedge$ zebra(y) $\wedge$ hantenv(z) $\Rightarrow$ hunt(x, y)

faster(Randy, Fajita)

$\forall x \forall y$ faster(x, y) $\wedge$ hunt(x, y) $\Rightarrow$ eat(x, y)

Here, we assume that hantenv(z) is true if z is an environment which favors the hunt. If hantenv(z) then from these premises, we can draw that conclusion eat(Randy, Fajita), where we can see an inferential mechanism that relies on combining factual and abstract knowledge.

In general, the argument $\{p_1, \ldots, p_n\} \models c$ is valid if and only if the associated proposition $\bigwedge_{\kappa=1}^{n} p_\kappa \Rightarrow c$ is a tautology. Now this corresponds with solving SAT, which is intractable. This is in fact the face of complexity recurrently emerging from the combinatorial search of satisfaction of Boolean constraints. In the following, we will see how the availability of related real-valued information from the environment leads to approximate reasoning processes that can be significantly more efficient than formal deduction.

What if some of the previous linguistic features are missing? Clearly, logic-based inferential processes are doomed to fail. More interestingly, suppose that instead of relying on the above collection of animal features, one is given a sentence in natural language that describes the animal. For example, one might want to answer the question:

$$\texttt{What is the animal which has hoofs,}$$
$$\texttt{as well as a long neck and dark spots?} \tag{2}$$

We are missing some of the features that describe the `giraffe` and, in addition, in this case we must be able to process the sentence to draw conclusions. Clearly, this is extremely critical, since we can easily reformulate the above question in such a way that the same features are present, whereas logic modifiers can indicate their absence, so as to indicate a different animal! As it will be shown in the next section, in other cases one might think of extending this animal identification problem to one in which, in addition to a linguistic query, we rely on the availability of a collection of animal pictures.

EXERCISES

1. [*16*] Based on the animal recognition problem whose rules are given in Table 6.2, consider the following argument:

$$\{\texttt{hair}, \texttt{hoofs}\} \models \texttt{giraffe} \vee \texttt{zebra}.$$

At first glance, one might conclude that it is true, since there are no other animals apart from zebra and giraffe which possess these two features. Is it really a valid argument? How can we enrich the KB so that this proposition is true?

2. [*14*] Consider the constraint

$$f_1^2 f_2^2 - f_1 f_2^2 - f_1^2 f_2 + 3 f_1 f_2 - f_1 - f_2 = 0.$$

Prove that it corresponds with the equivalence of f_1 and f_2.

6.2.2 Environments with symbols and subsymbols

As pointed out in Section 1.2, the separate view of models dealing with symbols and subsymbols does not turn out to be a very good strategy in case of intelligent agents that are exposed to environment where both symbolic and subsymbolic information is available. Of course, there are plenty of practical problems in which one can conceive excellent engineering solutions just by constructing an appropriate hybrid architecture, in which separate modules face different tasks and then communicate each other to achieve the goal. But that's not always the case! Let us consider the animal recognition problem that is reported in Table 6.2. This is in fact a nice prototype of those problems where an intelligent agent takes decision solely on the basis of a logic-based inferential process that relies on some triggering literals that support decision. Again, what happens if some of them are missing? For example, in the giraffe recognition task, the proposition R1 propagates evidence on the mammal category, once we know that the animal has got hair. Now let us assume to extend this

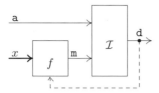

FIGURE 6.4

Decision taken on the basis of the feature vector x and of the available symbolic features a. The dashed line indicates more sophisticated processes where the prediction of the missing symbolic feature m is also affected by the final decision.

animal identification problem to one in which, while some of the triggering literals are missing, we rely on the availability of a collection of animal pictures. In this case the information for animal identification is integrated with a pattern $x \in \mathbb{R}^d$ composed of proper features extracted from the picture, but we can only rely on partial knowledge of the triggering literals, since some of them are missing. Clearly, hair is a triggering literal, since it allows us to infer abstract categories, like mammal (see R1) Interestingly, if it is missing, the decision can be activated by the acquisition of the feature-based representation of the pattern x. Beginning from animal pictures and, consequently, from their feature-based representation, one can also think of inducing directly the animal category using learning from examples. Hence one could also simply learn the functions which identify the animals; if we want to recognize giraffes, then we could focus on learning the function giraffe(x) and disregard the previous KB! This is exactly the opposite extreme of the pure logic inferential process: In this case, the decision does not involve rules, but it turns out to be an inductive process based solely on the content of the picture. Now, the prospected framework on the partial availability of literals (symbolic information) and feature-based pattern representations is quite common in practice and, obviously, it makes sense to use all the information available for the decision.

A direct approach for tackling the problem is that of benefiting from the joint availability of the KB and of the animal pictures simply by splitting the decision process into two modules based on learning from examples and logic inference, respectively. In so doing, the first module is charged of completing the information on the missing literals m by using the some available real-valued features x by learning the function f, thanks to the availability of a training set of labeled data $\mathcal{L} = \{(x_\kappa, m_\kappa) : \kappa = 1, \dots, \ell\}$. As a consequence, all the triggering literals are available to activate the formal logic inferential process that is carried out by the second module. While this piping scheme enriches the logic inference thanks to the availability of real-valued pattern features, it does not exploit the chance of revising the decision m on the missing literals on the basis of the final decision d. For example, if we rank the decision of f, each corresponding prediction of m can give rise to a different decision d, and when using only the piping flow of information (see Fig. 6.4), one implicitly relies on the assumption that the rank returned by f leads to the best decision. However, it might not fully resonate with the overall inference process and, therefore, it's interesting to explore how the two modules can cooperate more closely. How can this be done? Clearly, we need a way of expressing the overall behavior of the system, not only a criterion for finding m by learning from examples. No matter how it works, the decision d must be affected by m which, in turn, must also depend on d. This loopy process is referred to as *reasoning in the environment*, which is a nice emulation of some complex human reasoning schemes that can hardly be captured in the framework of formal logic.

Within this framework, while the prediction of m is an inductive process, the decision on d is based on deduction from the KB. The dashed line in Fig. 6.4 represents the loopy process which circulates the acquisition of evidence on m and d. The general discussion of the previous section suggests that reasoning based on formal logic only explores a corner of the planet of intelligent reasoning. While in formal logic triggering literals are symbols isolated from the environment, we can think of an extension that involves the feature space $X \subset \mathbb{R}$.

Now, we give additional insights on how to bridge logic-based descriptions of the environment with supervised examples. One of the most efficient ways of communicating with an intelligent agent is that of relying on logic formalisms. In a sense, if we ask the agent to be consistent with a collection of logic constraints, we are generalizing the act of teaching that is typical of supervised learning. We can think of any supervised pair (x_κ, y_κ) as a predicate $s(x_\kappa, y_\kappa)$, so in that classic setting the aim of learning is to return a *parsimonious* answer that is *soft-consistent* with the given predicates.

The natural extension of this scheme of learning is to maintain the principle of associating real-valued functions to symbolic entities. In order to catch the idea, it is instructive to come back to the learning tasks described in Fig. 6.4, where the KB is formally expressed by Eq. 6.1.4–(2). The supervised pairs can also be interpreted as a collection of predicates that corresponds with the classes a (square), b (diamond), c (circle), and d (triangle), respectively. Of course, the classes are not mutually exclusive and, therefore, for a given class, e.g., a, we cannot use the examples labeled with b, c, d as negative examples. As a consequence, a predicate like $\neg b(x_1)$ and others constructed to express the nonmembership are additional granules of information with respect to the given KB. Now let us associate the classes with corresponding real-valued functions collected in $f = (f_1, f_2, f_3, f_4)'$, so as to construct the pairs $(a, f_1), (b, f_2), (c, f_3), (d, f_4)$. When considering soft-satisfaction of the constraints, the supervised points can be associated with the hinge penalty

$$E_s = \sum_{j=1}^{n} \sum_{\kappa=1}^{\ell_s} s_{\kappa,j} \cdot \left(1 - y_{\kappa,j} \cdot f_j(x_\kappa)\right)_+, \tag{1}$$

where $s_{j,\kappa} = [y_{\kappa,j}$ is available$]$ and ℓ_s is the number of supervised examples. Now, since the logic constraints share the same structure, we can associate them with the same penalty term. For this reason, we only focus attention on the constraint $a \wedge b \Rightarrow c$. We start noting that any good candidate penalty must be consistent with the constraint. Ideally, the penalty is expected to return no error in case of truth, whereas

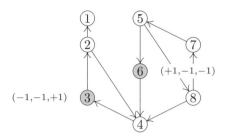

FIGURE 6.5

Simple relational environment, where any link means that the decision on the parents is propagated to the children. Notice that this can be very different with respect to the typical diffusion of recurrent neural networks.

it must penalize any triple for which the proposition is false, that is,

$$E(\mathsf{a} \wedge \mathsf{b} \Rightarrow \mathsf{c}) = \int_X \big(1 + f_1(x)\big) \cdot \big(1 + f_2(x)\big) \cdot \big(1 - f_3(x)\big)\, dx, \qquad (2)$$

where we assume that $f_j(x)$, $j \in \mathbb{N}_4$ takes on values in $[-1, +1]$. We can promptly see that this real-valued penalty is null when $f(x)$ is perfectly coherent with the constraint $\mathsf{a} \wedge \mathsf{b} \Rightarrow \mathsf{c}$. As already pointed out in the previous section, we can adopt different penalties for both the supervised examples and the logic constraints. The corresponding penalties present a relevant difference: While E_s acts on supervised examples, $E(\mathsf{a} \wedge \mathsf{b} \Rightarrow \mathsf{c})$ acts over X. A good tradeoff in this case is to enforce the penalization over all the available *unsupervised* data. This can be done by

$$E \mathbin{\dot{\frown}} \sum_{\kappa \in \mathbb{N}_{\ell_u}} \big(1 + f_1(x_\kappa)\big)\big(1 + f_2(x_\kappa)\big)\big(1 - f_3(x_\kappa)\big). \qquad (3)$$

Notice that one can naturally pose questions that go well beyond the membership of a, b, c, d. For example, the query $\forall x\ \mathsf{c}(x) \wedge \mathsf{d}(x) \Rightarrow \mathsf{a}(x)$ of Fig. 6.2 highlights a fundamental property of parsimonious inferential processes. While it is clearly false in a strict logic sense, it holds true in the given environment, which is also characterized by the presence of supervised examples that, once the parsimony principle is invoked, do contribute to restricting the inferential space.

Now let us consider the learning environment depicted in Fig. 6.5. We assume that we are given two types of vertex, namely $x_1, x_2, x_4, x_5, x_7, x_8 \in X_v \subset \mathbb{N}$ and $x_3, x_6 \in X_r \subset \mathbb{R}^3$. While x_3 and x_6 are feature-based patterns, the others in X_v are only characterized by their identifier, which is an integer. We assume that all the elements in $X = X_r \cup X_v$ are either of class a or b or c. Each class is associated with two unary predicates. For example, class a corresponds with a_r, a_v. The first one

$$\mathsf{a}_r : X_r \to \{\mathsf{T}, \mathsf{F}\}$$

is defined only on feature-based patterns in X_r, while

$$a_v \colon X_v \to \{\mathsf{T}, \mathsf{F}\}$$

on the vertices. Clearly, we need to impose consistency on a_r and a_v, that is, we need to impose $\forall x \in X_r \ a_r(x) \Leftrightarrow a_v(x)$. Functions that depend only on real-valued features are referred to as feature-based predicates, whereas those which depend on identifiers (integers) are labels referred to as predicates. The same holds for classes b and c. To sum up, suppose we have the following constraints:

$$
\begin{aligned}
&i && a_r(x_6),\, c_r(x_3); \\
&ii && r(x_2, x_1),\, r(x_3, x_2),\, r(x_4, x_3),\, r(x_2, x_4),\, r(x_8, x_4), \\
& && r(x_5, x_8),\, r(x_8, x_7),\, r(x_7, x_5),\, r(x_5, x_6); \\
&iii && \forall x, \forall y \quad r(x, y) \Rightarrow \big(b_r(x) \Rightarrow b_r(y) \vee b_v(y)\big); \\
&iv && \forall x, \forall y \quad r(x, y) \Rightarrow \big(c_r(x) \Rightarrow c_r(y) \vee c_v(y)\big); && (4) \\
&v && \exists x \quad b_r(x) \vee b_v(x); \\
&vi && \forall x \in X_r \quad a_r(x) \Leftrightarrow a_v(x); \\
&vii && \forall x \in X_r \quad b_r(x) \Leftrightarrow b_v(x); \\
&viii && \forall x \in X_r \quad c_r(x) \Leftrightarrow c_v(x).
\end{aligned}
$$

Constraints i come from the supervised pairs, while ii represents the relations amongst the vertices of the graph. Constraints iii and iv attach the meaning to the links: No matter what decision is taken on x, it "implies" the same decision on y on classes b and c. To some extent, as it will be seen in the following section, this is a way of carrying out a diffusion process that is commonly done in recurrent neural networks. However, there are remarkable differences, since in this case we control the diffusion mechanism on single vertices. In particular, while we propagate the evidence on b and c, we don't propagate the evidence on a.

Constraint v enforces solutions in which there is at least one individual that is classified as b and, finally, vi, vii, and $viii$ enforce the consistency between feature-based and hybrid predicates. Like in the previous example, we associate real-valued functions $f_{j,t}$ with $j = 1, 2, 3$ and $t \in \{r, v\}$ to the predicates a, b, c, respectively. Of course, the real-valued constraints to be associated with Eq. (4)–i are penalties like in Eq. (1). As for (4)–iii and (4)–iv, we can use the penalty

$$E_{iii, iv} = \sum_{j=2}^{3} \sum_{t \in \{r,v\}} \sum_{r(x,y)=\mathsf{T}} \big(1 + f_{j,r}(x)\big)\big((1 - f_{j,t}(y)\big). \qquad (5)$$

As already noticed, this penalty gives rise to evidence diffusion that is restricted to functions $f_{2,r}$, $f_{2,v}$, $f_{3,r}$, and $f_{3,v}$, since no evidence is propagated for class a. Notice that this penalty also enforces the diffusion by evidence coming from label predicates. The translation of the universal quantifiers is carried out by the accumulation over all the training data. As for constraint v, notice that, according to De Morgan laws, since it corresponds with $\neg(\forall\ f_{2,t}(x) = -1)$, we can carry out the translation by using again the universal quantifier translation. Hence consider the penalty

$$E_v = 1 - \prod_{t \in \{r,v\}} \prod_{x \in X} (1 - f_{2,t}(x))/2. \tag{6}$$

If $\forall f_{2,t}(x) = -1$ then $E_v = 0$. On the other hand, if at least for one of the factors $f_{2,t}(x) = +1$ then $E_v = 1$. Another possible translation is

$$E_v = \exp\left(-\gamma \sum_{x \in X} \sum_{t \in \{r,v\}} \left(1 + f_{2,t}(x)\right)\right). \tag{7}$$

We can promptly see that if $\forall x\ f_{2,t}(x) = -1$ then we have $E_v \simeq 1$, while $E_v \simeq 0$ if $\neg(\forall\ f_{2,t}(x) \neq -1)$, that is, if $\exists \bar{x} : f_{2,t}(\bar{x}) \simeq 1$. Clearly, for this to hold, we need to choose γ large enough. Of course, all this is made possible since X is finite. Finally, for the conversion of vi, vii, and $viii$ we can select the penalty

$$E_{vi,vii,viii} = \sum_{j=1}^{3} \sum_{x \in X_r} \left(f_{j,r}(x) \cdot f_{j,v}(x) - 1\right)^2. \tag{8}$$

As will be shown in the remainder of this chapter, the construction of the penalty functions can be driven by a general methodology that has been widely used in fuzzy systems. As an example, let us focus on the logic constraint $a \wedge b \Rightarrow c$ that can be rewritten as

$$\forall x \quad \neg\big((a(x) \wedge b(x)) \wedge \neg c(x)\big). \tag{9}$$

We notice that $a(x) \wedge b(x)$ can be naturally associated with $f(x) \cdot f_2(x)$, since whenever $f_1(x)$ and $f_2(x)$ take on values close to 0 and 1, it turns out that $\mathcal{T}(f(x)) = f_1(x) \cdot f_2(x)$ takes on values close to 0 and 1, which corresponds with the truth of $a(x) \wedge b(x)$. Likewise, given any Boolean variable like $c(x)$, clearly $1 - f_3(x)$ gets a related correspondence with $\neg c(x)$. As a consequence, we can associate $\neg\big((a(x) \wedge b(x)) \wedge \neg c(x)\big)$ with

$$\mathcal{T}(f(x)) = 1 - f_1(x) \cdot f_2(x) \cdot (1 - f_3(x)) = 1,$$

that is, with

$$f_1(x) \cdot f_2(x) \cdot (1 - f_3(x)) = 0. \tag{10}$$

Now we can go one step further to translate the universal quantifier. We start noting that since $1 - \mathcal{T}(f(x)) = f_1(x) \cdot f_2(x) \cdot (1 - f_3(x)) \geq 0$, whenever the associated isoperimetric constraint is verified,

$$E^{(p)} = \left(\int_{\mathbb{R}^2} (1 - \mathcal{T}(f(x)))^{2p}\right)^{1/2p} dx = 0, \tag{11}$$

the condition $\mathcal{T}(f(x)) = 1$ holds true. Of course, this holds $\forall p \in \mathbb{N}$. In particular, as $p \to \infty$ the above condition becomes

$$E^{\infty} \propto \sup_{x \in \mathbb{R}^2} \left(1 - \mathcal{T}(f(x))\right) = \sup_{x \in \mathbb{R}^2} \left(f_1(x) f_2(x)(1 - f_3(x))\right) = 0. \qquad (12)$$

In practice, the availability of a finite collection of (unsupervised) data, indexed by $\kappa \in \mathbb{N}_\ell$, suggests the approximation

$$E^{(p)} \propto \left(\sum_{\kappa \in \mathbb{N}_{\ell_u}} \left(f_1(x_\kappa) \cdot f_2(x_\kappa) \cdot (1 - f_3(x_\kappa))\right)^{2p} \right)^{1/2p}.$$

Clearly, if we want our functions f_j to return values in $[-1, 1]$ instead of $[0, 1]$, we can always transform the above equation into

$$E^{\infty}_{[-1,+1]} = \sup_{x \in \mathbb{R}^2} \left[f_1(x) f_2(x)(1 - f_3(x)) \right]$$

$$= \sup_{x \in \mathbb{R}^2} \left[\frac{1 + \overline{f}_1(x)}{2} \frac{1 + \overline{f}_2(x)}{2} \left(1 - \frac{1 + \overline{f}_3(x)}{2} \right) \right]$$

$$= \frac{1}{8} \sup_{x \in \mathbb{R}^2} \left(1 + \overline{f}_1(x)\right) \left(1 + \overline{f}_2(x)\right) \left(1 - \overline{f}_3(x)\right)$$

$$\sim \sup_{x \in \mathbb{R}^2} \left(1 + \overline{f}_1(x)\right) \left(1 + \overline{f}_2(x)\right) \left(1 - \overline{f}_3(x)\right),$$

which corresponds with the initial heuristic choice in the integrand function of Eq. (2). In order to satisfy all the m constraints stated for the learning environment, we can simply $\wedge$ them all, so the corresponding penalty is

$$E_s = \left(1 - \prod_{i=1}^{m} \mathcal{T}_i(f)\right)^2,$$

where $\mathcal{T}_i(f)$ is the truth value mapped to the corresponding real value. Another possible penalty is

$$E_w = \sum_{i=1}^{m} (1 - \mathcal{T}_i(f))^2,$$

Clearly, E_s is much more restrictive than E_w.

Now we want to show that there is a natural conversion into the language of real-valued constraints for the existential quantifier which is different from the heuristic adopted in Eqs. (6) and (7). We can promptly see that the penalty

$$E_\exists = \left(\inf_{x \in \mathbb{R}^2} \left(1 - f_{2,r}(x)\right) \right) \left(\inf_{x \in \mathbb{R}^2} \left(1 - f_{2,v}(x)\right) \right)$$

translates the existential quantifier. There is in fact a close connection with Eq. (12), which is disclosed when we consider that from

$$\lim_{p \to +\infty} \left(\int_{\mathbb{R}^2} \mathcal{T}(f(x))^{2p} \, dx \right)^{1/2p} = 1$$

we get

$$1 - \sup_{x \in \mathbb{R}^2} \mathcal{T}(f(x)) = 0 \Rightarrow \inf_{x \subset \mathbb{R}^2} (1 - \mathcal{T}(f(x))) = 0.$$

Hence, when the existential quantifier is involved, the dual penalty of (12) turns out to be

$$E_\exists = \inf_{x \in \mathbb{R}^2} \left(1 - \mathcal{T}(f(x)) \right). \tag{13}$$

Another way of translating the existential quantifier is to consider

$$E_\exists^p = \prod_{x \in X} \left(1 - \mathcal{T}(f(x)) \right).$$

This can promptly be checked, since $E_\exists^p = 0$ if there exist $\bar{x}$ such that

$$\mathcal{T}(f(\bar{x})) = 1.$$

Finally, let us discuss an important issue that is connected with the expression of truly relational information, that is naturally expressed by binary predicates. Suppose we are given the formula

$$\forall x \quad \texttt{Person}(x) \wedge \texttt{Person}(y) \wedge \texttt{FatherOf}(x, y) \Rightarrow \texttt{Male}(y). \tag{14}$$

Basically, this states that the father of any person is male. Such a knowledge granule can be translated into real-valued functions by using the above arguments. However, it's worth mentioning that $\texttt{fatherOf}(x, y)$ operates on X^2, which leads to doubling the dimensional space of the input! Of course, this shouldn't be neglected, since the task of learning the classifier $\texttt{fatherOf}(x, y)$, which is supposed to state whether y is the father of x, is likely to be remarkably more difficult than the other tasks involved in the above formula. However, we can do something different by translating the above knowledge granule into

$$\forall x \quad \texttt{Person}(x) \Rightarrow \texttt{Male}\big(\texttt{FatherOfFeatures}(x)\big). \tag{15}$$

The basic idea is to replace the binary predicate $\texttt{fatherOf}(x, y)$ with the function $\texttt{FatherOfFeatures}(x)$. Since the unary predicates $\texttt{Person}$ and $\texttt{Male}$ operate on $X = \mathbb{R}^d$, we can promptly see that the consistency of types in the above formula does require that $\texttt{FatherOfFeatures}$ returns a feature vector $f = \texttt{FatherOfFeatures} \in X$. This suggests that $\texttt{FatherOfFeatures}$ is in fact returning the features of the father of $x \in X$. For instance, if x is a vector of features that represents the face of a person then $f = $ FatherOfFeatures(x) is a vector of features that is expected to represent the face of the father of x. As a consequence, when (15) is one of the statements of a collection of predicates, their overall satisfaction requires the *generation* of the vector of features $f = $ FatherOfFeatures(x). Interestingly, this idea can easily be generalized with the purpose of giving an intelligent agent the task of creating patterns that are consistent with a given set of constraints. In a sense, this is not far away from the constructive mechanism that has been described for the N-queens task, where we also require the constraint satisfaction to construct a solution. However, in this case $f = $ FatherOfFeatures(x) constructs a feature vector $f \in X$, which is expected to be a feature-based representation of the face of the father. This is made possible by a neural network that is charged with approximating the function. It's quite obvious that the effectiveness of such a generative scheme is strongly dependent on the overall collection of constraints and on the chosen regularization scheme. In logic, the described idea of generating the feature vector $f \in X$ is referred to as *skolemization* — a few details are given in Section 6.6. Exercise 1 invites to reader to discuss the generality of skolemization, while Exercise 2 raises a question on the equivalence of formulas (14) and (15).

EXERCISES

1. [*20*]   Consider the generative process behind skolemization. Is true that it can always replace statement (14)?

2. [*15*]   Consider the formulas (14) and (15). Are they equivalent?

3. [*15*]   Consider Exercise 2 and suppose we enrich the set of constraints by stating that $\forall x$ FatherOf$(x, y) \Rightarrow$ Older(y, x), that is, the father is older than the son. Furthermore, suppose we have a regressor age(x), which estimates the age of a person from a picture $x \in X \subset \mathbb{R}^d$. Clearly, we have

$$[\text{age}(y) > \text{age}(x)] \Rightarrow \text{Older}(y, x).$$

Discuss qualitatively the differences between imposing

$$\forall x \forall y \quad \text{Person}(x) \wedge \text{Person}(y) \wedge \exists y \quad \text{FatherOf}(x, y) \Rightarrow \text{Male}(y) \wedge \text{Older}(y, x),$$

and simply considering the constraint of Exercise 2.

6.2.3 T-norms

As shown in above discussion, a possible expression of logic notions in terms of real-valued constraints can rely on an appropriate replacement of the $\wedge$ operator with the product and the $\neg$ with the complement to one. These transformations from the digital to the continuous setting of computation have been the subject of in-depth analyses, especially in fuzzy systems. There are in fact different ways of "and-ing" real-valued variables that can be deeply understood in the framework of the *triangular norms* (t-norms). A map

$$T : [0, 1]^2 \to [0, 1]$$

is a *triangular norm* if and only if $\forall x \in [0, 1]$, $\forall y \in [0, 1]$, $\forall z \in [0, 1]$ the following properties are satisfied:

$$
\begin{array}{lll}
T(x, y) = T(y, x) & \text{(symmetry);} & \\
T(x, T(y, z)) = T(T(x, y), z) & \text{(associativity);} & \\
(x \leq \overline{x}) \wedge (y \leq \overline{y}) \Rightarrow T(x, y)) \leq T(\overline{x}, \overline{y}) & \text{(monotonicity);} & \quad (1) \\
T(x, 1) = x & \text{(one identity).} &
\end{array}
$$

Whenever the "one identity" holds uniquely for 1 (that is, when 1 is the only idempotent), a function T with the above property is referred to as *Archimedean* t-norm. The following T are classic examples of t-norms:

$$
\begin{array}{l}
T_P(x, y) = x \cdot y, \\
T_G(x, y) = \min\{x, y\}, \qquad\qquad\qquad (2) \\
T_Ł(x, y) = \max\{x + y - 1, 0\}.
\end{array}
$$

T_P is referred to as the product t-norm (p-norm), T_G is the *Gödel* t-norm, while $T_Ł$ is the Łukasiewicz's t-norm. Exercise 1 invites the reader to check the properties (1). In the case of p-norm, the associativity makes it possible to map $\bigwedge_{\kappa=1}^{n} x_\kappa$ to $T(x_1, \ldots, x_n) = \prod_{\kappa=1}^{n} a_\kappa$. Likewise, in the case of the min-norm $T(x_1, \ldots, x_n) = \min_{\kappa \in \mathbb{N}_n}\{x_1, \ldots, x_n\}$. Now, suppose we want to translate $x \vee y$. We can use De Morgan's identities to translate $x \vee y$ into a formula with operators that can be directly translated by t-norms. Since we have $x \vee y = \neg x \wedge \neg y$, in the case of p-norm, the above formula can be mapped to

$$
x \vee y \mapsto to 1 - (1 - x) \cdot (1 - y) = x + y - x \cdot y,
$$

while in the case of the min-norm

$$
x \vee y \mapsto 1 - \min\{1 - x, 1 - y\} = \max\{x, y\}.
$$

This provides motivations to consider a map $S \colon [0, 1]^2 \to [0, 1]$ with the following properties:

$$
\begin{array}{lll}
S(x, y) = S(y, x) & \text{(symmetry);} & \\
S(x, S(y, z)) = S(S(x, y), z) & \text{(associativity);} & \\
(x \leq \overline{x}) \wedge (y \leq \overline{y}) \Rightarrow S(x, y)) \leq S(\overline{x}, \overline{y}) & \text{(monotonicity);} & \quad (3) \\
S(x, 0) = x & \text{(zero identity).} &
\end{array}
$$

It is referred to as a *triangular conorm* (t-conorm). Clearly, if T is a t-norm then

$$
S(x, y) = 1 - T(1 - x, 1 - y) \qquad\qquad (4)
$$

is a t-conorm. S is the t-conorm induced by the T t-norm.

Now we show that there is a very natural way of constructing t-norms that is based on the notion of *t-norm generator*. Suppose we are given a strictly decreasing function $\gamma : [0, 1] \to [0, +\infty]$ such that $\gamma(1) = 0$. Then define $T : [0..1]^2 \to [0, 1]$ as

$$T(x, y) = \gamma^{-1} (\gamma(x) + \gamma(y)). \tag{5}$$

When this construction is adopted, we say that γ is an additive generator of T. As it will be seen for the Łukasiewicz t-norm (8), this definition does require an extension of the notion of inverse function. In particular, γ^{-1} denotes the *pseudoinverse* of γ, that is,

$$\gamma^{-1}(y) = \sup \{ x \in [0, 1] : \gamma(x) > y \}. \tag{6}$$

In case of nondecreasing function we define $\gamma^{-1}(y) = \sup \{ x \in [0, 1] \mid \gamma(x) < y \}$. We can promptly see that the function $\gamma(x) = -\log(x)$ is an additive generator of the p-norm. According to (5), we have

$$T(x, y) = \exp\left(-(-\log(x) - \log(y)) \right) = \exp\left(\log(x \cdot y) \right) = x \cdot y. \tag{7}$$

Likewise, the function $\gamma(x) = 1 - x$ is an additive generator of the Łukasiewicz t-norm. Following (5), we have

$$T(x, y) = \gamma^{-1} (\gamma(x) + \gamma(y)) = \gamma^{-1} ((1 - x) + (1 - y)) = \gamma^{-1} (2 - (x + y)).$$

Now, if $x + y - 1 \geq 0$, we have

$$T(x, y) = \gamma^{-1} (2 - (x + y)) = 1 - (1 - 2 + x + y) = x + y - 1, \tag{8}$$

otherwise $T(x, y) = 0$, which is in fact the Łukasiewicz t-norm.

Now let us consider a t-norm generated by the nonincreasing function γ and consider the map

$$T(g(x), g(y)) = \gamma^{-1} (\gamma(g(x)) + \gamma(g(y))), \tag{9}$$

where we choose g such that

$$\gamma(g(x)) = \begin{cases} x - \frac{1}{2} & \forall x \in (1/2, 1], \\ 0 & \forall x \in [0, 1/2]. \end{cases}$$

When we restrict to functions T that are separately continuous with respect to each variable and to Archimedean t-norms, we can always find a strictly increasing function $f : [0, 1] \to [0, 1]$ such that

$$T(x, y) = f^{-1} (T_L(f(x), f(y))), \tag{10}$$

where T_L is the Łukasiewicz t-norm. This arises from Eq. (9) as follows. First, notice that

$$\begin{aligned} g^{-1} \circ T\big(g(x), g(y)\big) &= g^{-1} \circ \gamma^{-1}\big(\gamma \circ g(x) + \gamma \circ g(y)\big) \\ &= (\gamma \circ g)^{-1}\big(\gamma \circ g(x) + \gamma \circ g(y)\big) \\ &= \beta^{-1}\big(\beta(x) + \beta(y)\big) = T_L(x, y). \end{aligned}$$

Let $w := g(x)$ and $z := g(y)$, then we have $T(w, z) = gT_L(g^{-1}(w), g^{-1}(z))$.

Given a t-norm T, its residuum is defined by

$$(x \Rightarrow y) = \sup\{z : T(x, z) \leq y\}.$$

The residuum $(x \Rightarrow y)$ is also denoted as $T^*(x, y)$ and is a sort of adjoint of $T(x, y)$. In order to grasp its meaning, we start considering the case $x \leq y$, where we expect that the implication returns "high" values. This follows straightforwardly when considering that $T(x, z) \leq T(x, 1) = x$ and then

$$\sup\{z : T(x, z) \leq x \leq y\} = 1.$$

Clearly, the converse holds as well, and therefore $(x \Rightarrow y) = 1$ iff $x \leq y$. Now, let us consider its violation. We need to study $(x \Rightarrow y)$ with $x > y$. Now we have $T(x, z) \leq x$, that is, $T(x, z) \leq \min\{x, z\}$. Hence

$$(x \Rightarrow y) = \sup\{z : T(x, z) \leq \min\{x, z\} \leq y\} = y.$$

The converse holds true as well. That is, $(x \Rightarrow y) = y$ iff $x > y$. Clearly, $(1 \Rightarrow y) = y$. The difference in the translation of $\Rightarrow$ using the residuum and the direct application of the t-norms is very well illustrated in the case of the p-norm. The classic propositional calculus definition of the implication $\neg(x \wedge \neg y)$ would be translated into

$$x \Rightarrow y \mapsto \mathcal{T}(x, y) = 1 - x \cdot (1 - y).$$

For example, for $x = 0.55$ and $y = 0.6$ we get $\mathcal{T}(x, y) = 0.78$, whereas $(0.55 \Rightarrow 0.6) = 1$. Likewise, if $x = 0.55$ and $y = 0.2$, we have $\mathcal{T}(0.55, 0.2) = 0.56$. Clearly, this suggests that the implication holds true, which does not reflect its meaning. On the other hand, $(0.55 \Rightarrow 0.2) = 0.2$, which is definitely preferable for a coherent definition of $\Rightarrow$. To sum up, the given definition of residuum captures perfectly the meaning one would like to attach to the implication.

The residua of the three t-norms, Goguen, Gödel, and Łukasiewicz, defined by (2), are (see Exercise 2)

$$x \overset{P}{\Rightarrow} y = y/x, \tag{11}$$

$$x \overset{G}{\Rightarrow} y = y, \tag{12}$$

$$x \overset{Ł}{\Rightarrow} y = 1 - x + y. \tag{13}$$

When interpreting $\Rightarrow$ by the evaluation of $\neg x \vee y$, we notice that the Łukasiewicz implication, for $y = 0$, yields $\neg x \mapsto 1 - x$, which clearly offers a natural notion of negation.

EXERCISES

1. [20] Prove that $T_P, T_G, T_Ł$ satisfy the properties of t-norms.

2. [*18*] Prove Eqs. (11), (12) and (13) concerning t-norm residua.

3. [*18*] Suppose we want to redefine t-norms with truth values in $[l, h] \subset \mathbb{R}$. Given $t: [0, 1] \to [0, 1]$, what is the associated t-norm $\bar{t}: [l, h] \to [l, h]$?

4. [*15*] Based on the remapping defined in the previous Exercise 3, determine the value of the p-norm for the logical expression

$$a_1 \Leftrightarrow a_2 = a_1 \wedge a_2 \vee \neg a_1 \wedge \neg a_2.$$

6.2.4 Łukasiewicz propositional logic

The natural expression of negation in Łukasiewicz t-norm, as well as other relevant properties that will be described in this section, suggest focusing attention on Łukasiewicz propositional logic $[0, 1]_{\text{Ł}} = \{[0, 1], 0, 1, \neg, \wedge, \vee, \otimes, \oplus, \to\}$ that is defined as

$$
\begin{array}{lll}
(i) & \neg x = 1 - x & \text{(negation)}; \\
(ii) & x \wedge y = \min\{x, y\} & \text{(weak conjunction)}; \\
(iii) & x \vee y = \max\{x, y\} & \text{(weak disjunction)}; \\
(iv) & x \otimes y = \max\{0, x + y - 1\} & \text{(strong conjunction)}; \\
(v) & x \oplus y = \min\{1, x + y\} & \text{(strong disjunction)}; \\
(vi) & x \to y = \min\{1, 1 - x + y\} & \text{(implication)}.
\end{array}
$$

We notice in passing that $x \oplus y = \min\{1, x + y\}$ comes out as the Łukasiewicz conorm. This can be seen when using Eq. 6.2.3–(4). We have $S(x, y) = 1 - \max\{0, (1 - x) + (1 - y) - 1\} = 1 - \max\{0, 1 - x - y\}$. If $x + y < 1$ then we get $S(x, y) = 1 - (1 - x - y) = x + y = \min\{1, x + y\}$. If $x + y \geq 1$ then $S(x, y) = 1 - \max\{0, 1 - x - y\} = \min\{1, x + y\}$. The implication $\to$ comes from Eqs. 6.2.3–(11)–(13) when requiring the boundedness on $[0, 1]$. Łukasiewicz logic comes with two different concepts of conjunction and disjunction. We have the strong $\otimes$ and $\oplus$ connectives, and the weak $\wedge$ and $\vee$ connectives. An insightful discussion on the different nature of these connectives is stimulated in Exercise 1.

The class of $[0, 1]$-valued Łukasiewicz functions corresponding to formulas of propositional Łukasiewicz logic coincides with the class of McNaughton functions, namely the class of continuous finitely piecewise linear functions with integer coefficients. The algebra of formulas in the Łukasiewicz logic (Ł) is isomorphic to the algebra of McNaughton functions defined on $[0, 1]^n$. This is a crucial property that allows us to get a substantial advantage with respect other logics. In particular, we are interested in the fragment of Ł whose formulas correspond to concave functions. As it will be shown in the following, this has a strong consequence on the complexity issues of learning and reasoning with constraints. We start noticing that if ψ is a $[0, 1]$-valued function defined on $[0, 1]^n$ then ψ and $\neg \psi$ enjoy the nice property that f is convex iff $\neg \psi$ is concave. This is a straightforward consequence of the negation property (i), that is, $\neg \psi = 1 - \psi$. Now we can also see that $\wedge$ and $\oplus$ satisfy the property that the composition of concave functions leads to a concave function. Formally, if ψ_1, ψ_2 are concave then functions $\psi_1 \wedge \psi_2$ and

$\psi_1 \oplus \psi_2$ are still concave. Let us start with the weak conjunction $\wedge$. Let $0 \leq \mu \leq 1$ and $x_1, x_2 \in X$ and consider $x = \mu x_1 + (1 - \mu)x_2$. If ψ_i, $i = 1, 2$ is concave then $\psi_i(x) \geq \mu \psi_i(x_1) + (1 - \mu)\psi_i(x_2)$. Now we have

$$
\begin{aligned}
(\psi_1 \wedge \psi_2)(x) &= \min\{\psi_1(x), \psi_2(x)\} \\
&= \min\{\psi_1(\mu x_1 + (1 - \mu)x_2), \psi_2(\mu x_1 + (1 - \mu)x_2)\} \\
&\geq \min\{\mu \psi_1(x_1) + (1 - \mu)\psi_1(x_2), \mu \psi_2(x_1) + (1 - \mu)\psi_1(x_2)\} \\
&\geq \mu \min\{\psi_1(x_1), \psi_2(x_1)\} + (1 - \mu)\min\{\psi_1(x_2), \psi_2(x_2)\} \\
&= \mu(\psi_1 \wedge \psi_2)(x_1) + (1 - \mu)(\psi_1 \wedge \psi_2)(x_2).
\end{aligned}
$$

For the strong disjunction $\oplus$, we have

$$
\begin{aligned}
(\psi_1 \oplus \psi_2)(x) &= \min\{1, \psi_1(x) + \psi_2(x)\} \\
&= \min\{1, \psi_1(\mu x_1 + (1 - \mu)x_2) + \psi_2(\mu x_1 + (1 - \mu)x_2)\} \\
&\geq \min\{1, \mu \psi_1(x_1) + (1 - \mu)\psi_1(x_1) + \mu \psi_2(x_1) + (1 - \mu)\psi_2(x_2))\} \\
&= \min\{1, \mu(\psi_1(x_1) + \psi_2(x_1)) + (1 - \mu)(\psi_1(x_1) + \psi_2(x_1))\} \\
&\geq \mu \min\{1, (\psi_1(x_1) + \psi_2(x_1))\} + (1 - \mu)\min\{1, (\psi_1(x_1) + \psi_2(x_1))\} \\
&= \mu(\psi_1 \oplus \psi_2)(x_1) + (1 - \mu)(\psi_1 \oplus \psi_2)(x_2).
\end{aligned}
$$

Notice that while concavity of the McNaughton functions is closed under the pair $\wedge$, $\oplus$, the same does not hold for the pair $\wedge$, $\vee$. We can promptly see that the weak disjunction $\vee$ does not preserve concavity (see Exercise 2).

However, using similar arguments (see Exercise 3), we can see that $\vee$ does preserve convexity and that the same holds true for the strong conjunction $\otimes$. Again the closure is guaranteed from both conjunction and disjunction if we mix up weak and strong connectives. These closure results give suggestions on how to construct concave or convex function. We can mix up weak and strong connectives, but we can also use the negation so as to exploit both results on concavity and convexity closure.

Horn clauses are the most noticeable case of formulas that can be based on functions generated in this way. Let ψ_1 be convex and ψ_2 be concave and consider $\psi_1 \to \psi_2 = \neg\psi_1 \oplus \psi_2$. We end up with the weak disjunction of the concave functions $\neg\psi_1$ and ψ_2 which is a concave function. Hence $\psi_1 \to \psi_2$ is concave. Now suppose we are given a collection of convex functions φ_i, $i = 1, \ldots, p$. In the extreme case, they can be simply literals, since $\varphi_i(x) = x_i$ and $\varphi_i(x) = 1 - x_i$. If we construct

$$
\psi_1 = \bigotimes_{i=1}^{p} \varphi_i := \varphi_1 \otimes \ldots \otimes \varphi_p,
$$

then

$$
\neg\psi_1 \oplus \psi_2 = \neg\left(\bigotimes_{i=1}^{p} \varphi_i\right) \oplus \psi_2 = \left(\bigoplus_{i=1}^{p} \neg\varphi_i\right) \oplus \psi_2.
$$

From the closure of $\oplus$ with respect to concavity we conclude that the Horn clause, characterized by $\neg\psi_1 \oplus \psi_2$, is concave. Notice that the property arises when translating the implication with the strong Łukasiewicz connectives.

To sum up, the McNaughton functions corresponding to Ł-formulas in the fragment $(\otimes, \vee)^*$, composed of literals and not only of propositional variables, are convex, while the fragment $(\wedge, \oplus)^*$ is concave. Interestingly, as shown in Exercise 4, the fragment $(\wedge, \oplus)^*$ contains non-Horn clauses.

EXERCISES

1. [20] Discuss the relationships and the meaning of the strong and weak connectives of Łucasiewicz logic.

2. [17] Prove that $\vee$ does not preserve concavity, that is, if ψ_1, ψ_2 are concave then $\psi_1 \vee \psi_2$ is not concave.

3. [19] Prove that if ψ_1, ψ_2 are convex then $\psi_1 \otimes \psi_2, \psi_1 \vee \psi_2$ are convex.

4. [16] Prove that the McNaughton function associated with $(\neg\psi_1 \vee \psi_2) \to \neg\psi_3$ is concave, that is, $(\neg\psi_1 \vee \psi_2) \to \neg\psi_3 \in (\wedge, \oplus)^*$.

5. [15] Prove that $(\neg\psi_1 \wedge \psi_2) \vee \psi_3 \notin (\wedge, \oplus)^*$ and $(\neg\psi_1 \wedge \psi_2) \vee \psi_3 \notin (\otimes, \vee)^*$.

6.3 **Diffusion machines**

The notion of individual which lives in a graphical domain has enlightened the strict connections between inference and learning. It has been shown that a constraint machine carries out a computational scheme that, in general, can strongly depend on single individuals (vertices of the environmental graph). In many interesting cases the constraints reflect the structure of the environmental graph more that specific properties of the single tasks. We are now concerned with cases in which the constraints can be homogeneously expressed, that is, $\forall v \in V$:

$$z_v - s(z_{\text{ch}(v)}, x_v) = 0,$$
$$\gamma(z_v, x_v) = 0,$$

where $\text{ch}(v)$ is the ordered set of children of v. Here, there is a state variable z_v that contributes to expressing the values of $\gamma(z_v, x_v)$. In the rest of the section it will become clear that γ can conveniently be "factorized" as $\gamma = \psi_o \circ h$, where h is the function which returns the output of the machine by computing $h(z_v, x_v)$, while ψ_o is a constraint. In the simplest case, ψ_o yields supervised learning. Overall we can think of the function f that acts as an argument in the constraints as the pair $f = (s, h)'$. Unlike what has been seen so far, we are in the presence of functions where there is a hidden component — namely the transition function s. Function h is in fact the output function considered so far, on which the constraint ψ_o operates by $\psi_o(h(z_v, x_v)) = 0$. The hidden function s is in fact responsible for the information diffusion, whereas the output function h is used for enforcing the semantic knowledge of the environment. Here we overload the notation so as to denote by f also the function that maps all vertexes, that is, $f(x) = (f(x_1), \ldots, f(x_{|V|}))$. Similarly, we adopt the same notation for s and h. Hence, using Iverson's notation, the above constraints can be compactly rewritten as

$$\psi(f(x)) = \sum_{v \in V} \left([z_v - s(z_{\text{ch}(v)}, x_v) = 0] + [\psi_o(h(z_v, x_v)) = 0] \right) - 2|V| = 0. \quad (1)$$

We have just landed in a new territory where, because of the homogeneous structure of the constraints, we only need to learn the state s and the output h functions. The structure of constraint (1) is in fact induced from the environmental graph, but the "rules" are homogeneous throughout the graph. The specificity of this class of constraints is that they induce diffusion of information on the basis of the graph topology. For this reason the constraint machines that enforce the constraint (1) are referred to as *diffusion machines*. Notice that, while the satisfaction of complex constraints might not be easily given an intuitive interpretation, in this case the special structure of the constraints does suggest an appropriate computational model which carries out information diffusion in a way that can nicely be interpreted. In machine learning, recurrent networks for sequences are the most popular examples of diffusion machines.

6.3.1 Data models

Here we are concerned with the effects of the environmental data structure that is defined by the environmental graph $\mathcal{G} \sim (V, A)$, where V, as usual, is the set of vertices, and A is the set of arcs. In general, one can think of diffusion processes that are homogeneous only in portions of the graph. Hence we assume that the vertexes can be partitioned into the collection $V = \bigcup_{\alpha=1}^{p} V_i$, where $V_\alpha \cap V_\beta = \emptyset$ iff $\alpha \neq \beta$. The partition characterizes the type of data of the vertexes in V_α, which allows us to define different functions over the vertexes of $\mathcal{G}$. For any type α there is a pair of associated functions

$$
\begin{aligned}
s_\alpha &: Z_\alpha^{|\mathrm{ch}(v)|} \times X_\alpha \to Z_\alpha, \quad (z_{\mathrm{ch}(v)}, x_v) \mapsto s_\alpha(z_{\mathrm{ch}(v)}, x_v), \\
h_\alpha &: Z_\alpha \times X_\alpha \to Y, \quad (z_v, x_v) \mapsto h(z_v, x_v).
\end{aligned}
\tag{1}
$$

Function s_α is referred to as the *transition function*, while h_α is referred to as the *output function*. Here $X_\alpha \subset \mathbb{R}^{d_\alpha}$, $Z_\alpha \subset \mathbb{R}^{p_\alpha}$. The ordering relation on the $m_v = |\mathrm{ch}(v)|$ children of v is supposed to be same relation defined on V. When $v = \mathtt{nil}$, that is, when $\mathrm{ch}(v) = \emptyset$, we assume that $z_v = z_0$ is given. This is a way of restricting the class of functions that process information on $\mathcal{G}$ to take decisions on the vertices. The computation relies on the principle of summarizing the information needed for the decision in the state variable $z \in Z_\alpha \subset \mathbb{R}^{p_\alpha}$. We use the notation $z_{\mathrm{ch}(v)} \in Z_\alpha^{|\mathrm{ch}(v)|}$ to represent concisely the information in the children of v. Notice that the definition of s_α requires the correct identification of the arguments $x_{\mathrm{ch}(v)} = (x_{\mathrm{ch}(v),1}, \ldots, x_{\mathrm{ch}(v),m_v})$, which is made possible by the ordering relation on V. While the general computation given by Eq. (1) can be useful to deal with nonhomogeneous portions of the environmental graph, it's clear that we gain simplicity and efficiency when restricting the families of functions (s_α, h_α) only to (s, h), that is, when enforcing uniform computation over the graph. This also requires determining the degree of the graph, so as to consider for any vertex $m = \max_v \{m_v : v \in V\}$. Clearly, in so doing, the computation of $s(z_{\mathrm{ch}(v)}, x_v)$ requires padding with $\mathtt{nil}$ whenever $m_v < m$ — a child which is missing in a certain position can be replaced with a bud, with the associated $\mathtt{nil}$ symbol.

In the general case of Directed Ordered Acyclic Graphs (DOAG), we can promptly see that Eq. (1) corresponds with the data flow computational scheme that has been introduced in the previous chapter for feedforward neural networks (see Chapter 5, Algorithm F). This time the data flow computation arises because of constraints that are induced by the environmental graph, whereas the data flow taking place in feedforward networks arises because of the network structure, whose complexity is needed to produce complex maps. As far as the satisfaction of the constraints is concerned, as it will be clear in the following, we can employ exactly the same concepts used for learning with feedforward neural networks. When we deal with a general digraph, however, the presence of cycles requires an additional analysis, since the computation of $s_\alpha(x_{\mathrm{ch}(v)}, x_v)$ can become circular, that is, dependent on itself. However, regardless of the way $s_\alpha(x_{\mathrm{ch}(v)}, x_v)$ is determined, Eq. (1) is an expression of the constraint. While algorithmic issues will be covered in detail in the following, we notice that a possible way of discovering its satisfaction is to immerse it into a temporal basis by defining the associated discrete-time dynamical system

$$z_v(t+1) = s(z_{\text{ch}(v)}(t), x_v),$$
$$\gamma(\bar{z}_v, x_v) = \psi_o \circ h(\bar{z}_v, x_v) = 0.$$

Here, we assume that $z_v(t)$ is convergent, that is, $\bar{z}_v = \lim_{t \to \infty} z_v(t) < \infty$. In addition, the fixed point $\bar{z}_v$ of map s needs to satisfy the condition on the output functions $\gamma(\bar{z}_v, x_v) = 0$. This homogeneous formulation over the graph requires us to determine $f = (s, h)'$ that parsimoniously satisfy the constraints. Basically, the convergence of the above dynamical system yields relational representations in the environment, since for each vertex v we get the associated state value $\bar{z}_v$. It is equivalent to assuming that s is map which admits a fixed point. A case in which this happens is when s is a *contractive map*. Exercise 2 provides a simple example for discussing an appropriate choice of s which yields *relational consistency*. Like for neurons, we can use a threshold-linear aggregation of the states of the children, so that function s becomes

$$s(z_{\text{ch}(v)}, x_v) = \sigma \left(\sum_{u \in \text{ch}(v)} W_u z_u + U x_v \right). \tag{2}$$

We notice in passing that this special function keeps the general requirement on s of vertex invariance. This is the reason why the transition function is characterized by matrices W_u and U, which are independent of v. As a special case, Z could be the void space, which leads to dismissing s, while $h \colon Z \times X \to Y \mapsto h \colon X \to Y$.

An interesting case of information diffusion is the ranking of webpages by the PAGERANK algorithm, which arises when posing $\sigma = \text{id}$ in Eq. (2). Interestingly, this corresponds with random walk in graphs (see Exercise 1). PAGERANK yields the propagation of page authority through the graph according to the structure of the hyperlink only. The joint presence of diffusion and semantic-based constraints yields decisions based on $h(z_v, x_v)$, where function h is expected to weigh properly the two arguments. Depending on the learning task, diffusion and semantic-based constraints can jointly operate. Interestingly, not all vertexes are necessarily subjected to semantic-based constraints. In general, we denote by $\triangleright(v, \psi_o)$ a Boolean variable which is $\top$ if and only if the constraint ψ_o acts on the output functions for $v \in V$. This allows us to properly apply different environmental constraints on the given structured environment. If $\psi_o > 0$ then the empirical risk corresponding with the associated semantic-based constraint is created by accumulating the loss only over $\triangleright(v, \psi_o)$, that is,

$$E(f) = \sum_{v \in V} [\triangleright(v, \psi_o)] \psi_o(h(z_v, x_v)). \tag{3}$$

Now we want to see how learning tasks can naturally be framed in this general data model. The simplest case has been mostly addressed in the previous chapters of the book and includes the graph environment with individuals $J = \{(v_\kappa, x_\kappa) \in$

$V \times X : \kappa = 1, \ldots, \ell\}$ composed of a collection of distinct patterns, where individuals degenerate to patterns. If you are given three patterns then the environment is represented by the fully disconnected graph with three vertices drawn as boxes.

$$\boxed{1} \quad \boxed{2} \quad \boxed{\ell}$$

Basically, there is only one type of function s_α, h_α — homogeneous computation without partitioning of V — and the environmental graph $\mathcal{G}$ degenerates to a collection of vertexes without edges. This corresponds with what we have considered so far, namely environments without information diffusion. It's worth mentioning that, in so doing, we are missing an interesting processing of expressed relations amongst the vertexes of the graph. However, as put forward in the previous section, this doesn't mean that relations can only be expressed in this way. On the contrary, the diffusion process is only a special way of expressing relations, which comes from the presence of the hidden state variable. Of course, in this case both maps s and h degenerate since there is no state variable involved, so that output function h reduces to $h \colon X \to Y$. The training set $\mathcal{L}$ is not necessarily composed of supervised pair, since some labels may be missed.

Sequential homogeneous data is commonly encountered in machine learning. It is in fact the first true example of structured data. Again, V is not partitioned into different sets, that is, there is only one type of function (s, h). Moreover, the environmental graph $\mathcal{G}$ is simply a list of vertexes. An environment with three individuals organized as a sequence can be graphically represented by

$$\boxed{1} \longrightarrow \boxed{2} \longrightarrow \boxed{\ell}$$

Notice that we have chosen an ordering of the vertices that is coherent with the assumption that if $u, v \in V$ then $(u, v) \in A$ iff $u \prec v$. For any $v \in V$, in this case there's only one child in $\mathrm{ch}(v)$ and, moreover, $\mathrm{ch}(v) = \{v + 1\}$. Hence we can express the diffusion process taking place in the sequence and the output constraints by imposing $\forall v \in V$ that

$$
\begin{aligned}
z_v - s(z_{\mathrm{ch}(v)}, x_v) &= s(z_{v+1}, x_v) = 0, \\
\psi_o \circ h(z_v, x_v) &= 0.
\end{aligned}
\tag{4}
$$

While the first constraint yields information diffusion, the second one imposes restrictions on the values of $h(z_v, x_v)$. The satisfaction of this constraint requires one to convert the input sequence $\langle x \rangle$ to the output sequence $\langle y \rangle$, where $y_v = h(z_v, x_v)$. The semantics attached to the vertexes has important consequences on the processing. The equation on the transition function requires the initialization of the state corresponding with the `nil` vertex — denoted as z_0. Interestingly, in the special case of a *circular list*, we have no final `nil` symbol and, in addition, z_0 is not necessary. A related interesting case is that of an infinite list, which is suitable for agents performing truly online processing.

Like for plain environments, the supervision or other semantic constraints are activated whenever $\triangleright(v, \psi_o)$ is true. Notice that, unlike in the case of plain environments, the sequential structure plays a fundamental role in the computation over unsupervised vertexes, since the presence of the state makes it possible to perform a dynamical diffusion process. In addition to the selective supervision policy expressed by $\triangleright(v, \psi_o)$, there are two important distinct problems connected with sequential processing. In the first case, we are given a sequence, either finite or infinite — it might also be circular — and we want to impose constraints by considering a continuous diffusion process based on only one state reset (induced by $x_0 = $ nil). In this case, we are given a collection of sequences that are concatenated; unlike the previous case, the reset of the state is enforced at the beginning of every single sequence.

Many problems in machine learning are naturally framed in environments where data are represented by lists. Let us consider the classic problem of isolated word speech recognition. Speech signals are typically represented by sequences of frames collected every 10–20 ms, which are obtained by appropriate preprocessing. Each frame is a single vector that represents information attached with $v \in V$. The words have variable length, and there is supervision on each word. A training set of l supervised words can be represented as a graph, which degenerates to a list that is segmented into l portions separated by l nil vertexes. Here, we assume that there is no distinction between the initialization of the state for processing the overall list, and the initialization of on each word. The associated risk function coming from supervision requires the definition of $\triangleright(v, \psi)$, namely the identification of the speech frames where we attach the supervision label. The simplest solution is that of setting $\triangleright(v, \psi) = \top$ only for the last speech frame v of each word, but this might not necessarily be the best choice (see Exercise 3). Notice that the presence of nil in the list yields the reset of the state every time we start processing a new word, which is in fact what one expects from the computational model. When considering the temporal flow of the signal, the corresponding list has pointers going to the opposite direction (see Exercise 1 for a discussion on this issue).

While the isolated word recognition learning task requires the reset of dynamical system (4), the problem of phoneme recognition can be expressed again by a list of continuous speech frames, where there is no nil — apart the one which denotes the end of the list. Supervision is typically located on speech frames where there is strong reliability on the decision, whereas $\triangleright(p, \psi) = \mathsf{F}$ in other frames. The dynamical behavior is quite different with respect to that of isolated word recognition, since most of the emphasis on the classification is now in the current speech frame, though the context is also important in the classification.

Some learning tasks are naturally modeled by a collection of trees. As an example, consider the x–y trees that are used to model image document for their classification. In this case, the tree structure corresponds with the representation of the layout of the single documents, which are assumed to be independent of each other. Hence the processing based on Eq. (1) takes place on single trees. Like for lists, we need to reverse the links for state updating. Notice that we start from the leaves and then accumulate the states until we reach the root. Representations that resemble x–y trees are quite popular in document analysis and recognition; for example, graph-based representations are used for drawings like company logo. In these learning tasks $\triangleright(v, \psi) = \top$ is typically set on the root only, which is in fact the first node in the ordering relationship.

In some learning tasks, we need to go one step further and to make use of DOAGs. In other cases models based on cyclic graphs also naturally arise. For example, chemical molecules can be described by expressing the bonds between the atoms by arcs of a graph, which can clearly contain cycles. As pointed out in Section 1.1, the graphical representation is adequate also for tasks like Quantitative Structure Activity Relationship (QSAR), which explores the relationship between the chemical structure and the pharmacological activity in a quantitative manner, and Quantitative Structure-Property Relationship (QSPR), which is aimed at extracting general physical–chemical properties from the structure of the molecule. In these cases, the learning environment turns out to be a collection of undirected graphs, where each one is associated with a supervised value. Hence, these learning tasks are expected to return a decision on the single graphs. Interestingly, while in sequences, trees, and DOAGs $\triangleright(p, \psi) = \top$ typically holds only on a source, in a cyclic graph the condition is attached to any node of the graph. Clearly, since there is no ordering relation, there is no reason to prefer one node over others for receiving constraints. As already pointed out, in case of general digraphs, Eq. (1) makes sense only in case in which it admits a solution. An interesting case that deserves analysis is when functions s_α are linear. Let us consider only a single transition function on the nodes, so that we have

$$z_v = \sum_{u \in \mathrm{ch}(v)} W_u x_u + U v_v, \tag{5}$$

$$y_v = C z_v + D x_v. \tag{6}$$

Now define

$$z := (z_1', \ldots, z_\ell')',$$

$$x := (x_1', \ldots, x_\ell')',$$

$$U_G = \begin{pmatrix} U & 0 & \cdots & 0 \\ 0 & U & \cdots & 0 \\ 0 & 0 & \cdots & \vdots \\ 0 & 0 & \cdots & U \end{pmatrix} \in \mathbb{R}^{d \times p},$$

and consider the matrix $W_G \in \mathbb{R}^{n\ell \times n\ell}$ defined as follows. Let $A \in \mathbb{N}^{\ell \times \ell}$ be the adjacency matrix of the graph. Now consider the matrix $W_{\mathrm{ch}} := (W_1, \ldots, W_m)$, where m is the outdegree of $\mathcal{G}$. Furthermore, any vertex $v \in V$ is associated with its ordered parents $\mathrm{ch}(v) = \{c(v, 1), \ldots, c(v, m)\}$. Then consider the map which transforms $(A, W_{\mathrm{ch}}) \to W_G$, so that each coordinate of A' is mapped to a block of W_G as follows:

$$a_{v,u} \to [a_{v,u} = 0]0_m + [a_{v,u} = 1]W_{c(v,u)}. \tag{7}$$

Then Eq. (5) can be associated with the compact discrete-time linear system

$$z_{t+1} = W_G z_t + U_G x. \tag{8}$$

If the spectral radius of W satisfies the condition $\rho(W_G) < 1$ then we can prove that the above equation presents a BIBO (Bounded Input Bounded Output) behavior (see Exercise 2). Clearly, BIBO stability and the corresponding relational consistency depends on both matrix W and the adjacency matrix A of the graph. However, no matter what graph is given, we can always choose a matrix W such that $\rho(W_G) < 1$ (see Exercise 3).

While the collection of disconnected graphs is adequate for representing many problems where each graph provides a model of single structured objects, there are environments in which a connected graph is more adequate. In these cases we typically refer to *networked data*. The Web offers natural environments for these tasks. In this case we are interested in learning tasks where the machine might be expected to return a prediction also for every node. Ranking algorithms, like the mentioned PAGERANK, are a significant applicative example. Document classification is also another remarkable task. In this case the classification is based on both the document content and the structure of its hyperlinks. It's worth mentioning that the computation on connected networks typically leads to dismissing the ordering in the digraph. This is due to the fact that it is unreasonable to provide a different weight for different parents on large graphical domains. In these cases, there are natural symmetries which lead to the sharing of the parameters associated with the different parents. Whenever a different weight is attached to the parents, this is due to the need of expressing specific properties of finite structured objects.

The diffusion process evolving in the described machines leads us to discuss the role of the state in the modeling of relations amongst individuals. In order to disclose their connection, notice that the decision based on Eq. (1) requires the computation of the output by means of $h_\alpha(z, x_v)$. This indicates that the dimension p_α of the state space Z_α affects the weight of relational and semantic-based constraints in the decision. Clearly, as already pointed out, the absence of the state leads to learning agents that only involve the input for the decision. In the opposite case, the decision can also arise from only the presence of diffusion, which propagates relations amongst the vertexes. As already pointed out, this is in fact what happens in ranking algorithms like PAGERANK. Clearly, in the middle, there is a lot of room for clever heuristic choices to mix relational and semantic-based constraints.

Finally, we want to bring attention to the presence of multiple scalar functions (s_α, h_α) in a single vertex. For example, in web information processing, the views of a document can be substantially different. We can think of a vector-based representation of the content on which to define a function, but we can also consider link-based functions only. As previously pointed out, the introduction of different functions operating on a given individual opens important issues on coherent decision.

EXERCISES

1. [*M25*] Let $\langle a_n \rangle$ be a sequence. Let $A(z)$ be the generating function of $\langle a_n \rangle$. Prove that if $\langle a_n \rangle$ is convergent then

$$\lim_{n \to \infty} a_n = \lim_{\mathbb{R} \ni z \to 1} (1 - z) A(z).$$

2. [*28*] Given an environment represented by a graph with two fully-connected nodes, where $v_1, v_2 \in V$, provide an example of transition function s such that

$$z_1 = s(z_2, x_1); \quad z_2 = s(z_1, x_2)$$

yields a consistent relational constraint. Then give another example in which there is no solution to the above equations.

3. [*27*] Discuss appropriate choices of $\triangleright(p, \psi)$ for isolated word recognition.

6.3.2 Diffusion in spatiotemporal environments

In this section, we offer a complementary view of data models, which arises whenever spatiotemporal environments are involved. While many learning tasks can clearly be characterized by the graphical structure introduced in the previous section, all perceptual tasks require some additional thought. In order to shed light on this issue, let us consider the sequential environment characterized by Eq. 6.3.1–(4), which can be conveniently rewritten as $z_{t+1} = s(z_t, x_t)$. This corresponds with the temporal interpretation of the list, where we reverse the previously defined relation $\prec$, by setting $t = -v$. We can go one step further by providing a continuous interpretation of the previous difference equation by means of

$$\begin{aligned} \dot{z}(t) &= s(z(t), x(t)), \\ y(t) &= f(z(t), x(t)). \end{aligned} \tag{1}$$

Clearly, this differential equation inherits the basic ideas of Eq. 6.3.1–(4), and therefore it can still be regarded as a temporal constraint on the learning environment. However, Eq. (1) can also be given a different interpretation, which seems to be interwound with of laws of nature. This is suggested by replacing the discrete-time temporal basis with the ordinary continuous notion of time. In a sense, according to Eq. (1), any agent working in temporal environments obeys a law which describes the evolution of the state.

There are real-world problems which are naturally modeled by spatial constraints. For example, in computer vision, in addition to temporal coherence, there are spatial coherence requirements. Let us consider the case image processing and assume that images are represented by a trellis which corresponds to an opportune quantization. Each vertex v can be given a corresponding coordinate in the trellis, that is, $v \sim (i, j)$. Because of symmetry, if we consider only a single state variable for each vertex and assume a linear transition function then Eq. 6.3.1–(5) yields

$$z_{i,j} = w(z_{i-1,j} + z_{i+1,j} + z_{i,j-1} + z_{i,j+1}) + u x_{i,j}. \tag{2}$$

If the spatial quantization is chosen in such a way that the states on the neighbors are arbitrarily close, because of smoothness requirements, we have $z_{i-1,j} \simeq z_{i+1,j} \simeq z_{i,j-1} \simeq z_{i,j+1} \simeq z_{i,j}$. Therefore, from Eq. (2) we have $(z_{i-1,j} + z_{i+1,j} - 2z_{i,j}) + (z_{i,j-1} + z_{i,j+1}) - 2z_{i,j}) + \frac{u}{w}v_{i,j} = 0$. This leads us to conclude that this equation is in fact the discrete version of the partial differential equation

$$\nabla^2 z + \frac{w}{u}x = 0. \tag{3}$$

This turns out to be a translation of spatially symmetric relational constraints in the framework of image processing. Of course, the involvement of more complex trellises gives rise to higher-order equations and, moreover, we can enrich the state with vectorial representations. We notice in passing that the translation of relational constraints, along with the simple smoothness requirement, gives rise to a Poisson equation. Interestingly, the Laplacian also emerges from the minimization of the regularization term $\langle \nabla z, \nabla z \rangle$, which indicates that when stressing locality, the relational constraints collapse to regularization requirements. The only degree of freedom left by the relational constraint given by Eq. (3) is offered by the ratio u/w, which suggests that a richer state structure can significantly improve representational capabilities.

EXERCISES

▶ **1.** [*M32*]   According to the PAGERANK algorithm, the rank of a hyperlinked environment is determined by

$$z_v = d \sum_{u \in \text{pa}(v)} \frac{z_u}{|\text{ch}(u)|} + (1 - d)$$

where $0 < d < 1$. The value z_v is regarded as the *rank* of the page. Determine the rank of the nodes of the graph on the right-hand side. Under which condition $\sum_{v \in V} z_v = |V|$? Provide a probabilistic interpretation and give an example of a graph in which this normalization condition doesn't hold.

2. [*25*]   Prove that if $\rho(W) < 1$ then linear system (8) is BIBO stable.

3. [*13*]   Prove that given any real matrix A there exists $\alpha \in \mathbb{R}$ such that $\rho(\alpha A) < 1$.

4. [*20*]   Consider the Poisson equation (3). Discuss its relationship with Poisson editing in image processing [see P.M. Gangnet, A. Blake, *ACM Trans. Graph. (SIG- GRAPH'03)* **22** (2003), 313–318].

5. [*20*]   Discuss the Poisson editing algorithm (see Exercise 4) in the framework of relational and semantic-based constraints.

6.3.3 Recurrent neural networks

The previous discussion on relational domains and information diffusion provides an ideal framework for the introduction of *recurrent neural networks*. It has been pointed out that the data relations can be expressed by the functional equation 6.3.1–(1) that involves a state variable. Its translation in terms of linear equation has been already discussed, and it has been shown that in this case the model follows Eq. 6.3.1–(5).

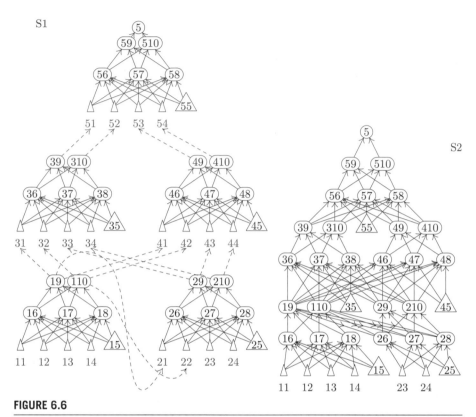

FIGURE 6.6

S1: Structure unfolding of the graph. The state of each node, represented by two variables, is propagated from its parents — at most two — and one input (triangle). Whenever incoming links are missing, the input is padded with the null value denoted by `nil`. While the transition function net, shared for each node, is drawn with solid links, the relational connections are drawn by dashed lines. S2: The compiling process of structure unfolding generates a feedforward net. We can see the five inputs and two neurons for each state variable that are used to determine the output at node 5.

Whenever the transition function s_α is chosen according to Eq. 6.3.1–(2), we are in the presence of a neural network that is referred to as a recurrent neural network. While in case of linear units the resulting linear dynamics given by Eq. 6.3.1–(5) can be summarized by the $\bowtie$ operation on A and W_{ch}, when involving the sigmoid functions such a strong algebraic support is lost! However, the shared graph-based computational scheme that is used in functional equation 6.3.1–(1) to expresses the relational constraints and in Eq. 6.3.1–(2) to propagate the activation in a feedforward network enables a compiling scheme to represent the required computation by a neural network. An example of compiling is shown in Fig. 6.6. In $S1$, we can see the structure unfolding of the graph, while $S2$ depicts the final construction of a feedforward neural network. Basically, given a DOAG $\mathcal{G}$ representing the environmental data and a consistent feedforward network $\mathcal{N}_{ch}$, the compiling process is an extension of the already defined $\bowtie$ operation, as we overload the symbol to denote by $\mathcal{G} \bowtie \mathcal{N}_{ch}$. The compiling described in Fig. 6.6 reports the construction of function s, while the output function is only shown in the graph supersource. The network associated with s yields a state of dimension $p = 2$ and processes at most two children by using three hidden units. Of course, the choice follows all design issues that have been faced during the discussion on feedforward networks. The output function h is simply modeled by a single neuron that takes a decision only as a function of the two state variables. For the sake of simplicity, in Fig. 6.6, the output network is only reported in the last expanded vertex. It's quite easy to realize that when we compose the DOAG data structure with the DOAG net architecture, the resulting graph is still a DOAG. As put forward in Section 6.4, this compiling scheme allows us to reuse most of the computational machinery used for feedforward networks. As already mentioned in Section 5.1 and in the previous section, the presence of cycles in the overall network needs state relaxation to an equilibrium point in order to provide a meaningful computation.

EXERCISES

1. [*15*]   Consider a learning environment based on lists which accumulate temporal information, like in speech recognition. Why are the list and temporal flow pointing to the opposite direction? How must we modify Eq. (1) if we want to enforce the same direction for the list and the temporal flow?

2. [*15*]   Given a graph with adjacency matrix A and weight matrices W_1 and W_2, construct matrix $W_G = A \bowtie (W_1, W_2)$, where

$$A = \begin{pmatrix} 0 & 1 & 0 & 0 \\ 0 & 0 & 1 & 1 \\ 0 & 0 & 0 & 1 \\ 1 & 0 & 0 & 0 \end{pmatrix}; \quad W_1 = \begin{pmatrix} w_{1,1,1} & w_{1,2,1} \\ w_{2,1,1} & w_{2,2,1} \end{pmatrix}; \quad W_2 = \begin{pmatrix} w_{1,1,2} & w_{1,2,2} \\ w_{2,1,2} & w_{2,2,2} \end{pmatrix}.$$

6.4 Algorithmic issues

This section is about solving the constraint satisfaction problems that arise in both learning and inference. As such, it's the core of constraint machines! In the most general case, one would very much like to determine the most parsimonious solution which satisfies the constraints expressed by Eq. 6.1.2–(1) for learning and inference, respectively. The formulation is simple and clear, yet the solution can be hard. The analysis in this section is based on the straightforward attack of the problem. In a sense, it's a generalization of batch-mode supervised learning. We are given data and constraints from which we want to extract regularities and take decisions. It will be shown that the general representation of the solution given in terms of the Green function of the differential operator L and of the reaction of the constraints supports the discovery of kernel-based solutions in a number of interesting problems. Intuitively, this simply means that while supervised learning corresponds with a distributional constraint reaction, where the Dirac deltas centered around the examples generate a plain kernel, in other interesting cases the convolution of the reaction of the constraints with the Green function of L yields a new kernel. Hence the marriage of the Green function with the reaction of the constraint leads to a specific kernel that is not only the outcome of smoothness requirements, but also somewhat incorporates the knowledge associated with that constraint. Basically, the cases in which the marriage of the plain kernel with the constraint reaction yields a new kernel enable the application of the mathematical and algorithmic machinery of kernel machines. However, you shouldn't neglect that we don't necessarily inherit the powerful convex structure of the classic kernel machine problems; when constraints are involved, in general, the corresponding optimization problem is not convex. There's more: The beauty of such a general framework, where everything comes back to kernels seems to be restricted to a few cases that are amenable of appropriate mathematical representations. However, when math doesn't help to gain kernel-based representations, it helps to always find a way out. While constraints are ideally enforced on data sets with the power of the continuum, an opportune sampling typically suffices to discover the solution. It is in fact shown that when we enforce pointwise constraints, we come back again to a kernel-based representation. More generally, the representation includes also free variables that are useful for both learning and inference. We can promptly get the idea behind pointwise constraint satisfaction over a finite collection of data, when considering that we simply need to construct an associated penalty that very much resembles the empirical risk function for supervised learning. When we assume that the constraint machine is based on a neural network, we can reformulate the learning problem by paralleling again supervised learning. This is where we can benefit from the power of deep learning. To sum up, the framework of constraint satisfaction shares with supervised learning the structure of the primal and of the dual solutions that are typical of neural networks and kernel machines, respectively.

The optimization schemes that are used for constraint machines can be used also for the special case of diffusion machines. However, the peculiarity of homogeneous constraints also enables specific algorithms that directly take into account the nature of diffusion processes.

6.4.1 **Pointwise content-based constraints**

In Section 6.1, we have already given the functional representation of the solution in case of pointwise constraints, which is expressed by Eq. 6.1.3–(6). Now suppose that X is sampled by $X^\sharp$ and that we are given a collection of soft-constraints $\mathcal{C}_\psi$. The results on functional representation indicate that we can treat unilateral and bilateral constraints in the same way. The optimal solution is found by using the superposition principle, that is,

$$f^*(x) = \sum_{\psi \in \mathcal{C}_\psi} \sum_{x_\kappa \in X^\sharp} \beta_{\psi,\kappa} \lambda_{\psi,\kappa} g(x, x_\kappa) = \sum_{x_\kappa \in X^\sharp} \left(\sum_{\psi \in \mathcal{C}_\psi} \lambda_{\psi,\kappa} \beta_{\psi,\kappa} \right) g(x, x_\kappa)$$

$$= \sum_{x_\kappa \in X^\sharp} \lambda_\kappa g(x, x_\kappa),$$

where $\lambda_{\psi,\kappa}$ is the κth Lagrangian multiplier associated with constraint ψ, which given by Eq. 6.1.3–(7), $\beta_{\psi,\kappa} := \beta_\psi(x_\kappa)$ is the given belief of ψ, and

$$\lambda_\kappa := \sum_{\psi \in \mathcal{C}_\psi} \lambda_{\psi,\kappa} \beta_{\psi,\kappa}.$$

As we can see, we end up with a kernel-based representation on $X^\sharp$, exactly like the one discovered for supervised learning by kernel machines. This time the unknown coefficients λ_κ need to be determined all over $X^\sharp$, but the unknowns associated with different constraints are grouped around each $x_\kappa \in X^\sharp$, so we needn't determine $\lambda_{\psi,\kappa}$ separately. For instance, in case of unilateral holonomic constraints, the collection $\mathcal{C}_\psi$ can be replaced with the isoperimetric constraint

$$\tilde{\Psi}(f) = \sum_{\psi \in \mathcal{C}_\psi} \sum_{x_\kappa \in X^\sharp} \beta_\psi(x_\kappa) \tilde{\psi}(x_\kappa, f(x_\kappa)).$$

This representation of f^* leads to the formulation of an optimization problem in the dual space. Hence, we can reuse all the mathematical and algorithmic apparatus of kernel machines to determine the coefficients λ_κ. A comment is in order concerning the passage to the dual space with λ_κ unknowns, which is related to the cardinality of $X^\sharp$. We've already seen that it's the presence of support vectors that somewhat limits the complexity of learning, but in any case, as the number of samples increases the problem becomes more and more difficult. In this new scenario, because of the abundance of (unsupervised) data, this number is typically very high, which means that we are typically in front of complex numerical problems connected with the dimension of the Gram matrix.

While supervised learning and any general pointwise constraint share the same solution, there is a significant difference in their concrete application, which is mostly due to the need, in the first case, to deal with supervised pairs. Unsupervised data are typically abundantly available and, therefore, they enable the massive enforcement of constraints different from supervised pairs. This leads to a new view of learning where one needn't distinguish between supervised and unsupervised learning! We only enforce constraints; it's the specific structure of supervised learning that requires us to provide supervised labels, while, in general, a constraint operates on the (unsupervised) training data. We can reinforce this concept by claiming that, *apart from exceptions, data are unsupervised*! When assuming this more general view, we can simply think of learning from constraints that operate on training data. When supervised and unsupervised data come together, we refer to *semisupervised learning algorithms*. Clearly, the general view of learning from constraints also leads to dismissing the need of semisupervised learning, it is common to mix up different types constraints. Hence semisupervised learning is just a very special case in which we enforce supervised pairs together with constraints that describe data distribution.

The analysis that we have carried out on pointwise constraints somewhat hides a fundamental problem that may arise depending on the class of environmental constraints. We must remind that discovery of the constraint reaction is generally hard. In particular, the corresponding optimization problem may admit many suboptimal solutions that are discovered by using the Euler–Lagrange equations. The corresponding collapse of dimensionality leads to a problem where the risk term associated with the constraints may not be convex, thus removing a fundamental guarantee for efficient optimization. This shouldn't be a surprise! The complexity of some learning environments is translated into the corresponding constraints, which may lead to complex optimization problems.

Of course, we can also formulate the learning problem in the primal space by using arguments put forward in Section 5.5.4. Let us consider Eq. 6.1.2–(1) stated for learning and inference. Instead of looking for solutions in a functional space, now we assume that $f(\chi)$ is given an explicit parametric representation. In case $\chi \to x$, we assume that $f(\chi) = f(x) = F(w, x)$, where $F \colon (\mathcal{W} \subset \mathbb{R}^m) \times (X \subset \mathbb{R}^d) \to (Y \subset \mathbb{R}^n)$. Then we need to satisfy

$$\forall x \in X \quad \psi(v, F(w, x)) = 0.$$

If $\psi \geq 0$, then this can be enforced on $X^\sharp \subset X$ by minimizing the corresponding penalty term

$$V_\psi(w) := \sum_{x_\kappa \in X^\sharp} \psi(v_\kappa, F(w, x_\kappa)).$$

Suppose we want to incorporate a collection $\mathcal{C}_\psi$ of constraints ψ. If $R_c(w)$ is a continuous regularization term then learning constraint $\mathcal{C}_\psi$ is reduced to

$$w^* \in \arg\min_w \left(\sum_{\psi \in \mathcal{C}_\psi} V_\psi(w) + \mu_c R_c(w) \right).$$

When we need to deal with individuals in a graphical environment, things get more involved. A clean formulation can be given in the special case — the opposite with respect to the previous case of $\chi = x$ — in which individuals reduce to the vertices, that is, $\chi = v$. We can immediately see that we end up with a finite optimization problem. Let $q_v := f(\chi) = f(v)$. Then one is concerned with the problem of discovering

$$q^* \in \arg\min_q \sum_{\psi \in \mathcal{C}_\psi} V_\psi(q) + \mu_d R_d(q),$$

where $R_d(q)$ is a discrete regularization term that is typically referred to as *graph regularization*. Like for continuous regularization, $R_d(q)$ favors smooth solutions, which means that q is expected to change smoothly from connected vertex. A classic choice is

$$R_d(q) = \frac{1}{2}\|\nabla q\|^2 = \langle q, \Delta q \rangle$$

(see Exercise 1 for details on a possible graph regularization approach). In most cases, because of the high dimension of the spaces involved, the search for q^* follows gradient descent. We have

$$\frac{\partial R_d(q)}{\partial q} = \Delta q,$$

and, therefore, stationary points of q are determined whenever

$$\Delta q + \frac{1}{\mu_d} \sum_{\psi \in \mathcal{C}_\psi} \frac{\partial}{\partial q} V_\psi(q) = 0. \tag{1}$$

Here Δ is the Laplacian of the graph. It can be proven that the solution of this equation consists of a *diffusion field* that nicely represents a number of interesting graph computational models, including random walk (see Exercise 2). Many tasks on a hyperlinked environment can be based on this model of graph regularization. The diffusion process turns out to be a sort of inference to determine the equilibrium. Notice that in other tasks, the emphasis is on the constraints of the environment, e.g., the n-queens problem. Most interesting situations involve general individuals in which dummy feature vertexes coexist with patterns (feature-based vertexes). In this general case we are concerned with finding

$$\begin{pmatrix} w^* \\ q^* \end{pmatrix} \in \arg\min_{w,q} \left(\sum_{\psi \in \mathcal{C}_{\psi(w)}} V_\psi(w) + \sum_{\psi \in \mathcal{C}_{\psi(q)}} V_\psi(q) + \mu R(w) \right).$$

A constraint machine, working in the primal space, which optimizes the above cost function, can be found in a number of different contexts. It can be used for classification in domains where features on the vertexes may be missing — collective classification in the literature. It can also be used to infer missing features by setting them as unknowns.

EXERCISES

1. [*M26*]　(Schoelkopf, 2004) Let $\mathcal{G} = \{V, A\}$ be a weighed graph, so that for any $e = (u, v) \in A$ there is a map $\omega\colon V^2 \to \mathbb{R}^+\ (u, v) \mapsto \omega_{u,v}$. Define

$$\frac{\partial q}{\partial e}\bigg|_u := \sqrt{\frac{\omega_{u,v}}{d_u}}\, q_u - \sqrt{\frac{\omega_{u,v}}{d_v}}\, q_v,$$

where $d_v = \sum_{u \sim v} \omega_{u,v}$. The *local variation* of q and the Laplacian of the graph are defined as

$$\|\nabla q\| := \sqrt{\sum_{e \vdash v}\left(\frac{\partial q}{\partial e}\bigg|_v\right)^2}, \quad (\Delta q)(v) := \frac{1}{2}\sum_{e \vdash v}\frac{1}{\sqrt{d}}\left(\frac{\partial}{\partial e}\sqrt{d}\frac{\partial q}{\partial e}\right)\bigg|_v,$$

and finally,

$$R_d(q) = \sum_{v \in V}\|\nabla q\|^2.$$

Prove that

$$\frac{\partial R_d(q)}{\partial q}\bigg|_v = (\Delta q)(v).$$

2. [*M26*]　Prove that the solution of Eq. (1) consists of a graph diffusion process and determine the conditions under which we reduce to random walk.

3. [*M28*]　Extend the results of Exercise 2 to DAGs.

6.4.2 Propositional constraints in the input space

The discussion on pointwise constraints has highlighted a fundamental aspect behind the special functional representation based on finite kernel-based expansion. Pointwise constraints can be given a distributional interpretation, so the optimal functional solution is the output of a linear system characterized by the regularization operator. Hence, from the superposition principle, the output is determined by expanding the Green function over the points of the training set. Interestingly, this is simply the outcome of the definition of the Green function, so a point x_κ, distributionally expressed by $\propto x_\kappa \delta(x - x_\kappa)$, yields $\propto g(x, x_\kappa)$. Clearly, for a generic constraint ψ with reaction ω_ψ, the convolution $g * \omega_\psi$ cannot enjoy such a simplification. On the contrary, in general such a computation is trapped in the curse of dimensionality. The complexity in the computation of the convolution has an intriguing counterpart when considering the related local equation 6.1.3–(10). In both cases, the need to compute the integral or to solve the partial differential equations is intractable for high dimensional spaces. However, the knowledge on the type of constraints may allow us to discover more efficient representations that are specifically inherited by their structure.

Here, we address an important case, which consists of propositional description, just like the example of Table 6.1–(ix). The idea is that we have prior knowledge that can be expressed on some of the coordinates of $x \in X$. Such knowledge involves the values of the coordinates that are assumed to belong to an interval that might be eventually infinite. Basically, for some of the d coordinates, we are given primitive propositions of the form $x_i \geq \bar{x}_i$, where $\bar{x}_i$ is a given threshold. In the example (ix) of Table 6.1 the proposition $[(m \geq 30) \wedge (p \geq 126)] \Rightarrow d$ formally defines a knowledge granule for the diagnosis of diabetes, which is positive whenever the body mass m and the level of fasting plasma glucose p exceed some thresholds. Here m and p are two coordinates of the input, and the proposition combines the primitives $m \geq 30$ and $p \geq 126$. Some patterns $x \in X$ might come with a number of associated propositions to define their membership. In principle, these propositions are expressed in propositional calculus and can be expressive. In simple cases, like the above example on diabetes diagnosis, these propositions can naturally be associated with corresponding multiinterval constraints, each defining a certain set X_i. In so doing, we naturally extend learning from supervised examples to the situation in which the points are replaced by sets X_i that are defined by the corresponding characteristic functions $[x \in X_i]$. Hence, instead of the classic pairs (x_κ, y_κ), we can regard propositional descriptions as a collection $\{(X_i, y_i)\}$, which turns out to be a generalized view of the notion of training set.

We assume that the open sets X_i can also degenerate to single points and that we are given ℓ_d points and ℓ_o ordinary sets, so as $\ell = \ell_d + \ell_o$. In order to perform soft-satisfaction of the supervised learning constraints, we need to provide a probability density of the data. A strong simplification arises when choosing:

$$p(x) = \frac{1}{\ell} \sum_{i=1}^{\ell_o} \frac{1}{|X_i|}[x \in X_i] + \frac{1}{\ell} \sum_{\kappa=1}^{\ell_d} \delta(x - x_\kappa), \tag{1}$$

which corresponds with assuming uniform distribution over the sets X_i and Dirac distributional degeneration for points x_κ. Here we tacitly assume that all constraints are given the same belief. Alternatively, the same weight in the formulation of the problem arises if we think of an environment with uniform probability distribution $1/\ell$, where the agent distinguishes the belief in the constraints as follows: $\beta_{\psi_i}(x) = (1/|X_i|)[x \in X_i]$ for propositional constraints and $\beta_{\psi_\kappa}(x) = \delta(x - x_\kappa)$ for supervised pairs. The association by propositional rules with multiintervals X_i leads naturally to reformulating supervised learning as a problem of soft-constraint satisfaction characterized by the functional index

$$\begin{aligned} E(f) = &\frac{1}{\ell} \sum_{j=1}^{n} \sum_{\kappa=1}^{\ell_d} V(x_\kappa, y_{j,\kappa}, f_j(x_\kappa)) \\ &+ \frac{1}{\ell} \sum_{j=1}^{n} \sum_{i=1}^{\ell_o} \int_X \frac{[x \in X_i]}{|X_i|} V(X_i, y_{i,\kappa}, f_j(x)) \, dx, \end{aligned} \tag{2}$$

which corresponds with the general index $\int_X p(x)\check{\psi}(x, f(x)) \, dx$ when expressing $\check{\psi}(x, f(x))$ and the probability distribution of Eq. (1). In the following analysis we make the further assumption that

$$\forall x_1, x_2 \in X_i \quad (1 - y_{i,j} f_j^*(x_1))(1 - y_{i,j} f_j^{\star}(x_2)) > 0, \tag{3}$$

where $y_{i,j} \in \{-1, 1\}$ is the target for function f_j on set X_i. Here, we assume $1 - y_{i,j} f_j^*$ is nonzero on the sets and that it does not change its sign within each such set (*sign consistency hypothesis*). For every $x \in X$, the reaction of the generic ith constraint is different depending on whether we are considering a degenerate set (point) or an ordinary one. In the first case from the previous analysis we get for every $\kappa \in \mathbb{N}_{\ell_d}$ and $j \in \mathbb{N}_n$

$$\omega_{\kappa, j}(x) = y_{\kappa, j} \lambda_{\kappa, j} \delta(x - x_\kappa).$$

For nondegenerate sets, under the sign consistency hypothesis, we have that $\forall x \in X_i$ $(1 - y_{i,j} f_j(x))_+ = 1 - y_{i,j} f_j(x)$. Hence we get

$$\omega_{i,j}(x) = -\frac{[x \in X_i]}{|X_i|} \frac{\partial}{\partial f_j} (1 - y_{i,j} f_j(x))_+ = y_{i,j} \lambda_{i,j} \frac{[x \in X_i]}{|X_i|}.$$

Let 1_{X_i} be the characteristic function of X_i, that is, $1_{X_i}(x) = [x \in X_i]$. Now if we define

$$\beta(x, X_i) := \frac{1}{|X_i|} \big(g * 1_{X_i}\big)(x)$$

as the *set kernel* associated to the pair (x, X_i), we end up with the representational form

$$f_j^*(x) = \sum_{\kappa \in S_d} y_{\kappa, j} \lambda_{\kappa, j} g(x - x_\kappa) + \sum_{i \in S_o} y_{i,j} \lambda_{i,j} \beta(x, X_i), \tag{4}$$

where S_d and S_o denote the support sets of the constraints.

In order to get a glimpse on the class of representations that arise from Eq. (4), let us consider $X = \mathbb{R}$ and restrict ourselves to the case of a single propositional constraint characterized by an interval $[a..b] \subset \mathbb{R}$, where we know that a certain function has given the target $+1$. When we use the Gaussian regularization operator (see Exercise 4.4–4)

$$L = \sum_{k=0}^{\infty} (-1)^k \frac{\sigma^{2k}}{2^k k!} \nabla^{2k}, \tag{5}$$

the reaction of the constraint $[x \in [a, b]](f(x) - 1) = 0$ is

$$g * 1_{[a,b]}(x) = \mathrm{erf}\left(\frac{x - a}{\sigma}\right) - \mathrm{erf}\left(\frac{x - b}{\sigma}\right),$$

which is drawn in Fig. 6.7.

The functional representation given by Eq. (4) allows us to approach the problem in finite dimensions. We can just parallel what has been done in Section 4.4 concerning supervised learning, which leads to collapse of dimensionality. When plugging

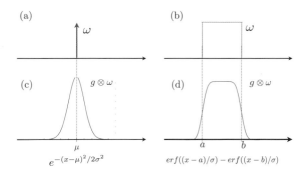

FIGURE 6.7

Constraint reactions corresponding (a) to a classic supervised pair and (b) to the soft constraint $\forall x \in [a, b]$ $f(x) = 1$ (box constraint) when choosing the regularization operator that generates Gaussian kernels. In (c) and (d) we can see the emergence of the plain and box kernel, respectively. The pseudo-differential regularization operator leads to a new kernel-based solution and prescribes its form (box kernel).

the solution expressed by Eq. (4) into the error function, we mostly need to understand the reduction of the regularization term $\langle Pf, Pf \rangle$. We can promptly see that we need to compute the three terms $\langle Pg(\cdot - x_h), Pg(\cdot - x_\kappa)\rangle$, $\langle Pg(\cdot - x_\kappa), P\beta(\cdot, X_i)\rangle$, and $\langle P\beta(\cdot - X_i), P\beta(\cdot, X_j)\rangle$. Now we have already seen that

$$\langle Pg(\cdot - x_h), Pg(\cdot - x_\kappa)\rangle = G(h, \kappa). \tag{6}$$

Then we have

$$\begin{aligned}
\langle Pg(\cdot - x_\kappa), P\beta(\cdot, X_i)\rangle &= \langle g(\cdot - x_\kappa), P^\star P\beta(\cdot, X_i)\rangle \\
&- \langle g(\cdot - x_\kappa), Lg(\cdot) * 1_{X_i}(\cdot)\rangle \\
&= \langle g(\cdot - x_\kappa), \delta(\cdot) * 1_{X_i}(\cdot)\rangle \\
&= \langle g(\cdot - x_\kappa), 1_{X_i}\rangle \\
&= \beta(x_\kappa, X_i).
\end{aligned} \tag{7}$$

Notice that the degeneration of X_i to a single point x_i can be handled by replacing $1_{X_i} \to \delta(\cdot - x_i)$, which restores Eq. (6). For the third term we have

$$\begin{aligned}
k(X_i, X_j) &:= \langle P\beta(\cdot, X_i), P\beta(\cdot, X_j)\rangle \\
&= \langle L\beta(\cdot, X_i), \beta(\cdot, X_j)\rangle \\
&= \langle 1_{X_i}, \beta(\cdot, X_j)\rangle \\
&= \langle 1_{X_i}, g * 1_{X_j}\rangle.
\end{aligned} \tag{8}$$

Notice that the degeneration of both X_i and X_j to single points x_i and x_j can be handled by replacing $1_{X_i} \to \delta(\cdot - x_i)$ and $1_{X_j} \to \delta(\cdot - x_j)$, which restores Eq. (6). If only one of the two sets degenerates to a point then we recover Eq. (7). We can prove that k is symmetric and nonnegative definite (see Exercise 1). Moreover, when using Eqs. (6), (7), and (8), we can easily conclude that the overall index (2) can be converted into a finite-dimensional quadratic function (see Exercise 2 for details). This is in fact the consequence of the kernel-based representation given by Eq. (4), which enables us to use all the mathematical and algorithmic kernel machine apparatus.

The functional representation (4) comes from assuming that the probability distribution is given by Eq. (1). This assumption is just an extension of the idea that is typically followed in supervised learning: Data are statistically distributed according to the supervised training set. This is quite good when the data collection is pretty big, but in many real-word problems we have only a few labels available and the supervised training set distribution doesn't represent a good approximation of the probability distribution! While we enrich a traditional supervised set with propositional rules, the underlying probability distribution dictated by Eq. (1) only makes a somewhat arbitrary assumption on the distributions inside multiintervals, which is supposed to be uniform. Clearly, this is quite good for small sets X_i, but we are still biasing the distribution with data on which a sort of supervision has been attached, thus neglecting the distribution of unsupervised data. What if we decide to approach the problem of learning with supervised data and propositional rules without making any assumption on p? The most challenging approach could be that of reformulating learning as an optimization problem with the additional unknown p that is supposed to an element of a functional space. In order to simplify the formulation, one can make the assumption

$$p(x) = \sum_{\kappa=1}^{\bar{\ell}} \pi_\kappa g(x - x_\kappa), \tag{9}$$

where the expansion with the Green function is over all the collected (supervised and unsupervised) data and the coefficients π_κ are supposed to be unknown. The probabilistic normalization requires us to impose

$$\sum_{\kappa=1}^{\bar{\ell}} \pi_\kappa = \gamma^{-1}, \tag{10}$$

where $\gamma := \int_X g(x - x_\kappa)$. Unlike in the previous case, we needn't distinguish the expression of the reaction of the constraints, which simply turns out to be

$$\omega_{\kappa,j}(x) = \frac{y_{\kappa,j}}{\mu} \sum_{h=1}^{\bar{\ell}} \pi_h g(x - x_h).$$

We notice in passing that $y_{\kappa,j}$ is directly given for supervised point by means of the pair $(x_\kappa, y_{\kappa,j})$. In case of propositional rules modeled by multiintervals, given x_κ as one of the ℓ examples of the training set, we can determine its category simply by checking whether it satisfies the or-ing of the given propositions. If we consider the normalization condition of Eq. (10) then the optimal solution can be expressed as

$$f_j^*(x) = \sum_{\kappa=1}^{\bar{\ell}} g(\cdot) * \omega_{\kappa,j}(\cdot) =$$

$$\frac{1}{\mu} \sum_{\kappa=1}^{\bar{\ell}} y_{\kappa,j} \left(\sum_{h=1}^{\bar{\ell}} \pi_h \right) g(\cdot) * g(\cdot - x_h)$$

$$= \sum_{\kappa=1}^{\bar{\ell}} \frac{y_{\kappa,j}}{\mu\gamma} \bar{g}(x - x_h)$$

$$= \sum_{\kappa=1}^{\bar{\ell}} \lambda_{\kappa,j} \bar{g}(x - x_h)$$

where $\lambda_{\kappa j} := \frac{y_{\kappa,j}}{\mu\gamma}$ and $\bar{g}(\cdot) := g(\cdot) * g(\cdot - x_h)(x) = \int_X g(x-u)g(u-x_h)\,du$. Hence we are back to a kernel expansion. Again we can reuse the mathematical and algorithmic kernel apparatus for approaching the problem.

EXERCISES

1. [22] Prove that k defined by Eq. (8) is symmetric and nonnegative definite.

2. [18] Prove that functional (2) can be converted into a finite-dimensional quadratic form.

3. [HM47] The derivation of Eq. (4) concerning the representation of f in case of supervised learning over point- and multiintervals is based on the sign consistency hypothesis (3). The analysis is based on the paper published in [S. Melacci, M. Gori, *IEEE Trans. Pattern Anal. Mach. Intell.* **35** (2013), 2680–2692], which also makes use of this hypothesis. Can we get rid of it?

4. [28] Reproduce the experimental results for the artificial data set in [S. Melacci, M. Gori, *IEEE Trans. Pattern Anal. Mach. Intell.* **35** (2013), 2680–2692] (Fig. 5).

5. [HM50] Reformulate learning from examples and propositional rules addressed in Section 6.4 as an optimization problem with the additional unknown p that is supposed to be an element of a functional space. Express the reaction of the probabilistic normalization constraint and write the differential equation for p. Is the problem well-posed? What if we have supervision on p available?

6. [25] Prove that differential operator $L = \sum_{\kappa=0}^{k}(-1)^\kappa \rho_\kappa \nabla^{2\kappa}$ is invertible when assuming that P is given by Eq. (1).

6.4.3 Supervised learning with linear constraints

Now let us consider another example in which we can give useful expressions of the constraint reaction for the reduction of learning to the framework of kernel machines. We consider the case of bilateral linear hard constraints like those referred to in Table 6.1–(vi) and (vii). The structural difference is that in the first case the constraints act on the finite dimensional space of the learning parameters, while in the second case they apply on a functional space. Exercise 1 stimulates the solution in the case of finite dimensional case, whereas here we cover the more general formulation. Suppose, as usual, that our perceptive space is $X = \mathbb{R}^d$ and that $\forall i \in \mathbb{N}_m$, $\forall x \in X \ \psi_i(f(x)) := a_i' f(x) - b_i = 0$, where $a_i \in \mathbb{R}^n$ and $b_i \in \mathbb{R}$ are given. Basically, these are holonomic bilateral constraints that can be rewritten in a compact form as $Af(x) = b$, where $b \in \mathbb{R}^m$ and a_i' is the ith row of a given constraint matrix $A \in \mathbb{R}^{m \times n}$. In the following, we assume $n > m$ and rank $A = m$. We discuss the solution for the class of so-called rotationally symmetric differential operators P that are expressed as

$$P := (\sqrt{\rho_0} D_0, \sqrt{\rho_1} D_1, \dots, \sqrt{\rho_\kappa} D_\kappa, \dots, \sqrt{\rho_k} D_k)'. \tag{1}$$

Such operators correspond via $L = (P^\star)' P$ to $L = \sum_{\kappa=0}^{k} (-1)^\kappa \rho_\kappa \nabla^{2\kappa}$, which is an invertible operator (see Exercise 6).

We first address the case $\rho_0 \neq 0$, for which we show by a counterexample that being a solution of the Euler–Lagrange equations is not a sufficient condition to solve the associated problem of learning from hard constraints, even though such a problem is convex in this case. We can promptly see that the Euler–Lagrange equations (10) are satisfied for a constant function $\bar{f}$. We have $L\bar{f} = \sum_{\kappa=0}^{k} (-1)^\kappa \rho_\kappa \nabla^{2\kappa} \bar{f} = \rho_0 \bar{f}$ and $\nabla_f \phi_i(\bar{f}) = a_i$. Hence we get $\rho_0 \bar{f} + A'\lambda = 0$, where the Lagrangian multipliers are collected in the constant vector $\lambda \in \mathbb{R}^m$. Now all the constant solutions $\bar{f}$ of the above algebraic equation are also solutions of $\rho_0 A \bar{f} + A A'\lambda = 0$ and therefore of, $\rho_0 b + A A'\lambda = 0$. Let $\det(AA') \neq 0$, which is implied by the assumptions $n > m$ and rank $A = m$. We can determine the vector of Lagrange multipliers by

$$\lambda = -\rho_0 (AA')^{-1} b, \tag{2}$$

and consequently, denoting by λ_i the ith component of the vector λ, the reaction of the ith constraint is $\omega_i = -a_i \lambda_i$ which, in turn, yields

$$\omega_i = \rho_0 a_i \left((AA')^{-1} b \right)_i. \tag{3}$$

Hence, recalling that the overall reaction of the constraints is $\omega = \sum_{i=1}^{m} \omega_i$, the solution is given by

$$\bar{f} = g * \omega = \rho_0 A'(AA')^{-1} b \int_X g(\zeta) d\zeta.$$

Due to $Lg = \delta$, we get $\sum_{\kappa=0}^{k} (-1)^\kappa \rho_\kappa \nabla^{2\kappa} g = \delta$. When using the Fourier transform $\hat{g}(\xi)$ of g, we get $\rho_0 \hat{g}(\xi) + \sum_{\kappa=1}^{k} \rho_\kappa |\xi|^{2\kappa} \hat{g}(\xi) = 1$. For $\xi = 0$ we get $\rho_0 \hat{g}(0) = \rho_0 \int_X g(\zeta) d\zeta = 1$. Finally,

$$\bar{f} = A'(AA')^{-1} b. \tag{4}$$

Now we check that the function $\bar{f}$ found above is *not* a solution of the problem of learning from hard constraints, for which we have just solved the associated Euler–Lagrange equations. First of all, since it is a constant, each nonzero component $\bar{f}_j$

of f does not belong to the Sobolev space $W^{k,2}(\mathbb{R}^d)$, so $\bar{f}$ does not belong to the ambient space F. At a first look, this may be considered a minor issue since one may still replace the ambient space by a different one, for which the same form of the Euler–Lagrange equations still holds. The real issue is that, for the obtained $\bar{f}$, the value $E(\bar{f}) = \| \bar{f} \|^2$ assumed by the objective functional is not even finite. However, the solution (4) has a concrete meaning that can be grasped if we associated the given problem with one in which the constant b is replaced with the function b such that $b(x) = b$ for $\| x \| < R$ and $\lim_{\|x\| \to \infty} f(x) = 0$, where $R > 0$. Of course, it a closely related problem, since we can promptly see that this new problem admits the solution $\bar{f}(x) = A'(AA')^{-1}b(x)$, which is the same as (4) within the sphere $\| x \| < R$. Unlike the original problem, we can always choose the decay function b so as $\bar{f} \in W^{k,2}(\mathbb{R}^d)$.

We now consider the case $\rho_0 = 0$. In this case one can easily verify that $\bar{f} = A'(AA')^{-1}b$ solves again the associated Euler–Lagrange equations with the constant choice $\lambda = 0$. Although such $\bar{f}$ does not belong to F (so it is not an optimal solution of the original problem of learning from hard constraints when it is set on F), its components $\bar{f}_j$ belong to the generalized Sobolev space $H_P(\mathbb{R}^d)$, i.e., the set of functions $f_j : \mathbb{R}^d \to \mathbb{R}$ for which $\| f_j \|_P^2$ is finite. Finally, since $E(f) = \| f \|^2 \geq 0$ for any admissible f and $E(\bar{f}) = \| \bar{f} \|^2 = 0$, one can conclude a-posteriori that $\bar{f}$ is indeed an optimal solution $f^\star$ to the problem of learning from hard constraints above when it is set on

$$\bar{F} = \underbrace{H_P(\mathbb{R}^d) \times \cdots \times H_P(\mathbb{R}^d)}_{n \text{ times}}. \tag{5}$$

A relevant case is the one in which a generalization of the previously discussed linear constraints is combined with classic learning from supervised examples. We suppose that $\rho_0, \rho_k > 0$. We are given the learning set

$$\left\{ (x_\kappa, y_\kappa) : x_\kappa \in \mathbb{R}^d, \quad y_\kappa \in \mathbb{R}^n, \quad \kappa \in \mathbb{N}_{m_d} \right\}$$

and the linear constraint $Af(x) = b(x)$, where $b \in \mathcal{C}_0^{2k}(X, \mathbb{R}^m)$ is a smooth vector-valued function with compact support. Such a constraint is still intended in hard sense, as usual, whereas the given supervised pairs induce soft constraints expressed in terms of the quadratic loss. This is reasonable in practice. For example, in the task of Table 6.1–(vii), while the single asset functions can be learned in a soft way from supervised examples, a constraint on the assets, like the overall money available, must be intended in a hard sense.

Since this is a problem with mixed hard and soft constraints, we search for a solution $\bar{f}$ of the Euler–Lagrange equations

$$L\bar{f}(x) + A'\lambda(x) + \frac{1}{m_d} \sum_{\kappa=1}^{m_d} (\bar{f}(x) - y_\kappa)\delta(x - x_\kappa) = 0. \tag{6}$$

Let us determine the vector of Lagrange multipliers λ. We start noting that

$$
ALf = A\sum_{\kappa=0}^{k}(-1)^{\kappa}\rho_{\kappa}\nabla^{2\kappa}f
$$

$$
=\sum_{\kappa=0}^{k}(-1)^{\kappa}\rho_{\kappa}A\nabla^{2\kappa}f
$$

$$
=\sum_{\kappa=0}^{k}(-1)^{\kappa}\rho_{\kappa}\nabla^{2\kappa}Af
$$

$$
=\sum_{\kappa=0}^{k}(-1)^{\kappa}\rho_{\kappa}\nabla^{2\kappa}b=Lb,
$$

where $Lb\in\mathcal{C}_0^0(X,\mathbb{R}^m)$ has compact support. Hence we get

$$
Lb(x)+A\left(A'\lambda(x)+\frac{1}{m_d}\sum_{\kappa=1}^{m_d}(\bar{f}(x)-y_\kappa)\delta(x-x_\kappa)\right)=0,
$$

from which we find that the Lagrange multiplier distribution λ is given by

$$
\lambda(x)=-(AA')^{-1}\left(\gamma Lb(x)+\frac{1}{m_d}\sum_{\kappa=1}^{m_d}A(\bar{f}(x)-y_\kappa)\delta(x-x_\kappa)\right). \tag{7}
$$

Now, if we plug this expression for λ into the Lagrange equations (6), we get

$$
\gamma L\bar{f}(x)=c(x)+\frac{1}{m_d}\sum_{\kappa\in\mathbb{N}_{m_d}}Q(y_\kappa-\bar{f}(x))\delta(x-x_\kappa),
$$

where $c(x):=\gamma A'(AA')^{-1}Lb(x)$ and $Q:=I_n-A'(AA')^{-1}A$. Let $\alpha_\kappa^{(ql)}:=(1/m_d)(y_\kappa-\bar{f}(x_\kappa))/\gamma$. By inverting the operator L, we get

$$
\bar{f}(x)=\frac{\int_X g(\zeta)c(x-\zeta)d\zeta}{\gamma}+\sum_{\kappa=1}^{m_d}Q\alpha_\kappa^{(ql)}g(x-x_\kappa). \tag{8}
$$

EXERCISES

1. [28] Given a vectorial linear machine $f:X\to Y, x\mapsto Wx$, where $W\in\mathbb{R}^{d\times n}$ and the set of bilateral holonomic linear constraints defined by $\forall x\in X\ Af(x)=b$, where $A\in\mathbb{R}^{n\times m}$ and $b\in\mathbb{R}^m$, formulate learning in both the case of soft and hard constraints by assuming the parsimony criterion which consists of minimizing the objective function $P(W)=\|W\|^2$. What if we are also given a collection of supervised examples?

2. [23] Consider the problem of supervised learning from examples with the probabilistic normalization constraint $\forall x\in X\ \sum_{j=1}^{n}f_j(x)=1$. Find the optimal solution in case of a parsimony index defined by differential operator P.

6.4.4 **Learning under diffusion constraints**

Now it's time to explore diffusion-based learning and inference schemes. Apparently, these look like other constraints, so one might end up with the conclusion that there is nothing new to say here. However, as already pointed out in Section 6.3, the special uniform propagation of evidence suggests a specific analysis. Furthermore, diffusion machines introduce hidden state variables into the arguments of the constraints, which is of remarkable importance. The presence of hidden variables enriches constraint machines in terms of relational modeling. It has been pointed out that we can indeed construct relations without involving state variables, but diffusion yields in fact a different type of relations. A closer look at constraints with hidden state variables leads us to conclude that there is an underlying process of diffusion. Notice that it's only the semantics attached to the output functions which makes it possible to express specific knowledge domain. Even though one can likely find exceptions, when hidden variables are involved, the constraints cannot be built up on information that is not semantically labeled and, therefore, in this case they turn out to be homogeneous! As such, diffusion seems to be the outcome of lack of specific knowledge. Unlike most of the discussed semantic-based constraints, which propagate evidence according to their specific meaning, diffusion machines carry out uniform propagation.

We start discussing the case of graphical environments modeled by directed acyclic graphs that, as already seen, dramatically simplifies the information flow. Many tasks are in fact characterized by such a flow. The case of a sequence is the simplest one, but it can be considered as a good representative of the entire class of DAGs. An interesting case is that in which the entire sequence needs to be classified. Here supervised learning takes place from the collection of supervised sequences

$$\{(\langle x_\kappa \rangle, y_\kappa), \quad \kappa = 1, \dots, \ell\},$$

where the generic sequence $\langle x_\kappa \rangle$ of length t_κ is given the supervision y_κ. The concrete position of the problem does require specifying the computing style. Two different schemes are possible: First, the response of the machine is taken, at t_κ, on the last element x_{t_κ}. Second, one waits for the machine to settle on a stable state, and takes the output by processing the value of the state. Of course, this relaxation dynamics makes sense whenever the recurrent network exhibits stable behavior. While in principle one might consider the response after a given number of steps after the end of the sequence, marked by t_κ, it's quite obvious that such a response is brittle and mostly meaningless. Similar tasks can involve prediction that, however, are more often formulated in a different way. Instead of waiting for the end of the sequence, the decision can take place online, which is a computing style that is clearly interesting for also classification. A given set of sequences can always be connected so as to create one single sequence. In so doing, the training set turns out to be

$$\{(\langle x_{\kappa, v(\kappa)} \rangle, \langle y_{\triangleright(v(\kappa), \psi_o)} \rangle), \quad \kappa = 1, \dots, \ell\}$$

where the index $\kappa = 1, \ldots, \ell$ denotes the κth sequence, while $v(\kappa)$ is the index associated with a generic atom of the sequence — it is in fact a vertex of the overall environmental graph. Here the constraint ψ_o is active on $v(\kappa)$ whenever $\triangleright(v(\kappa), \psi_o) = \top$. We rely on the underlying assumption that the ℓ sequences are independent of each other, so that we assume to begin the computation with the reset of the state. For any $\kappa = 1, \ldots, \ell$ we assume that for the last vertex $v_o(\kappa)$ of the list we have $z_{v_o(\kappa)} = z_o$. Learning with diffusion models requires the satisfaction of constraints expressed by Eq. 6.3–(1), with the additional specification that we might disregard the application of the constraint on h on some vertices. In general, in the case of a DOAG $\mathcal{G} = \{V, A\}$, one can formulate learning in diffusion machines as the problem of discovering the functions s and h such that

$$\binom{s}{h} \in \underset{s,h}{\arg\min} \frac{1}{2}\left(\mu_s \langle P_s s, P_h s\rangle + \mu_h \langle Ph, Ph\rangle + \sum_{v \in V_{\psi_o}} \psi_o(h(z_v, x_v))\right)$$

$$[\forall v \in V_o](z_v - z_o = 0) + [\forall v \in V \setminus V_o](z_v - s(z_{\mathrm{ch}(v)}, x_v) = 0).$$

Here, the problem is formulated as parsimonious constraint satisfaction by using the parsimony index based on regularization operators P_s and P_h. As we can see, the hidden state function s is in fact constrained to obey the diffusion model, while the output function h must satisfy the constraint expressed by ψ_o. This formulation assumes that we are strongly-enforcing $z_v = s(z_{\mathrm{ch}(v)}, x_v)$, while the constraint on the output function $\psi_o(h(z_v, x_v))$ is only soft-enforced. This is quite a natural assumption that can be regarded as a strong belief in the graphical model expressed by the transition function s. Exercise 3 proposes a discussion on the relaxation of this constraint. It's worth mentioning that $[\triangleright(v, \psi_o)]$ can dramatically change the nature of the learning problem. Basically, $[\triangleright(v, \psi_o)]$ defines the set $V_{\psi_o} = \{v \in V : [\triangleright(v, \psi_o)]\}$ on which the constraint is active. It's quite clear that if ψ_o is enforced only on a few vertexes then the hidden variables become more important, so the machine is strongly driven by its state transition function s. On the other hand, in the extreme case, when all vertexes receive the constraint ψ_o then the role of the state variable becomes less important. Another remark is in order concerning the state transition function s, which is forced to the reset state in the set V_o. Like for sequences, the presence of these vertices can be useful when creating a collection of distinct graphs that are separated by vertices in V_o. For example, the learning tasks on Quantitative Structure Activity Relationship described in Section 1.1 require the construction of a training set which is composed by a collection of graphs. In other cases, the set V_o is reduced to the set of buds of the environmental graph $\mathcal{G}$ that is used to fill out the missing children.

When dealing with directed ordered graphs, the learning algorithms can parallel those for feedforward neural networks, since we are still in the presence of a data flow computational scheme. The analysis carried out Section 6.3.3 indicates that we can in fact perform *structure unfolding*, so as to reduce the learning algorithm to backpropagation. Since it comes from the unfolding of the environmental graphical structure, in this case the algorithm is referred to as *Backpropagation Through Structure* (BPTS).

The compiling process illustrated in Fig. 6.6 shows clearly that we are simply using backpropagation over the $S2$ neural network. In the case of lists, because of the presence of a temporal basis, the gradient computation algorithm is referred to as *Backpropagation Through Time* (BPTT). The algorithm keeps the strong property of *spatial locality* during the weight updating of the network, which is inherited from backpropagation. Basically, when learning the weight $w_{i,j}$, only information local to the weight is used for its updating. However, in the case of data organized in long lists, while spatial locality is still an important computational requirement, one might decide to dismiss it by preferring the satisfaction of *temporal locality*. The problem with long list is that the associated time-unfolded network becomes a very deep network. In the extreme case, if we need a truly online processing, we cannot store a virtually infinite network! The lack of temporal locality makes clear the limitation of this approach. It's quite obvious that this is not restricted to the case of sequences. When graphs are involved, the critical issues connected with temporal locality arise in environmental tasks defined on a network environment that can be regarded as an infinite structure, e.g., think of the Web.

Now let's see another algorithm for gradient computation that, in case of sequences, possesses the temporal locality property. For a general graph, we'll see that this turns into a *vertex locality property*, and that the same algorithmic idea can still be applied. To make things easy, suppose we are dealing with a recurrent neural network characterized by

$$z_{t+1,i} = \sigma\left(\sum_{j \in \mathrm{pa}(i)} w_{i,j} z_{t,j} \right), \tag{1}$$

where $i \in H \bigcup O$ and $j \in I \bigcup H \bigcup O$, with I, H, O being the sets of input, hidden, and output units, respectively. As we can see, this is in fact a way of expressing the constraint $z_v = s(z_{\mathrm{ch}(v)}, x_v)$. As already noticed, we need to set $t = -v$ and consider the aggregation function s simply as a combination of the state $z_{\mathrm{ch}(v)} = z_t$ and of the input $x_v = x_t$. Notice that Eq. (1) is in fact an explicit way of expressing the coordinates of the state z and the sum over $\mathrm{pa}(i)$ is a classic forward step over the network architecture. When referring to the above general learning formulation, here we simply have $h = \mathrm{id}$. Suppose we work in the primal space and, furthermore, assume that the regularization terms expressed by differential operators are simply replaced with quadratic expressions, so the gradient yields the classic weight decay. The discovery of the minimum by gradient descent does require focusing on

$$\frac{\partial}{\partial w_{j,k}} \sum_{t \in \mathcal{T}_{\psi_o}} \psi_o(z_t(w), x_t) = \sum_{t \in \mathcal{T}_{\psi_o}} \sum_{i \in O} \frac{\partial \psi_o(z(w), x_t)}{\partial z_{t,i}} \frac{\partial z_{t,i}(w)}{\partial w_{j,k}}, \tag{2}$$

where $\mathcal{T}_{\psi_o} = V_{\psi_o}$ is the set of time indexes on which the constraint ψ_o applies. The above factorization can be used as a basic step for gradient computation. Basically, the first factor can always be determined by the forward steps. For the second one, we can exploit a recursive scheme that relies on the definition of $\zeta_{t,i,j,k} :=$

$\partial z_{t,i}(w)/\partial w_{j,k}$. This tensor needs to be determined $\forall i \in O, j,k \in I \bigcup H \bigcup O$. From Eq. (1) we get

$$\zeta_{t+1,i,j,k} = \sigma'\left(\sum_{p \in \mathrm{pa}(i)} w_{i,p} z_{t,p} \right) \sum_{p \in \mathrm{pa}(i)} w_{i,p} \zeta_{t,p,j,k} + \delta_{i,j} x_{t,k}, \qquad (3)$$

where the sum is extended over $p \in H \bigcup O$. Notice that the terms associated with the Kronecker delta $\delta_{i,j}$ arise because of the input terms $x_{t,k}$, and that the previous temporal recursive equation is initialized by $z_{0,i,j,k} = 0$. The gradient computation based on this algorithmic scheme is referred to as *Real Time Recurrent Learning* (RTRL). There is a remarkable difference with respect to BPTT. As we can see RTRL is not local in space, whereas it is local in time. This last property makes it appropriate for truly online computation, but it's quite obvious that its lack of spatial locality in many cases leads to a complex computational scheme (see Exercise 4). To sum up, the gradient computation learning algorithms BPTT and RTRL are somewhat complementary: While one is local in space, the other one is local in time. Interestingly, when making restrictions on the diffusion constraints we can gain both spatial and time locality. This arises in particular when the recurrent connections are restricted to neural self-loops (see Exercise 8). The idea which drives the gradient computation according to Eqs. (2) and (3) goes beyond sequences, since it can naturally be extended to DOAGs (see Exercise 6), where the notion of temporal locality is translated into *vertexlocality.*

When abandoning the DOAG structure for the environmental graph, things become more involved. If we lose the ordering relation on the children, so that the environmental graph is reduced to a DAG, then function s is dramatically restricted to be a symmetric function, which is independent of any permutation of $\mathrm{ch}(v)$. However, nothing changes on the algorithmic structure discussed for DOAG, the only difference being that we must keep the coherence of developing a symmetric function. The simplest example of enforcing such a coherence is that in which, in Eq. (1), we enforce the equality constraints on the weights of the matrices associated with the children. When abandoning the DAG structure, we enter another computational world! The case of circular lists is the simplest example of a cyclic structure that is not suited for the described learning algorithms. The processing on circular lists corresponds with the relaxation-based architectures described in Section 5.1.3. The reason why we cannot rely on the above algorithmic framework is that, because of the cyclic structure that arises when losing the DAG property, we may visit the same vertex many times! Hence setting supervision over the vertexes requires us to specify how many times they have already been visited. This doesn't seem to correspond with meaningful computational processes, unless we wait for an eventual relaxation to an equilibrium point. The compiling scheme described in Section 6.3.1 is still valid, but the result of combining a generic cyclic graph $\mathcal{G}$ with a feedforward network $\mathcal{N}$ yields a graphical neural network $\mathcal{G} \bowtie \mathcal{N}$ that inherits cyclic connections. How does learning change in case of cyclic connections? In order to grasp the main idea, let us go back to recurrent network with cyclic connections discussed in Section 5.1.3 that processes single inputs — just like a common feedforward network. If the dynamical evolution ends up in convergence then we can still use the backpropagation algorithm. We only need to consider that the computation of the gradient requires the values of the outputs of the neurons that have been reached at the end of the relaxation to the equilibrium point. Some more details on this issue are given in Section 6.6, including a discussion in the case of graphs.

The massive experimentation with recurrent neural networks based on the model described so far — especially for sequences — early led to identifying that they are not very effective in learning tasks that require capturing long-term symbolic dependencies. We can get an insight on the reason of this limitation by considering the following simple example with one neuron only:

$$z_{t+1} = \sigma(wz_t + ux_t). \tag{4}$$

To make things easy, suppose that $\sigma = \tanh$, and start to analyze the case in which $x_t = 0$. We can promptly see that there are two different dynamical behaviors. First, if $w < \sigma'(0) = 1$ then $\lim_{t \to \infty} z_t = 0$, whereas if $w > \sigma'(0) = 1$ then there are two symmetric stable points z^* and $-z^*$ (see Exercise 10). If the recurrent network is used to recognize a sequence that is supervised only at the end, then there are in fact serious issues of gradient vanishing. This immediately arises when noticing that the application of BPTT yields $\delta_t = w\sigma'(a_t)\delta_{t+1}$. Now, for t large enough, we have $z_t \simeq z^*$, and $w\sigma'(a^*) \leq 1$, with $w\sigma'(a^*) = 1$ holding only in case in which in the graph of the functions $z(a) = a/w$ and $z(a) = \sigma(a)$ the line is tangent to the sigmoidal curve. Of course, the case $w\sigma'(a) < 1$ leads us to conclude that δ_t vanishes as we get far away from the supervision, which propagates the same property to the gradient! The case $w\sigma'(a) = 1$ leads to marginal stability. When the input $x_t \neq 0$, we can promptly see that there is no robust information latching, since any small noise can lead to switching to the equilibrium $z^* = 0$. Hence, whenever one wants to latch information, the condition $w\sigma'(a) \leq 1$ must be satisfied in the strict sense. As a consequence, the recurrent network is doomed to experience gradient vanishing and is unable to capture long-term dependencies. Additional details are given in Exercise 11. It can be proven that this dramatic limitation is not restricted to this simple example, but that it is in fact connected with the limited computational capabilities expressed by the chosen map s. At each layer, the time unfolding process in recurrent networks with sigmoidal units leads to a reduction of error δ, which results in an exponential decay. The core of the problem is in the limited computational capability of the map s defined by Eq. (4), a limitation that still applies in the case of vectorial variables — more neurons to represent the state. How can we circumvent this problem? Since this deflating behavior of δ_t is due to the limited expressiveness of s, we need to introduce more complex state transition functions.

The additional complexity of the state transition function s is expected to sustain an inflating/deflating behavior for δ_t that must be stimulated by the presence of specific events in the sequence. Long Short Term Memory (LSTM) recurrent networks represents a successful example of constructing a state transition with the required inflating/deflating property. Additional details are given in Section 6.6. Interestingly, Exercise 12 suggests a potential alternative to LSTM.

EXERCISES

1. [*HM45*] Solve the learning problem with holonomic bilateral quadratic constraints and supervised pairs.

2. [*HM50*] In case of hard-constraints the Lagrangian multipliers are functions in the perceptive space. Based on the proof of Theorem 2 in Appendix C, propose an algorithm for finding the Lagrangian multipliers and the constraint reactions. *Hint:* Use a cycling scheme that initializes the multipliers to a constant, and then determine an update by using Eq. C–(13).

3. [*HM47*] Discuss the solution of learning under DAG graphical environment when the constraints on the hidden state variables $x_v = s(x_{\mathrm{ch}(v)}, x_v)$ is relaxed (soft constraint formulation).

4. [*18*] Analyze the computational complexity of the BPTT and RTRL algorithms (space and time). Give an example in which BPTT is preferable and another example in which the opposite holds.

5. [*20*] The idea behind RTRL can be used also for learning in feedforward neural networks. Is the corresponding algorithm more efficient than backpropagation?

6. [*27*] RTRL works for lists. How can it be extended to graphs? RTRL possesses the time locality property. How can this property be translated into the graphical domain?

7. [*45*] RTRL equations (3) might explode. Can you find any condition for preventing explosion? More technically, can you find conditions for guaranteeing BIBO (Bounded Input Bounded Output) stability? Discuss qualitatively the role of $\mathcal{T}_{\psi_o}$ in stability, as well as the set $\mathcal{T}_o = V_o$ in which we reset the state.

8. [*32*] Consider a recurrent neural network for sequence where the recurrence is restricted to the set H of hidden units that only contain self-loops — connections are only accepted provided that $i \leq j$. Write down an algorithm for gradient computation which is local in both space and time.

9. [*HM46*] Let us consider a recurrent network in the continuous setting of computation. We formulate learning as the process which yields a stationary point of

$$E = \int_0^t \sum_\alpha V(z(\tau))\delta(\tau - t_\alpha)\,d\tau$$

under the holonomic constraint

$$\dot{z}_i(t) = \sigma\left(\sum_j w_{i,j}(t)z_j(t) + \sum_\kappa u_{i,\kappa}(t)x_\kappa(t)\right).$$

Derive the solution $t \mapsto w_{ij}(t)$ in the framework of variational calculus.

10. [*M23*] Consider the recurrent network define by Eq. (4). Prove that the claimed statement that if $w \leq \sigma'(a)$ then $\lim_{t\to\infty} z_t = 0$, whereas if $w > \sigma'(a)$ then there are two symmetric stable points z^* and $-z^*$.

11. [*M24*] Consider the recurrent network defined by Eq. (4). Prove the claimed statement, namely that $w\sigma'(a) \leq 1$.

12. [*HM48*]   Consider the unfolded networks created by a given recurrent network where

$$z_i = \gamma \left(\sum_j w_{i,j} z_j \right),$$

where $i \in O \bigcup H$ and $j \in O \bigcup H \bigcup I$. In order to deal with the problem of long-term dependencies, suppose that γ is not chosen in advance, but that it is taken from a functional space and learned from examples. In particular, formulate a learning algorithm based on the framework introduced in Exercises 5.5–1 and 5.5–2.

6.5 Life-long learning agents

The path followed so far has led us to formulate machine learning mostly as an optimization problem with parameters to be learned over an available training set. The performance is assessed by using an appropriate statistically significant test set. The discussion on the perceptron algorithm and the translation of batch-mode learning algorithms for neural nets into online schemes do not really address the view point put forward in Section 1.2.3. In spite of a number of appreciable attempts, also the functional representation of kernel machines, along with its efficient mathematical and algorithmic apparatus, does not offer a natural scheme for an online formulation of learning. The formulation proposed in Section 4.4.4, which is based on Euler–Lagrange differential equations, is quite different since it's inherently local in the feature space. Hence, instead of looking for an online equation by rearranging the global representation based on kernel expansion, we might think of using directly the local model provided by the differential equations. While at a first look this seems the right way of driving online learning schemes, the problem is that the Euler–Lagrange equations are defined in $\mathbb{R}^d$, that is, they are typically in high-dimensional spaces. This is a very serious warning from a computational complexity view point. Something similar holds for the learning from constraints framework introduced in the previous section. The given result on the functional representation of the agent is indeed nice, since it discloses the deep structure of the "body" of the agent. However, as pointed out in the previous section, we can use concretely these functional statements only for some special classes of constraints, for which the constraint reactions are expressed in such a way to transform functional optimization into a finite-dimensional problem (collapse of dimensionality). Eqs. 6.1.3–(4), 6.1.3–(5), 6.1.3–(11), and 6.1.3–(12), in their general form, express f^* as a function of the constraint reaction which, in turn, depends on f^*. While this circularity can be broken from simple constraints, it is in fact a sign of complexity in more general tasks. Consider the case of hard constraints. Basically, the construction of the dual form, which leads to expressing $\theta(\lambda)$, cannot follow the straightforward path of kernel machines, in which the minimization with respect to the primal variables makes it possible to directly express $\theta(\lambda)$ by appropriate elimination of the primal variable w. One possibility is to use alternate optimization steps in the primal and dual space, but there are critical issues of convergence. Moreover, it looks like there is something wrong with the current indisputable dominating school of thought in machine learning, which biases any approach with the concept of a given training set. First, it ignores time! Of course, every mathematical and algorithmic framework which relies on the notion of training set can, in principle, be made online. However, this seems to be extremely unnatural.

When looking at the formulated view of learning from constraints, we are still mostly inside the framework of learning as a process to be constructed over a given data collection. As a result, also the local models offered by the corresponding Euler–Lagrange equations suffer from this strong assumption. A radical paradigm shift does require thinking of data as a temporal stream and reformulating learning as a process fully immersed in time. When shifting our view on time, we can also easily interpret the behavior of the teenager nerd mentioned in Section 1.2.4, who leaves the computer science lab meeting to protect himself from information overloading! Time naturally stimulates and drives the focus of attention and the active interaction with the environment. Just like humans, artificial agents need to pass through developmental steps that incorporate appropriate focusing mechanisms. Overall, a good theory on the processes of learning and inference that might be successful in machine learning neither needs to emulate human cognition nor to capture biological issues, but likely needs to provide an efficient computational perspective. When blessing this view point, we are essentially subscribing to the principle that there are in fact information-based laws that govern learning mechanisms regardless of human or artificial agents. In this section, we introduce a *natural learning theory* based on the principle of *least cognitive action*, which is inspired by the related mechanical principle, and by the Hamiltonian framework for modeling the motion of particles. The introduction of the kinetic and potential energy leads to a surprisingly natural interpretation of learning as a dissipative process. The kinetic energy reflects the temporal variation of the synaptic connections, while the potential energy is a penalty that describes the degree of satisfaction of the environmental constraints.

6.5.1 Cognitive action and temporal manifolds

In most challenging and interesting learning tasks taking place in humans, unlike machines, the underlying computational processes does not seem to offer a neat distinction between the training and the test set. As time goes by, humans react surprisingly well to new stimuli, while keeping past acquired skills, which seems to be hard to reach with nowadays intelligent agents. This suggests us to look for alternative foundations of learning, which are not necessarily based on statistical models of the whole agent life. We can think of learning as the outcome of laws of nature that govern the interactions of intelligent agents with their own environment, regardless of their nature. We reinforce the underlying principle that the acquisition of cognitive skills by learning obeys information-based laws on these interactions, which hold regardless of biology.

The notion of time is ubiquitous in laws of nature. Surprisingly enough, most studies on machine learning have relegated time to the related notion of iteration step. On the one hand, the connection is pertinent and apparently sound, since it involves the computational effort that is also observed in nature. From one side notice that time acts as an index of any human perceptual input. Now, while the iteration steps in machine learning somewhat parallel this idea, most algorithms neglect the smoothness of the temporal information flow. As a consequence, we pass through the inputs of the training set that, however, turn out to be an unrelated picture of artificial life, since the temporal relations are lost. This might be one the main reasons of current performance gap in challenging tasks of vision understanding with respect to humans.

Here we discuss how to express and incorporate time in its truly continuous nature, so that the evolution of the weights of the neural synapses follows equations that resemble laws of physics. The environmental interactions are modeled under the general framework of constraints on the learning tasks. In the simplest case of supervised learning we are given the collection $\mathcal{L} = \{(t_\kappa, x(t_\kappa)), y_\kappa\}_{\kappa \in \mathbb{N}}$ of supervised pairs $(x(t_\kappa), y_\kappa)$ over the temporal sequence $\{t_\kappa\}_{\kappa \in \mathbb{N}}$. We assume that those data are learned by a feedforward neural network characterized by the function $F(\cdot, \cdot) : \mathbb{R}^m \times \mathbb{R}^d \to \mathbb{R}^n$, so the input $x(t)$ is mapped to $y(t) = F(w(t), x(t))$. Learning affects the synapses by changing the weights $w(t)$ during the agent life taking place in the horizon $\mathcal{T} = [0, T]$, where T can become pretty large, so that the condition $T \to \infty$ might be reasonable.

The environmental interactions take place over a temporal manifold, so the perceptual input space $X \in \mathbb{R}^d$ is "traversed" by the map $\mathcal{T} \to X \subset \mathbb{R}^d \ t \mapsto x(t)$. The following analysis holds for different types of constraints, including the previous pointwise constraints of supervised learning where each example x_κ can be associated with the loss $\left(V(t, w(t)) = F(w(t), x(t)) - y_\kappa\right)^2 \cdot \delta(t - t_\kappa)$. For example, the soft-enforcement of a general unilateral constraint $\check{\psi}$ can be imposed by keeping as low as possible the following associated *potential function*:

$$V(t, w(t)) = \tilde{\psi}\left(x(t), F(w(t), x(t))\right) \tag{1}$$

where $\tilde{\psi} = (-\check{\psi})_+$. Basically, the potential is the temporal counterpart of the penalty function of the constraint. When looking at the neural network, we can regard $w \in \mathbb{R}^m$ as the *Lagrangian coordinates* of a virtual mechanical system, so the potential of the system is defined in terms of the Lagrangian coordinates w. In this perspective, as already noticed, we look for trajectories $w(t)$ that possibly lead to configurations with small potential energy. Following the duality with mechanics, we also introduce the notion of kinetic energy. Now we parallel the notion of velocity by considering how quickly the weights of the connections are changing. We can also dualize the notion of mass $m_i > 0$ of a certain particle by introducing the *mass* of a connection weight. In so doing, the overall system is characterized by the conjugate variables that correspond with the *position* $w_i(t)$ and *velocity* $\dot{w}_i$. Then we define the *kinetic energy* as

$$K(\dot{w}) = \frac{1}{2} \sum_{i=1}^{m} m_i \dot{w}_i^2. \tag{2}$$

Now let us consider a Lagrangian inspired from mechanics, which is split into the potential V and the kinetic energy as

$$F_m(t, w, \dot{w}) = \sum_{i=1}^{m} F(t, w_i, \dot{w}_i) = \sum_{i=1}^{m} K(t, \dot{w}_i) - V(t, w). \tag{3}$$

The Lagrangian F_m has an intriguing meaning that arises when we explore the underlying cognitive processes. Let us consider

$$F(t, w, \dot{w}) := \zeta\left(t, \dot{x}(t)\right) F_m(t, w, \dot{w})$$

and define $\zeta\left(t, \dot{x}(t)\right)$ as the *developmental function* of the agent. We can think of $\zeta([0, T], \dot{x}([0, T]))$ as an infinite-dimensional vector so that $\zeta\left(t, \dot{x}(t)\right)$ properly filters out the environmental constraints. Hence small values of $\zeta\left(t, \dot{x}(t)\right)$ reduce the corresponding penalty of the constraint. In so doing, we somehow "protect" the agent from information overflow! The value of $\zeta\left(t, \dot{x}(t)\right)$ is supposed to be "small" at the beginning of learning, while as time goes by its growth enforces the satisfaction of the constraint. This somehow emulates human life, where babies are not forced to react to complex stimulus. They pass through developmental stages by properly focusing attention on the tasks. While the explicit temporal dependence in $\zeta(\cdot, \dot{x}(\cdot))$ is basically needed to impose monotonicity, so as to carry out protection from information overload, the dependence on $\dot{x}$ models the need of focusing attention. At a given time t and certain level of development, the agent reacts differently to the external stimulus. It must be able to filter out and focus attention on what is expected to be more relevant in its current stage of development. The developmental function $\zeta\left(t, \dot{x}(t)\right)$ can be expressed by the natural factorization

$$\zeta\left(t, x(t)\right) = \rho(t)\varphi(\dot{x}(t)). \tag{4}$$

Function $\rho \colon [0, T] \to \mathbb{R}$, which is supposed to be monotonic, is referred to as the *dissipation function*. The reason of the name will be captured later on when discussing energetic issues behind the emergence of learning. Function φ is the referred to as the *focus function*. Its role is that of selecting the inputs $x(t)$ that are relevant for learning. The dependency on $\dot{x}$ is due to the principle that the agent needs to focus attention on highly changing inputs.

We are now ready to reformulate learning in the general framework of environmental constraints. The living agent is characterized by a neural network whose weight $w(t)$ vector is that for which $\delta A|_{w^*} = 0$, where

$$A(w) = \frac{1}{T} \int_0^T F(t, w, \dot{w}) \, dt \tag{5}$$

is the *cognitive action* of the system. The intriguing analogies with analytic mechanics are summarized in Table 6.3. A comment on the optimization of the cognitive action (5) is in order. In general, the problem makes sense whenever the Lagrangian is given all over $[0, T]$. The very nature of the problem results in the explicit time dependence of F, which, in turn, depends on the information coming from the interactions of the agent with the environment. The underlying assumption in learning processes is that after the agent has inspected a certain amount of information, it will be able to make predictions on the future.

Hence, whenever we deal with a truly *learning environment*, in which the agent is expected to capture regularities in the inspected data, we can make the assumption that the weight trajectory converges to an end-point that somewhat expresses the saturation of the agent learning capabilities. Hence we make the following border assumptions:

$$\lim_{t \to 0} \dot{w}_i(t) = 0, \qquad \lim_{t \to T} \dot{w}_i(t) = 0. \tag{6}$$

Table 6.3 Links between natural learning theory and classical mechanics.

Learning	Mechanics	Remarks
w_i	q_i	Weights are interpreted as generalized coordinates.
$\dot{w}_i$	$\dot{q}_i$	Weights variations are interpreted as generalized velocities.
v_i	p_i	The conjugate momentum to the weights is defined by using the machinery of Legendre transforms.
$A(w)$	$S(q)$	The cognitive action is the dual of the action in mechanics.
$F(t,w,\dot{w})$	$L(t,q,\dot{q})$	The Lagrangian F is associated with the classic Lagrangian L in mechanics.
$H(t,w,v)$	$H(t,q,p)$	When using w and v, we can define the Hamiltonian, just like in mechanics.

As it will be shown later, these conditions can be guaranteed if we assume that long-life learning undergoes a *day–night rhythm* scheme. Such a scheme follows the corresponding human metaphor: The perceptual information is only provided during the day, while the agent "sleeps" at night without receiving any perceptual information, which is translated into the condition $\dot{x} = 0$. We assume to undergo a long-life learning scheme which repeats days of life according to the above rhythmic scheme. Before discussing this assumption, we start noticing that in perceptual tasks, the day-night rhythm doesn't alter the semantics that can be captured from the environmental information flow. Hence, just like an uninterrupted flow, this rhythmic interaction keeps the semantics, but favors the simplicity and the effectiveness of learning processes, since the lack of night stimulus facilitates the verification of the condition (6) on the right border. Unlike an uninterrupted flow, the day–night rhythm allows us small weight updates from consecutive days. Hence, if $w(t_\kappa)$ is the weight vector at the end of day κ, the day after, the weight $w(t_{\kappa+1}) \simeq w(t_\kappa)$, which facilitates the approximation of the condition $\dot{w}(t = t_{\kappa+2}) = 0$. Now let's write the equations of the weights $w_i(t)$ of the synaptic connections.

If we pose $D = d/dt$ then the Euler–Lagrange equation $DF'_{w_i} - F'_i = 0$ becomes

$$m_i \ddot{w}_i + \frac{\dot{\zeta}}{\zeta} \dot{w}_i + V'_{w_i} = 0. \tag{7}$$

As we can see, the developmental function ζ, which is always positive, strongly affects the neurodynamics. In order to start understanding its effect, suppose that there is no focus of attention, that is, $d\varphi(\dot{x}(t))/dt := \dot{\varphi} = 0$. In this case the above equation reduces to

$$m_i \ddot{w}_i + \frac{\dot{\rho}}{\rho} \dot{w}_i + V'_{w_i} = m_i \ddot{w}_i + \theta \dot{w}_i + V'_{w_i} = 0, \tag{8}$$

where the last reduction comes from choosing the dissipation function $\rho(t) = e^{\theta t}$. Exercise 1 proposes a different dissipation function, which results in a remarkably different asymptotic behavior. The integration of the Euler–Lagrange equation does require knowing, for each weight, a couple of conditions. One could simply assume that the agent has a "gestation period" in which it does not receive any stimulus, so $\dot{w}_i = 0$, and that we depart from random values of w_i. This results in a Cauchy problem which is well-posed: The Euler–Lagrange equations can be directly integrated by classic numerical analysis algorithms.

The two border day–night environmental conditions are quite different, but they require quite an unusual analysis with respect to related classic developments in physics. Whenever the Euler–Lagrange equations are asymptotically stable, if we start with the Cauchy initialization scheme and enforce the day–night environment, the condition $\dot{w}_i(T) = 0$ can be arbitrarily approximated at night. As a matter of fact, its fulfillment is facilitated when using "short days." For the daily incremental learning to make sense, we need to retain, during the night, what has been learned during the day. Now the potential switches from day $t \in \mathcal{T}_d$ to night $t \in \mathcal{T}_n$ is expressed by

$$V(t, w(t)) = \begin{cases} \tilde{\psi}\big(x(t), F(w(t), x(t))\big), & \text{for } t \in \mathcal{T}_d \\ \tilde{\psi}\big(0, F(w(t), 0)\big). & \text{for } t \in \mathcal{T}_n \end{cases}$$

Hence at night we have

$$M\ddot{w} + \frac{\dot{\zeta}}{\zeta}\dot{w} + \tilde{\psi}\big(0, f(w, 0)\big) = 0, \tag{9}$$

where $M = \text{diag}\{m_i\}$. This equation models the retainment process that takes place at night! As a matter of fact, the constraints need to be satisfied also when turning off the stimulus. At the end of each day, this is not necessarily verified, so as we need "night relaxation" to satisfy Eq. (9). Notice that while we can always regard the enforcing of the day–night environment as a consistent modification of the learning environment with the Cauchy initialization, we are in front of a dynamics that can be significantly different with respect to the one which arises from processing an uninterrupted information flow by using Cauchy initialization. The day–night rhythmic assumption still relies on random initial weights, but night relaxation contributes to the fulfillment of the required boundary conditions. However, we can promptly see that the theory relies on the underlying assumption that there are a combinatorial number of solutions. For example, as shown in Section 5.3.3, this holds true for feedforward neural networks. Exercise 1 proposes a convergence analysis on the day–night rhythmic conditions.

The modeling of intelligent agents, which react to the environment asformulated by Eq. (8), somewhat escapes the borders of most statistical machine learning approaches, which rely on the appropriate definition of the learning, validation, and test set. These agents live in their own environment, so any crisp separation between the learning and the test phase turns out to be artificial. While this seems to be unreasonable in the framework of classic benchmark-based measurements of the performance, there are a number of arguments to claim that in many real-world problems this approach, based on information-based laws, can be more appropriate (see Section 1.5).

EXERCISES

▶ **1.** [*M25*] Write down the differential equations of learning the connection weights w_i when choosing the dissipation function $\rho(t) = \rho_0 t + \rho_1 \text{logistic}(\theta^{-1}t)$, where $\rho_0, \rho_1 > 0$. Suppose there is no focus of attention. What is the dissipation energy $Z(T)$? Discuss in particular the cases $\rho_0 = 0$ and $\rho_1 = 0$ and the convergence of the dynamical system.

 2. [*M20*] Consider the trivial dissipation function which is identically constant, that is, $\forall t \in [0, T]$ $\rho(t) = 1$. Discuss the dynamics in the special case of supervised learning.

3. [*M20*] Consider the dissipation function $\rho(t) = e^{\theta t}$ with $\theta < 0$. Discuss the dynamics in the special case of supervised learning.

4. [*HM45*] One could provide an interpretation of the cognitive action that very much resembles the regularized risk in machine learning. In this case, the kinetic energy can be regarded as the parsimony term, while the potentials are the penalties associated with the constraints. However, in the cognitive action defined by the Lagrangian of Eq. (3), the potential comes with a flipped sign. If we use higher-order differential operators for defining the kinetic energy things might become different. Why?

5. [*HM50*] Reformulate the cognitive action and the corresponding learning problem by considering a generic class of functions $f \in F$ instead of a neural network.

6. [*M22*] Write down the equations of the agent behavior for supervised learning in the case $\rho(t) = e^{\theta t}$.

7. [*M30*] Discuss the dynamics for a developmental function in which $\varphi(\dot{x}) = \dot{x}^2$.

8. [*M30*] Discuss the dynamics for a developmental function in which $\varphi(\dot{x}) = 1/(1 + \dot{x}^2)$.

6.5.2 Energy balance

In order to get a qualitative picture of the neurodynamics behind Eq. 6.5.1–(7), we can carry out an analysis based on energy-based invariants. This discussed more deeply in Section 6.6. Here we use a more straightforward approach. From Eq. 6.5.1–(7) we get

$$\sum_{i=1}^{m} m_i \ddot{w}_i \dot{w}_i + \frac{\dot{\zeta}}{\zeta} \sum_{i=1}^{m} \dot{w}_i^2 + \sum_{i=1}^{m} V'_{w_i} \dot{w}_i = 0. \tag{1}$$

We have $DK(\dot{w}(t)) = D\left(1/2 \sum_{i=1}^{m} m_i \dot{w}_i^2\right) = \sum_{i=1}^{m} m_i \ddot{w}_i \dot{w}_i$ and $DV(t, w(t)) = \sum_{i=1}^{m} V'_{w_i} \dot{w}_i$. If we plug these identities into the above balancing equation and integrate over the domain $[0 .. T]$, we get

$$V(0, w(0)) + K(\dot{w}(0)) = V(T, w(T)) + K(\dot{w}(T)) + Z(T) \tag{2}$$

where

$$Z(T) := \sum_{i=1}^{m} \int_0^T \frac{\dot{\zeta}}{\zeta} \dot{w}_i^2 \, dt. \tag{3}$$

Eq. (2) is clearly an energy balance. However, before making statements on the overall energy balance, we need to understand better the term $Z(T)$ which involves the developmental function. We start to see what happens when the agent does not perform any focus of attention. In that case $D\varphi(x(t)) = 0$, and we get

$$Z(T) = Z_d(T) = \sum_{i=1}^{m} \int_0^T \frac{\dot{\rho}}{\rho} \dot{w}_i^2 dt > 0, \tag{4}$$

where the positiveness comes from the assumption that ρ is monotone. If we consider the developmental function expressed by Eq. 6.5.1–(4) then we have

$$\frac{\dot{\zeta}}{\zeta} = \frac{\dot{\varphi}}{\varphi} + \frac{\dot{\rho}}{\rho} = D(\ln \varphi + \ln \rho). \tag{5}$$

As a consequence, the developmental energy $Z(T)$ takes into account both the dissipation energy and the energy coming from focusing of attention. From Eq. (5) we get

$$Z(T) := \sum_{i=1}^{m} \int_0^T \frac{\dot{\zeta}}{\zeta} \dot{w}_i^2 dt = \sum_{i=1}^{m} \int_0^T \left(\frac{\dot{\varphi}}{\varphi} + \frac{\dot{\rho}}{\rho} \right) \dot{w}_i^2 dt = Z_d(T) + Z_f(T),$$

$$Z_f(T) := \sum_{i=1}^{m} \int_0^T \frac{\dot{\varphi}}{\varphi} \dot{w}_i^2 dt.$$

Now this energy is clearly connected with the process of focusing attention. Of course, $Z_f(T)$ depends on the policy followed for focusing of attention, and can be either positive or negative. But a good choice is that overall, $\dot{\zeta}/\zeta \geq 0$. From Eq. (2), the overall energy balance becomes[2]

$$V(0) + K(0) = V(T) + K(T) + Z_d(T) + Z_f(T). \tag{6}$$

Suppose that there is no focus-driven computation, so $Z_f(T) = 0$. Furthermore, since on the border of $[0, T]$ we have $K(0) = K(T) = 0$, we get $V(0) = V(T) + Z_d(T)$. This balancing equation states that the initial potential $V(0)$, which gives us an idea of the difficulty of the corresponding task (constraint), drops to $V(T)$, which is hopefully as small as possible, by dissipating the energy $Z_d = V(T) - V(0) > 0$. Interestingly, this means that learning processes takes place by dissipating energy, which seems to be reasonable regardless of the body of the agent. A straightforward interpretation in this direction comes out when inspecting Eq. 6.5.1–(8), which is in fact a generalized damped oscillator. The attracting behavior depends in fact on the structure of the potential V, while the dissipation strongly characterizes the dynamics. For large values of θ with respect to the virtual masses m_i, Eq. 6.5.1–(8) reduces to classic gradient descent trajectories

$$\dot{w}_i = -\frac{m_i}{\theta} V'_{w_i}. \tag{7}$$

The dynamics dictated by Eq. 6.5.1–(8) is richer than gradient descent. The damped oscillation might result in a virtuous process for getting around local minima or any configuration with small gradient. We must bear in mind that we are addressing the

---

[2] Here, with some abuse of notation, for the sake of simplicity we overloaded the symbols V and K by setting $V(0, w(0)) \leftarrow V(0)$, $V(T, w(T)) \leftarrow K(T)$ and $K(0, w(0)) \leftarrow K(0)$, $K(T, w(T)) \leftarrow K(T)$.

dynamics deriving from the choice $\rho(t) = e^{\theta t}$. For different choices the dynamics and the corresponding energy balancing equations can be remarkably different. In Exercise 1 we also parallel the analysis for the dissipation function $\rho(t) = e^{\theta t}$ to other interesting cases.

EXERCISES

▶ **1.** [*HM50*] Provide conditions under which the day–night rhythmic conditions guarantee the satisfaction of $\dot{w}(0) = \dot{w}(T) = 0$.

2. [*HM50*] Consider the dynamics arising from the day/night boundary conditions in (6). What is the dependence on the initial values w_i? Under which conditions after many days of life the agent behavior becomes independent of w_i? *Hint:* Consider that the day–light boundary conditions do not affect the values of $w_{i,j}$. The explicit presence on the Lagrangian of a penalizing term on $w_{i,j}$ might change the dynamics significantly.

6.5.3 Focus of attention, teaching, and active learning

The developmental function introduced in the previous section returns values $\zeta(t, \dot{x})$ that are independent of x. We can extend it to return a value $\zeta(t, x, \dot{x})$ that also depends on the specific input. A discussion on this issue is invited in Exercise 1. This extension still relies on the factorization $\zeta(t, x, \dot{x}) = \rho(t)\phi(x, \dot{x})$, thus keeping the same role for the dissipation function, while enriching the focus function ϕ. The developmental function presents some intriguing analogies with Lagrangian multipliers and probability density. However, one shouldn't be betrayed by the shared mathematical structure, since there's something that deeply characterizes function ζ. Basically, it is not simply the outcome of the class of optimization problems considered so far. In case of hard constraints, the Lagrangian multipliers come out from the problem at hand, while for soft-constraints we typically assume that the probability density comes with the availability of the training set. Function ζ requires us to choose the pair ρ, ϕ. Apart from the dissipation issues connected with ρ function, the choice of $\phi(x, \dot{x})$ requires some more careful analyses. As already pointed out, it makes sense to exploit $\dot{x}$ to drive attention on quickly changing events. In addition, the dependence on x can drive the focus of attention on "easy" patterns. Basically, we can face the nerd paradox sketched in Section 1.2.4 by adopting the *easy first* focusing policy. This is extremely useful in highly-structured tasks. In these cases, one expects the focusing function $\phi(x, \dot{x})$ to be pretty small at the beginning of learning on "difficult patterns" x. This favors learning on easy patterns that allows the agent to conquer intermediate structures that will likely help in subsequent inferences on more complex patterns. Interesting, the developmental function ζ might not only be the translation of dissipation and focus of attention. One could regard $\zeta(t, x, \dot{x})$ as a task carried out by an agent different from the one that is learning — for example, a teacher could be involved! This suggests a dramatic change on the environmental interaction of the agent with the environment. This paradigm shift, which begins with replacing λ with ζ in the cognitive action, consists of regarding ζ as an independent agent that lives in the same environment. Interestingly, in this new context, the agent is not necessarily a teacher. It could just be another agent with its own objectives, which consists of satisfying specific constraints. Overall, we gain a new picture where there are *social constraints*. While the presence of ζ as a factor in the cognitive action is a natural choice, social constraints deliver a more general view where a society of agents interact with their own purposes. Social constrains somewhat govern the interactions of the agents by appropriate rules. For instance, a social rule might enforce education, where some agents are expected to learn from a teacher. Others might stimulate agents to follow those which operate better. In this new context, the ideas put forward in Section 1.1 naturally arises; in addition to respecting the constraints, the intelligent agent is also expected to interact actively with the environment. This is a fundamental component of any learning process that is too often neglected in machine learning. In general, the active interaction can be very rich and also extremely complex. It has to do with semantic issues on the communication of the agent with the environment. In general, we need a language for expressing the interactions. The simplest one consists of actively asking the category of certain patterns that are typically those for which the agent is unable to establish a certain classification. How can this be done? We need an intelligent selection of critical patterns. Now the constraint satisfaction over the available data depends significantly on both the constraint and the point $x \in X$. In case of supervised learning this distinction does not apply, so that it is convenient to focus only on points where the applied constraints are not satisfactorily verified. We can parallel what happens in humans learning: They ordinarily experiment the difficulty of correctly explaining some critical examples within their current model of the environment, and operate actively looking for an explicit support to their lack of understanding. The selection of appropriate examples on which to issue queries on their category can nicely be discovered in the framework of learning from constraints. Since the learning process is strongly driven by supervised examples, it might be the case that in some regions of the feature space there is lack of labels with corresponding smaller accuracy. The active support finalized to the supervision of those examples contributes to provide evidence on their categories. Suppose we sort $X = \{x_\kappa : \kappa = 1, \ldots, \ell\}$ in descending order according to the value of $\check{\psi}(x_\kappa, f(x_\kappa))$. The agent interacts with the environment with the purpose of improving its accuracy over the worst elements of X, which can directly be selected from the first positions in $X^{\downarrow}$.

Algorithm Q (*Learning from queries and examples*).   Given a set of constraints Ψ, data X, and a precision ϵ, together with the routines LEARN, SORT, WORST AVERAGE-WORST and SUPERVISION-ON, the algorithm learns the function f. The algorithm also needs a number n_w of questions that I am willing to answer.

Q1. [Assign.] Make the following assignments: $f \leftarrow \text{LEARN}(\Psi, X)$, $X^{\downarrow} \leftarrow \text{SORT}(X)$, $X_w \leftarrow \text{WORST}(X^{\downarrow}, n_w)$, $\overline{\psi}_w \leftarrow \text{AVERAGE-WORST}(X_w)$.

Q2. [Finished yet?] If $\overline{\psi}_w \geq \epsilon$, the algorithm stops and returns f, otherwise it goes on to the next step.

Q3. [Update.] Get the supervisions $Y_w \leftarrow \text{SUPERVISION-ON}(X_w)$ and add the new pointwise constraints $\Psi \leftarrow \Psi \cup \psi|_{X_w, Y_w}$. Then go back to step Q1.   ▮

The cardinality n_w of X_w is chosen to restrict the number of questions. Small values of n_w inject a shy behavior in the agent, which refrains from posing too many questions. On the contrary, large values of n_w push the agent towards the opposite behavior. Clearly, this interaction does depend also on the difficulty of the task, and on the appropriateness of the learning from constraint model that is used. Questions are posed whenever the average value of the error on the worst examples AVERAGE-WORST(X_w) exceeds a certain threshold ϵ. As the agent receives supervision on those examples, they enrich Ψ (step Q3), so the corresponding learning is facilitated because of the specificity of the supervised pairs. The strength of this learning scheme relies on the full exploitation of the learning from constraint approach. Basically, we assume that learning is modeled by many different constraints that mostly operate on unsupervised data. The joint satisfaction of the constraints offers the mathematical model for the understanding of the environment. The critical examples turn out to be those on which the agent asks for support. The actual interaction that arises very much depends on the environment, as well as on the modeled constraints, whose satisfaction stimulates the questions. Human interactions with the environment do not restrict to simple queries on pattern category. They can pose complex questions that involve relations amongst data, as well as abstract properties that involve more categories. Again, just as the satisfaction of single pointwise constraints involving the category $y_{\kappa,j}$, both type of interactions can be modeled by constraints! This time, however, the agent is expected to carry out a more abstract process which involves learning of constraints.

6.5.4 Developmental learning

Learning from constraints has been shown to possess a number of desirable features for the construction of a general theory of learning. In particular, it makes possible to focus on the unifying notion of constraint, while disregarding the distinction between supervised, semisupervised, and unsupervised learning. The shift to spatiotemporal environments also suggests dismissing the difference between training, validation, and test set. While all this gives strong motivations for the adoption of the unifying notion of constraint, the actual declination in real-world tasks is not always simple and natural. One might wonder whether the interaction with the environment can allow us to learn the constraints. This is indeed a crucial issue, since it opens to a new intriguing scenario: An agent can start learning from given constraints, thus developing the corresponding optimal tasks defined by f^*. As soon as f^* is available, one can consider the exploration of "rules" that link different components f_j^* of the optimal solution (content-based constraints) and the relations amongst the data (relational constraints). Any such rule can be regarded as a constraint that can be added to the previous collection used for learning. As a consequence, this enables us to determine an updated value for f_j^*. Of course, cycling is quite natural, so one can start thinking of a development plan of the agent, which passes through the conquest of the skills of some tasks to the abstraction of new rules that, in turn, lead to the acquisition of new skills, and so on and so forth. This opens doors to learning with constraints, that is, to the unified view of learning from — and of — constraints. To sum up, learning with constraints is interwound with developmental learning. How can we develop the abstraction which leads to the learning of constraints? The previous discussion on the alternate development of tasks and constraints does help. However, no abstraction is possible until symbolic descriptions of data are available. This suggests that learning is fired by the acquisition of specific concepts that are provided with supervision on the environment. The learning of any constraint does relies on the availability of data created from the maps f_j. As soon as these are available, we need to discover rules on f_j. We can use an information-based approach based on the MMI principle. The idea is similar with respect to what has been shown in Section 2.3.3 for the case of hidden variables — unsupervised learning. Beginning from vectorial function f, we perform MMI so as to transform the tasks f into a vector ρ, which is expected to provide an abstract description of the relationships between the coordinates of f. In terms of random variables, Y (produced by f) is used to construct R. Each coordinate of R is expected to express the presence of a certain rule over Y. The penalty $-I(Y, R \mid \rho)$ must be minimized (it's the flipped-sign of mutual information). As the convergence is reached, the penalty term $-I(Y, R \mid f)$ produces the vector of rules ρ. Their number is chosen in advance and depends on the degree of symbolic compression that we want to achieve. When learning with this symbolic layer, we construct a random variable R that somewhat expresses the structural dependencies in the output Y. In so doing, Y is associated with a constraint that has the general form of the mutual information between Y and R. Clearly, because of its nature, such a constraint is properly adapted so as to transfer the maximum information in R. Notice that, while the output variable is typically constrained by other environmental constraints, it is also involved in the joint minimization of $-I(Y, R \mid \rho)$. In so doing, $-I(Y, R \mid \rho)$ has a twofold role: It is learned from the structure of f, which is itself learned from the $-I(Y, R \mid \rho)$ that has been developing, in a cycling process. Depending on the probability distribution, the MMI yields different constraints amongst the f coordinates. Again notice that the MMI is maximized while we minimize the error with respect to the target via supervised learning. The reason why we need developmental learning is that we better work in stages: Supervised learning facilitates the development of the MMI rule — and vice versa — because of circularity.

EXERCISES

▸ **1.** [*HM47*]   Consider the extended class of developmental functions defined as $\zeta : \mathcal{T} \times X \times X \; (t, x, \dot{x}) \mapsto \zeta(t, x, \dot{x})$. Reformulate the theory of Section 6.5.1 at the light of this extended definition, where $\zeta(t, x, \dot{x}) = \rho(t)\phi(x, \dot{x})$.

6.6 Scholia

Section 6.1 Constraint machines are one of the dominant keywords in this book. Machine learning has been formulated in the framework of constraint satisfaction in many papers, yet not much attention has been paid on a possible deep unification around the notion of constraint. The idea of enforcing constraints and exploring parsimonious satisfaction has been primarily explored in the field of kernel machines. As already pointed out, however, in this framework, mostly pointwise constraints have been explored. Early attempts to enrich the principle of parsimonious soft-satisfaction has been cultivated by the author in a number of papers [see, e.g., S. Melacci, M. Maggini, M. Gori, *Proc. of the 19th Int. Conf. on Artif. Neural Netw.* (2009), 653–662, M. Diligenti, M. Gori, M.M. Maggini, L. Rigutini, *Proc. of the 20th Int. Conf. on Inductive Logic Programm.* (2010), M. Diligenti, M. Gori, M. Maggini, L. Rigutini, *Proc. of the 19th European Conf. on Artif. Intell.* (2010), 433–438, M. Gori, S. Melacci, Learning with convex constraints, *Artificial Neural Networks—ICANN 2010* (2010), 315–320, S. Melacci, M. Gori, *Proc. of the 12th Int. Conf. on Artif. Intell. Around Man and Beyond* (2011), 21–32. The promotion of the idea of soft-satisfaction of constraints for carrying out inferential processing can be clearly seen in early studies on Hopfield networks [J.J. Hopfield, D.W. Tank, *Biol. Cybern.* **52** (1985) 141–152]. While this kind constraint satisfaction results in inferential steps of a recurrent neural net, the satisfaction of pairwise constraints is somehow a counterpart in supervised learning. The discussion in Section 6.1 provides reasons for treating multilayer and Hopfield nets in the same framework, since it revolves around the unified computational mechanisms behind learning and inference. One can find the roots of this unification principle in human learning. In his theory on "learning by doing," American philosopher John Dewey theorized that learning should be relevant and practical, not just passive and theoretical [see his book *Democracy and Education: An Introduction to the Philosophy of Education* (Macmillan, 1916)]. Hence making inferences and doing are expected to be processes that need to be deeply related with learning, so a good theory should cover them in the same framework. The classification of constraints along with the variational setting of learning based on regularization operators is discussed with plenty of details in G. Gnecco, M. Gori, S. Melacci, M. Sanguineti, *Neural Comput.* **27** (2015) 388–480. The paper leads to the introduction of the notion of *constraint reaction*, which is a generalized view of the parameters learned in SVM. Early traces of the idea of carrying out inference in the framework of learning with constraints can be found in M. Gori, S. Melacci, *IEEE Trans. Neural Netw. Learn. Syst.* **24** (2013), 825–831. The theory borrows its conceptual backbone from variational calculus and, particularly on optimization under subsidiary conditions [see, e.g., I.M. Gelfand, S.V. Fomin, *Calculus of Variations* (Dover Publications, 1963), and *Calculus of Variations* (Springer, 1996) vol. 1 and 2 by M. Giaquinta and S. Hildebrand]. In case of hard constraints the solution of the formulated learning problem does require discovering both the tasks f and the adjoint Lagrangian multiplier functions λ. As we introduce the slack variables, this holds true also for soft-constraints. Hence the process of learning turns out to assume the intriguing feature of developing at the same time the tasks and the weight to be assigned to the environmental constraints. In a sense, this corresponds with the mechanism of focus of attention. Basically, if you look at the Lagrangian function, the multipliers — and, therefore, the reactions — enable us to express the tasks, but at the same time give the degree of importance of the constraints; the vanishing of the multipliers is in fact indicating that we are in the presence of a straw constraint.

The taxonomy given in G. Gnecco, M. Gori, S. Melacci, M. Sanguineti, *Neural Comput.* **27** (2015) 388–480 borrows concepts that are classic in related disciplines. Any holonomic bilateral constraint can be transformed by differentiation into a constraint of the form 6.1.2–(3), and one can regain the original constraint by integration. However, the converse does not hold true [see M. Giaquinta and S. Hildebrand *Calculus of Variations* vol. 1 p. 98]. The name "isoperimetric" derives from the classical isoperimetric problem of the calculus of variations, which consists in determining a plane figure of the largest possible area, whose boundary is constrained to have a specified length. Like for kernel machines, the regularization can be introduced in the framework of regularization operators, which have intriguing connections with the notion of kernel [see A.J. Smola, B. Schoelkopf, K.R. Mueller, *Neural Netw.* **11** (1998), 637–649, G. Gnecco, M. Gori, M. Sanguincti, *Neural Comput.* **25** (2013), 1029–1106, Q. Ye, G.E. Fasshauer, *Numer. Math.* **119** (2011), 585–611.]

The representational results given in Section 6.1 can be extended to a collection of constraints that are assumed to be functionally independent. This means that we can always find two permutations σ_f and σ_ψ of the indexes of the n functions f_j and of the ℓ constraints ψ_κ, respectively, such that $\psi_{\sigma_\psi(1)}, \ldots, \psi_{\sigma_\psi(\ell(x_0))}$ refer to the constraints actually defined in x_0, and the Jacobian matrix

$$J_\psi = \frac{\partial(\psi_{\sigma_\psi(1)}, \ldots, \psi_{\sigma_\psi(\ell(x_0))})}{\partial(f^o_{\sigma_f(1)}, \ldots, f^o_{\sigma_f(\ell(x_0))})} \tag{1}$$

evaluated in x_0, is not singular. In this case, the Lagrangian multipliers are uniquely determined. Notice the functional dependence of a given set is not associated with the property that some of them are straw constraints, since in that case, the Lagrangian multipliers are basically distributed amongst the constraints. In case of soft-constraints, straw constraints typically arise because of strong regularization. As already noticed, the curse of dimensionality sends us a severe warning on the effectiveness of the theory: The given functional theorem turns out to be useful for inspiring new representation theorems, but, unfortunately, they don't allow us to determine a direct solution by solving the Euler–Lagrange differential equations. This is in fact a strong message for the immersion of the agent into a temporally ordered environment. An in-depth understanding of the interplay between learning and inference in this framework is still and open research problem.

Section 6.2 For centuries, the study of logic has been regarded as an attempt to understand and improve thinking, reasoning, and argument as they occur in real life.

While formal logic restricts attention to the structure of the arguments, informal logic aims at building logic models suited to this purpose. Methods in informal logic are expected to properly combine the classic reasoning with arguments with an assessment of their effectiveness in real life. This is in fact an effort which is rooted in many older attempts to understand arguments occurring daily in our life. There are intriguing connections between formal and informal logic, yet their relations are in some ways controversial. In the long run, informal logic is expected to embrace the challenge of providing a complete theory of reasoning to unify formal deductive and inductive logic [see R.H. Johnson, *The Rise of Informal Logic* (Vale Press, 1996) on page 11]. The interdisciplinary study of real life argument is often called *argumentation theory*. It incorporates studies and insights from cognitive psychology, rhetoric, dialectics, computational modeling, semiotics, communication studies, artificial intelligence, and other related disciplines. The interpretation of arguments in natural language needs to face issues of informal reasoning due to the semantics that emerges from the single words embedded in the sentences. Hence, while formal logic relationships are still useful and expressive, they mostly miss the strength of inductive processes which attach a meaning depending on the context. The deep meaning of argument informal logic has been broadened even further with the purpose of including nonverbal elements, like picture, art, cartoons, graphs, and diagrams.

According to Judea Pearl [*Probabilistic Reasoning in Intelligent Systems: Networks of Plausible Inference* (Morgan Kauffman, 1988)], scientists working on approaches to uncertainty can be classified into three formal schools, where they are referred to as *logicist*, *neo-calculists*, and *neo-probabilist*. The logicist school faces uncertainty using truly symbolic techniques that don't involve numerical techniques, e.g., nonmonotonic logic. Scientists in the neo-calculist school propose new numerical-based calculi, like the Dempster–Shafter calculus and fuzzy logic, while neo-probabilists adhere to the traditional framework of probability theory. However, he maintains that a more fundamental taxonomy can be drawn along the dimensions of *extensional* vs *intensional* approaches. In his own words:

> *The extensional approach, also known as production systems, rule-based systems, and procedure-based systems, treat uncertainty as a generalized truth value attached to formulas and (following the tradition of classical logic) compute the uncertainty of any formula as a function of the uncertainty of its sub-formulas. In the intensional approach, also known as declarative or model-based, uncertainty is attached to "state of affairs" or subsets of "possible worlds." Extensional systems are computationally convenient but semantically sloppy, while intensional systems are semantically clear but computationally clumsy.*

For example, the certainty of the conjunction $a \wedge b$, expressed in some t-norm, depends on the certainty of the literals a and b. By contrast, intensional approaches express a certainty value that cannot be simply computed when knowing the certainty value of the single literals. For example, in probability theory the probability of the event characterized by the conjunction of a and b cannot be expressed by combining the probabilities of the single events. Learning with constraints breaks this distinction by joining the possibility of a knowledge-based representation of the environment with a truly intensional process.

The integration of symbolic and subsymbolic representations is the subject of the NeSy International Workshops.[3] The corresponding proceedings are sources of a rich literature. The experiment on animal recognition has been popularized in the field of artificial intelligence by Patrick Winston [*Artificial Intelligence* (Addison-Wesley, 1984)].

Early studies where we can clearly see trace of the bridging between logic constraints and machine learning that reflect what is presented in the book are given in M. Diligenti, M. Gori, M. Maggini, L. Rigutini, *Proc. of the 19th European Conf. on Artif. Intell.* (2010), 433–438. The theory is based on the adoption of kernel machines; the basic idea has been given a more formal statement in M. Diligenti, M. Gori, M. Maggini, L. Rigutini, *Mach. Learn.* **86** (2012), 57–88. Early trace on the inferential processes in the environment was published in M. Gori, S. Melacci, *IEEE Trans. Neural Netw. Learn. Syst.* **24** (2013), 825–831. The idea of an individual, which consists of joining labels and real-valued features, was given in M. Diligenti, M. Gori, C. Saccà, *IEEE Trans. Neural Netw. Learn. Syst.* **27** (2016), 1322–1332. thanks to the introduction of the principle of learning in variable dimensional spaces. An overall view of the approach, which includes the process of learning and reasoning as described in the book, can be found in M. Diligenti, M. Gori, C. Saccà, *Artif. Intell.* **244** (2017), 143–165. The credit for the complexity of learning with appropriate fragments of Ł goes to Francesco Giannini, who mostly discovered the corresponding closure of concavity of McNaughton functions F. Giannini, M. Diligenti, M. Gori, M. Maggini, *Proceedings of ECML PKDD 2017* (2017). The principles behind parsimonious logic are the subject of current investigation in the AI lab of the University of Siena. The description of the semantic-based regularization language for setting up experiments with logic constraints and supervised examples was early given a first release in C. Saccà, M. Diligenti, Marco Gori, *Recent Advances of Neural Netwo. Models and Appl.*, in *Smart Innov., Sys. and Tech.* **26** (2014) 15–23 edited by S. Bassis, A. Esposito, F.C. Morabito. Related studies that somewhat resemble the idea of using the math apparatus of constraint satisfaction for learning and reasoning can be found in Z. Hu, X. Ma, Z. Liu, E. Hovy, E. Xing arXiv:1603.06318, D. Lowd, C. Meek, *Proc. of the Eleventh ACM SIGKDD Int. Conf. on Knowl. Discovery in Data Mining* (2005), 641–647. and, particularly, in I. Donadello, L. Serafini, *European Conf. on Comput. Vision* (2014), 283–298, L. Serafini, A.S. d'Avila Garcez, *AI* IA 2016 Advances in Artif. Intell.* (2016), 334–348, L. Serafini, A.S. d'Avila Garcez, arXiv:1606.04422, 2016. Other related remarkable results can be found in S. Teso, R. Sebastiani, A. Passerini, *Artif. Intell.* **244** (2015), 166-187, C. Saccà, S. Teso, M. Diligenti, A. Passerini, *BMC Bioinform.* **15** (2014) A. Passerini, *Intell. Sys. Reference Library* **49** (2013), 283–333 edited by M. Bianchini, M. Maggini, L.C. Jain, and E. Cilia, N. Landwehr, A. Passerini, *Fundam. Inform.* **113** (2011), 151–177. Excellent

---

[3] Neural-Symbolic Integration, http://www.neural-symbolic.org/.

sources on foundations of triangular norms can be found in P. Hájek, L. Godo, F. Esteva, arXiv:1302.4953. and E.P. Klement, R. Mesiar, E. Pap, *Triangular Norms* (Kluwer Academic Publisher, 2000). In mathematical logic, the Skolem normal form (SNF) refers to a logic formula without existential quantifiers (named after Thoralf Skolem) in prenex normal form. Every first-order formula can be given a corresponding representation in Skolem normal form. Insights on appropriate selection of the weight of different constraints can be found in R. Alcalá, P. Ducange, F. Herrera, B. Lazzerini, F. Marcelloni, *IEEE Trans. Fuzzy Syst.* **17** (2009), 1106–1122, M. Cococcioni, P. Ducange, B. Lazzerini, F. Marcelloni, *Soft Comput.* **11** (2007), 1013–1031, which promote the adoption multiobjective evolutionary approaches, and in F. Esposito, D. Malerba, G. Semeraro, *Appl. Artif. Intell.* **8** (1994), 33–84. Related studies in cases in which the domain knowledge involves textual documents are presented in M. Ceci, D. Malerba, *J. Intell. Inf. Syst.* **28** (2007), 37–78.

A number of significant results on both foundations and applications have been achieved in the community of statistical relational learning [see, e.g., *Logic for Learning: Learning Comprehensible Theories from Structured Data* by John W. Lloyd (Springer-Verlag, 2003), A. Nareyek, *Constraint-Based Agents: An Architecture for Constraint-Based Modeling and Local-Search-Based Reasoning for Planning and Scheduling in Open and Dynamic Worlds* (Springer-Verlag, 2001), S. Jabbari, R.M. Rogers, A. Roth, S.Z. Wu, *Advances in Neural Inf. Process. Syst.* (2016), 1570–1578, A. Lallouet, M. Lopez, L. Martin, C. Vrain, *Proc. of Int. Conf. on Tools with Artif. Intell.* (2010), 45–52, C.M. Cumby, D. Roth, *Proc. of the 20th Int. Conf. on Machine Learning* (2003), 107–114, S. Muggleton, H. Lodhi, A. Amini, M.J.E. Sternberg, *Int. Conf. on Discovery Science* (2005), 163–175. G. Farnadi, S.H. Bach, M.F. Moens, L. Getoor, M. De Cock, *Mach. Learn.* **106** (2017) 1971–1991. Special attention has been devoted to random Markov fields S.H. Bach, M. Broecheler, B. Huang, L. Getoor, arXiv:1505.04406 and Markov logic networks M. Richardson, P. Domingos, *Mach. Learn.* **62** (2006) 107–136.

Section 6.3 The view of recurrent neural networks herein presented as diffusion machines is quite atypical. They are mostly regarded as a specific chapter of machine learning used primarily for processing sequences, and were introduced at the dawn of the first connectionist wave at the end of the 1980s. A good coverage on recurrent networks and on also on their processing on graphs can be found in J.F. Kolen, S.C. Kremer, *A Field Guide to Dynamical Recurrent Networks* (IEEE Press, 2001).

Many studies have been focused on the identification of appropriate vector-based features also in problems like QSAR, in which one can think of using graphs for pattern representation [see, e.g., V.V. Kovalishyn, I.V. Tetko, A.I. Luik, V.V. Kholodovych, A.E.P. Villa, D.J. Livingstone, *J. Chem. Inf. Comput. Sci.* **38** (1998), 651–659 as a successful example of this approach]. In pattern recognition, for many years, scientists were thinking of structural representations and related recognition algorithms [see, e.g., H. Bunke, "Structural and syntactic pattern recognition" in *Handbook of Pattern Recognition and Computer Vision* (World Scientific Publishing Co., 1993), 163–209, L.P. Cordella, P. Foggia, C. Sansone, M. Vento, *IEEE Trans. Pattern*

Anal. Mach. Intell. **26** (2004), 1367–1372]. The idea of using neural networks for the processing of any data structure appeared at the beginning of the 1990s mostly thanks to Jordan Pollack [*Artif. Intell.* **46** (1990), 77–106], who introduced the recursive auto-associative memories (RAAM). This stimulated related studies by Alessandro Sperduti and Antonina Starita [*IEEE Trans. Neural Netw.* **8** (1997), 714–735.]. Later on, Paolo Frasconi et al. [*IEEE Trans. Neural Netw.* **9** (1998), 714–735] provided a more general framework for the processing of data structures that contributed to identify different classes of graphs, depending on the corresponding computational model. It was early clear that the computational style of recurrent networks for sequences is basically retained in case of directed ordered graphs, where the data flow model doesn't encounter any loops. Recurrent neural nets with cycles were investigated in M. Bianchini, M. Gori, L. Sarti, F. Scarselli, *IEEE Trans. Neural Netw.* **17** (2006) 10–18, but a systematic treatment appeared a few year later in F. Scarselli, M. Gori, A.C. Tsoi, M. Hagenbuchner, G. Monfardini, *IEEE Trans. Neural Netw.* **20** (2009), 61–80. Related studies were also carried out in the case unsupervised learning B. Hammer, A. Micheli, A. Sperduti, M. Strickert, *Neural Netw.* **17** (2004), 1061–1085. and B. Hammer, A. Micheli, A. Sperduti, M. Strickert, *Neurocomputing* **57** (2004), 3–35, while in M. Gori, A. Petrosino, *IEEE Trans. Neural Netw.* **15** (2004) 1435–1449 the structured information was represented by nondeterministic fuzzy frontier-to-root tree automata.

The computational mechanisms behind all these models very much resemble what is done in models of page rank in hyperlinked environment. In particular, there are intriguing similarities with the `PageRank` algorithm L. Page, S. Brin, R. Motwani, T. Winograd, *Proc. of the 7th Int. World Wide Web Conf.* (1998), 161–172., which posses a number of remarkable properties reported in M. Bianchini, M. Gori, F. Scarselli, *ACM Trans. Internet Technol.* **5** (2005), 92–128. The diffusion in continuous spatiotemporal domains, which leads to the Poisson equation (3), has been successfully adopted in the problem of *image editing* P. Pérez, M. Gangnet, A. Blake, *ACM Trans. Graph.* **22** (2003), 313–318, where a function is constructed under the guidance of a vector field, which may or may not be the gradient field of a source function. This is done for interactive cut-and-paste with cloning tools for replacements.

Section 6.4 The concrete development of algorithms in the framework of learning from constraints has followed paths that can nicely be retraced. The underlying primal/dual dichotomy is ubiquitous: Like in the classic framework given for neural networks and kernel machines, we can learn the weights in the primal or in the dual space, respectively. The discovery of the Lagrangian multipliers — and then of the reactions — is the main route to gain a solution that, at the same time, returns a score of the constraint. On the other hand, this is often hard to compute, especially in the case of hard constraints in which the Lagrangian is a function with domain in the perceptual space X. A viable solution is that of constructing a recursive computational structure to discover the fixed points of $f^*(\cdot) = g(\cdot) * (-\lambda(\cdot)\nabla_f \psi(\cdot, f^*(\cdot)))$. An insight on this approach is also given in Exercise 6.4-2.

However, while in case of supervised learning the Green function of the regularization operator can correspond with a traditional kernel, for other constraints the solution of the above recursive functional equation might result in new kernels that somewhat incorporate the knowledge associated with the constraint and the regularization operator. The noticeable case of *box kernels* S. Melacci, M. Gori, *IEEE Trans. Pattern Anal. Mach. Intell.* **35** (2013), 2680–2692 is of remarkable importance whenever we deal with granules of knowledge that are expressed by propositions on the feature space. The incorporation of convex constraints is discussed in M. Gori, S. Melacci, *Proc. of the 20th Int. Conf. in Artif. Neural Netw.* (2010), 315–320. The discovering of new marriages between differential operators and constraints, giving rise to new appropriate kernels, is an open research problem.

The discovery of learning algorithms for the case of diffusion, in which we deal with hidden variables, has been investigated with much more emphasis in the primal, under the umbrella of recurrent neural networks. The idea of time-unfolding can be found in the seminal of the Parallel Distributed Processing group [*Parallel Distributed Processing* vol. 1 by J.L. McClelland and D.E. Rumelhart (MIT Press, 1988)]. The RTRL algorithm, along with its temporal locality property, was introduced in R.J. Williams, D. Zipser, *Neural Comput.* **1** (1989) 270–280, R.J. Williams, D. Zipser, *Connect. Sci.* **1** (1989) 87–111. The basis of self-loop recurrent networks, which arises in a learning algorithm that is local both in time and space, was proposed in M. Gori, Y. Bengio, R. De Mori, *Proc. of the Int. Joint Conf. on Neural Netw.* (1989), 643–644 and subsequently refined in M. Gori, *Proc. of Neuro-Nimes* (1989), 83–93, Y. Bengio, R. De Mori, M. Gori, *Special Issue on Artificial Neural Networks, Pattern Recognit. Lett.* **13** (1992) 375–386, P. Frasconi, M. Gori, G. Soda, *Neural Comput.* **4** (1992), 120–130. Related investigations were carried out in P. Campolucci, A. Uncini, F. Piazza, B.D. Rao, *IEEE Trans. Neural Netw* **10** (1999), 253–271.

Recurrent networks that process with relaxation so as to reach an equilibrium were proposed in the discrete setting in L.B. Almeida, *Proc. of the Int. Conf. on Neural Netw.* (1987), 609–618, L.B. Almeida, *Neural Computers, Neuss 1987* (1988), 199–208 edited by R. Eckmiller, Ch. von der Malsburg. A related analysis for time-continuous processing, where the recurrent network is modeled as a differential equation, was introduced in L.B. Almeida, *Proc. of the Int. Conf. on Neural Netw.* (1987), 609–618, and successively refined in F.J. Pineda, *J. Complex.* **4** (1988), 216–245. and F.J. Pineda, *Neural Comput.* **1** (1989), 161–172. A nice general view, which embraces most interesting recurrent network, can be found in B.A. Pearlmutter, *Int. Joint Conf. on Neural Networks* (1989), 365–372, B.A. Pearlmutter, *Neural Comput.* **1** (1989), 263–269.

At the beginning of the 1990s, a few people began realizing that the process of learning sequences, where one wants to detect long-term dependencies, was very expensive from a computational point of view and, even worse, it was quite ineffective! This was quite clear during experiments on grammatical inference, in which recurrent networks were asked to recognize whether a given sequence has been generated by a given grammar [see, e.g., C.W. Omlin, C.L. Giles, *J. ACM* **43** (1996), 937–972., P. Frasconi, M. Gori, M. Maggini, G. Soda, *Mach. Learn.* **23** (1996), 5–32]. The seminal paper by Bengio et al. [*IEEE Trans. Neural Netw.* **5** (1994), 157–166.] had the merit of clearly detecting the problem connected with the difficulties of capturing long-term dependencies. They provided a solid argument to prove that the gradient vanishes with the length of the sequence and also stated negative results on a wide class of recurrent nets. The limitations behind the classic recurrent nets that circulated in the literature were recognized by other scientists and, particularly, by Hochreiter and Schmidhuber, who early came out with the proposal of Long-Short Time Memory (LSTM) models [see S. Hochreiter, J. Schmidhuber, *Proc. of Advances in Neural Inf. Process. Syst.* **9** (1997), 473–479 and S. Hochreiter, J. Schmidhuber, *Neural Comput.* **9** (1997), 1735–1780]. We like to think of the gating LSTM structure as a clever way of facing the gradient vanishing at every state transition, which relies on a transition function that exhibits different "modes" at different times. This is in fact due to the presence of the gates that drive the structure of the transition function on the basis of the input. In so doing, we get rid of the homogeneous application of the transition function, thus constructing a state updating mechanism that can alternatively reduce or increase the gradient. It is in fact the nonhomogeneous transition function that allows us to escape the trap of the exponential decay. A challenging alternative to the gating approach is proposed in Exercise 6.4–12. Additional insights on the problem of long-term dependencies appears in S. Hochreiter, Y. Bengio, P. Frasconi, J. Schmidhuber, *A Field Guide to Dynamical Recurrent Neural Networks* (IEEE Press, 2001).

Section 6.5 The necessity of lifelong agent learning was early advocated by Sebastian Thrun and Tom Mitchell in mid-1990s [*Robot. Auton. Syst.* **15** (1995), 25–46]. They began to notice that robots can likely benefit from the experience acquired during their entire lifetime by transferring knowledge. They argued that knowledge transfer is plays a crucial role if robots are to learn control with moderate learning times in complex scenarios. The topic raised interest in the research community, where people began thinking also of selective mechanisms to focus attention. In *Proc. Int. Conf. on Machine Learning* (2009), 41–48, Yoshua Bengio et al. formalized new training strategies that they referred to as *curriculum learning*, where they proposed a method for accessing the examples that are driven by information-based principles. In A. Carlson, J. Betteridge, B. Kisiel, B. Settles, E.R. Hruschka Jr., T.M. Mitchell, *Proc. of the Twenty-Fourth AAAI Conf. on Artif. Intel.* (2010), 1306–1313 the idea of lifelong learning was proposed for attacking natural language. An agent was supposed to extract information from a web text to populate a growing knowledge base of facts and knowledge, and to learn to improve the reading capabilities. Recently, the topic of lifelong learning has received growing attention. In particular, it has been deeply studied by involving the interplay with transfer learning and multitask learning [Z. Chen, B. Liu, *Lifelong Machine Learning* (Morgan & Claypool Publishers, 2016)].

The approach discussed herein follows the principle of formulating the emergence of intelligence processes by information-based laws. This was advocated in S. Frandina, M. Gori, M. Lippi, M. Maggini, S. Melacci, *Proc of Int. Conf on Artif. Neural Netw. and Machine Learning* (2013), 82–89. and, later on, formalized by the introduction of the principle of least cognitive action A. Betti, Marco Gori, *Theor. Comput.*

Sci. **633** (2016), 83–99. The proposed formalization goes beyond the notion of kinetic energy reported herein by offering an intriguing connection with regularization operators. The corresponding energy balancing becomes more involved, though a new invariant is discovered which generalizes the Poisson brackets. It's worth mentioning that the formulation of the principle of least cognitive action doesn't really require us to determine a minimum of the functional, but simply a stationary point. This leads apparently to losing connection with the classic framework of regularization, where one wants to minimize the sum of the loss with the regularization term. The adoption of high order temporal differential operators to generalize the kinetic energy also yields an additional awkward sign-flip that depends on the order [A. Betti, Marco Gori, *Theor. Comput. Sci.* **633** (2016), 83–99] However, when interpreting these temporal processes in terms of energy balance, no matter what the chosen order of the operator is, everything becomes clear. A closer look on the role of dissipation was also given in M. Gori, M. Maggini, A. Rossi, *Neural Netw.* **81** (2016) 72–80, where the emphasis is posed on reframing supervised learning in this new context. The immersion in the temporal manifold along with the idea of dissipation incorporated in the Hamiltonian framework, where the whole Lagrangian contains a temporally-growing exponential term, can be found in L. Herrera, L. Nunez, A. Patino, H. Rago, *Am. J. Phys.* **53** (1985), 273–277.

The importance of providing a symbolic interpretation of the internal developed representations has been regarded as an important topic beginning with early 1990s [see, e.g., C.L. Giles, C.W. Omlin, *Proc. of the 1992 IEEE Workshop* (1992), 13–22, S. Das, C.L. Giles, G.Z. Sun, *Advances in Neural Inf. Process. Syst.* **5** (1993), C.W. Omlin, C.L. Giles, *J. ACM* **43** (1996), 937–972, P. Frasconi, M. Gori, M. Maggini, G. Soda, *Mach. Learn.* **23** (1996), 5–32, P. Frasconi, M. Gori, M. Maggini, G. Soda, *Int. Joint Conf. on Neural Netw.* (1991), 811–816, P. Frasconi, M. Gori, M. Maggini, G. Soda, *IEEE Trans. Knowl. Data Eng.* **7** (1995), 340–346, M. Gori, M. Maggini, G. Soda, *Workshop on Combining Symbolic and Connectionist Process., ECAI '94*, (1994), 78–87, P. Frasconi, M. Gori, M. Maggini, G. Soda, *Knowl.-Based Syst.* **8** (1995), 313–332 and B. Apolloni, A. Esposito, D. Malchiodi, C. Orovas, G. Palmas, J.G. Taylor, *IEEE Trans. Neural Netw.* **15** (2004), 1333–1349]. In order for an agent to fully conquer most interesting human cognitive skills, a sort of symbolic representation needs to be developed to exploit the communication with humans and other agents. This feature is likely to be very useful in the framework of developmental learning.

The spread of the ideas cultivated in the framework of lifelong learning likely needs to undergo a complex process that has been encountering explicit and implicit obstacles raised by members of the scientific community. Unlike what one could believe, this is not so odd! First, statistics is a noble and illustrious ancestor of machine learning. As such, even though some people in machine learning might not realize to have come from such a family, they regard benchmark tests as the indisputable method to assess the performance! If you were grown up with the idea of sampling data of probability distributions then benchmark-based assessment is quite a natural outcome. Hence the systematic accumulation and careful organization of training data becomes of central importance, and it quite often regarded as an important asset of research teams and companies. In addition, today's widespread practice of evaluating machine learning systems over predefined benchmark datasets have certainly contributed to the tremendous progress made on several specific problems. Second, even though smart people likely suspect that benchmark radicalization, in the long run, might not necessarily be a clever choice, at the moment it might reinforce their visibility. In addition, when turning to lifelong learning, one might need to discuss its current foundational framework and change dramatically towards new unexplored paths. As nicely pointed in D. Geman, S. Geman, *Proc. Natl. Acad. Sci.* **113** (2016), 9384–9387, such research directions are not very well promoted, so as the paradigm of lifelong learning doesn't currently shift!

As pointed out by Marcello Pelillo in an informal personal communication, the benchmark-oriented attitude, which nowadays dominates the machine learning community, bears some resemblance to the influential testing movement in psychology which has its roots in the turn-of-the-century work of A. Binet and T. Simon *The Development of Intelligence in Children: The Binet–Simon Scale* (Williams & Wilkins, 1916) on IQ tests. In both cases, in fact, we recognize a familiar pattern: A scientific or professional community, in an attempt to provide a rigorous way of assessing the performance or the aptitude of a (biological or artificial) system, agrees on a set of standardized tests which, from that moment onward, becomes the ultimate criterion for validity. As is well known, though, the IQ testing movement has been severely criticized by many scholars not only for the social and ethical implications arising from the idea of ranking human beings on a numerical scale but also, more technically, on the grounds that, irrespective of the care with which these tests are designed, they are inherently unable to capture the multifaceted nature of real-world phenomena. As David McClelland put it in a seminal paper [*Am. Psychol.* **28** (1973) 1–14] which set the stage for the modern competency movement in the US, the criteria for establishing the "validity" of these new measures really ought to be not grades in school, but "grades in life" in the broadest theoretical and practical sense. We maintain that the time is ripe for the machine learning community to adopt a similar "grade-in-life" attitude towards the evaluation of its systems and algorithms. We do not, of course, intend to diminish the importance of benchmarks, as they are indeed invaluable tools to make the field devise better and better solutions over time, but we propose using them in much the same way as we use school exams for assessing our children abilities: Once they pass the final one, and are therefore supposed to have acquired the basic skills, we allow them to find a job in the real world. A closer interaction with real life seems to be extremely welcome in the current scientific context. Is it really necessary to collect millions of images to test computer vision abilities? Likewise, do we really need huge speech corpora for checking the understanding capabilities of conversational agents? How long does it take for humans for assess concretely visual and language understanding capabilities? A few minutes of close interaction are typically enough to come up with a very clear assessment of visual and language skills! Hence it seems to be quite obvious to let people assess intelligent agents by a crowd-sourcing scheme, where they are invited to rank the agent skills. When restricting the task of agent assessment to scientists who must use their own credentials, one might reasonably regard the corresponding rank as the true evaluation index of an intelligent agent working in real life!

As pointed out in M. Gori, M. Lippi, M. Maggini, S. Melacci, M. Pelillo, *Proc. of the 18th Int. Conf. in Image Analysis and Process.* (2015), 697–709, we urge machine learning labs to open their doors and to go "en plein air," thereby allowing people all over the world, from the researcher to the layman, to freely play and interact with their grown-up systems. This will very likely result in a paradigm shift in the way in which methodologies and algorithms are evaluated in computer vision, and the launch of this "open lab paradigm" will certainly stimulate critical investigations on new methods and algorithms in an attempt to effectively deal with totally unrestricted visual environments.

Epilogue

No epilogue, I pray you...
Shakespeare, *A Midsummer-Night's Dream* scene *V, i, 363* (1605)

Machine Learning. https://doi.org/10.1016/B978-0-32-389859-1.00014-3

443

NOWADAYS, we can get in touch with machine learning in many different ways, especially because of the explosion of web resources, that are extremely useful for supporting the development of applications. We can quickly gain a nitty-gritty explanation of basic concepts by professional multimedia material (see e.g. http://www.popularmechanics.com/science/math/amp28539/what-is-a-neural-network/), and you can also find books for assisting you during the design of applications (see e.g. the great WEKA environment [I.H. Witten, Data Mining: Practical Machine Learning Tools and Techniques, *Addison-Wesley*, Reading MA, (2011)]).

This book is more oriented to foundations of machine learning and reasoning in intelligent agents. It gives the reader a unified view of constraint-based environments by bridging their symbolic and sub-symbolic representations. Now it's time to unroll the carpet. While we have covered different topics in machine learning, most of the foundational principles presented in Chapters 4 and 5 turn out to be strongly connected with the (primal vs dual) representation of the function associated with the agent. In the first case the computational models rely on a set of learnable parameters that somewhat synthesize all the training examples, whereas in dual representations each learnable parameter is directly associated with a correspondent example.

About ten years ago, a technical report circulated that was later published in [Y. Bengio, Learning deep architectures for AI, *Found. Trends Mach. Learn.* **2** (1) (2009) 1–127] where the shallow architectures associated with the dual representation were strongly criticized because of the lack of compositional capabilities and of other limitations, including the expression of symmetries. The paper advocated the power of deep architectures in the primal space. Others soon joined and reinforced the idea, which was mostly driven by Geoffrey Hinton, Yoshua Bengio, and Yann Le Cun. This generated a second wave of interest in connectionist models[1] with a corresponding spectacular impact not only in science [Y. Le Cun, Y. Bengio, G.E. Hinton, Deep learning, *Nature* **521** (7553) (2015) 436–444], but also in the real life (see, e.g., *Comment le "deep learning" révolutionne l'intelligence artificielle* published at http://www.lemonde.fr/, *Researcher dreams up machines that learn without humans*, https://www.wired.com, and *Artificial Intelligence Swarms Silicon Valley on Wings and Wheels*, which appeared on https://www.nytimes.com). In addition to a number of significant theoretical improvements with respect to studies in the 1990s, those people early realized that it makes sense to fully exploit the impressive representational capabilities offered by deep architectures, since learning could rely on impressive parallel computation capabilities that had become easily affordable. The picture was completed when it was of clear that, even though deep learning is hungry of data, we could think of collecting really huge training sets — a dream during the first wave of interest in connectionist models! In summer 2008, Fei-Fei Li and associates discovered the crowdsourcing Amazon Mechanical Turk and came up with

---

[1] Here we don't consider the foundations of the end of 1950s, with perceptrons and related machines, so as the first wave of interest is located at the end of the 1980s.

a related approach to collect millions of labeled images extracted from the net. No graduate student could have accepted such a crazy task, but crowdsourcing succeeded (see [J. Deng, W. Dong, R. Socher, L.-J. Li, K. Li, L. Fei-Fei, ImageNet: a large-scale hierarchical image database, in *CVPR*, 2009], [J. Deng, A.C. Berg, K. Li, L. Fei-Fei, What does classifying more than 10,000 image categories tell us? in *Proc. of the 11th European Conf. on Computer Vision* (2010), 71–84], [O. Russakovsky, et al., ImageNet large scale visual recognition challenge, *Int. J. Comput. Vis.* **115** (3) (2015) 211–252])! Related work was carried out in other labs with impressive data collections [A. Torralba, R. Fergus, W.T. Freeman, 80 million tiny images: a large data set for nonparametric object and scene recognition, *IEEE Trans. Pattern Anal. Mach. Intell.* **30** (11) (2008) 1958–1970], while an interesting view on image tagging, which is framed in the context of content-based image retrieval, is given in [X. Li, T. Uricchio, L. Ballan, M. Bertini, C.G.M. Snoek, A. Del Bimbo, Socializing the semantic gap: a comparative survey on image tag assignment, refinement, and retrieval, *ACM Comput. Surv.* **49** (1) (2016)]. Nowadays, results are under the eyes of everyone. In addition to the impact in vision, deep learning has been successfully adopted in many different contexts, including natural language processing (see, e.g., [R. Socher, C.C.-Y. Lin, A.Y. Ng, C.D. Manning, Parsing natural scenes and natural language with recursive neural networks, in *Proc. of ICML* (2011), 129–136], [X. Glorot, A. Bordes, Y. Bengio, Domain adaptation for large-scale sentiment classification: a deep learning approach, in *Proc. of ICML* (2011)], [B. Hu, Z. Lu, H. Li, Q. Chen, Convolutional neural network architectures for matching natural language sentences, *NIPS* (2014), 2042–2050]) and bioinformatics (see, e.g., [P. di Lena, K. Nagata, P. Baldi, Deep architectures for protein contact map prediction, *Bioinformatics* **28** (19) (2012) 2449–2457], [A. Lusci, M.R. Browning, D. Fooshee, S.J. Swamidass, P. Baldi, Accurate and efficient target prediction using a potency-sensitive influence-relevance voter, *J. Cheminformatics* **7** (2015)], [S. Min, B. Lee, S. Yoon, Deep learning in bioinformatics, *CoRR*, arXiv:1603.06430 (2016)]).

While this book provides basic foundations on deep learning and kernel machines, it mostly promotes the notion of constraint for composing a unified view to understand intelligent agents. It is claimed that learning and inference naturally arise as problems of constraint satisfaction. The continuous based formulation turns out to embrace perceptual tasks, as well as logic-based knowledge granules. This mixture somewhat dismisses the need for hybrid models and opens the doors to truly novel challenges. The constrained-based approach to learning and reasoning of Chapter 6 somewhat integrates the deep learning view by shifting the emphasis to mechanisms for accurate descriptions of the environment. In addition to in-depth discussions, the proposed rethinking of the environmental interactions opens the doors to deep generative schemes that are not only driven by supervision, but also significantly benefit from abstract descriptions. To some extent we can think of agents that "creatively" generate new patterns! The process of constraint satisfaction can generate solutions of N-queens, but can also generate the face of my father or produce a picture of the next handwritten number The theory of parsimonious satisfaction of constraints is given its own formulation along with the relationships with statistics and the minimum description length principle. Interestingly, it is a natural extension of the regularization concept that is used in kernel machines, and it's self-contained. It suggests dismissing the difference between supervised, unsupervised, and semisupervised learning, since we only need to express the consistency of data with the given constraints.

When intelligent agents are immersed into highly structured environments, one soon realizes that there is still a long way to gain many elusive cognitive skills. A sound theory on the emergence of intelligence should hopefully explain how an agent, which lives in an environment characterized by continuous-based perceptual information, can develop an internal symbolic representation of the world, which can be used for communication in the society. Interestingly, as the agent conquers this feature, it can acquire knowledge by compact symbolic expressions, perform deductions that turn out also to be useful in gaining additional skills also at perceptual level. This seems to follow a sort of induction–deduction loop that leads to bridging perception with symbolic reasoning. However, other spectacular cognitive skills are not gained by simply observing the world! As pointed out in Chapter 6, a good theory of learning cannot neglect the immersion in the environment, where the agent is expected to act. The basic idea behind recurrent neural networks is that those actions are fully supervised, but in many cases, the agent can simply expect to receive a reward/punishment signal from the environment — which corresponds with the framework of reinforcement learning.

No matter what computational model we bear in mind, we need to realize that learning by doing [J. Dewey, Democracy and Education: An Introduction to the Philosophy of Education, *Macmillan*, New York (1916).] opens the doors to the acquisition of new cognitive skills that were very well established in what Marvin Minsky called the "society of mind" [M. Minsky, The Society of Mind, *Simon & Schuster, Inc.*, New York (1986)]. In a sense, this is connected with the basic assumption that while doing the agent modifies the conditions under which it operates. This typically happens in human dialogues, when one is given a certain task to solve. As we disclose of importance of giving the agents a purpose, it looks like that everything becomes more clear. In many interesting cases there are in fact intermediate steps in which it can receive feedback from the environment. In reinforcement learning, this happens during the agent life, but one can also think of cases where the statement of the purpose gives the agent a precise goal to achieve. An agent involved in the N-queens task could benefit from a reinforcement signal during its actions but, as already shown, it also clearly benefit from the formal statement of the overall task to be solved.

As symbolic knowledge is induced from environmental interactions, it becomes immediately important to start exploring social behaviors. Social agents can enjoy the privilege of receiving knowledge granules from others — also humans — which possess a symbolic representation of their world. As already mentioned, their perceptual interaction must also be expressed in abstract symbolic representations. As shown in Chapter 6, the interaction with graphical domains — think of sequences as a special case — suggests the introduction of constraints that operate on hidden variables that can be interpreted as expression of the state of the agent. The importance of diffusion processes becomes clearer and clearer as we consider the immersion in time and consider long-life learning processes. They take place in reinforcement learning and in recurrent networks, where the state defines the agent behavior. Of course, we can directly inject the purpose, so that the agent gains a clear picture of its mission. This is what happens in most machine learning tasks like classification and regression. However, we can restrict the interaction with reward/punishment statements on the performed actions, and we can think of a truly process of "learning the purpose" without providing its specific statement. In a social network of agents, one can explore the meaning of receiving reward from many agents that is the consequence of gaining a sort of popularity. While all this seems to be strictly connected with human social behavior, one shouldn't neglect the importance of dealing with the related concept of multiagent systems [J. Ferber, Multi-Agent Systems: An Introduction to Distributed Artificial Intelligence, *1st edition, Addison-Wesley Longman Publishing Co., Inc.*, Boston, MA, USA (1999)], which is very popular in artificial intelligence. Stressing the idea of agent cooperation could lead to important progresses. As stated by Yuval Noah [Y.N. Harari, Sapiens and Homo Deus, *Harper Collins Publishers* (2017)], the emphasis on cooperation, along with the ability of prospecting nice stories, was in fact one of the fundamental evolutionary ingredients for the success of homo sapiens.

The book also promotes a view on the emergence of intelligent processes that is somewhat inspired to information-based laws of nature. This reflects the previous discussion on the actual immersion in the environment. When dealing with perceptual tasks, the principle of least cognitive action can be used to discover the correspondent agent behavior. Interestingly, the laws that arise from the principle rely on the classic notion of time that is shared with humans. One shouldn't confuse its role with the iteration index of many online learning algorithms. The time has come for thinking of software agents that live on the web and perform in principle nearly all human tasks. Classic problems in computer vision and in natural language processing might take a very different face when reframing the agent evaluation according to the long-life learning perspective. This new assessment scheme might not be simply interesting for emulating humans, but to favor the development of approaches for improving efficiency and performances. People could be reluctant to explore such a path, simply because of the great results that have already been achieved in the last few years under the blessing of the benchmark assessment scheme. However, the blessing relies on crowdsourcing, something new for statisticians, and others. What if we fully trust crowdsourcing by stressing its role in a truly dynamic way? Couldn't it be more appropriate to use crowdsourcing during all its life? This kind of question gave rise to the *en plein air movement* [M. Gori, M. Lippi, M. Maggini, S. Melacci, M. Pelillo, En plein air visual agents, in *Proc. of the Int. Conf. on Image Analysis and Processing* (2015), 697–709]. The workshop "A cena con i pattern," Convegno GIRPR 2014, and the ICPR tutorial "Data-driven pattern recognition: Philosophical, historical, and technical issues" began promoting this approach in machine learning and pattern recognition. It might resemble the act of painting outdoors, which contrasts with studio painting under rules that might contribute to creating a predetermined look. Painting outside in natural light has been of crucial importance in the Barbizon school, in Hudson River School, and for Impressionists. What if also scientists open the doors of their labs under the en plein air scheme?

Answers to exercises

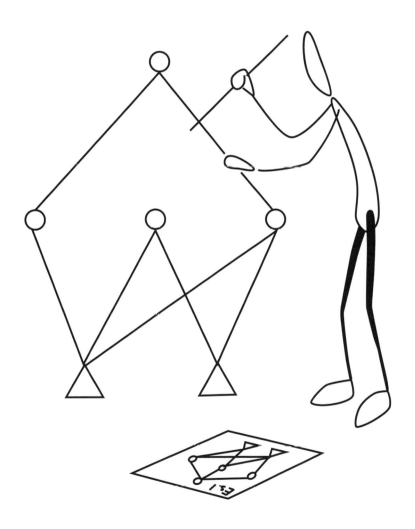

Machine Learning. https://doi.org/10.1016/B978-0-32-389859-1.00015-5

Section 1.1
Section 1.1.1

1. Let $\epsilon > 0$ be a proper threshold, and let $M_y := \max\{y_1, \dots, y_n\}$ be the maximum value of y and $S_y := \max\{y_1, \dots, y_n\} \setminus M_y$ be the second maximum of y. Furthermore, assume that $h(y) = 0$ indicates the absence of decision (rejection). Then for all $i = 1, \dots, n$ choose $h_i(y) = [M_y - S_y > \epsilon][i = \arg\max y]$. This gets rid of the robustness issue. Clearly, the choice of ϵ, which isn't addressed in this exercise, isn't a negligible issue.

2. If we have an algorithm for learning f_2 then we can avoid learning f_1, since we have $f_1 = (h_1^{-1} \circ h_2) \circ f_2 \circ (\pi_2 \circ \pi_1^{-1}) = h \circ f_2 \circ \pi$. On the contrary, if we have an algorithm for f_1, it cannot be used for learning f_2. In case in which also π_2 and h_2 are invertible the above equation enlightens a symmetry equivalency property of learning χ by f_1 and f_2, since we have $f_2 = h^{-1} \circ f_1 \circ \pi^{-1}$, where $\pi^{-1} = (\pi_2 \circ \pi_1^{-1})^{-1} = \pi_1 \circ \pi_2^{-1}$ and $h^{-1} = (h_1^{-1} \circ h_2)^{-1} = h_1 \circ h_2^{-1}$. In this case, learning f_1 and f_2 is essentially equivalent.

3. We can regard $e \in E$ as the trees represented according to Fig. 1.1. A possible construction of π consists of simply collecting the information on the leaves, which is done by pasting together `buying`, `maint`, `doors`, `person`, `lug-boot`, and `safety`. Since we are dealing with features having scalar attributes, if we use one-hot encoding for all features then we can use Boolean vector $x \in \{0, 1\}^{22}$. For instance, $x = (1, 0, 0, 0; 0, 1, 0, 0; 0, 0, 0, 1, 0; 0, 1, 0; 0, 1, 0; 0, 0, 1)$ is coding a car described by `buying` $=$ `vhigh`, `maint` $=$ `high`, `doors` $= 5$, `comfort` $= 4$, `lug` $-$ `boot` $= 4$, and `safety` $=$ `big`. We notice in passing that the constructed internal representation $x \in X$ disregards the structural information inherently associated with the tree of Fig. 1.1.

4. Patterns coming from camera typically come with resolutions and, consequently, details that aren't necessary for the purpose of the recognition. If we keep all this information then the inferential process is more complex. In particular, we do require more examples for the learning process.

5. The inferential process carried by the agent is defined by function $\chi = h \circ f \circ \pi$, where the functions h and π are defined in advance on the basis of coding issues. As a consequence, the learning process, which consists of discovering f, is affected by the choices of h and π. Let us compare the one-hot and the classic binary codes. Clearly, the same information is supported with more compact codes when using the classic binary representation. However, while we save space for the representation of the information, its encryption by the binary code makes induction remarkably more difficult. Clearly, because of the exclusive firing of one out of ten outputs, the one-hot representation is simpler in terms of induction.

Section 1.1.4

6. The overall number of classes is $26 + 10 = 36$ (letters plus digits). The five hidden units of Fig. 1.4 make it possible to code, at most, $32 = 2^5$ classes and, therefore, the task cannot be learned. It is worth mentioning that if the hidden units take on real values such a negative result ceases to apply.

Section 1.1.5

1. We can replace function (4) with $C(f_1, f_2) = \sum_{x \in X^\sharp} \left(-(2f_1(x_1) - 1)(2f_2(x_2) - 1) \right)_+ = \sum_{x \in X^\sharp} \left(2(f_1(x_1) + f_2(x_2)) - 4f_1(x_1)f_2(x_2) - 1 \right)_+$. Notice that while this penalty function serves the purpose of forcing coherent decisions, it doesn't exhibit the robustness of penalty (4). How could robustness be incorporated in this case?

2. The penalty $C(f_1, f_2) = \sum_{x \in X^\sharp} \frac{1}{2}(f_2(x_1) - f_1(x_1))^2$ serves the purpose.

3. They are not equivalent. The first one likely yields more precision, whereas the second one likely yields more recall. Notice that an additional interpretation of the retina pattern occupancy principle is to replace vectors m^c with $\bar{m}^c$, where this time we consider the allowed region of the retina that can be occupied.

Section 1.2
Section 1.2.1

2. Let us look for a solution of the form $F_n = \phi^n$. Then ϕ^n must satisfy $\phi^n = \phi^n - 1 + \phi^n - 2$, which admits the two solutions $\phi = (1 + \sqrt{5})/2$ and $\widehat{\phi} = (1 - \sqrt{5})/2$. Let $c_1, c_2 \in \mathbb{R}$. Then $F_n = c_1 \phi^n + c_2 \widehat{\phi}^n$ is still a solution. Finally, we need to satisfy the conditions $F_0 = 0$ and $F_1 = 1$, which require to verify the linear system $c_1 + c_2 = 0$ and $c_1 \phi + c_2 \widehat{\phi} = 1$, which has solutions $c_1 = 1/\sqrt{5}$ and $c_2 = -1/\sqrt{5}$.

3. The proof is given by induction on n. The basis of induction is trivial. For the induction step suppose the claim holds true for n. Then we have

$$\begin{pmatrix} 1 & 1 \\ 1 & 0 \end{pmatrix}^{n+1} = \begin{pmatrix} F_{n+1} & F_n \\ F_n & F_{n-1} \end{pmatrix} \begin{pmatrix} 1 & 1 \\ 1 & 0 \end{pmatrix} = \begin{pmatrix} F_{n+2} & F_{n+1} \\ F_{n+1} & F_n \end{pmatrix}.$$

4. We can easily check that the sequence is correctly interpreted. Moreover we can prove that this generative scheme does produce Fibonacci numbers. The proof comes out directly if we use the identity in Exercise 3 and calculate the determinant of both sides, while using Binet's theorem.

Section 1.2.4

6. Giotto di Bondone, a famous artist in the late Middle Ages, is typically regarded as an example of violation of the statement "the student cannot overcome the teacher." In the book *Lives of the Artists* by Giorgio Vasari (Oxford University Press, Oxford, 1550) it is pointed out that "Giotto truly eclipsed Cimabue's fame just as a great light eclipses a much smaller one." Giotto was in fact a student of Cimabue. Interestingly, machines can overcome the supervisor. This is clearly a slippery issue, until we make it clear that in machine learning, machines don't necessarily receive supervision information from the potential human competitor. As an example, a banknote acceptor can be based on neural networks that learn from examples to detect counterfeits. An in-depth discussion, along with the presentation of a real-world banknote acceptor, is given by Frosini, Gori and Priami [*IEE Trans. Neural Netw.* **6** (1996), 1482–1490]. If the machine and the human competitor are expected to recognize counterfeits on the basis of the sensorial information produced by the machine then we are outside the domain of the student-teacher relationship. In this specific case, according to the author's experience, machines significantly outmatch human competitors. Like in other examples, machines can in fact rely on certified supervision information, which is not affected by errors of the teacher.

Section 1.3

1. No, we cannot! The accuracy also depends on $|N_t|$, which is neither involved in the precision nor in the recall.

3. The proof can promptly be given when noticing that $2/F_1 = 1/p + 1/r$. Hence we have $2/\max\{p, r\} \le 1/p + 1/r \le 2/\min\{p, r\}$, and, finally we get $\max\{p, r\} \ge F_1 \ge \min\{p, r\}$.

Section 1.4

1. Speech signals come in nature with a modulation. The message that is transmitted from the brain to the speech articulators is in fact the one which performs the modulation of the vibration produced by the vocal folds. Hence, the isolation of speech portion must involve the absence of speech production (no air pumped from the lungs). This is the reason why we need to impose $s(t) \simeq 0$ instead of $s'(t) \simeq 0$. Although, visual information is the outcome of high frequency signals, the reflection of light from the objects, returns video signals from the camera that already report the gray level (or color). Hence, the segmentation of homogeneous visual portions in the single frames, requires us to check whether the spatial gradient is nearly null — $|\nabla_x v(t, x)|^2 \simeq 0$.

Section 2.1
Section 2.1.1

1. We can balance the error function by defining

$$E(f) = 6 \sum_{\kappa=1}^{10} V(x_\kappa, y_\kappa, f(x_\kappa)) + 2 \sum_{\kappa=11}^{30} V(x_\kappa, y_\kappa, f(x_\kappa)) + \sum_{\kappa=31}^{60} V(x_\kappa, y_\kappa, f(x_\kappa)).$$

In general, if $\mathcal{N} = \{n_1, \ldots, n_c\}$ is the number of examples for each class, we can define $\forall i = 1, \ldots, n_c$, $\mu_i := \mathrm{LCM}(\mathcal{N})/n_i$, where $\mathrm{LCM}(\mathcal{N})$ is the Least Common Multiple of elements in $\mathcal{N}$ and choose $E(f) = \sum_{i=1}^{n_c} \sum_{\kappa=1}^{n_i} V(x_\kappa, y_\kappa, f(x_\kappa))$.

2. Consider the loss function defined by Eq. (6). We can promptly see that symmetry does not hold. If we choose $y = 1$ and $f(x) = 0$ then $V(x, y, f) := \log 2$, while if $y = 0$ and $f(x) = 1$ then $V(x, y, f(x)) = 0$.

3. Let $p(x) := yf(x)$. We can promptly see that the cross-entropy of Eq. (6) can be written as $V(x, y, f) = \tilde{V}(y, p(x)) = -[y - 1 = 0]\log((1 + p(x))/2) - [y + 1 = 0]\log((1 - p(x))/2)$. Hence loss depends both on $p(x)$ and y. Moreover, when checking the expression on $y \in \{-1, +1\}$, we conclude that it is impossible to express $V(x, y, f)$ only as a function of $p(x)$.

5. We consider the case $Y = \{0, 1\}$. We have $V(x, y, f) = \sum_{j=1}^{n} y_j \log f_j(x) + (1 - y_j)\log(1 - f_j(x))$. In case we use the softmax output, because of the inherent probabilistic constraint, we can simply use $V(x, y, f) = \sum_{j=1}^{n} y_j \log f_j(x)$ as suggested in Chapter 1, Section 1.3.4.

6. Let $p(x) := yf(x)$. We have $V(x, y, f) = 1 - \exp(-\exp(-p(x)))$. In order to see whether this function is an appropriate loss for classification, let us consider how it works in case of sign agreement/disagreement. For $p(x) = 0$ we have $V(x, y, f) := 1 - e^{-1}$. As $p(x) \to \infty$ (very strong sign agreement) we have $V(x, y, f) = 0$, whereas if $p(x) \to -\infty$ (very strong sign disagreement) we get $V(x, y, f) = 1$. To sum up, we are in front of an appropriate loss with the property that its values are somewhat related to the number of errors.

9. From Eq. (14), we get

$$
\begin{aligned}
V(x, y, f) &= -\frac{1+y}{2} \ln p(1 \mid x; f) - \frac{1-y}{2} \ln p(-1 \mid x; f) \\
&= -\frac{1+y}{2} \ln \frac{\exp(f(x))}{1 + \exp(f(x))} - \frac{1-y}{2} \ln \frac{1}{1 + \exp(f(x))} \\
&= \ln(1 + \exp(f(x))) - \frac{1+y}{2} f(x) \\
&= \ln(1 + \exp(f(x))) - \ln \exp \frac{1+y}{2} f(x) = \ln \frac{1 + \exp(f(x))}{\exp \frac{1+y}{2} f(x)}.
\end{aligned}
$$

If we distinguish the cases $y = -1$ and $y = +1$, we get

$$
\ln \frac{1 + \exp(f(x))}{\exp \left(\frac{1+y}{2} f(x)\right)} = \begin{cases} \ln(1 + \exp(f(x))), & \text{if } y = -1, \\ \ln(1 + \exp(-f(x))), & \text{if } y = +1. \end{cases}
$$

Hence $V(x, y, f) = \ln(1 + \exp(-yf(x)))$, which restores the logistic loss defined by Eq. (5) with $\theta = 0$. Notice that the softmax probabilistic model that has been used expresses the probability by using only one function. Alternatively, we could have chosen $\exp(f_j(x))/\sum_{i=1}^{n} \exp(f_i(x))$ with $j = 1, 2$. What are the differences? *Hint:* See Exercise 11.

11. We follow the analysis of Exercise 9 by using the softmax probabilistic assumption $P(Y_j = +1 \mid f(x)) = \exp(f_j(x))/\sum_{i=1}^{n} \exp f_i(x)$ where $j = 1, \ldots, n$. Now let X_j be the class of patterns of class j and let $y_{j,\kappa} = 2[x_\kappa \in X_j] - 1$. The log-likelihood is

$$\mathcal{L}(f) = \frac{1}{\ell} \sum_{\kappa=1}^{\ell} -\ln p(y_\kappa \mid x_\kappa, f) = \frac{1}{n\ell} \sum_{\kappa=1}^{\ell} \sum_{j=1}^{n} -\frac{1+y_{j,\kappa}}{2} \ln P(Y_j = +1 \mid f_j(x_\kappa))$$

$$= \frac{1}{n\ell} \sum_{\kappa=1}^{\ell} \sum_{j=1}^{n} -\frac{1+y_{j,\kappa}}{2} \ln \frac{\exp(f_j(x))}{\sum_{i=1}^{n} \exp f_i(x_\kappa)}$$

$$= \frac{1}{n\ell} \sum_{\kappa=1}^{\ell} \sum_{j=1}^{n} \frac{1+y_{j,\kappa}}{2} \ln \frac{\sum_{i=1}^{n} \exp f_i(x)}{\exp(f_j(x_\kappa))}$$

$$= \frac{1}{n\ell} \sum_{\kappa=1}^{\ell} \sum_{j=1}^{n} \frac{1+y_{j,\kappa}}{2} \ln \sum_{i=1}^{n} \exp(f_i(x_\kappa) - f_j(x_\kappa)).$$

The associated loss is $V(x_\kappa, y_\kappa, f) = (1/(2n)) \sum_{j=1}^{n} (1+y_{j,\kappa}) \ln \sum_{i=1}^{n} \exp(f_i(x_\kappa) - f_j(x_\kappa))$.

Section 2.1.3

1. The lack of knowledge in medicine applications might suggest refraining from applying a certain treatment, modeled by $f(x)$ over pattern x, in cases that were never detected, for which we nominally get $p(x) = 0$.

3. While the method is very straightforward and, at first glance, yields a good approximation of f^*, unfortunately, it does not work in many real-world problems where there are not enough data for the local approximation by $\hat{f}$. The other problem is the curse of dimensionality which, when using classic metrics, leads to odd neighbors.

4. We distinguish the case $\ell = |Y| = 3$ and $\ell = 4$.

$\ell = 3$ Here, we prove that $\arg\min_{s \in \mathbb{R}} v(s) = \{y_2\}$. The proof is based on the preliminary remark that the minimum must necessarily be on one of the points of Y. If it were not the case, we would have

$$\begin{aligned}
s < y_1, & \quad v'(s) = -3; \\
y_1 < s < y_2, & \quad v'(s) = -1; \\
y_2 < s < y_3, & \quad v'(s) = +1; \\
s > y_3, & \quad v'(s) = 3,
\end{aligned}$$

which indicates that there is no minimum in $\mathbb{R} \setminus Y$. Then we get

$$v(s) = |y_1 - s| + |y_2 - s| + |y_3 - s| = \begin{cases} s = y_1 & v(s) = |y_2 - y_1| + |y_3 - y_1|, \\ s = y_2 & v(s) = |y_1 - y_2| + |y_3 - y_2|, \\ s = y_3 & v(s) = |y_1 - y_3| + |y_2 - y_3|. \end{cases}$$

Since Y is sorted in ascending order, we can promptly see that, from $|y_3 - y_2| < |y_3 - y_1|$ we have $v(y_2) < v(y_1)$. Likewise, from $|y_1 - y_2| < |y_1 - y_3|$ we have $v(y_2) < v(y_3)$. Finally, $\{y_2\} = \arg\min_{s \in \mathbb{R}} v(s)$.

$\ell = 4$ Following similar arguments, we can easily see that the minimum cannot be an element of Y. Moreover, this time any $s \in \mathbb{R}$ such that $y_2 < s < y_3$ is such that $v'(s) = 0$. The function v is convex and any $s \in \{y_2, y_3\}$ yields the minimum of v.

Hence med(Y) is a minimum in both the cases, but for $\ell = 4$ there are infinitely many minima. Notice that the min v is independent of the specific values of y_κ and only depends on their ordering! This is a strong property, which has an interesting consequence: In a skewed distribution, the mean is farther out in the long tail than the median.

5. We give the proof for the case of single output functions. We have $\min_f \mathbb{E}_{XY}(\|y - f(x)\|_1) = \min_f \mathbb{E}_X \mathbb{E}_{Y|X}(\|y - f(x)\|_1) = \mathbb{E}_X \min_{f(x)} \mathbb{E}_{Y|X}(\|y - f(x)\|_1)$, so the problem is reduced to determining $\min_{f(x)} \mathbb{E}_{Y|X}(\|y - f(x)\|_1)$. Let $\epsilon > 0$ and define $\phi(f(x)) := (1/n_\epsilon) \sum_{\alpha=1}^{n_\epsilon} |y_\alpha - f(x)|$. Then there exists $n_\epsilon \in \mathbb{R}$ such that $|\phi(f(x)) - \mathbb{E}_{Y|X}(\|y - f(x)\|_1)| < \epsilon$. Hence we have

$$\min_{f(x)} \phi(f(x)) - \epsilon < \min_{f(x)} \mathbb{E}_{Y|X}(\|y - f(x)\|_1) < \min_{f(x)} \phi(f(x)) + \epsilon.$$

Because of the continuity of $\mathbb{E}_{Y|X}(\|y - f(x)\|_1)$, this also means that n_ϵ can be chosen in such a way that

$$\arg\min_{f(x)} \phi(f(x)) - \epsilon < \arg\min_{f(x)} \mathbb{E}_{Y|X}(\|y - f(x)\|_1) < \arg\min_{f(x)} \phi(f(x)) + \epsilon. \quad (\clubsuit)$$

Now we calculate $\min_{f(x)} \phi(f(x))$. Notice that $\phi(f(x))$ can be rewritten as

$$\phi(f(x)) = \sum_{\alpha=1}^{m} (y_\alpha - f(x))[y_\alpha - f(x) > 0] + \sum_{\alpha=m+1}^{n_\epsilon} (f(x) - y_\alpha)[y_\alpha - f(x) < 0]$$

$$(\diamondsuit)$$

In case n_ϵ is even, from $(\diamondsuit)$, we can immediately conclude that

$$D_{f(x)} \phi(f(x)) = n_\epsilon - 2m \neq 0,$$

that is, the median is the output. In addition, we can always consider an even n_ϵ. Suppose that n_ϵ is odd. We can easily see that $\arg\min_{f(x)} \sum_{\alpha=1}^{n_\epsilon} |y_\alpha - f(x)|$ is one of supervised targets that we denote as $y_{\overline{\alpha}}$. This can be proved by contradiction. Let us assume that $f(x)$ is a minimum that is not one of the supervised points y_α. Then $\phi(f(x))$ is differentiable and $D_{f(x)}\phi(f(x)) = 0$. However, when using ($\diamondsuit$), we get $D_{f(x)}\phi(f(x)) = n_\epsilon - 2m \neq 0$, which yields $n_\epsilon = 2m$, thus contradicting the hypothesis that n_ϵ is odd. Let us denote by $y_{\overline{\alpha}}$ the supervised point that is the argument of the minimum, that is, $f(x) = y_{\overline{\alpha}}$. If we add $|y_{\overline{\alpha}} - f(x)|$ to $\phi(f(x))$, clearly the minimum does not change, that is,

$$\arg\min_{f(x)} \phi(f(x)) = \arg\min_{f(x)} \left(\sum_{\alpha=1}^{n_\epsilon} |y_\alpha - f(x)| + |y_{\overline{\alpha}} - f(x)| \right)$$

$$= \arg\min_{f(x)} \phi(y_{\overline{\alpha}}, f(x)). \qquad (\heartsuit)$$

Now $\phi(y_{\overline{\alpha}}, f(x))$ is composed of an even number of terms. Furthermore, suppose we move $y_{\overline{\alpha}} \to y_{\overline{\alpha}} + \delta$. Clearly, $\forall \epsilon_\phi > 0$ there exists $\delta > 0$ such that $|\phi(y_{\overline{\alpha}+\delta}, f(x)) - \phi(y_{\overline{\alpha}}, f(x))| < \epsilon_\phi$, that is,

$$\phi(y_{\overline{\alpha}+\delta}, f(x)) - \epsilon_\phi < \phi(y_{\overline{\alpha}}, f(x)) < \phi(y_{\overline{\alpha}+\delta}, f(x)) + \epsilon_\phi. \qquad (\spadesuit)$$

Now $\phi(y_{\overline{\alpha}+\delta}, f(x))$ is constructed over an even number of distinct nodes, so we have reduced the analysis to the case of even nodes. Following arguments similar to the case of odd n_ϵ, we can easily see that the argument of the minimum is not on the targets and therefore can be determined from ($\diamondsuit$) by nullifying the derivative with respect to $f(x)$. We have $D_{f(x)}\phi(y_{\overline{\alpha}}, f(x)) = n_\epsilon - 2m = 0$, from which $m = n_\epsilon/2 = \mathrm{med}(\phi(y_{\overline{\alpha}+\delta}, f(x)))$. Hence from ($\spadesuit$) and ($\heartsuit$) we get $\mathrm{med}(\phi(y_{\overline{\alpha}+\delta}, f(x))) - \epsilon_\phi < \arg\min_{f(x)} \phi(y_{\overline{\alpha}}, f(x)) = \arg\min_{f(x)} \phi(f(x)) < \mathrm{med}(\phi(y_{\overline{\alpha}+\delta}, f(x))) + \epsilon_\phi$. By subtracting ϵ and using ($\clubsuit$), we get $\mathrm{med}(\phi(y_{\overline{\alpha}+\delta}, f(x))) - \epsilon_\phi - \epsilon < \arg\min_{f(x)} \phi(f(x)) - \epsilon < \arg\min_{f(x)} \mathbb{E}_{Y|X}(\|y - f(x)\|_1)$. Likewise, $\mathrm{med}(\phi(y_{\overline{\alpha}+\delta}, f(x))) + \epsilon_\phi + \epsilon > \arg\min_{f(x)} \phi(f(x)) + \epsilon < \arg\min_{f(x)} \mathbb{E}_{Y|X}(\|y - f(x)\|_1)$. Let $\epsilon_t := \epsilon + \epsilon_\phi$. For any $\epsilon_\phi > 0$ and $\epsilon > 0$ we can always find $\delta > 0$ and $n_\epsilon \in \mathbb{N}$ such that $\mathrm{med}(Y \mid X = x) - \epsilon_t < \arg\min_{f(x)} \mathbb{E}_{Y|X}(\|y - f(x)\|_1) < \mathrm{med}(Y \mid X = x) + \epsilon_t$. Finally, the case of even n_ϵ is a special case of the analysis carried out for odd n_ϵ.

Section 2.1.4

1. We have $\mathbb{E}_{XY}(Y - \mathbb{E}_{Y|X}(Y \mid X)) = \mathbb{E}_{XY}Y - \mathbb{E}_{XY}(\mathbb{E}_{Y|X}(Y \mid X))$. By definition of $\mathbb{E}_{XY}$,

$$\mathbb{E}_{XY}(\mathbb{E}_{Y|X}(Y \mid X)) = \int_{X \times Y} \mathbb{E}_{Y|X}(Y = y \mid X = x) p_{XY}(x, y) dx dy$$

$$= \int_X \left(\int_Y \left(\int_Y y p_{Y|X}(y \mid x) dy \right) p_{Y|X}(y \mid x) dy \right) p_X(x) dx$$

$$= \int_X \int_Y y p_{Y|X}(y \mid x) dy \left(\int_Y p_{Y|X}(y \mid x) dy \right) p_X(x) dx$$

$$= \int_Y \int_X y p_{XY}(x, y) dx dy = \mathrm{E}_{XY}(Y).$$

Finally, $\mathrm{E}_{XY}(Y - \mathrm{E}_{Y|X}(Y \mid X)) = \mathrm{E}_{XY}(Y) - \mathrm{E}_{XY}(\mathrm{E}_{Y|X}(Y \mid X)) = \mathrm{E}_{XY}(Y) - \mathrm{E}_{XY}(Y) = 0.$

Section 2.2
Section 2.2.1

1. We use the classical Laplacian analysis, suggesting that if the trials are independent and all possible values of p are assumed equally likely, then

$$\frac{\int_0^1 p^{r+1}(1-p)^{m-r}}{\int_0^1 p^r(1-p)^{m-r}} = \frac{r+1}{m+2}.$$

Notice that if we always observe successes, then the formal replacement $r \to m$ yields $(m+1)/(m+2)$. Now the sun won't rise tomorrow with probability $2.3 \cdot 10^{-10}$. A detail analysis on Laplace's rule of succession can be found in S. L. Zabell, *Erkenntnis* **31** (1989), 283–321.

3. Let us consider the case in which $X = [0, m] \subset \mathbb{R}$, where we have $p_X(x) = [0 \leq x \leq m]/\theta$. The likelihood function is $L(\theta) = [0 \leq x \leq m]/\theta^\ell$ where $\forall x_\kappa \in X^\sharp$ we have $x_\kappa \leq m$. Now $\hat{\theta} = \sup_{\theta > x_\kappa} L(\theta) = \max_\kappa x_\kappa = m$. What if $X = [a, b] \subset \mathbb{R}$? Of course, we can always map these samples to $(0, m]$, which leads to conclude that $\hat{\theta} = 1/(b - a)$. Finally, if $X \in \mathbb{R}^d$, we can apply the same idea. Any $x \in X$ can be mapped to a ball $B_\rho(0)$ of radius ρ such that $\mathcal{L}^d(B_\rho(0)) = \mathcal{L}^d(X)$. Clearly, the analysis for the ball corresponds with the single-dimensional case, and we conclude that $\theta = \rho$.

4. The log-likelihood is $l(\sigma) = \sum_{\kappa=1}^\ell (-\ln 2 - \ln \sigma - |x_\kappa|/\sigma)$. Then

$$\frac{dl(\sigma)}{d\sigma} = \sum_{\kappa=1}^\ell \left(-\frac{1}{\sigma} + \frac{|x_\kappa|}{\sigma^2} \right) = -\frac{\ell}{\sigma} + \frac{1}{\sigma^2} \sum_{\kappa=1}^\ell |x_\kappa|.$$

Hence we get $\hat{\sigma} = (1/\ell) \sum_{\kappa=1}^\ell |x_\kappa|$.

5. We are given a collection of Bernoulli trials $X = \{x_1, \ldots, x_\ell\}$, where the generic $x_\kappa \in \{0, 1\}$ is the outcome of trial (e.g., heads or tails). Now we can promptly see that $p^{x_\kappa}(1-p)^{1-x_\kappa} = [x_\kappa = 1]p + [x_\kappa = 0](1 - p)$, and therefore the log-likelihood is

$$l(p) = \sum_{\kappa=1}^{\ell} \ln p^{x_\kappa}(1-p)^{1-x_\kappa} = \sum_{\kappa=1}^{\ell}(x_\kappa \ln p + (1-x_\kappa)\ln(1-p))$$

$$= \ln p \sum_{\kappa=1}^{\ell} x_\kappa + \ln(1-p)\sum_{\kappa=1}^{\ell}(1-x_\kappa)$$

$$= \ell\bar{x}\ln p + \ell(1-\bar{x})\ln(1-p).$$

Now from

$$\frac{dl(p)}{dp} = \frac{\bar{x}}{p} + \frac{\bar{x}-1}{1-p} = 0,$$

we get $\hat{\theta} = \hat{p} = \bar{x}$. This critical point is in fact a maximum, since we have

$$\frac{dl^2(p)}{dp^2} = -\frac{\bar{x}}{p^2} + \frac{\bar{x}-1}{(1-p)^2} < 0.$$

We notice in passing that if instead of knowing $X = \{x_1, \ldots, x_\ell\}$ only the resulting number h of heads is available, then the likelihood would have been $L(p) = p^h(1-p)^{\ell-h}$. It's easy to see that we get the same estimate as in the case in which all the information X is available. We have

$$\frac{dl(p)}{dp} = \frac{d}{dp}(h\ln p + (\ell-h)\ln(1-p)) = \frac{h}{p} - \frac{\ell-h}{1-p} = 0,$$

from which we get $\hat{p} = h/\ell$. Finally, we can promptly conclude that this is a maximum by checking that $dl^2(p)/dp^2 < 0$.

Section 2.2.3

2. We use the recursive Bayes learning stated by Eq. (1). We consider the updating of $p(\theta \mid L)$ for $L = \{4, 7, 2, 8\}$. When $x_1 = 4$ comes, we have

$$p(\theta \mid L_1) = \alpha p(x = 4 \mid \theta)p(\theta \mid L_0) = \alpha_1 \frac{[4 \le \theta \le 10]}{\theta}.$$

In order to determine α_1, we impose $\alpha_1 \int_4^{10} d\theta/\theta = 1$, from which we get $\alpha_1 = 1/(\ln 5 - \ln 2)$. When $x_2 = 7$ comes, we have

$$p(\theta \mid L_2) = \alpha_2 p(x = 7 \mid \theta)p(\theta \mid L_1) = \alpha\alpha_1 \frac{[7 \le \theta \le 10][4 \le \theta \le 10]}{\theta^2}$$

$$= \alpha_2 \frac{[7 \le \theta \le 10]}{\theta^2},$$

where $\alpha_2 := \alpha\alpha_1$ is determined by imposing $\alpha_2 \int_7^{10} d\theta/\theta^2 = 1$, from which we get $\alpha_2 = 70/3$. When $x_3 = 2$ comes, something different happens. We have

$$p(\theta \mid L_3) = \alpha_3 p(x = 2 \mid \theta) p(\theta \mid L_2) = \alpha\alpha_2 \frac{[2 \leq \theta \leq 10][7 \leq \theta \leq 10]}{\theta^3}$$

$$= \alpha_3 \frac{[7 \leq \theta \leq 10]}{\theta^3}.$$

Basically, this time the interval in which the density $p(\theta \mid L_3)$ is not updated, but its structure changes, since it's more peaked. If we impose the normalization condition, we get $\alpha_3 = 9800/51$. Now let $x_m := \max_\kappa x_\kappa$. Induction on $n > 1$ leads us to conclude that

$$p(\theta \mid L_n) = \frac{n-1}{x_m^{1-n} - 10^{1-n}} \frac{[x_m \leq \theta \leq 10]}{\theta^n}.$$

As $n \to \infty$ the density $p(\theta \mid L_n)$ becomes highly peaked. For $\theta = x_m$ we get

$$p(\theta = x_m \mid L_n) = \frac{n-1}{(x_m^{1-n} - 10^{1-n})x_m^n} = \frac{n-1}{x_m - 10(x_m/10)^n},$$

and $p(\theta = x_m \mid L_n) = (n-1)/(x_m^{1-n} - 10^{1-n}) \to \infty$ as $x_m \to 10$. This distributional degeneration of the posterior is in fact indicating that Bayesian learning has been completed. Finally, we can compute $p(x \mid L)$ according to Eq. (4). When plugging the above expression for $p(\theta \mid L_n)$ and using the expression in Excrcise 2.2.1–3 for $p(x \mid \theta)$, we get

$$p(x \mid L_n) = \int_0^{10} p(x \mid \theta) p(\theta \mid L_n) d\theta$$

$$= \frac{n-1}{x_m^{1-n} - 10^{1-n}} \int_0^{10} \frac{[0 \leq x \leq \theta][x_m \leq \theta \leq 10]}{\theta^{n+1}} d\theta$$

$$= \frac{n-1}{x_m^{1-n} - 10^{1-n}} \left([x < x_m] \int_{x_m}^{10} \frac{d\theta}{\theta^{n+1}} + [x \geq x_m] \int_x^{10} \frac{d\theta}{\theta^{n+1}} \right)$$

$$= \frac{n-1}{n(x_m^{1-n} - 10^{1-n})} \left([x < x_m] \left(\frac{1}{x_m^n} - 10^{-n} \right) + [x \geq x_m] \left(\frac{1}{x^n} - 10^{-n} \right) \right).$$

Notice that as $n \to \infty$, if the random variable X yields a maximum $x_m \to 10$ then, as one could expect, $p(x \mid L_n) \to 1/10$. Moreover, it's worth mentioning that according to Exercise 2.2.1–3, the MLE in this case also yields $p(x \mid L_n) \to 1/10$. Clearly, the difference between these estimators arises for small n.

4. We consider the likelihood of having m_H heads in m trials: $p(L_m \mid \theta) = \theta^{m_H}(1 - \theta)^{1-m_H}$. The prior on θ is

$$p(\theta) \sim B(\theta, \alpha, \beta) = \frac{1}{B(\alpha, \beta)} \theta^{\alpha-1}(1 - \theta)^{\beta-1},$$

where $B(\alpha, \beta) := \int_0^1 t^{\alpha-1}(1-t)^{\beta-1}\,dt$. The posterior $p(\theta \mid L_m)$ turns out to be

$$p(\theta \mid L_m) = \frac{\theta^{m_H}(1-\theta)^{m-m_H}\theta^{\alpha-1}(1-\theta)^{\beta-1}}{\int_0^1 \theta^{m_H}(1-t)^{m-m_H}t^{\alpha-1}(1-t)^{\beta-1}dt}$$

$$= \frac{\theta^{m_H+\alpha-1}(1-\theta)^{m-m_H+\beta-1}}{\int_0^1 t^{m_H+\alpha-1}(1-t)^{m-m_H+\beta-1}dt},$$

so that $p(\theta \mid L_m) \sim B(\theta, m_H + \alpha, m - m_H + \beta)$. Since the prior and the posterior exhibit the same probability distribution, we conclude that the prior Beta distribution on θ is conjugate with the Bernoulli data distribution. Since we have discovered the expression of the posterior in a closed analytic form, we can directly determine the optimal MAP estimation by determining the value $\hat\theta_{MAP}$ which maximizes $p(\theta \mid L_m)$. If we differentiate w.r.t. θ, we can easily see that

$$\hat\theta_{MAP} \in \arg\max_\theta p(\theta \mid L_m) = \frac{m_H + \alpha - 1}{m + \alpha + \beta - 2}.$$

Hence, the MAP estimate is equivalent to MLE when considering a virtual data set with $\alpha - 1$ additional heads and $\beta - 1$ additional tails. For $\alpha = \beta = 1$, MAP reduces to MLE. For example, the hyperparameters $\alpha = 9$ and $\beta = 1$ model the case of a tossed coin strongly biased on the heads. Hence the coefficients α and β nicely model our belief on the Bernoulli distribution. They are referred to as *prior hyperparameters*, whereas $m_H + \alpha$ and $m - m_H + \beta$ are called *posterior hyperparameters*.

5. For a fixed σ^2, we have $p(x \mid \theta) = p(x \mid \mu) \sim \mathcal{N}(\mu, \sigma^2)$, and our prior is $p(\theta) = p(\theta \mid L_0) \sim \mathcal{N}(\mu_0, \sigma_0^2)$. Now, because of Exercise 3, the product of Gaussians is still a Gaussian, and therefore if we use the recursive Bayes learning stated by Eq. (1), we get

$$p(\theta \mid L_1) = \alpha_1 p(x_1 \mid \theta)p(\theta) = \mathcal{N}\left(\frac{\sigma_0^2 x_1 + \sigma^2 \mu_0}{\sigma_0^2 + \sigma^2}, \frac{\sigma_0^2 \sigma^2}{\sigma_0^2 + \sigma^2}\right),$$

$$p(\theta \mid L_2) = \alpha_2 p(x_2 \mid \theta)p(\theta \mid L_1).$$

Let $\hat\mu_n = (1/n)\sum_{\kappa=1}^n x_\kappa$. By induction on n, we have

$$p(\theta \mid L_n) = \mathcal{N}\left(\frac{\sigma_0^2 n\hat\mu_n + \sigma^2 \mu_0}{n\sigma_0^2 + \sigma^2}, \frac{\sigma_0^2 \sigma^2}{n\sigma_0^2 + \sigma^2}\right). \tag{*}$$

Now let us define

$$\mu_n = \frac{n\sigma_0^2}{n\sigma_0^2 + \sigma^2}\hat\mu_n + \frac{\sigma^2}{n\sigma_0^2 + \sigma^2}\mu_0, \qquad \sigma_n^2 = \frac{\sigma_0^2 \sigma^2}{n\sigma_0^2 + \sigma^2}.$$

As $n \to \infty$ the distribution becomes highly peaked since $\lim_{n\to\infty}(\sigma_0^2\sigma^2)/(n\sigma_0^2 + \sigma^2) = 0$. This is the indication that Bayesian learning has converged. Now we can compute $p(x \mid L)$ according to Eq. (4). When using Eq. (*), we get

$$p(x \mid L_n) = \int_{-\infty}^{\infty} p(x \mid \theta) p(\theta \mid L_n) \, d\theta = \int_{-\infty}^{\infty} g_{\sigma^2}^{x}(x - \theta) g_{\sigma_n^2}^{\mu_n}(\theta - \mu_n) \, d\theta,$$

where $g_{\sigma^2}^{x}$ and $g_{\sigma_n^2}^{\mu_n}$ are Gaussian functions, along with their parameters. If we take $\vartheta = x - \theta$, we get

$$p(x \mid L_n) = \int_{-\infty}^{\infty} g_{\sigma_n^2}^{x - \mu_n}(x - \mu_n - \vartheta) g_{\sigma^2}^{0}(\vartheta) \, d\vartheta = \left(g_{\sigma_n^2}^{\mu_n} * g_{\sigma_n^2 + \sigma^2}^{0} \right)(x),$$

where we have used the property stated in Exercise 3 on the convolution of Gaussians. To sum up, the Bayesian estimate is a Gaussian $\mathcal{N}(\mu_n, \sigma_n^2 + \sigma^2)$.

Section 2.2.4

1. We start with the basis of induction and assume that all the variables are independent of each other, that is $\forall v_i \in V$, $\mathrm{pa}(i) = \emptyset$. Then $p(v_1, \ldots, v_\ell) = \prod_{i=1}^{\ell} p(v_i)$, which collapses to Eq. (1), when considering that $p(v_i \mid v_{\mathrm{pa}(i)}) = p(v_i \mid v_\emptyset) = p(v_i)$. Now suppose by induction that $p(v_1, \ldots, v_{\ell-1}) = \prod_{i=1}^{\ell-1} p(v_i \mid v_{\mathrm{pa}(i)})$. The new vertex V_n receives connections from the vertexes in $i \in \mathrm{pa}(\ell)$; then we have

$$p(v_1, \ldots, v_\ell) = p(v_\ell \mid v_1, \ldots, v_{\ell-1}) p(v_1, \ldots, v_{\ell-1})$$

$$= p(v_\ell \mid v_1, \ldots, v_{\ell-1}) \prod_{i=1}^{\ell-1} p(v_i \mid v_{\mathrm{pa}(i)}) = \prod_{i=1}^{\ell} p(v_i \mid v_{\mathrm{pa}(i)}).$$

2. We use the MLE, then we must maximize $L = \prod_{\kappa=1}^{\ell} p(y_\kappa, x_{\kappa,1}, \ldots, x_{\kappa,d}) = \prod_{\kappa=1}^{\ell} q(y_\kappa) \prod_{i=1}^{d} q_i(x_{\kappa,i} \mid y_\kappa)$, or equivalently,

$$l = \sum_{\kappa=1}^{\ell} \sum_{i=1}^{d} \left(\ln q(y_\kappa) + \ln q_i(x_{\kappa,i} \mid y_\kappa) \right) = d \sum_{\kappa=1}^{\ell} \ln q(y_\kappa) + \sum_{\kappa=1}^{\ell} \sum_{i=1}^{d} \ln q_i(x_{\kappa,i} \mid y_\kappa),$$

where $q(y) \geq 0$, $\sum_{y=1}^{c} q(y) = 1$, $q_i(x \mid y) \geq 0$ and $\sum_{x \in \{-1,+1\}} q_i(x \mid y) = 1$. The maximization can be handled by using the Lagrangian

$$\mathcal{L} = d \sum_{\kappa=1}^{\ell} \ln q(y_\kappa) + \sum_{\kappa=1}^{\ell} \sum_{i=1}^{d} \ln q_i(x_{\kappa,i} \mid y_\kappa) + \sum_{y=1}^{c} \lambda_{1,y} q(y) + \lambda_{2,y} \left(\sum_{y=1}^{c} q(y) - 1 \right)$$

$$= \lambda_{1,x} \sum_{x \in \{-1,+1\}} q_i(x \mid y) + \lambda_{2,x} \left(\sum_{x \in \{-1,+1\}} q_i(x \mid y) - 1 \right).$$

Differentiation w.r.t. $q(y)$ yields

$$\frac{\partial \mathcal{L}}{\partial q(y)}(q(\hat{y})) = d \sum_{\kappa=1}^{\ell} \frac{[y = y_\kappa]}{\hat{q}(y)} + \lambda_{1,y} + \lambda_{2,y} = 0,$$

from which we get

$$\hat{q}(y) = -\frac{d}{\lambda_{1,y} + \lambda_{2,y}} \sum_{\kappa=1}^{\ell} [y = y_\kappa].$$

When imposing the normalization conditions, we get

$$-\sum_{y=1}^{c} \frac{d}{\lambda_{1,y} + \lambda_{2,y}} \sum_{\kappa=1}^{\ell} [y = y_\kappa] = -\frac{d}{\lambda_{1,y} + \lambda_{2,y}} \sum_{\kappa=1}^{\ell} \sum_{y=1}^{c} [y = y_\kappa] = -\frac{d\ell}{\lambda_{1,y} + \lambda_{2,y}} = 1.$$

Hence $-d/(\lambda_{1,y} + \lambda_{2,y}) = 1/\ell$, and finally, $\hat{q}(y) = (1/\ell) \sum_{\kappa=1}^{\ell}[y = y_\kappa]$. Differentiation w.r.t. $q_i(x \mid y)$ yields

$$\frac{\partial \mathcal{L}}{\partial q_i(x \mid y)}(\hat{q}_i(x \mid y)) = \sum_{\kappa=1}^{\ell} \frac{[(x = x_{\kappa,i}) \wedge (y = y_\kappa)]}{\hat{q}_i(x \mid y)} + 2(\lambda_{1,x} + \lambda_{2,x}) = 0,$$

from which we get $\hat{q}_i(x \mid y) = -(1/(2(\lambda_{1,x} + \lambda_{2,x}))) \sum_{\kappa=1}^{\ell}[(x = x_{\kappa,i}) \wedge (y = y_\kappa)]$. Finally, from the normalization conditions, we get $\sum_{x \in \{-1,+1\}} \hat{q}_i(x \mid y) = -1/(2(\lambda_{1,x} + \lambda_{2,x})) \sum_{\kappa=1}^{\ell}[y = y_\kappa] = 1$, from which we get

$$\hat{q}_i(x \mid y) = \frac{\sum_{\kappa=1}^{\ell}[(x = x_{\kappa,i}) \wedge (y = y_\kappa)]}{\sum_{\kappa=1}^{\ell}[y = y_\kappa]}.$$

Section 3.1

1. We have $(\nabla f)_i = \partial_i(c_j x_j) = c_i$ and $(\nabla g)_i = \partial_i(x_j A_{jk} x_k) = (A_{ik} + (A')_{ik}) x_k$.

2. The index (1) becomes $E(W, b) = \sum_{\kappa=1}^{\ell} \sum_{i=1}^{n} (y_{\kappa i} - (W x_\kappa)_i - b_i)^2$, where $y_{\kappa i}$ is the i-th component of the κ-th target y_κ. In order to find normal equations define

$$\hat{W} = \begin{pmatrix} w_{11} & w_{12} & \cdots & w_{1n} \\ w_{21} & w_{22} & \cdots & w_{2n} \\ \vdots & & & \vdots \\ w_{d1} & w_{d2} & \cdots & w_{dn} \\ b_1 & b_2 & \cdots & b_n \end{pmatrix},$$

and let $(Y)_{ij} = y_{ij}$ with $Y \in \mathbb{R}^{\ell,n}$. Then $E(\hat{W}) = \text{tr}\big((Y - \hat{X}\hat{W})(Y' - \hat{W}'\hat{X}')\big) = \text{tr}(YY') - 2\,\text{tr}(Y\hat{W}'\hat{X}') + \text{tr}(\hat{X}\hat{W}\hat{W}'\hat{X}')$, and normal equations are obtained imposing $\partial_{\hat{W}} E = 0$. Since $\partial_{\hat{W}} \text{tr}(Y\hat{W}'\hat{X}') = \hat{X}'Y$ and $\partial_{\hat{W}} \text{tr}(\hat{X}\hat{W}\hat{W}'\hat{X}') = 2(\hat{X}'\hat{X}\hat{W})$ normal equations are $\hat{W}^* = (\hat{X}'\hat{X})^{-1}\hat{X}'Y$.

3. The properties follow from the linearity of E plus the fact that $E\,1 = 1$. In fact, we have

$$\hat{\sigma}^2_{xy} = E(XY - X\,EY - Y\,EX + EX\,EY) = E(XY) - EX\,EY$$

$$= \frac{1}{\ell}\sum_{\kappa=1}^{\ell} x_\kappa y_\kappa - \overline{x}\cdot\overline{y},$$

if we take $X = Y$, we obtain the desired expression also for $\hat{\sigma}^2_{xx}$.

Section 3.1.1

1. Assume you have at least two examples, then using normal equations (4) we have (the index κ always varies from 1 to ℓ):

$$\begin{pmatrix} w \\ b \end{pmatrix} \simeq \frac{1}{\ell \sum_\kappa x_\kappa^2 - (\sum_\kappa x_k)^2} \begin{pmatrix} \ell & -\sum_\kappa x_\kappa \\ -\sum_\kappa x_\kappa & \sum_\kappa x_\kappa^2 \end{pmatrix} \begin{pmatrix} p\sum_\kappa x_\kappa^2 + q\sum_\kappa x_\kappa \\ p\sum_\kappa x_\kappa + \ell q \end{pmatrix}$$

$$= \begin{pmatrix} p \\ q \end{pmatrix}.$$

3. i. If rank $\hat{X} = d + 1 < \ell$ then 0 is an eigenvalue, since rank $P_\perp = d + 1$, otherwise rank $P_\perp = \ell$, and therefore 0 is not an eigenvalue. Now we have $P_\perp^d \hat{X} = (\hat{X}(\hat{X}'\hat{X})^{-1}\hat{X})\hat{X} = \hat{X}$, and therefore 1 is the only other eigenvalue.

ii. Concerning $Q_\perp^d$ we can draw the same conclusions on the spectrum, whereas rank$(Q_\perp^d) = \ell - \text{rank}(P_\perp^d)$. The case of $P_\perp^u$, $Q_\perp^u$ is left to the reader.

5. Suppose we use the scaling $x \to \alpha x$ with $\alpha > 0$. Then $\hat{w}_\alpha = (w'/\alpha, b)'$ clearly constructs a function such that $\hat{X}\hat{w} = \hat{X}_\alpha \hat{w}_\alpha$. Now we prove that also $\hat{X}\hat{w}^* = \hat{X}_\alpha \hat{w}_\alpha^*$ holds true, that is $\hat{w}_\alpha^* := ((w^*)'/\alpha, b)'$ yields the optimal solution once $\hat{w}^*$ is given. We notice that $\min_{\hat{w}} \|y - \hat{X}_\alpha \hat{w}\|^2 = \|y - P_\perp^\alpha y\|^2 = \|Q_\perp^\alpha y\|^2 \leq \|Q_\perp^\alpha\|^2 \|y\|^2 = \|Q_\perp\|^2 \|y\|^2$, where the last equality comes from the property stated in Exercise 3. Hence α-scaling does not change the $\|Q_\perp\|^2 \|y\|^2$ bound on the minimum.

6. Since we consider inversion in the classic sense, we can limit the discussion to the case $d + 1 \leq \ell$. In this case, M is in fact a fat matrix, and consequently the solution of $M\hat{X}\hat{w} = My$ corresponds with the solution of $\hat{X}\hat{w} = y + z$, where $Mz = 0$ — basically z is in the kernel of M. This means that the solution of the abnormal equations neither perfectly fits the conditions nor represents the best fitting — unless $M = \hat{X}$.

7. If $\det T \neq 0$, then the spectrum of $Q_\perp(T)$ does not change and the considerations of Exercise 5 hold. Basically, $w \cdot Tx + b = \omega \cdot x + b$, where $\omega := T'w$, and the problem is reduced to learning with $(\omega', b)'$. Clearly, if T is not a full rank matrix, we lose information in the preprocessing.

10. The data are represented by $\hat{X} = \begin{pmatrix} 0 & 0 & 1 \\ 0 & 1 & 1 \end{pmatrix}$ and $y = \begin{pmatrix} -1 \\ +1 \end{pmatrix}$. From the normal equations, we get the solution $\hat{w} = (\alpha, 2, -1)'$ with $\alpha \in \mathbb{R}$. For any $w_1 = \alpha$, the corresponding line $x_2 = 0.5$ separates the training set. A more straightforward formulation consists of disregarding the coordinate w_1 and considering the one-dimensional problem defined by $\hat{X} = \begin{pmatrix} 0 & 1 \\ 1 & 1 \end{pmatrix}$ and $y = \begin{pmatrix} -1 \\ +1 \end{pmatrix}$ which directly yields the same separating line $w_2 = 0.5$.

11. The data are represented by

$$\hat{X} = \begin{pmatrix} 0 & 0 & 1 \\ 0 & 1 & 1 \\ 0 & \alpha & 1 \end{pmatrix}, \qquad y = \begin{pmatrix} -1 \\ -1 \\ +1 \end{pmatrix}.$$

From the normal equations we get $w_2 = (2\alpha - 1)/(\alpha^2 - \alpha + 1)$ and $b = -\alpha^2/(\alpha^2 - \alpha + 1)$. For example, for $\alpha = 2$ we get the separating line $x_2 = 4/3$. Clearly, one could simply reduce the dimensionality of the representation, just like in Exercise 10, that is, one can drop the coordinate w_1. Under this restriction, for $\alpha = 1/2$ we have $w_2 = 0$ and $b = 1/3$. The interpretation is quite interesting: We have $w'x = 0$ and the problem of learning is reduced to that of determining the barycenter of the given set of points, where the mass can either be $+1$ or -1.

Section 3.1.2

1. Take $A^+ = A'(AA')^{-1}$ with A real ($A^* = A'$), the first property is $AA'(AA')^{-1}A = A$, the second $A'(AA')^{-1}AA'(AA')^{-1} = A'(AA'^{-1})$, the third is trivial since in our special case $AA^+ = I$, while the last one is $(A'(AA')^{-1}A)' = A'(AA')^{-1}A$.

5. The problem can be approached by minimizing $L(\hat{w}, \lambda) = \hat{w}^2 + \lambda \cdot (\hat{X}\hat{w} - y)$. This yields $\nabla_{\hat{w}} L = 2\hat{w} + \hat{X}'\lambda = 0$ and $\nabla_\lambda L = \hat{X}\hat{w} - y = 0$. From the first equation we get $\hat{w} = -\hat{X}'\lambda/2$, and therefore $\lambda = -2(\hat{X}\hat{X}')^{-1}y$. Finally, $\hat{w} = \hat{X}'(\hat{X}\hat{X}')^{-1}y$. According to the definition given in Eq. (4), this is the pseudoinverse of fat full-rank matrix $\hat{X}$.

Section 3.1.3

1. We begin proving that $II E_r > 0$. We have that $\forall u \in \mathbb{R}^{d+1}$ with $u \neq 0$, $u \cdot (\mu I_d + \hat{X}'\hat{X})u = \mu \sum_{i=1}^d u_i^2 + \|\hat{X}u\|^2$. Now we can consider two cases depending on the structure of u. If $u_i \neq 0$ for at least one $i \in \{1, \ldots, d\}$ then $\sum_{i=1}^d u_i^2 > 0$, and therefore $u'(\mu I_d + \hat{X}'\hat{X})u > 0$. Since $u \neq 0$ the only other possibility is $u = \gamma e_{d+1} = \gamma(0, \ldots, 0, 1)'$. In this case $\hat{X}u = \gamma \hat{X}(0, \ldots, 0, 1)' = \gamma(1, \cdots, 1)$, and $u \cdot (\lambda I_d + \hat{X}'\hat{X})u = \gamma^2(\ell + 1) > 0$. Finally, the claim follows when considering that positive matrices, which have positive eigenvalues, are nonsingular.

5. Notice that if the penalty term is convex then adding the regularization term preserves convexity. This arises from the property that the sum of any two convex functions is convex. When the penalty term is only local minima free, the problem is more difficult. To the best of my knowledge, it's an open research problem.

6. The matrix $\hat{X}$ is

$$\hat{X} = \begin{pmatrix} x_1^d & x_1^{d-1} & \cdots & x_1 & 1 \\ x_2^d & x_2^{d-1} & \cdots & x_2 & 1 \\ \vdots & \vdots & \cdots & \vdots & \vdots \\ x_\ell^d & x_\ell^{d-1} & \cdots & x_\ell & 1 \end{pmatrix}.$$

If $d + 1 \geq \ell$ we can use the normal equations, whereas if $d + 1 > \ell$ we can use the pseudoinverse or the ridge regularization. Notice that the problem with $\hat{X}$ (Vandermonde matrix) is that it is ill-conditioned, which creates serious numerical problems for large dimensions.

Section 3.2

1. No, it isn't. A counterexample comes out directly when choosing the following learning task $\{((1, 1), -1), ((2, 2), +1)\}$ and $\alpha > 2$.

2. For the normalization we have $\sum_i \text{smx}_i = \sum_i \exp(a_i)/\sum_j \exp(a_j) = 1$. Now suppose that $a_i \gg a_j$, then $\text{smx}_k \approx \exp(a_k - a_i) \approx [k = i]$.

Section 3.2.1

1. We have $x_1 \wedge x_2 \wedge \cdots \wedge x_n = [x_1 + x_2 + \cdots + x_n \geq n]$, and $x_1 \vee x_2 \vee \cdots \vee x_n = [x_1 + x_2 + \cdots + x_n \geq 1]$.

2. If f is linearly separable, it means that there exists a hyperplane $w_1 x_1 + w_2 x_2 + \cdots + w_n x_n = t$ that separates the points of the Boolean hypercube on which the function assumes value 1 from those on which the function assumes value 0. Then $f(x_1, x_2, \ldots, x_n) = [w_1 x_1 + w_2 x_2 + \cdots + w_n x_n \geq t]$.

3. We already know that $x \oplus y = (x \vee y) \wedge (\neg x \vee \neg y)$, then the only thing we have for sure is that it is not of order 1. Then assume that is of order 1, this means that $x \oplus y = [w_1 x + w_2 y \geq t]$; because of the symmetry $x \oplus y = y \oplus x$ we have that whenever $w_1 x + w_2 y \geq t$ it must also be $w_1 y + w_2 x \geq t$, and hence $\frac{1}{2}(w_1 + w_2)x + \frac{1}{2}(w_1 + w_2)y \geq t$; similarly, when $w_1 x + w_2 y < t$, it is $\frac{1}{2}(w_1 + w_2)x + \frac{1}{2}(w_1 + w_2)y < t$. Now let $\frac{1}{2}(w_1 + w_2) \equiv \gamma$, then $x \oplus y = [\gamma(x + y) \geq t]$. For this expression to be consistent with the definition of the XOR function, it must be that when $x = y = 0$, $\gamma \cdot 0 \geq t$, when $x = 1$ and $y = 0$, $\gamma \cdot 1 < t$, and when $x = y = 1$ it must be $\gamma \cdot 2 \geq t$. These inequalities cannot be satisfied simultaneously, hence we proved that XOR cannot be of order 1 by contradiction.

4. Let $|R^\sharp| = n$, then Φ is finite since there are exactly 2^{2^n} Boolean functions of n variables. Furthermore, one can safely assume that there is no image $X^\sharp$ for which $\sum_{i=1}^{D} w_i \varphi_i = t$ and that t is rational since in both cases we can just change $t \to t + \delta$, where δ is less than the smaller nonzero value that $|\sum_i w_i \varphi_i - t|$ can assume on any image $X^\sharp$ and $t + \delta$ is rational.

Now suppose that all w_i are rational, then $w_i = p_i/q_i$ with $q_i > 0$ for all $i = 1$, $\dots$, D and if we say that $t = \tau/q_{D+1}$ $(q_{D+1} > 0)$ we can define $w_i' := (\sum_{j=1}^{D+1} q_j) w_i$ and $t' = (\sum_{j=1}^{D+1} q_j) t$. In this way w_i' and t' are integers and $[\sum_{i=1}^{D} w_i \varphi_i \geq t] = [\sum_{i=1}^{D} w_i' \varphi_i \geq t']$ for all the possible images on the retina. Suppose now that there exists an index k for which w_k is irrational; then one can replace this number with any rational number w_k' that lies in the interval $(w_k .. w_k + \delta/2^{2^n})$. In this way the sum $\sum_{i=1}^{D} w_i \varphi_i$ cannot change by more than δ and then the value of $[\sum_{i=1}^{D} w_i \varphi_i \geq t]$ remains unchanged.

5. For any negative weight w_j set $x_j \leftarrow \bar{x}_j$, $w_j \leftarrow -w_j$ and $t \leftarrow t + |w_j|$.

6. (a) By construction there are as many vertexes as pixels, then R_n has exactly n^2 vertexes. Now the "vertical" edges are $n - 1$ for each row and for each column of vertexes then they are $2n(n - 1)$ in total; then we have to add $2(n - 1)^2$ "diagonal" edges.

(b) All we have to do is to exhibit a way to build a spanning cycle for each R_n. It is convenient to distinguish between n odd and n even. Then these two spanning cycles for R_6 and R_7 show how we can do this in general

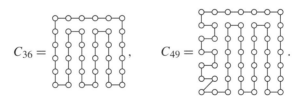

The idea here is that when n is odd one can zigzag between the first two columns and then complete the pattern as in the even case.

7. Strictly speaking, it is not possible to define a midpoint function that given two vertices finds a single midpoint for essentially two reasons: The first one is that given two vertices i and j of a graph there is more than one shortest path that connects them; the second reason is that for paths of odd length there is no vertex that is equally distant from both ends of the path so that instead we have two midpoints. Then it is more natural to assume that in this case $i \rightarrowtail j$ is a set of points rather than a single point as in the continuous case — in particular, it will be the set of all midpoints (for even length paths) or midpoints (for odd length paths) of all shortest paths between i and j. Then it is natural to say that a set of vertexes C of a graph is midpoint-convex if for all i and j of such set $i \rightarrowtail j \subseteq C$. Notice, to conclude, that all this discussion does not change the statement that $[X^\sharp$ is convex$]$ is of order 3, since each of the $\varphi_i = \lfloor p_i \rightarrowtail q_i \subseteq X^\sharp \rfloor$ can be written as an AND of predicates that depend only on 3 points.

Section 3.2.3

1. (a) For $d = 1$ we can use the normal equations with $\hat{X} = \left(\begin{smallmatrix} x_1 & 1 \\ x_2 & 1 \end{smallmatrix}\right)$ and $y = \left(\begin{smallmatrix} 1 \\ -1 \end{smallmatrix}\right)$; thus we have

$$\begin{pmatrix} w \\ b \end{pmatrix} = \begin{pmatrix} \frac{2}{(x_1-x_2)^2} & -\frac{x_1+x_2}{(x_1-x_2)^2} \\ -\frac{x_1+x_2}{(x_1-x_2)^2} & \frac{x_1^2+x_2^2}{(x_1-y_1)^2} \end{pmatrix} \begin{pmatrix} x_1 - x_2 \\ 0 \end{pmatrix} = \begin{pmatrix} \frac{2}{x_1-x_2} \\ -\frac{x_1+x_2}{x_1-x_2} \end{pmatrix}.$$

Notice that in the case of the symmetric configuration with respect to the origin, $x_1 = -1$ and $x_2 = 1$, we get $w = -1$ and $b = 0$. In this case, $f(x) = -x$ that clearly fits the data.

(b) In this case the normal equations assume the form

$$\begin{pmatrix} \hat{x}_1 & \hat{x}_2 \end{pmatrix} \begin{pmatrix} \hat{x}_1' \\ \hat{x}_2' \end{pmatrix} (\alpha \hat{x}_1 + \hat{x}_2) = \begin{pmatrix} \hat{x}_2 & \hat{x}_2 \end{pmatrix} \begin{pmatrix} 1 \\ -1 \end{pmatrix},$$

and since x_1 and x_2 are linearly independent this reduces to $\alpha\hat{x}_1^2 + \beta\hat{x}_1'\hat{x}_2 = 1$ and $\alpha\hat{x}_1\hat{x}_2 + \beta\hat{x}_2^2 = -1$. From here solving for α and β we immediately get $\alpha = (\hat{x}_2^2 + \hat{x}_1'\hat{x}_2)/(\hat{x}_1^2\hat{x}_2^2 - (\hat{x}_1'\hat{x}_2))$, $\beta = -(\hat{x}_1^2 + \hat{x}_1'\hat{x}_2)/(\hat{x}_1^2\hat{x}_2^2 - (\hat{x}_1'\hat{x}_2))$. Plugging these results back into $\hat{w} = \alpha\hat{x}_1 + \beta\hat{x}_2$, we finally obtain $w = (2 + x_2^2 + x_1'x_2)x_1 - (2 + x_1^2 + x_1'x_2)x_2$ and $b = x_2^2 - x_1^2$. Now since $\hat{X}'\hat{X}(\alpha\hat{x}_1 + \beta\hat{x}_2) = \hat{x}_1(\alpha\hat{x}_1^2 + \beta\hat{x}_1'\hat{x}_2) + \hat{x}_2(\alpha\hat{x}_1'\hat{x}_2 + \beta\hat{x}_2^2) = \hat{x}_1 - \hat{x}_2$, in the case $\hat{x}_1^2 = \hat{x}_2^2$ we have that $\hat{x}_1 - \hat{x}_2$ is an eigenvector of $\hat{X}'\hat{X}$.

2. Define

$$\hat{X} = \begin{pmatrix} c_1x_{11} & c_1x_{12} & \cdots & c_1x_{1d} & c_1 \\ c_2x_{21} & c_2x_{22} & \cdots & c_2x_{2d} & c_2 \\ \vdots & & & \vdots & \vdots \\ c_\ell x_{\ell 1} & c_\ell x_{\ell 2} & \cdots & c_\ell x_{\ell d} & c_\ell \end{pmatrix}, \qquad y = \begin{pmatrix} c_1y_1 \\ \vdots \\ c_\ell y_\ell \end{pmatrix}.$$

Then it is clear from Eq. (1) that $E = \frac{1}{2}\|y - \hat{X}\hat{w}\|^2$ whose minimization gives the normal equations in the standard form 3.1.1–(4).

3. Using normal equations with

$$\hat{X} = \begin{pmatrix} 0 & 0 & m \\ 1 & 0 & 1 \\ 1 & 1 & 1 \\ 0 & m & m \end{pmatrix} \quad \text{and} \quad y = \begin{pmatrix} -m \\ -1 \\ -1 \\ m \end{pmatrix},$$

one obtains $\hat{w}^* = (1 + m^2)^{-1}(-1 - m^2, 2m^2, -m^2)'$, thus the separating line is $x_2 = (1 + m^2)/(2m^2)x_1 + 1/2$. For $m > 0$ the angular coefficient of such a line is a monotonically decreasing function with a horizontal asymptote at $x_2 = 1/2$; this proves that for any finite m this line actually is a separating line for these examples and that for $m = \infty$ the line goes through an example, thus failing to separate. For $m = 0$ (that is, when there are no more examples 1 and 4) the angular coefficient becomes ∞, and the solution is $x_1 = 0$; this is a good solution since the remaining examples 2 and 3, lying on the right of the line, belong to the same category.

4. Take m examples of the kind $\bullet$ in $(1, 0)$ and m examples of type $\times$ in $(0, 1)$. Since here we have more unknowns than (clusters of) examples, we use the Penrose pseudoinverse in order to find the solution with $\hat{X} = \begin{pmatrix} m & 0 & m \\ 0 & m & m \end{pmatrix}$ and $y = \begin{pmatrix} m \\ -m \end{pmatrix}$. The solution is $\hat{w}^* = (1, -1, 0)$, which gives as a separating line for every m the bisector $x_2 = x_1$.

5. The problem formulated in this exercise has been discussed in a more general framework in M. Brady, R. Raghavan, J. Slawny, *IEEE Trans. Circuits Syst.* **36** (1989), 665–674 where you can also find its solution.

Section 3.3

2. In this case, we have $\hat{X} = \begin{pmatrix} x_1 & 1 \\ x_2 & 2 \end{pmatrix}$ and $y = \begin{pmatrix} +1 \\ -1 \end{pmatrix}$. Then we have

$$\begin{pmatrix} x_1^2 + x_2^2 & x_1 + x_2 \\ x_1 + x_2 & 2 \end{pmatrix} \begin{pmatrix} w \\ b \end{pmatrix} = \begin{pmatrix} x_1 - x_2 \\ 0 \end{pmatrix}$$

and get $w = 2/(x_1 - x_2)$, $b = (x_2^2 - x_1^2)/(x_1 - x_2)^2$. Notice that in the case of the symmetric configuration w.r.t., the origin $x_1 = -1$ and $x_2 = -1$ we get $w = -1$ and $b = 0$. In this case $f(x) = -x$, which clearly fits the data.

3. The solution can be found by applying normal equations: $(\hat{x}_i \hat{x}_i' + \hat{x}_j \hat{x}_j')(\alpha \hat{x}_i + \beta \hat{x}_j) = \hat{x}_i - \hat{x}_j$. Hence $\alpha = (\hat{x}_j^2 + \hat{x}_i' \hat{x}_j)/(\hat{x}_i^2 \hat{x}_j^2 - (\hat{x}_i' \hat{x}_j)^2)$ and $\beta = -(\hat{x}_i^2 + \hat{x}_i' \hat{x}_j)(\hat{x}_i^2 \hat{x}_j^2 - (\hat{x}_i' \hat{x}_j)^2)$. From this we directly have that $w = (2 + x_j^2 + x_i' x_j)x_i - (2 + x_i^2 + x_i' x_j)x_j$ and $b = x_i^2 - x_j^2$.

4. When studying the eigenvectors of $\hat{X}' \hat{X}$, we notice that $\hat{X}' \hat{X} = \sum_{\kappa=1}^{\ell} x_\kappa x_\kappa'$. Now let us restrict to the case of two examples only, namely x_i and x_j, where $\|x_i\| = \|x_j\| := \rho$. Then from the normal equations we saw that there exists $\gamma \in \mathbb{R}$ such that $w = \gamma(x_i - x_j)$. As a consequence,

$$\hat{X}' \hat{X} \gamma(x_i - x_j) = \gamma \left(x_i x_i' + x_j x_j' \right)(x_i - x_j) = \gamma x_i x_i' x_i - \gamma x_j x_j' x_j = \gamma \rho(x_i - x_j).$$

Section 3.4

1. Step P2 normalizes the examples. Then if we do not perform such replacement before the output when a new example arrives in order to compute $f = \hat{w}' \hat{x}$, we need to properly normalize the example by inserting R as the last coordinate instead of 1.

2. It doesn't change since the only property that we use in the proof is the fact that the inner product is symmetric.

3. As it has been done in order to derive Eq. 3.4.3–(5), we just have to bound $1/\cos\varphi_i$. Eq. 3.4.3–(3) still holds true, while Eq. 3.4.3–(4) becomes $\|\hat{w}_t\|^2 \le \eta^2(R^2+1)t$, since

$$\|\hat{w}_{\kappa+1}\|^2 \le \|\hat{w}_\kappa\|^2 + \eta^2(\|x_i\|^2+1) \le \|\hat{w}_\kappa\|^2 + \eta^2(R^2+1).$$

Then we have

$$1 \ge \cos\varphi_t \ge \delta\sqrt{\frac{t}{R^2+1}}, \quad \text{that is,} \quad t \le \left(1+\frac{1}{R^2}\right)\left(\frac{R}{\delta}\right).$$

This means that when we know that $R > 1$ — with a fixed ratio R/δ — the algorithm without the normalization is more efficient.

4. Yes, in fact by assumption there exists a hyperplane represented by the vector a such that for all the examples $\hat{x}_i$ in the training set we have $y_i a'\hat{x}_i > 0$. Now call d_i the distance of each point $\hat{x}_i$ from the hyperplane and define $\delta = (1/2)\min_i d_i$, then for each point of the training set we have $y_i a'\hat{x}_i > \delta$.

5. In this case, one cannot conclude that linear separability implies strong linear separability. Suppose, for example, that a is the hyperplane that correctly separates the examples of the training set but assume that the distances d_i of the points $\hat{x}_i$ from a form a sequence such that $\lim_{i\to\infty} d_i = 0$, in this case it is clear that one cannot find a $\delta > 0$ such that for all the examples $y_i a'\hat{x}_i > \delta$. In this case the bound 3.4.3–(5) has to be modified to take into account the way in which d_i approaches 0; let us discuss this in some details.

It is clear that the sequence $\langle d_i \rangle$ induces a sequence $\langle \delta_i \rangle$ where the generic term is defined by $\delta_i = \frac{1}{2}\min_{j<i} d_j$. Then the bound in Eq. 3.4.3–(5) becomes $t \le 2(R/\delta_i)^2$, where the index i is the number of examples that have been processed by Agent Π and t is the number of times that a weights update occurred during these i examples (clearly, $t \le i$). Now suppose that the oracle gives examples such that $\delta_i \approx \Delta/i$ as i becomes bigger. Then the bound reduces to $t \le 2(R/\Delta)^2 i^2$, which is not meaningful since we already knew that $t \le i$. On the other hand, suppose that $\delta_i \approx \Delta/\log i$, this time the bound says that $t \le 2(R/\Delta)(\log i)^2$, which is a meaningful statement about the convergence of the Agent. All this discussion indicates that "effectiveness" of the Agent is largely determined by the benevolence of the oracle that presents the examples.

9. In order to prove the convergence, we use the same scheme of Section 3.4.3. We can simply modify the bounds 3.4.3–(3) and 3.4.3–(4) by considering that $\hat{w}_o \ne 0$. We have the new bounds $a \cdot \hat{w}_t > a \cdot \hat{w}_o + \eta\delta t$, and $\|\hat{w}_t\|^2 \le \hat{w}_o^2 + 2\eta^2 R^2 t$. Hence we get $(a \cdot \hat{w}_o + \eta\delta t)/\sqrt{\hat{w}_o^2 + 2\eta^2 R^2 t} \le 1$ from which $t \le (\eta R^2 - a \cdot \hat{w}_o\delta + \sqrt{(\eta R^2 - a \cdot \hat{w}_o\delta)^2 + (\hat{w}_o^2 - (a \cdot \hat{w}_o)^2)\delta^2})/(\eta\delta^2)$. In case of $\hat{w}_o = 0$ this returns the already seen bound. Notice that $a \cdot \hat{w}_o \le \hat{w}_o$, where $\|a\| = 1$. Then $\hat{w}_o^2 - (a \cdot \hat{w}_o)^2 \ge 0$ so that the roots of the quadratic equation are both real, but we only consider the positive one. Interestingly, when $\hat{w}_o \ne 0$ the learning rate affects the bound.

11. Here, we only provide a sketch of the solution. Any input $\check{x} \in \mathbb{R}^d$ with missing data contains at least one coordinate $\check{x}_i = \check{x}_o$, where $\check{x}_o \in \mathbb{R}$ is the code the missing data. In so doing, we can express the generation of the input by the linear machine $x = M\check{x}$, where $M \in \mathbb{R}^{d \times d}$. When considering all patterns of the training set, we have to satisfy the constraint $X = \check{X}'M$, where we need to impose consistency on the known features. Then the predicted feature vector x is used to compute $f(x) = \hat{X}w$, which yields the constraint $\hat{X}w = y$. This enables us to formulate learning as the parsimonious satisfaction of the above two constraints.

Section 4.1
Section 4.1.2

1. Suppose we restrict to $p = 2$, which prevents us from the mentioned problem of ill-conditioning. When d increases, the number of monomials is given by $|H| = \binom{p+d-1}{p} = \binom{1+d}{2} = d(d+1)/2$. Hence the number of features only grow polynomially with the input dimension.

Section 4.1.4

1. From (3), we have

$$p_{2d,d} = \frac{1}{2^{2d-1}} \sum_{\kappa=0}^{d-1} \binom{2d-1}{\kappa} = \frac{1}{2^{2d}} \left(\sum_{\kappa=0}^{d-1} \binom{2d-1}{\kappa} + \sum_{\kappa=d}^{2d-1} \binom{2d-1}{\kappa} \right) = \frac{1}{2}.$$

The asymptotic behavior can be obtained when noticing that $c_{\ell d} = 2^{1-\ell} \sum_{\kappa=0}^{d-1} \binom{\ell-1}{\kappa}$.

2. We notice that half of all dichotomies can be associated with the code

$$\overbrace{1,1\ldots,1}^{i \text{ bits}}, \overbrace{0,0,\ldots,0,}^{\ell-i \text{ bits}}$$

where $i = 1, \ldots, \ell$, while the other half comes out when considering the dichotomies associated with a code obtained by flipping all bits.

3. The proof comes out when using the result from Exercise 2 and Eq. (1).

(basis) We have $c_{1d} = 2$ and $c_{\ell 1} = 2d$. The first one is trivial, while the second one follows from Exercise 2.

(inductive step) Using induction, we have $c_{(\ell-1)d} = 2 \sum_{h=0}^{d-1} \binom{\ell-2}{h}$, $c_{(\ell-1)(d-1)} = 2 \sum_{\kappa=0}^{d-2} \binom{\ell-2}{\kappa}$. If we replace h with $h = \kappa - 1$ in the first equation and sum up, when considering Eq. (1) we get

$$c_{\ell d} = c_{(\ell-1)d} + c_{(\ell-1)(d-1)}$$

$$= 2 \sum_{\kappa=1}^{d} \binom{\ell-2}{\kappa-1} + 2 \sum_{\kappa=0}^{d-2} \binom{\ell-2}{\kappa}$$

$$= 2\sum_{\kappa=1}^{d-2}\left(\binom{\ell-2}{\kappa-1}+\binom{\ell-2}{\kappa}\right)+2\binom{\ell-2}{d-1}+2\binom{\ell-2}{d-2}+2\binom{\ell-2}{0}$$

$$= 2\sum_{\kappa=1}^{d-2}\binom{\ell-1}{\kappa}+2\binom{\ell-1}{d-1}+2\binom{\ell-1}{0}=2\sum_{\kappa=0}^{d-1}\binom{\ell-1}{\kappa}.$$

4. Shifting the index ℓ by one, Eq. (3) becomes $c_{\ell d} = c_{(\ell-1)d} + c_{(\ell-1)(d-1)}$. Together with the initial conditions $c_{1d} = 2$ and $c_{\ell d} = 0$ if $d \leq 0$ this relation defines the coefficients $c_{\ell d}$. Now consider the generating functions $G_\ell(z) = \sum_d c_{\ell d} z^d$. Using initial conditions, we have $G_1(z) = 2z/(1-z)$; and from the recursion relation we have $G_\ell(z) = G_{\ell-1}(z) + zG_{\ell-1}(z)$. Then

$$G_\ell(z) = (1+z)G_{\ell-1}(z) = (1+z)^2 G_{\ell-2}(z) = \cdots = (1+z)^{\ell-1}G_1(z)$$
$$= \frac{2z(1+z)^{\ell-1}}{1-z}.$$

This means that the generating function can be decomposed as $G_\ell(z) = 2zA(z)B_\ell(z)$ where $A(z) = 1/(1-z) = \sum_{d\geq 0} z^d$ and $B_\ell(z) = \sum_{d\geq 0}\binom{\ell-1}{d}z^d$. Knowing that the overall multiplication by z shifts the coefficients of $A(z)B_\ell(z)$ by one, and using the well known formula for the coefficients of the multiplication of two generating functions, we conclude immediately that $c_{\ell d} = 2\sum_{k=0}^{d-1}\binom{\ell-1}{k}$, which is indeed equivalent to Eq. (3).

5. The proof naturally arises when we assume that one coordinate i of $x_{\ell+1,i} = 0$. In that case all the dichotomies can be determined just like in case when we are working with $d-1$ variables. Now any general position can be obtained by rotating the axis so that one coordinate of $x_{\ell+q}$ is null, which clearly doesn't change the geometry. Hence $D = C(\ell, d-1)$.

6. From the first equality in Eq. (2) we have $c_{\ell d} = 2\left(\sum_{k\geq 0}\binom{\ell-1}{k} - \sum_{k\geq d}\binom{\ell-1}{k}\right) = 2^\ell - 2\sum_{k\geq 0}\binom{\ell-1}{k+d}$. So let us focus on the term $\sum_{k\geq 0}\binom{\ell-1}{k+d}$, and let $t_k = \binom{\ell-1}{k+d}$. The exercise suggests that this series is actually a hypergeometric series. To prove this property and to find the parameters of the corresponding hypergeometric function, since $t_0 = \binom{\ell-1}{d} \neq 0$, we just need to compute the ratio t_{k+1}/t_k: If this ratio is a rational function of k then we can conclude that our series is hypergeometric. We have

$$\frac{t_{k+1}}{t_k} = \frac{(k+1)(k+d-\ell+1)(-1)}{(k+d+1)(k+1)}.$$

We can then conclude that $c_{\ell d} = 2^\ell - 2\binom{\ell-1}{d} F\left(\begin{matrix}1, d-\ell+1\\d+1\end{matrix}\Big| -1\right)$. Now, by using the *reflection law* of the hypergeometric functions,

$$F\left(\begin{matrix}a, c-b\\c\end{matrix}\Big| z\right) = \frac{1}{(1-z)^a} F\left(\begin{matrix}a, b\\c\end{matrix}\Big| \frac{-z}{1-z}\right),$$

with $z = -1$, $a = 1$, $b = \ell$, and $c = d + 1$, we eventually have the required representation $c_{\ell d} = 2^\ell - \binom{\ell-1}{d} F\left(\begin{smallmatrix} 1, \ell \\ d+1 \end{smallmatrix} \middle| \frac{1}{2}\right)$. In general this sum does not have a simple closed form, unless for values of d near 1, $\ell/2$, and ℓ.

Section 4.2
Section 4.2.1

1. We can easily see that $w^* = (0, 0)$. Consider the level curve $w_1 + w_2 = \alpha$. Since $w_1 \geq 0$ and $w_2 \geq 0$, we have $\alpha \geq 0$. Then the minimum is reached for $\alpha = 0$, which means $w^* = (0, 0)$. If we use the Lagrangian approach, we have $L(w_1, w_2, \lambda_1, \lambda_2) = w_1 + w_2 - \lambda_1 w_1 - \lambda_2 w_2$. From $\nabla_w L = 0$ we get $\lambda_1 = \lambda_2 = 1$, leading us to conclude that $L \equiv 0$. Finally, from $\nabla_\lambda L = 0$ we get $w^* = (0, 0)$.

2. The analysis of the constraints makes it possible to conclude immediately that $w^* = (1, 0)$. The KKT conditions are

$$\binom{1}{0} + \lambda_1 \binom{3(1 - w_1)^2}{1} + \lambda_2 \binom{0}{-1} = \binom{0}{0}, \quad \lambda_1 (w_2 - (1 - w_1)^3 = 0, \tag{$*$}$$

$$\lambda_2 w_2 = 0, \quad \lambda_1 \leq 0, \quad \lambda_2 \leq 0.$$

From $\lambda_2 w_2 = 0$, we know that either $w_2 = 0$ or $\lambda_2 = 0$. In the first case, the only point which satisfies $g_1(w_1, w_2) \leq 0$ is $w^* = (1, 0)'$. However, this point doesn't satisfy $(*)$, since for finite values of λ_1 we cannot get $\lambda_1 (w_2 - (1 - w_1)^3) = 0$. On the other hand, if consider $\lambda_2 = 0$ we end up with the same violation of $(*)$, since the condition also requires $\lambda_1 = 0$. It's interesting to construct function θ. We can easily see that

$$\theta(\lambda_1, \lambda_2) = 1 - \frac{1}{\sqrt{-3\lambda_1}} - \frac{\lambda_1}{(\sqrt{-3\lambda_1})^3},$$

which is maximized for $\lambda_1 \to -\infty$. Notice that we can easily understand the reason why $\lambda_1 \to -\infty$ when considering that on the optimal solution $\partial_{w_1} g_1|_{w^*} = 0$. Moreover, notice that there is no duality gap, since $\sup_\lambda \theta(\lambda) = \min_w p(w) = 1$.

3. The first statement is correct, and one can further convince himself that the solution of the problem exists and it is in fact $w^* = (1, 0)$ by drawing the constraint $h = 0$. The problem with the second statement is that the substitution is done naively and the condition $w_1 \geq 1$ (which is implicit in the original constraint) is lost.

4. The objective function is not convex. We can promptly see that there is a manifold of minima, that is,

$$\arg\min_w p(w) = \{w : w = (\alpha, 0), 0 \leq \alpha \leq 1\} \cup \{w : w = (0, \beta), \ 0 \leq \beta \leq 1\}.$$

Now the Lagrange dual function is $\theta(\lambda) = \inf_w \left(w_1 w_2 + \lambda_1 w_1 + \lambda_2 w_2 + \lambda_3 (1 - w_1^2 - w_2^2) \right)$. We can easily see that $\sup_\lambda \theta(\lambda) = -1/2$. In any point of the manifold of minima $p(w^*) = 0$, and therefore there is a duality gap of $-1/2$. It's interesting to capture the reasons behind the gap. It arises because of the presence of multiple minima. If we choose a minimum such that $w^* = (\alpha, 0)$, with $0 \le \alpha \le 1$, then we have $\lambda_2 w_2^* = 0$, with $\lambda_2 < 0$. However, in the same point $\lambda_1 w_1^* < 0$ and $\lambda_3 (1 - (w_1^*)^2 - (w_2^*)^2) < 0$. These two terms give rise to the duality gap. Clearly, this holds regardless of the minimum in which we check the KKT conditions.

5. 1. In case of linear kernel we have $k(x, z) = x \cdot z$. Hence $k(x_h, x_\kappa)$ can be compactly placed in the Gram matrix

$$k = \begin{pmatrix} 0 & 0 & 0 & 0 \\ 0 & 1 & 1 & 0 \\ 0 & 1 & 2 & 1 \\ 0 & 0 & 1 & 1 \end{pmatrix}$$

and the solution to the learning equations (8) is $\lambda_1 = 4$, $\lambda_2 = 2$, $\lambda_3 = 0$, $\lambda_4 = 2$, which can easily be understood when looking at the symmetry of the problem. The solution tells us that $S_= = \{x_1, x_2, x_4\}$ while example x_3 is a straw vector. The quickest way of computing the intercept b is from (11) by choosing x_1 as a support vector, since we immediately get $b = 1$. In the primal, we have $w = \sum_{\kappa=1}^{\ell} \lambda_\kappa y_\kappa \phi(x_\kappa)$. Since $\phi(x_\kappa) = x_\kappa$, $w = -2\binom{1}{0} - 2\binom{0}{1} = -\binom{2}{2}$. Finally, the maximum margin problem is solved by the separation line

$$x_{.,1} + x_{.,2} = 1/2.$$

2. Any relabeling does not affect the quadratic programming scheme associated with Eq. (8).

3. By direct substitution in Eq. (8) of the roto-translation $x \mapsto x_0 + Px$, where P is a rotation matrix

4. In the primal, according to Eq. (2), we need to minimize $1/2(w_1^2 + w_2^2)$ under the constraints $b \ge 1$, $-w_1 - b \ge 1$, $-w_1 - w_2 - b \ge 1$, $-w_2 - b \ge 1$. Employing quadprog Matlab® instruction, we get $w_1 = w_2 = -2$ and $b = 1$, which corresponds with the separation line that was found in the dual representation.

6. If $k = k'$ then its spectral decomposition is $k = U \Sigma U'$. Since $k \ge 0$, for any $\omega \in \mathbb{R}^\ell$ we have $\omega' k \omega \ge 0$, that is, $\omega' U \Sigma U' \omega = \zeta' \Sigma \zeta \ge 0$, where $\zeta := U' \omega$. Now we can always choose ω such that $U' \omega = e_i$, with $i \in \mathbb{N}_\ell$, since $U U' \omega = U e_i$ yields $\omega = U e_i$. Finally, for all $i \in \mathbb{N}_\ell$, by such a choice of ω, we have $\omega' U \Sigma U' \omega = e_i' \Sigma e_i \ge 0$, which yields $\sigma_i \ge 0$.

8. In the following, we use the matrix k whose coordinates are defined by $\tilde{k}(x_i, x_j) := y_j k(x_i, x_j)$. Now let us define $\hat{\alpha} := (\alpha', b)'$ let $\hat{K}$ be the block matrix

$$\hat{K} := \begin{pmatrix} K & 1 \\ y' & 0 \end{pmatrix},$$

and $\hat{y} = (y', 0)'$. The problem of learning is reduced to solving the linear equation $\hat{K}\hat{\alpha} = \hat{y}$.

10. We can generate these extreme cases beginning from the example of Exercise 5. First, we consider the case of degeneration $|S_=| = \ell = 4$. This can be obtained by moving $x_3 = (1, 1) \rightarrow (1/2, 1/2)$ and keeping the rest unchanged. The case $|S_=| = 2$ can be discovered when moving slightly $x_3 = (1/2, 1/2) \rightarrow (1/2 - \epsilon, 1/2 - \epsilon)$ with $0 < \epsilon < 1/2$.

12. First, we notice that the objective function $1/2w^2$ leads an optimization problem with the same parsimony term as for ridge regression. When considering $f(x) = w \cdot \phi(x) + b$, this parsimony term turns out to be $1/2(\nabla f(x))^2$. Amongst others, we can think of this term as of a sensitivity index, which is clearly dependent only on variables which control the variation of the inputs, whereas it is independent of b.

14. If $k(x_h, x_\kappa) > 0$ then it is not singular, and therefore from $\nabla\theta = 0$ we can promptly see that there is only one solution. Moreover, the Hessian of θ corresponds with $-k < 0$, with k being the Gram matrix. Hence the only critical point is a maximum. Concerning the qualitative meaning of the constraints, we notice that $\lambda_\kappa \geq 0$ indicates that the contribution of all similarities between x and the generic x_κ of the training set has to be considered in the overall decision. The constraint $\sum_{\kappa=1}^{\ell} \lambda_\kappa y_\kappa = 0$ can nicely be interpreted as a balancing of positive and negative Lagrangian multipliers (constraint reactions).

16.

$$k(x_h, x_\kappa) = \langle \phi(x_h), \phi(x_\kappa) \rangle^2 \leq \|\phi(x_h)\|^2 \|\phi(x_\kappa)\|^2 = \langle \phi(x_h), \phi(x_h) \rangle \langle \phi(x_\kappa), \phi(x_\kappa) \rangle$$
$$= k(x_h, x_h)k(x_\kappa, x_\kappa).$$

Section 4.2.2

3. As $C \rightarrow 0$ we have $\lambda_\kappa = 0$, and therefore $f(x) = b$. The bias can be determined just like for hard-constraints, and therefore we can use Eq. (12), so $b = \sum_{\bar{i} \in S_=} y_{\bar{\kappa}}/n_s$.

Section 4.2.3

5. Support vectors emerge from numerical solution of quadratic programming, but one always needs to set up a threshold that specifies the border between support and straw vectors. The choice of different thresholds results in a different number of support vectors.

Section 4.3
Section 4.3.1

1. We can provide a counterexample for $X = \mathbb{R}^2$. Let $s = \sqrt{2}/2$, $x = (1, 0)'$, $y = (1, 1/2)'$, $z = (1, -1/2)$, and define $\phi(u) := u$, so that $k_\varphi(u, v) = u'v/(\|u\| \cdot \|v\|)$. We have $k_\varphi(x, y) = 2/\sqrt{5} \Rightarrow x \sim y$, $k_\varphi(x, z) = 2/\sqrt{5} \Rightarrow x \sim z$ and $k_\varphi(y, z) = 3/5 \Rightarrow y \nsim z$. Hence we conclude that $y \sim x$ and $x \sim z$, but $y \nsim z$, hence there is no transitivity.

2. A possible factorization of $k(x_h, x_\kappa) = (x'_h x_\kappa)^2$ is the one given by Eq. (3). However, the feature map $(x_1, x_2) \xrightarrow{\phi} (x_1^2, x_1 x_2, x_2 x_1, x_2^2)$, yields clearly another factorization of the same kernel. Furthermore, notice that at any dimension D there are infinitely many equivalent features that arise from a rotation with matrix $R \in \mathbb{R}^{D \times D}$ of $\phi(x)$. For $h(x) = R\phi(x)$, we have $k(x, z) = \phi(x)'\phi(z) = \phi(x)'R'R\phi(z) = h'(x)h(z)$, which indicates that any $h(x)$ produced by rotation of $\phi(x)$ is a valid feature map for k.

3. A possible feature factorization is $(x_1, x_2) \xrightarrow{\phi} \left(x_1^2, \sqrt{2}x_1 x_2, x_2^2, \sqrt{2c}x_1, \sqrt{2c}x_2\right)$.

Section 4.3.2

1. The singular value decomposition of $k_4 = k(X_4^\sharp)$ is

$$
U = \begin{pmatrix} 0 & 0 & 0 & 1 \\ -0.2610 & 0.7071 & 0.6572 & 0 \\ -0.9294 & 0 & -0.3690 & 0 \\ -0.2601 & -0.7071 & 0.6572 & 0 \end{pmatrix}, \quad \Sigma = \begin{pmatrix} 4.5616 & 0 & 0 & 0 \\ 0 & 1 & 0 & 0 \\ 0 & 0 & 0.4384 & 0 \\ 0 & 0 & 0 & 0 \end{pmatrix},
$$

from which we see that the given Gram matrix is nonnegative definite. Hence k is a kernel. The Mercer's features corresponding to $X^\sharp$ are

$$
\begin{pmatrix} x_1 \\ x_2 \\ x_3 \\ x_4 \end{pmatrix} = \begin{pmatrix} 0 & 0 \\ 1 & 0 \\ 1 & 1 \\ 0 & 1 \end{pmatrix} \xrightarrow{\phi} \begin{pmatrix} \phi'(x_1) \\ \phi'(x_2) \\ \phi'(x_3) \\ \phi'(x_4) \end{pmatrix} = \begin{pmatrix} 0 & 0 & 0 & 0 \\ -0.5573 & 0.7071 & 0.4352 & 0 \\ -1.9850 & 0 & -0.2444 & 0 \\ -0.5573 & -0.7071 & 0.4352 & 0 \end{pmatrix} = \phi_4.
$$

It's worth mentioning that these features refer to the classic inner product of $\mathbb{R}^4$, whereas if we use $\langle \cdot, \cdot \rangle_\sigma$, as assumed in Eq. (5), then Mercer's features are the eigenvectors of k, namely the columns of k_4. Notice that on the given domain $X_4^\sharp$ we have $K(x, z) = k(x, z) = (x \cdot z)^2$ and, therefore, in addition to the discovered Mercer's features, one can exhibit also the feature space given by (3) and the one discussed in Exercise 4.3.1–2.

2. We know that[1] $k = k^* = k'$. Now if σ is an eigenvalue then $\overline{\sigma}$ is also an eigenvalue. Let u and v be the eigenvectors associated with σ and $\overline{\sigma}$, respectively, that is, $ku = \sigma u$ and $kv = \overline{\sigma} v$. Then we have $\sigma \langle u, v \rangle = \langle ku, v \rangle = \langle u, k^* v \rangle = \langle u, kv \rangle = \overline{\sigma} \langle u, v \rangle$, from which we immediately draw the conclusion, since $\langle u, v \rangle (\sigma - \overline{\sigma}) = 0 \Rightarrow \sigma = \overline{\sigma}$.

3. Let σ_1, σ_2 be two distinct (real) eigenvalues of $k = k'$ and denote by u_1, u_2 the corresponding eigenvectors. Then $ku_1 = \sigma_1 u_1$ and $ku_2 = \sigma_2 u_2$. Now we have $\sigma_2 \langle u_1, u_2 \rangle = \langle u_1, ku_2 \rangle = \langle k^* u_1, u_2 \rangle = \langle ku_1, u_2 \rangle = \sigma_1 \langle u_1, u_2 \rangle$. Then we get $(\sigma_2 - \sigma_1) \langle u_1, u_2 \rangle = 0$. Since $\sigma_1 \neq \sigma_2$ we conclude that $\langle u_1, u_2 \rangle = 0$, that is, the eigenvectors u_1 and u_2 are orthogonal.

4. If $K = K'$ then its spectral decomposition is $K = U \Sigma U'$. Since $K \geq 0$, for any $\omega \in \mathbb{R}^\ell$, we have $\omega' K \omega \geq 0$, that is, $\omega' U \Sigma U' \omega = \zeta' \Sigma \zeta \geq 0$, where $\zeta := U' \omega$. Now we can always choose ω such that $U' \omega = e_i$, with $i = 1, \ldots, \ell$, since $UU' \omega = Ue_i$ yields $\omega = Ue_i$. Finally, for all $i = 1, \ldots, \ell$, by such a choice of ω, we have $\omega' U \Sigma U' \omega = e_i' \Sigma e_i = \sigma_i \geq 0$. The opposite implication is straightforward. If $\forall \kappa = 1, \ldots, \ell, \sigma_\kappa \geq 0$ then, $\forall \omega \in \mathbb{R}^\ell$, we have $\omega' K \omega = \zeta' \Sigma \zeta = \sum_{\kappa=1}^\ell \sigma_\kappa \zeta_\kappa^2 \geq 0$.

5. We provide two proofs:

1. Let $t \in \mathbb{R}$ and $x, y \in \mathbb{R}^\ell$. Now let us consider the consequences of $p(t) := (tx - z)^2 \geq 0$. We have $\forall t \in \mathbb{R}$ that $t^2 x^2 - 2t \langle x, z \rangle + z^2 \geq 0$. The minimum of p is achieved for $\overline{t} = \langle x, z \rangle / \|x^2\|$. If we plug $\overline{t}$ back into $p(t) \geq 0$ we get the Cauchy–Schwarz inequality.

2. From $(\|x\|z \pm \|z\|x)^2 \geq 0$ we get $2\|x\|^2 \|z\|^2 + 2\|x\| \|z\| \langle x, z \rangle \geq 0$, from which we get the Cauchy–Schwarz inequality.

We can promptly see that the only case in which the equality holds is when x is collinear to z.

6. Since k is a kernel, then Gram matrix satisfies $K = \begin{pmatrix} k(x,x) & k(x,z) \\ k(z,x) & k(z,z) \end{pmatrix} \geq 0$. Hence $\forall \kappa = 1, \ldots, \ell, \sigma_\kappa \geq 0$ (see Exercise 4) $\det K = \prod_{\kappa=1}^\ell \sigma_\kappa \geq 0$. Finally, this yields $\det K = k(x, x)k(z, z) - k(x, z)k(z, x) \geq 0$.

8. Since $\forall p \in \mathbb{N}_0, a_p \geq 0$ we can define $\sqrt{a_p}$, we have

$$\sum_{n=0}^\infty a_n \cos(n(x - z)) = \sum_{n=0}^\infty (a_n \cos(nx) \cos(nz) + a_n \sin(nx) \sin(nz))$$
$$= \langle \phi(x), \phi(z) \rangle_\sigma,$$

where the feature map is defined by $\phi(y) := (1, \cos(y), \sin(y), \ldots, \cos(py), \sin(py) \ldots)'$. This leads us to conclude that k is a kernel. Notice that this is just an instance of the expansion defined in general by Eq. (7), where $\forall n \in \mathbb{N}_0, \sigma_n = \sqrt{a_n}$. The feature $\phi(y)$ is an eigenvector of K_∞.

---

[1] Gram matrices are defined on the real field, so that the notion of Hermitian and transpose matrices coincide.

9. From the definition of kernel

$$\langle \sum_{i=1}^{n} \alpha_i \phi_i, \sum_{j=1}^{n} \alpha_j \phi_j \rangle = \sum_{i=1}^{n} \sum_{j=1}^{n} \alpha_i \alpha_j \langle \phi_i, \phi_j \rangle = \sum_{i=1}^{n} \sum_{j=1}^{n} \alpha_i \alpha_j \delta_{i,j} k_i = \sum_{i=1}^{n} \alpha_i^2 k_i,$$

which completes the proof when setting $\mu = \alpha_i^2$. As a consequence, the new kernel is a conic combination of the kernel collection $\{k_i : i = 1, \ldots, n\}$.

11. Let $\alpha_\kappa := y_\kappa \lambda_\kappa$, so Eq. 4.2.1–(9) can be rewritten as $f^*(x) = \sum_{\kappa=1}^{\ell} \alpha_\kappa k(x, x_\kappa) + b$, with $k(x, x_\kappa) = \langle \phi(x), \phi(x_\kappa) \rangle$. Now let $\widehat{\phi}(x) := (\phi'(x), 1)'$ and define $\hat{k}(x, x_\kappa) := \langle \widehat{\phi}(x), \widehat{\phi}(x_\kappa) \rangle_+ = \langle \phi(x), \phi(x_\kappa) \rangle + 1$. We can easily see that $\langle \cdot, \cdot \rangle_+$ is an inner product, which is in fact a straightforward consequence of the hypothesis that $\langle \cdot, \cdot \rangle$ is an inner product. Hence $f^*(x) = \sum_{\kappa=1}^{\ell} \alpha_\kappa k(x, x_\kappa) + b = \sum_{\kappa=1}^{\ell+1} \alpha_\kappa \hat{k}(x, x_\kappa)$, where $\alpha_{\ell+1} := b$. Finally, this makes it possible to conclude that the presence of the bias corresponds with an additional virtual example $x_{\ell+1}$ such that $\hat{k}(x_{\ell+1}, x_{\ell+1}) = 1$.

12. The problem with using $\alpha = K_\ell^{-1} y$ for supervised learning is that K_ℓ is generally ill-conditioned, so the corresponding problem is ill-posed. This is immediately evident in the case of linear kernels, where $K_\ell = XX'$, or $\hat{K}_\ell = \hat{X}\hat{X}'$ in case in which there is also a bias term (see Exercise 11). In particular, in this case $K_\ell = XX' > 0$ only if we have enough examples, otherwise K_ℓ is not invertible. The passage to a feature space makes it possible to get $K_\ell > 0$, but this is typically interwound with ill-conditioning. In case of polynomial kernel, this is related to the issue of ill-conditioning of Vandermonde matrix (see also the example on regression concerning the dependence of the air pressure from the height).

Section 4.3.3

1. We need to show, symmetry, bilinearity and positive-definiteness. Symmetry immediately follows from the properties of the kernel. Also, homogeneity is trivial. Now consider $w(x) = \sum_{j=1}^{\ell_w} \alpha_j^w k(x, x_j^w)$, then we can express $v + w$ by $v + w = \sum_{r=1}^{\ell_v + \ell_w} \alpha_r k(x, z_r)$, where $r = 1, \ldots, \ell_v + \ell_w$ is a renumbering of the ordered sequence of indexes h, κ, so that $z_r^{v+w} = [r \le \ell_v] x_\kappa^v + [r > \ell_v] x_j^w$ and $\alpha_r^{v+w} = [r \le \ell_v] \alpha_h^v + [r > \ell_v] \alpha_j^w$. Hence it follows

$$\langle u, v + w \rangle_k^0 = \sum_{h=1}^{\ell_u} \sum_{r=1}^{\ell_v + \ell_w} \alpha_h^u \alpha_r^{v+w} k(x_h^u, z_r^{v+w})$$

$$= \sum_{h=1}^{\ell_u} \sum_{r=1}^{\ell_v + \ell_w} \alpha_h^u ([r \le \ell_v] \alpha_h^v + [r > \ell_v] \alpha_j^w) k(x_h^u, [r \le \ell_v] x_\kappa^v + [r > \ell_v] x_j^w)$$

$$= \sum_{h=1}^{\ell_u} \sum_{\kappa=1}^{\ell_v} \alpha_h^u \alpha_\kappa^v k(x_h^u, x_\kappa^v) + \sum_{h=1}^{\ell_u} \sum_{j=1}^{\ell_v} \alpha_h^u \alpha_j^w k(x_h^u, x_j^w) = \langle u, v \rangle_k^0 + \langle u, w \rangle_k^0.$$

Finally positive definiteness follows from the positive definiteness of the Gram matrix and from the fact that $\langle u, u \rangle_k^0 = 0 \Rightarrow u = 0$. Indeed the special reproducing property of $\langle \cdot, \cdot \rangle_k^0$ together with the Cauchy–Schwarz inequality (see Exercise 4.3.2–6), $\forall x \in X$ we get

$$|u(x)|^2 = |\langle u, k_x \rangle_k^0|^2 \leq k(x, x)|\langle u, u \rangle| = 0 \Rightarrow u = 0.$$

2. Given $x_1 \neq 0$ and $x_2 \neq 0$ in X, consider the functions $f_1(x) = (x_1 \cdot x)^2$ and $f_2(x) = (x_2 \cdot x)^2$. We can construct an orthonormal set by using Gram–Schmidt orthonormalization method. We have $h_1 = f_1$, $u_1 = h_1/\|h_1\|$, $h_2 = f_2 - \langle f_2, u_1 \rangle_k^0 u_1$ and $u_2 = h_2/\|h_2\|$. Therefore

$$u_1 = \frac{(x_1 x)^2}{x_1 \cdot x_1}, \quad u_2 = \frac{(x_1 \cdot x_1)^2 (x_2 \cdot x)^2 - (x_1 \cdot x_2)^2 (x_1 \cdot x)^2}{(x_1 \cdot x_1)^2 (x_2 \cdot x_2)^2 - (x_1 \cdot x_2)^4},$$

are orthonormal.

4. (P1) Using Cauchy–Schwarz inequality, we have $|\delta_x(f) - \delta_x(g)| = |\langle f - g, k_x \rangle_k^0| \leq \sqrt{k(x, x)}\|f - g\|_k^0$.

(P2) Take a Cauchy sequence $\langle f_n \rangle$ that converges pointwise to 0; since $\langle f_n \rangle$ is a Cauchy sequence, we can always find a constant C such that $\|f_n\|_k^0 < C$ for all $n \in \mathbb{N}$. Then for all $\epsilon > 0$ we can always find an N_1 so that for all $n, m > N_1$ $\|f_n - f_m\|_k^0 < \epsilon/2C$; let $f_{N_1}(x) = \sum_{i=1}^r \alpha_i k(x, x_i)$. Choose N_2 so that for all $i = 1, \ldots, r$ and all $n \geq N_2$ we have $|f_n(x_i)| < \epsilon/(2r|\alpha_i|)$. So for all $n \geq \max\{N_1, N_2\}$ we conclude that $\|f_n\|_k^0 \leq |\langle f_n - f_{N_1}, f_n \rangle_k^0| + |\langle f_{N_1}, f_n \rangle_k^0| \leq \|f_n - f_{N_1}\|_k^0\|f_n\|_k^0 + \sum_{i=1}^r |\alpha_i f_n(x_i)| < \epsilon$.

5. Consider a Cauchy sequence $\langle f_n \rangle$ in H_k; the evaluation functional is continuous also in H_k thus the sequence $\langle f_n(x) \rangle$ is convergent to a point $f(x) = \lim_{n \to +\infty} f_n(x)$ for all $x \in X$. We ask ourselves whether the function $f(x)$ is still in H_k. To see this, consider, for every n, an approximating function $g_n \in H_k^0$ such that $\|f_n - g_n\|_k < 1/n$; this can always be done since H_k^0 is dense in H_k. This approximating sequence converges pointwise to f since $|g_n(x) - f(x)| \leq |g_n(x) - f_n(x)| + |f_n(x) - f(x)| = |\delta_x(g_n - f_n)| + |f_n(x) - f(x)|$, where the first term goes to 0 because of the continuity of δ_x on H_k. Moreover, $\langle g_n \rangle$ is a Cauchy sequence: $\|g_m - g_n\|_k^0 = \|g_m - g_n\|_k \leq \|g_m - f_m\|_k + \|f_m - f_n\|_k + \|f_n - g_n\|_k \leq 1/m + 1/n + \|f_m - f_n\|_k$. The last thing that we have to show is that f defined as the limiting function of $\langle g_n \rangle$ is a limiting value also in the $\|\cdot\|_k$ norm: This is the case thanks to the denseness of H_k^0 in H_k. At this point we can conclude that f_n converges to f in the $\|\cdot\|_k$ norm: $\|f_n - f\|_k \leq \|f_n - g_n\|_k + \|g_n - f\|_k \leq 1/n + \|g_n - f\|_k$.

Section 4.3.4

3. We restrict the proof to the case of $X = \mathbb{R}$. We start noticing that $B_0(u) := [|u| \leq 1/2]$ is a kernel (see Exercise 4.3.2–10). The Fourier transform of $B_n(\cdot)$ is

$$\hat{B}_n(\omega) = \mathcal{F}\left(\bigotimes_{i=1}^{n}[|u| \le 1/2]\right) = \prod_{i=1}^{n}\mathcal{F}([|u| \le 1/2]) = \mathrm{sinc}^{n+1}(\omega/2),$$

where $\mathrm{sinc}(\alpha) := \sin\alpha/\alpha$. Now if we restrict to odd values of n then we conclude that $\hat{B}_n(\omega) \ge 0$, since $n+1$ is even, and therefore $\mathrm{sinc}^{n+1}(\omega/2) \ge 0$. From $\hat{B}_n(\omega) \ge 0$ we conclude that $B_n(\|x - z\|)$ is a kernel (see Exercise 2).

5. Since k_1 and k_2 are kernels, we have $k_1(x, z) = \langle\phi_1(x), \phi_1(z)\rangle$ and $k_2(x, z) = \langle\phi_2(x), \phi_2(z)\rangle$.

- $k(x, z) = k_1(x, z) + k_2(x, z)$. Since k_1 and k_2 are kernels, $\forall \ell \in \mathbb{N}$, $\forall u \in \mathbb{R}^\ell$, $u \cdot K_1 u \ge 0$ $u \cdot K_2 u \ge 0$, and then we have $u \cdot Ku = u \cdot (K_1 + K_2)u = u \cdot K_1 u + u \cdot K_2 u \ge 0$, from which we conclude that k is a kernel.
- $k(x, z) = \alpha k_1(x, z)$. Using a similar argument, $u \cdot k(x, z)u = u \cdot (\alpha k_1)u = \alpha u \cdot k_1 u$.
- $k(x, z) = k_1(x, z)k_2(x, z)$. The Hadamard product of two non-negative matrices is also non-negative.

Section 4.4

1. According to what is stated in the solution of Exercise 4.3.2–11, we can always redefine the kernel $\hat{k}(x, x_\kappa) = 1 + k(x, x_\kappa)$, so as to use directly the definition of $\|f\|^2$ given in Eq. 4.4.1–(2). In this case, we have $f_{\hat{k}} = f_k + \sum_{\kappa=1}^{\ell}\sum_{h=1}^{\ell}\alpha_h\alpha_\kappa$.

3. From Euler–Lagrange equations, we get that the stationary points must satisfy the differential equation

$$2f(x)f'(x)(1 - f'(x)) - f^2(x)f''(x) + f(x)(1 - f'(x))^2 = 0.$$

Here, we can promptly see that $\forall\alpha \in (-1, +1)$ the family of functions $\{f_\alpha : f_\alpha(x) := [x \ge \alpha]x\}$ satisfies the Euler–Lagrange equations in $(-1, 0) \cup (0, 1)$. However, the only one which also meets the boundary condition is clearly $f(x) = [x \ge 0]x$. Now we want to see whether there are any solution that never cross the zero level, that is, such that $\forall x \in [-1, +1]$ $(x) \ne 0$. The differential equation reduces to $f(x)f''(x) + (f')^2 - 1 = 0$. The general solution is $f(x) = \pm\sqrt{2c_2x + c_2^2 - c_1 + x^2}$. When imposing the boundary conditions, we find $\bar{f}(x) = \sqrt{(1/2)x - 1/2 + x^2}$, this function cannot be defined in $[-1, +1]$, since $(1/2x - 1/2 + x^2 < 0$ for $x \in [-1, 1/2)$. Hence we conclude that there is no solution in $C^1([-1, 1]; \mathbb{R})$. This minimization problem sheds some light on machine learning tasks. Suppose that x is a temporal variable. The minimization of functional E consists of discovering a signal f, with small "energy" $f^2(x)$ and a value of $f'(x)$ which is as close as possible to 1. Interestingly, the discovery of the minimum $f(x) = [x \ge 0]x$ arises when noticing that the positive function E, which is positive, vanishes for $f(x) = 0$ or $f(x) = x$.

4. We consider the case $\sigma = 1$. The derivative of the Gaussian is interwound with Hermite polynomials, since we have

$$H_n(x) = (-1)^n e^{x^2} \frac{d^n}{dx^n} e^{-x^2}.$$

Since the Hermite numbers are $H_n = H_n(0) = \left(1 - [n \mod 2 = 0]\right)(-1)^{n/2} n! / ((1/2n))$, we have

$$\lim_{x \to 0} Lg = \sum_{\kappa=0}^{\infty} (-1)^{\kappa} \frac{1}{\kappa! 2^{\kappa}} \frac{d^{2\kappa} g}{d^{2\kappa}}\bigg|_{x=0} = \sum_{\kappa=0}^{\infty} \frac{(-1)^{\kappa}}{\kappa! 2^{\kappa}} H_{2\kappa}(0)$$

$$= \sum_{\kappa=0}^{\infty} \frac{(-1)^{\kappa}}{\kappa! 2^{\kappa}} \frac{(-1)^{\kappa}(2\kappa)!}{\kappa!} = \sum_{\kappa=0}^{\infty} \frac{(2\kappa)!}{(\kappa!)^2 2^{\kappa}} = \sum_{\kappa=0}^{\infty} \frac{(2\kappa)^{\kappa}}{\kappa! 2^{\kappa}}$$

$$= \sum_{\kappa=0}^{\infty} \frac{2\kappa(2\kappa - 1)\cdots\kappa}{2\kappa(2\kappa - 2)\cdots 2} = \infty.$$

This is consistent with the fact that g is the Green function of L.

6. From Appendix B we can easily see that $P^* = P = D^2$, and therefore $L = D^4$. Finally, we have $D^4 |x|^3 = \delta(x)$.

Section 5.1
Section 5.1.1

1. The contribution of the two equal neurons in the hidden layer can always be expressed by $w_{o,1}\sigma(wx) + w_{o,2}\sigma(wx) = (w_{o,1} + w_{o,2})\sigma(wx) = w_{o,e}\sigma(wx)$. This identity tells us that the two hidden units 1 and 2 can be replaced with e, where the weight $w_{o,2}$ connecting the output to the *equivalent neuron* is determined by $w_{o,e} = w_{o,1} + w_{o,2}$.

2. Since W_1 and W_2 are simultaneously diagonalizable, there exists P such that $W_1 = P\,\mathrm{diag}(\omega_{1,i})P^{-1}$ and $W_2 = P\,\mathrm{diag}(\omega_{2,i})P^{-1}$. Hence $y = W_2 W_1 x = P\,\mathrm{diag}(\omega_{2,i})P^{-1}P\,\mathrm{diag}(\omega_{1,i})P^{-1} = P\,\mathrm{diag}(\omega_{2,i})\,\mathrm{diag}(\omega_{1,i})P^{-1} = P\,\mathrm{diag}(\omega_{2,i}\omega_{1,i})P^{-1}$. Clearly, W_1 and W_1 are commutative since $\mathrm{diag}(\omega_{2,i}\omega_{1,i}) = \mathrm{diag}(\omega_{1,i}\omega_{2,i})$. Since W_1 and W_2 are invertible, we have $W_2^{-1}y = W_1 x$ and $W_1^{-1}y = W_2 x$. Finally, when summing them up, we get $y = (W_1^{-1} + W_2^{-1})^{-1}(W_1 + W_2)x$.

Section 5.1.2

1. When we rewrite the cascade as

$$y(x) = \ln\left(1 + \beta\exp\left(1 - \ln\left(1 + \beta\exp\left(1 - x\right)\right)\right)\right)$$
$$= \ln\left(1 + \beta\exp\left(\ln e - \ln\left(1 + \beta\exp\left(1 - x\right)\right)\right)\right)$$

$$= \ln\left(1 + \beta \exp\left(\ln\frac{e}{1 + \beta \exp(1 - x)}\right)\right) = \ln\left(1 + \beta \frac{e}{1 + \beta \exp(1 - x)}\right)$$
$$= \ln\frac{1 + \beta e + \beta \exp(1 - x)}{1 + \beta \exp(1 - x)},$$

we have $\lim_{x \to -\infty} y(x) = 0$ and $\lim_{x \to +\infty} y(x) = \ln(1 + \beta e)$. Now if $\beta = (e - 1)/e$ we get $\lim_{x \to +\infty} y(x) = \ln(1 + \beta e) = 1$. Hence the asymptotic behavior of $y(\cdot)$ is the same as that of the sigmoid. For $x = 0$ we get $y(x) = \ln(2 - (1/e)) \simeq 0.49$, which is only close to 0.5 of the squash function. Now for small x, we have

$$y(x) = \ln\frac{1 + \beta e + \beta \exp(1 - x)}{1 + \beta \exp(1 - x)} = \ln\left(1 + \frac{\beta e}{1 + \beta \exp(1 - x)}\right)$$
$$\simeq \frac{\beta e}{1 + \beta \exp(1 - x)} = \frac{e - 1}{1 + \frac{e}{e-1}\exp(1 - x)}.$$

Its derivative in $x = 0$ is $y'(0) = (e^2(e - 1)^2)/((e^2 + e - 1)^2) \simeq 0.263$, that is quite close to 0.25, that is, the value of the derivative of the sigmoid for $x = 0$. This value, as well as $y(0)$, indicates a lack of symmetry with respect to the case of the sigmoid. What is the reason of this symmetry breaking?

3. The proof that there is no collapse in the cascade has already been given in Section 5.1.2. Let us consider its interpretation in $\mathbb{C}$. It has been proven that the equation $w_2 \cdot e^{w_1 x} = w_3 x$ expresses the lack of collapse of the exponential function. This equation can be rewritten as $w_2 \cdot (\cos(w_1 x) + i \sin(w_1)x) = w_3 x$ that, at least, requires us to impose $\cos(w_2 \cdot (\cos(w_1 x)) = \cos(w_3 x)$. This condition explores the collapse of the cascade in case of sinusoidal functions. Clearly, the condition does not hold for all $x \in \mathbb{R}$.

4. Let us consider the cascade $y = \sigma(w_2 \sigma(w_1 x + b_1) + b_2)$. We have

$$\frac{\partial}{\partial x}\sigma(w_2 \sigma(w_1 x + b_1) + b_2) = w_2 w_1 \sigma'(a_2)\sigma'(a_1),$$

which never flips sign with $x \in \mathbb{R}$. Hence we conclude on the monotonicity of the cascading structure.

5. A similar cascade can be recursively described as $y_1 := (w_1 x + b_1)^2 + c_1$ and $y_{n+1} = (w_{n+1} y_n + b_n)^2 + c_n$. We have $y_1 = w_1^2 x^2 + 2w_1 b_1 x + (b_1^2 + c_1)$, from which we can promptly see that any quadratic polynomial with $a_2 \geq 0$ can be realized. Let $p_2(x) := a_2 x^2 + a_1 x + a_0$ be the given polynomial. Then we need to solve $w_1^2 = a_2$, $2w_1 b_1 = a_2$, and $b_1^2 + c_1 = a_2$, which yields $w_1 = \pm\sqrt{a_2}$, $b_1 = 1/2$, and $c_1 = a_2 - 1/4$. Clearly, there is no cascade of units that can realize a polynomial with $a_2 < 0$. Notice that, as we increase the degree of the polynomial to be realized, we see that the computational capabilities decrease. Given any polynomial $p_n(x) = \sum_{\kappa=0}^{n} a_\kappa x^\kappa$, we have $y_{n+1}(x) = w_{n+1}^2 y_n^2(x) + 2w_{n+1} b_n y_n(x) + (b_n^2 + c_n)$. When moving from n to $n + 1$, the number of coefficients of the cascade increases by 3, while the corresponding degree doubles! Hence, we conclude that the space of polynomials that can be realized by this quadratic cascading is pretty limited.

Section 5.1.3

1. We know that Eq. (1) converges to any of the solutions of $(W - I)\hat{x} = 0$. Now we have $\hat{x}_1 = \alpha u, \hat{x}_2 = \alpha W u, \ldots, \hat{x}_{t+1} = \alpha \hat{W}^t u$. Hence, in case of convergence, we have $\hat{x}^* = \lim_{t \to \infty} \hat{x}_t \propto u$.

Section 5.1.4

1. In order to assess the difficulty of learning we discuss the separation surfaces when using the two different encodings of the outputs. When we use one-hot encoding, for each class, the determination of separation surface is reduced to discovering the region defined by (5). Hence, the aim of learning is that of constructing the separation surfaces which introduce a degree of complexity that is induced by the given single classes. Let $x_{1,o}$ and $x_{2,o}$ be the two outputs of the network. When Boolean coding is adopted, the separation surfaces are defined by

$$\forall x \in C_3 \bigcup C_4 : x_{1,o} = f_1(w, x) > \delta,$$
$$\forall x \in C_2 \bigcup C_4 : x_{2,o} = f_2(w, x) > \delta.$$

Since the separation surfaces involve the union of sets, the degree of complexity clearly increases. We can easily see that as the number of classes increases, this issue becomes more and more relevant.

Section 5.2
Section 5.2.1

1. We can rewrite Eq. (1).

$$\begin{array}{c} a \\ b \\ c \\ d \end{array} \begin{pmatrix} 0 & 0 & 1 \\ 1 & 0 & 1 \\ -1 & -1 & -1 \\ 0 & 1 & 1 \end{pmatrix} \begin{pmatrix} w_1 \\ w_2 \\ b \end{pmatrix} < 0.$$

Let $\hat{w}, \tilde{w} \in \mathcal{W}_\wedge = \{w : Xw < 0\}$ belong to the space of solutions and let $\alpha \in [0, 1]$. Then we have $X\hat{w} < 0$ and $X\tilde{w} < 0$. (Notice that $X \neq \hat{X}$, since the signs of the positive examples are properly set in such a way to write the above inequality.) Now consider $w = \alpha \hat{w} + (1 - \alpha)\tilde{w}$, so Xw can be written as $Xw = X(\alpha \hat{w} + (1 - \alpha)\tilde{w}) = \alpha X\hat{w} + (1 - \alpha)X\tilde{w} < 0$, from which we conclude that $\mathcal{W}_\wedge$ is clearly a convex set.

2. The OR function is linearly separable. We can follow the analysis of Exercise 1.

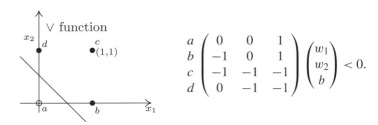

Clearly, $w_1 = w_2 = 1$, $b = -\frac{1}{2}$ is a solution, and the proof on $\mathcal{W}_\vee$ convexity follows exactly the same arguments as for $\mathcal{W}_\wedge$.

8. We can replace the Heaviside function with the sign function as follows: $H(a) = (1 + \text{sign}(a))/2$. Suppose we have $a_2 = H(wx_1 + b)$. It can be replaced with

$$a_2 = \frac{1 + \text{sign}(wx_1 + b)}{2} = \frac{1}{2} + \text{sign}\left(\frac{w}{2}x_1 + \frac{b}{2}\right) = \text{sign}\left(\frac{w}{2}x_1 + \frac{1+b}{2}\right).$$

Hence the weights of the equivalent sign net are $w_s = \frac{1}{2}$ and $b_s = (1 + b)/2$. The Heaviside net (left) and the associated sign net are in the figure below.

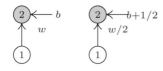

Section 5.2.3

2. 1. The location of region X is a consequence of the definition of separation surfaces and of the given condition $f(w, x) \geq 0$. We start noting that the configuration $w_{6,3} = w_{6,4} = w_{6,5} = 1$ and $b_6 = -5/2$ assigns the $\wedge$ function to neuron 6. Hence the delimitation of X comes from the signs of w_3, w_4, and w_5.

2. The number of lines in the plane is given by $L_3 = 1 + 3(3 + 1)/2 = 7$.

3. We start noticing that the perfect realization of the LTU of Fig. 5.7 clearly results in a failure to construct the parity predicate as soon as we slightly change the inputs. For instance, units 5 and 6 are jointly fired only if $\sum_{\kappa=1}^{4} x_\kappa = 1$, which leads to conclude that only the perfect coding of the inputs into real numbers returns a correct solution. (Of course, this is impossible in practice!) This has a side effect on the space of the solution weights, which has null measure in $\mathbb{R}^d$. In order to confer robustness to the solution, w.l.o.g. let us focus on units 5 and 6. Let $\delta \in (0, 1/2)$. If we replace $x_5 = H\left(\sum_{\kappa=1}^{4} x_\kappa - 1\right) \rightarrow H\left(\sum_{\kappa=1}^{4} x_\kappa - (1 - \delta)\right)$, $x_6 = H\left(1 - \sum_{\kappa=1}^{4} x_\kappa\right) \rightarrow H\left(1 - \delta - \sum_{\kappa=1}^{4} x_\kappa\right)$, then the network still returns the same solution. This solution faces the robustness on the input, and the weight space is no longer of null measure in $\mathbb{R}$. The answer to the question on the convexity is left to the reader.

Section 5.3
Section 5.3.2

1. We assume that the domain X is partitioned into $n = \lceil \mathcal{L}^d(X)/p^d \rceil$ hypercubes. Since f satisfies Lipschitz condition, if we consider the middle point x_m of a generic box, we have $\| f(x) - f(x_m)\| < K\|x - x_m\|$ where K is the Lipschitz constant. Let $\epsilon > 0$ be the accuracy that we want to achieve, that is, the brute force approximation scheme must satisfy the condition $\| f(x) - f(x_m)\| < \epsilon$ for all hypercubes with edge p. If we impose $\| f(x) - f(x_m)\| < K\|x - x_m\| \leq Kp\sqrt{d} = K\sqrt{d}\left(\mathcal{L}^d(X)/n\right)^{1/d} < \epsilon$ then we get

$$n > \mathcal{L}^d(X)\left(\frac{K\sqrt{d}}{\epsilon}\right)^d.$$

Hence, the number of nodes (hypercubes) grows exponentially with the dimension of X, which corresponds with stating that the accuracy is $O(1/n)^d$.

2. We prove that $f^*(x_k^m) = f(x_k^m)$. For $n = 1$ we can promptly see that $f^*(x_1^m) = \left(f(x_1^m) - f(x_0^m)\right)H(x_1^m - x_1) = f(x_1^m)$. Now suppose by induction that $f^*(x_{k-1}^m) = \sum_{i=1}^{k-1}(f(x_i^m) - f(x_{i-1}^m))H(x_k^m - x_i)$. Then we have

$$f^*(x_k^m) = \sum_{i=1}^{k}(f(x_i^m) - f(x_{i-1}^m))H(x_k^m - x_i)$$

$$= f(x_k^m) - f(x_{k-1}^m))H(x_k^m - x_k) + \underbrace{\sum_{i=1}^{k-1}(f(x_i^m) - f(x_{i-1}^m))H(x_k^m - x_i)}_{f^*(x_{k-1}^m)}$$

$$= f(x_k^m) - f^*(x_{k-1}^m) + f^*(x_{k-1}^m) = f(x_k^m).$$

The accuracy analysis is left to the student.

Section 5.3.3

1. We give an insight on the proof and let the reader provide a formal proof. Suppose we are using $\sigma(a) = e^{-a^2/2}$. Furthermore, let us assume that we use two different hidden units, so the corresponding lines are not parallel. In that case, any direction, which is not parallel to one of the separating lines, yields an activation $a_i = w_{i,1}x_1 + w_{i,2}x_2 + b_i$ which grows up with no upper bound. Hence the corresponding output $\sigma(a)$ soon vanishes, thus defining a bounded domain.

3. If we set $h = p \cdot q$, then we are reduced to proving that

$$(q!)^p \leq (pq)!.$$

The property is trivially verified for $p = 1$. Moreover, we easily prove that it also holds for any choice of q and $p = 2$. In this case the inequality becomes $(q!)^2 \le (2q)!$. This can promptly be verified when noticing that[2]

$$\underbrace{q^2 \cdot (q-1)^2 \cdots 2^2 \cdot 1^2}_{(q!)^2} \le \underbrace{\overbrace{2q \cdot (2q-1) \cdots (q+1)}^{>q!} \cdots \overbrace{q \cdots 2 \cdot 1}^{q!}}_{(2q)!}$$

Now, in order to prove the property for any q and $p > 2$, we use the inequality below [from page 52 of The Art of Computer Programming vol. 1 (Addison-Wesley)]

$$\forall q \ge 1, \quad \frac{q^q}{e^{q-1}} \le q! \le \frac{q^{q+1}}{e^{q-1}}$$

and get

$$(q!)^p \le \frac{q^{pq+p}}{e^{pq-p}} \le \frac{q^p}{e^{1-p}p^{pq}} \frac{(pq)^{pq}}{e^{pq}-1} = \alpha(p,q) \frac{(pq)^{pq}}{e^{pq}-1}, \tag{$*$}$$

where $\alpha(p,q) := q^p/(e^{1-p}p^{pq})$. Now we prove that $\alpha < 1$ for $p > 2$. Let $Q := [3,\infty) \times [1,\infty)$ and consider the extension of α

$$\varphi(x,y) := y^x/(e^{1-x}x^{xy}) = \exp(x \ln y - 1 + x - xy \ln x) = e^{\psi(x,y)}$$

where $\varphi: Q \to \mathbb{R}$ and $\psi: Q \to \mathbb{R}$, $\psi(x,y) := x \ln y - 1 + x - xy \ln x$. We have $\partial_x \psi = \ln y + 1 - y \ln x - y$ and $\partial_y \psi = x/y - x \ln x$. Now for $x \ge 3$, $\partial_x \psi = \ln y + 1 - y \ln x - y \le \ln y + 1 - y \le 0$ and $\partial_y \psi = x/y - x \ln x \le x - x \ln x < 0$ For $(x,y) = (3,1)$ we have $\psi(3,1) = -1 + 3 - 3\ln 3 < 0$, and therefore $\phi(3,1) < 1$. Since $\forall (x,y) \in Q$, $\partial_x \varphi < 0$, $\partial_y \varphi < 0$ and $\varphi(3,1) < 1$, we conclude that $\forall (x,y) \in Q \varphi(x,y) < 1$. When restricting to α, from $(*)$ we get

$$(q!)^p \le \frac{q^{pq+p}}{e^{pq-p}} \le \frac{q^p}{e^{1-p}p^{pq}} \frac{(pq)^{pq}}{e^{pq}-1} = \alpha(p,q)\frac{(pq)^{pq}}{e^{pq}-1} \le \frac{(pq)^{pq}}{e^{pq}-1} \le (pq)!$$

which completes the proof.

Section 5.4
Section 5.4.1

1. We give a proof for $\Omega = \mathbb{R}$. We have $(u * v)(t) = \int_{-\infty}^{+\infty} u(t-\tau)v(\tau)d\tau$. Let $\alpha = t - \tau$ be. Then

---

[2] The reader is invited to prove the property by induction on p by considering any value of q.

$$(u * v)(t) = \int_{-\infty}^{+\infty} u(t - \tau)v(\tau)d\tau = -\int_{+\infty}^{-\infty} v(t - \alpha)u(\alpha)d\alpha = (v * u)(t).$$

An elegant proof, which is general, relies on the isomorphism between $*$ and $\cdot$, which is induced by the Fourier transform. We have $\widehat{(u * v)} = \hat{u} \cdot \hat{v} = \hat{v} \cdot \hat{u} = \widehat{(v * u)}$.

2. If $\mathcal{D}_\Omega = [1, \ldots, n_i] \times [1, \ldots, n_j]$ then for all $(m, n) \in \mathcal{D}_\Omega$ we define $y_{m,n} = \sum_{h=1}^{n_i} \sum_{\kappa=1}^{n_j} h_{m-h,n-\kappa} v_{h,\kappa}$. This is motivated by the restriction on the definition of $v_{h,\kappa}$ from $\mathbb{N}^2$ to $\mathcal{D}_\Omega$. Notice that $y_{m,n} = \sum_{h=1}^{n_i} \sum_{\kappa=1}^{n_j} h_{m-h,n-\kappa} v_{h,\kappa} = \sum_{h=1}^{\infty} \sum_{\kappa=1}^{\infty} h_{m-h,n-\kappa} v_{h,\kappa}^\infty$, where $v_{h,\kappa}^\infty = v_{h,\kappa}$ for $(h, \kappa) \in \mathcal{D}_\Omega$ and $v_{h,\kappa}^\infty = 0$ otherwise.

In order to establish the relation between the continuous and discrete setting of computation, we simply notice that the sampling of Ω to $[\Omega]$ yields $z = \Delta \odot (m, n)$ and $u = \Delta \odot (h, \kappa)$, where $\Delta := (\Delta_i, \Delta_j)$ and $\text{side}(\Omega) = (n_i \Delta_i, n_j \Delta_j)'$. Hence we have

$$y(z) = \int_\Omega h(z - u)v(u)du$$

$$\simeq \Delta_i \Delta_j \sum_{i_u=1}^{n_i} \sum_{j_u=1}^{n_j} \underbrace{h((m - h)\Delta_i, (n - \kappa)\Delta_j)}_{h_{m-h,n-\kappa}} \underbrace{v(h\Delta_j, \kappa\Delta_j)}_{v_{h,\kappa}}$$

$$\propto \sum_{i_u=1}^{n_i} \sum_{j_u=1}^{n_j} h_{m-h,n-\kappa} v_{h,\kappa}$$

4. We have $u * v|_{[-1..1]}(z) = \int_{-1}^{+1}(z - \alpha)(1 - \alpha)d\alpha = 2z + 2/3$. Likewise, we have $v * u|_{[-1..1]}(z) = \int_{-1}^{+1}(z - 1 + \alpha)\alpha d\alpha = 2/3$. Hence $u * v|_{[-1..1]}(z) \neq v * u|_{[-1..1]}(z)$.

Section 5.5
Section 5.5.2

3. The proof of Eq. (2) can be given by using Taylor expansion with Lagrange remainder. There exist $a_1 \in (a, a + h)$ and $a_2 \in (a - h, a)$ such that

$$\sigma(a + h) = \sigma(a) + h\sigma^{(1)}(a) + \frac{h^2}{2}\sigma^{(2)}(a) + \frac{h^3}{3!}\sigma^{(3)}(a_1),$$

$$\sigma(a - h) = \sigma(a) - h\sigma^{(1)}(a) + \frac{h^2}{2}\sigma^{(2)}(a) - \frac{h^3}{3!}\sigma^{(3)}(a_2).$$

Hence we have

$$\frac{\sigma(a + h) - \sigma(a - h)}{2h} = \sigma^{(1)}(a) + \frac{h^2}{6}\frac{\sigma^{(3)}(a_1) + \sigma^{(3)}(a_2)}{2}.$$

Under the condition of continuity of $\sigma^{(3)}$, there exists $\tilde{a} \in [a - h, a + h]$ such that $(\sigma^{(3)}(a_1) + \sigma^{(3)}(a_2))/2 = \sigma^{(3)}(\tilde{a})$, and therefore we end up with Eq. (2). The second part of the exercise if left to the reader, who is expected to prove that the asymmetric approximation of the derivative yields an error which is $O(h)$, which is remarkable higher than the one obtained by symmetric approximation!

7. We discuss the cases of $\sigma = \tanh$, $\sigma = $ logistic, and $\sigma = (\cdot)_+$.

(*i*) If $\sigma = \tanh$ then we can promptly see that all neurons return the null output, that is, $\forall i \in \mathbb{N}$, $x_i = 0$. From (5) we promptly see that $g_{\kappa i j} = 0$, where we replaced $g_{\kappa i j}^o = 0$ with $g_{\kappa i j} = 0$, since there is only one output.

(*ii*) If $\sigma = $ logistic then $x_{\kappa j} = 1/2$. Let us consider the weights of the output layer. Again, since we only have one output unit, we can replace $\delta_{\kappa i}^o$ with $\delta_{\kappa i}$. We have $\delta_{\kappa o} = \sigma'(1/2)(1/2 - y_\kappa) = 1/4(1/2 - y_\kappa)$. Hence if $y_\kappa = 0$ then $\delta_{\kappa o} = 1/8$ else $\delta_{\kappa o} = -1/8$. Finally, $g_{\kappa o i} = \delta_{\kappa o} x_{\kappa i} = 1/16 \operatorname{sign}(1/2 - y_\kappa)$. For all the other layers, $g_{\kappa i j} = 0$ since, from (7), we immediately see that $\delta_{\kappa i j} = 0$.

(*iii*) If $\sigma = (\cdot)_+$, we can draw the same conclusion as for (*i*).

Section 5.5.3

4. The stationarity conditions are

$$\frac{\partial E}{\partial w_1} = \delta_b + \delta_c = 0,$$

$$\frac{\partial E}{\partial w_2} = \delta_d + \delta_c = 0,$$

$$\frac{\partial E}{\partial b} = \delta_a + \delta_b + \delta_c + \delta_d = 0$$

Now we can promptly see that $w_1 = w_2 = 1$ and $b = -1$ is a stationary configuration. This can be seen as follows. Consider any line defined by $w_1 = w_2 = 1$. Clearly, because of the symmetry, we have $\delta_b = \delta_d$. From the stationarity conditions, we promptly get that also $\delta_a = \delta_c$ holds true. This condition, paired with $\delta_b = \delta_d$, characterizes the stationary configuration $w_1 = w_2 = 1$ and $b = -1$. Is it a minimum, a saddle point, or a maximum? What about the configuration $w_1 = -w_2$ and $b = 0$? Are there any other stationary points?

5. Let the output of the neuron be expressed by $y = \sigma(w_1 x_1 + w_2 x_2 + b)$. The perfect loading of the training set requires the solution of $\sigma(b) = \underline{d}$, and $\sigma(w_1 + b) = \overline{d}$. Clearly, the loading problem is independent of w_2. Moreover, we need to choose $b < 0$, $w_1 > 0$, and $|w_1| > |b|$. The choice $w_2 = 0$ respects the parsimony principle and gives rise to the separating line $w_1 x_1 + b = 0$. Now the MMP (maximum margin problem) is clearly solved by the separating line $2x_1 - 1 = 0$, which corresponds with the 1D infinite space $\mathcal{W} = \{\alpha \in \mathbb{R} : (w_1, b) = \alpha(2, -1)\}$. This yields $\underline{d} = \sigma(-\beta)$ and $\overline{d} = \sigma(\beta)$ with $\beta \in \mathbb{R}^+$. As a consequence, for the MMP to be solved, the targets must be chosen according to that symmetric structure with respect to the threshold 0.5.

Section 5.7
Section 5.7.1

2. Suppose we change $w_{3,1}$ and $w_{3,2}$ so as to get the rotation of the separating line w, associated with unit 3, indicated below.

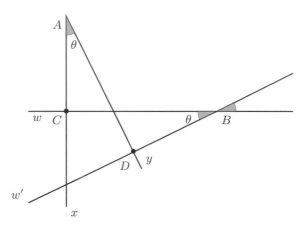

We want to see the change of the activation a_3 corresponding to this change. The activation is proportional to the distance of points to w. Let us consider the point A. When rotating w by the angle θ, the distance moves from $\overline{AC}$ to $\overline{AD}$, so the variation is

$$\overline{AD} - \overline{AC} = \frac{\overline{AC} + \overline{BC}\tan\theta}{\cos\theta} - \overline{AC} = \underbrace{\overline{AC}(\frac{1}{\cos\theta} - 1)}_{u(\theta)} + \underbrace{\overline{BC}\sin\theta}_{v(\theta)}.$$

Now the variation $\overline{AD} - \overline{AC}$ depends on the two terms $u(\theta)$ and $v(\theta)$. The proof is completed when noting that

$$\lim_{\theta \to 0} \frac{u(\theta)}{v(\theta)} = 2\frac{\overline{AC}}{\overline{BC}} \cdot \lim_{\theta \to 0} \frac{1 - \cos\theta}{\sin 2\theta} = 0.$$

4. Suppose we have an OHL net with multiple outputs, where each output is connected to its own hidden units. This corresponds with c separated networks with one output unit, for which we know that the error function is local minima free. This simplified hypothesis yields an error function which is the sum over the functions associated with the single local minima free networks. Now, as we can promptly see for the analysis of the gradient and of the Hessian function, the sum of local minima free error functions is a local minima free error function.

6. Let us consider the rectifier paired with the quadratic function (similar conclusions can be driven when pairing it with different loss functions). Suppose the following condition holds true: $(a_{\kappa i})'_{+}((a_{\kappa i})_{+} - d_{\kappa}) = 0$. If $a_{\kappa i} > 0$ then we get

$(a_{\kappa i})_+ = d_\kappa$ (perfect match). If $a_{\kappa i} \leq 0$ then the above condition does not impose any constraint on the activation of the neuron. This is remarkably different with respect to the logistic (and hyperbolic tangent) transfer functions, since in that case the condition $\Delta = 0$ characterizes stationary points such that the only stable is the global minimum. Although there is a small movement in saturated configurations, there is still the heuristic power to escape from saddles and maxima. This is an important remark, since it indicates that some initializations of the weights might stuck the learning process also for linearly separable examples. This clearly happens also in the case of a single neuron! This property indicates that there is an important difference between ridge neurons characterized by sigmoidal functions and rectifiers, which reflect their different possibilities of recovering from erroneous initializations.

7. The gradient with respect to the bias for the output units is

$$\frac{\partial E}{\partial b_o} = 1'\Delta_2 = \sum_{k=1}^{\ell} \delta_{o,\kappa} = \frac{1}{4}\left(\sum_{\kappa \in P} 0 - \overline{d} + \sum_{\kappa \in N} 0 - \underline{d}\right) = -\overline{d}|P| - \underline{d}|N|.$$

Hence $\frac{\partial E}{\partial b_o} = 0$ iff we are dealing with balanced training sets, where $|N| = |P|$ and $\underline{d} = -\overline{d}$. If this is not the case then the output bias term contributes to the gradient. Now all the neural outputs are null, apart from the output that, however, is independent of the input. Let x_o be the common output that is reached in the stationary configuration. Then we have $(x_o - \overline{d})|P| + (x_o - \underline{d})|N| = 0$, that is,

$$x_o = -\overline{d}|P| + \underline{d}|N||P| + |N|.$$

8. Eq. (5), along with the corresponding analysis for the case of a single neuron, suffices to draw the conclusion.

Section 5.7.2

3. We distinguish the case of uniform and Gaussian distribution.

(*i*) Uniform distribution: We calculate σ_W^2 directly from its definition $\sigma_W^2 = \int_{-w}^{+w} \omega^2 d\omega = (2/3)w^3$.

(*ii*) Normal distribution: Let $w = \beta\sigma_W$ and $\gamma > 4$. Then a Gaussian distribution can be generated coherently with the normalization condition

$$(1/(\sqrt{2\pi}))\int_{-w}^{+w} e^{-\omega^2/2\sigma_W^2}d\omega = 2\,\mathrm{erf}(w) \simeq 1.$$

Section 6.1

Section 6.1.1

1. Instead of directly solving the problem, an abstraction interpretation based on vectorial formalism dramatically simplifies the notation and solution. We can restate linear consistency by considering the three functions, $f_1(x) = \hat{W}_1'\hat{x}$, $f_2(y) = \hat{W}_2'\hat{y}$, and $f_3(x) = \hat{W}_3'\hat{x}$, where matrices $\hat{W}_i$, $i = 1, 2, 3$ match the coherence condition $\hat{W}_3 = \hat{W}_1\hat{W}_2$. The best fitting can be expressed, as discussed in Exercise 3.1–2, by minimizing $(1/2)(\|Y - \hat{X}\hat{W}_1\|_F^2 + \|Z - \hat{Y}\hat{W}_2\|_F^2 + \|Z - \hat{X}\hat{W}_3\|_F^2$ subject to the constraint on the matrices $\hat{W}_3 = \hat{W}_1\hat{W}_2$. Here $\|A\|_F^2 = \text{tr}(A'A)$ is the Frobenius norm for the real matrix A. Because of the simple nature of the constraint it can be readily incorporated into the objective function, thus reducing the problem to the unconstrained minimization of

$$E(\hat{W}_1, \hat{W}_2) = \frac{1}{2}\left(\|Y - \hat{X}\hat{W}_1\|_F^2 + \|Z - \hat{Y}\hat{W}_2\|_F^2 + \|Z - \hat{X}\hat{W}_1\hat{W}_2\|_F^2\right).$$

The gradients of this function can be evaluated using the results from Exercise 3.1–2, so that after some calculations we find that

$$\partial_{\hat{W}_1} E(\hat{W}_1, \hat{W}_2) = \hat{X}'\hat{X}\hat{W}_1 - \hat{X}'Y + (\hat{X}'\hat{X}\hat{W}_1\hat{W}_2 - \hat{X}'Z)\hat{W}_2',$$

$$\partial_{\hat{W}_2} E(\hat{W}_1, \hat{W}_2) = \hat{Y}'\hat{Y}\hat{W}_2 - \hat{Y}'Z + \hat{W}_1'(\hat{X}'\hat{X}\hat{W}_1\hat{W}_2 - \hat{X}'Z).$$

3. Finding $f_1(x_1)$ and $f_2(x_2)$ is equivalent to solving the equation $\alpha^2 - s\alpha + p = 0$, where $s = f_1(x_1) + f_2(x_2)$, $p = f_1(x_1) \cdot f_2(x_2)$, and $s = 2p$. Hence we get $\alpha = p \pm \sqrt{p^2 - p}$. Now, since $f_1(x_1), f_2(x_2) \in [0, 1]$, we get real solutions only if $p = 0$ or $p = 1$, which corresponds with $f_1(x_1) = f_2(x_2) = 0$ or $f_1(x_1) = f_2(x_2) = 1$, respectively.

4. Clearly, they both express the notion of equality when the functions take on values close to Boolean targets. Basically, whenever $f_1(x_1) \simeq 0$ and $f_2(x_2) \simeq 0$ or $f_1(x_1) \simeq 1$ and $f_2(x_2) \simeq 1$, we have $f_1(x_1) - 2f_1(x_1)f_2(x) + f_2(x_2) \simeq (f_1(x_1) - f_2(x_2))^2$. However, notice that as we depart from Boolean-like values, the behavior of the constraints is very different. As an extreme case for $f_1(x_1) = f_2(x_2) = 1/2$, the constraint of Eq. (7) is not verified, since we get $f_1(x_1) - 2f_1(x_1)f_2(x) + f_2(x_2) = 1/2$. On the contrary, the constraint $(f_1(x_1) - f_2(x_2))^2 = 0$ is clearly met. The constraint $f_1(x_1) = f_2(x_2)$ has is in fact the most general statement of equivalence, which yields for any pair of real numbers, whereas the constraint of Eq. (7) has a decision-like flavor!

Section 6.1.2

2. It suffices to notice that, given $f \in F$, considering the isoperimetric constraint $\Psi(f) = 0$, where

$$\Psi(f) := \int_X \sum_{\kappa=1}^{\ell} \delta(x - x_\kappa) V(x_\kappa, y_\kappa, f(x)) \, dx = \sum_{\kappa=1}^{\ell} V(x_\kappa, y_\kappa, f(x_\kappa))$$

was once softly enforced, corresponds with supervised learning.

3. To see this, let $(u)_+ := \max\{0, u\}$. Then the unilateral constraint $\check{\phi}_i(x, f(x)) \geq 0$ is equivalent to $(-\check{\phi}_i(x, f(x)))_+ = 0$ and also to $\left((-\check{\phi}_i(x, f(x)))_+\right)^2 = 0$. This equivalence should be treated with care when applying the classical theory of Lagrange multipliers (see Section 6.1.3), since it requires some properties that might be lost in making such a transformation. For instance, the reduction of a unilateral constraint $\check{\phi}_i(x, f(x)) \geq 0$ to the corresponding bilateral $(-\check{\phi}_i(x, f(x)))_+ = 0$ may cause the loss of the differentiability. So the direct extension of theoretical results from the case of bilateral constraints to unilateral ones is not always feasible in a plain way.

5. The overall 2-nd degree of mismatch is

$$\frac{1}{4}\left((y_1 - f_1(\overline{x}))^2 + (y_2 - f_2(\overline{x}))^2\right) + \frac{1}{2}\int_X (f_1(x)(1 - f_2(x)))^2 \, dx.$$

While the first part involves supervised pairs on the same $\overline{x}$, the second models a logic-type constraint. Clearly, their soft fulfillment requires expressing corresponding belief, since it may be qualitatively different on different points of the domain. Basically, the belief can be thought of as a weight to judge the subsequent constraint verification.

6. We have

$$\omega_\psi = -2\lambda\left(\psi_1(x, f(x))(1, 1)' + \psi_2(x, f(x))(1, 2)' + \psi_3(x, f(x))(2, 3)'\right). \quad (*)$$

For high values of λ, we have $\psi_3(x, f(x)) = \psi_1(x, f(x)) + \psi_2(x, f(x))$, and therefore we get $\omega_\psi = -2\lambda\left(\psi_1(x, f(x))(1, 1)' + \psi_2(x, f(x))(1, 2)' + (\psi_1(x, f(x)) + \psi_2(x, f(x))(2, 3)'\right) = -2\lambda\left(\psi_1(x, f(x))(3, 4)' + \psi_2(x, f(x))(3, 5)'\right)$. Notice that for high values of λ, the solution associated with $\{\psi_1, \psi_2\}$ is nearly the same as the solution associated with $\{\psi_1, \psi_2, \psi_3\}$. Even tough, the last constraint can be derived by summing up the previous two, in the last case, the reaction is shared amongst all the constraints. The way it is shared is expressed by Eq. $(*)$.

Section 6.1.3

3. Clearly, the constraints ψ and ψ_α define the same functional space, and therefore they lead to the same optimal solution of the learning problem. In particular, the Euler–Lagrange equations:

$$Lf(x) + \lambda(x)\nabla_f \psi(x, f(x)) = 0,$$
$$Lf(x) + \lambda_\alpha(x)\nabla_f(\alpha(x)\psi(x, f(x))) = 0,$$

are satisfied, from which we get $\forall x \in X \ (\lambda(x) - \alpha(x)\lambda_\alpha(x))\nabla_f \psi(x, f(x)) = 0$. Now, since $\nabla_f \psi(x, f(x)) \neq 0$, we conclude that $\lambda_\alpha(x) = \lambda(x)/\alpha(x)$.

Section 6.1.4

1. Suppose $r \in \mathbb{N}_n$ denotes the row index of the chessboard, while $x, y \in \mathbb{N}_n$ are used to locate the position of the queens (e.g., in the classic chessboard with $n = 8$, $x = 13$ corresponds with the second row and fifth column). The case in which the queen in position x belongs to row r is denoted by $x \triangleright r$. Furthermore, let us assume that $P(x) = \top$ whenever a queen is in position located at x and define $\text{IN}(x, r) \Leftrightarrow x \triangleright r$ and $\text{DIFF}(x, y) \Leftrightarrow x \neq y$. Now, $\neg \exists y \ \text{IN}(y, r) \wedge \text{DIFF}(x, y) \wedge P(y) \equiv \forall y \ \neg \text{IN}(y, r) \vee \neg \text{DIFF}(x, y) \vee \neg P(y)$, and therefore

$$\forall r \quad \exists x \ \text{IN}(x, r) \wedge P(x) \wedge [\neg \exists y \ \text{IN}(y, r) \wedge \text{DIFF}(x, y) \wedge P(y)]$$
$$\Leftrightarrow \forall r \quad \exists x \ \text{IN}(x, r) \wedge P(x) \wedge [\forall y \ \neg \text{IN}(y, r) \vee \neg \text{DIFF}(x, y) \vee \neg P(y)]$$
$$\Leftrightarrow \forall r \quad \exists x \forall y \ \text{IN}(x, r) \wedge P(x) \wedge [\neg \text{IN}(y, r) \vee \neg \text{DIFF}(x, y) \vee \neg P(y)].$$

Finally, we get the logic constraint

$$\forall r \forall y \quad \exists x \ \text{IN}(x, r) \wedge P(x) \wedge [\neg \text{IN}(y, r) \vee \neg \text{DIFF}(x, y) \vee \neg P(y)],$$

which is in CNF. In addition, all the quantifiers have been moved to the beginning of the formula, that is, it is in *prenex normal form*. Interestingly, not only any formula can be written in CNF, but its quantifiers can always be moved to the beginning. Clearly, the nonattacking condition also requires similar constraints on the columns and on the diagonal of the chessboard.

2. In order to claim equivalence, it suffices to take the nonnegative functions $\alpha_{2,1} = (f_3^2)/(f_1^2 + f_2^2)$ and $\alpha_{3,1} = (f_3/(f_1 f_2))^2$, since we immediately get $\check{\psi}_2 = \alpha_{2,1}\check{\psi}_1$ and $\check{\psi}_3 = \alpha_{3,1}\check{\psi}_1$.

3. If we replace $f_1^2 - f_2^2 = (f_1 + f_2)(f_1 - f_2)$ and $f_1 - f_2 = 1 + f_4$ in ψ, we get $f_1 + f_2 + f_1 f_4 + f_2 f_4 - f_3 = 0$, that is, $\{\psi_1, \psi_2\} \models \overline{\psi}$.

Section 6.2
Section 6.2.1

1. The argument is not valid. We use refutation trees. In the refutation tree the beginning chain from hair to ¬zebra comes from joining the premises with the negation of the conclusion after having used De Morgan theorem. We can easily see that some paths cannot be closed with falsification. For example, the paths on ¬longlegs, ¬longneck, ¬giraffe, ¬tawny, and ¬darkspots are open and they can only be closed when ungulate ∧ longlegs ∧ longneck ∧ ¬tawny ∧ darkspots is true, which is in fact informing us on the missing features to identify a giraffe. In order for the proposition to be true, we can enrich the KB so as to restrict the class of admissible animals.

Section 6.2.2

2. *Sketch:* No, they aren't. While formula (15) assumes that every person has got a father, the same doesn't hold for formula (14). Of course, we can modify (14) in such a way to state such an existential property as follows: $\forall x \forall y$ Person$(x) \wedge$ Person$(y) \wedge$ $\exists y$ FatherOf$(x, y) \Rightarrow$ Male(y). The translation of this formula into a real-valued constraint can be simplified, again, by skolemization. Alternatively, we can convert the universal and the existential classifier into the search for the inferior and superior of the function, respectively.

Section 6.2.3

1. We check T_P, T_G, and $T_Ł$ separately.

- T_P is clearly symmetric and associative. As for monotonicity we have that if $(x \leq \bar{x}) \wedge (y \leq \bar{y})$ then $T_P(x, y) = xy \leq \bar{x}\bar{y} = T_P(\bar{x}, \bar{y})$. Finally, p-norm satisfies the One-identity property, since we have $T_P(x, 1) = x \cdot 1 = x$.
- T_G is clearly symmetric and associative. As for monotonicity we have that if $(x \leq \bar{x}) \wedge (y \leq \bar{y})$ then $T_G(x, y) = \min\{x, y\} \leq \min\{\bar{x}, \bar{y}\}$. Finally, $T_G(x, 1) = x$.
- $T_Ł$ is clearly symmetric and associative. As for monotonicity we have that if $(x \leq \bar{x}) \wedge (y \leq \bar{y})$ then $x + y \leq \bar{x} + \bar{y}$ from which we immediately draw the conclusion. Finally, $T_Ł(x, 1) = \max\{1 + x - 1, 0\} = x$.

2. We start with the $T_p(x, y) = x \cdot y$. If $x \neq 0$, we have $(x \Rightarrow y) = \sup\{z : T(x, z) \leq y\} = \sup\{z : x \cdot z \leq y\} = y/x$. If $x = 0$ then $\sup\{z : x \cdot z \leq y\} = 1$. Notice that this holds true regardless of y. In particular, if $y = 0$ then $x \Rightarrow y$ returns $\neg x = 1$. If $x = 1$ and $y = 0$ then $\sup\{z : x \cdot z \leq y\} = 0$. To sum up, the p-norm returns $\neg 0 = 1$ and $\neg 1 = 0$. Basically, it is involutive when dealing with crisp values. However, the involution property $\neg\neg x$ fails for $x \in (0, 1)$. In that case $\neg x = 0$ and then $\neg\neg x = 1$. Concerning the Gödel t-norm, $(x \Rightarrow y) = \sup\{z : T(x, z) \leq y\} = \sup\{z : \min\{x, z\} \leq y\} = y$. Like for the p-norm we can easily check that involution fails. Finally, for the Łukasiewicz t-norm $(x \Rightarrow y) = \sup\{z : T(x, z) \leq y\} = \sup\{z : \max\{x + z - 1, 0\} \leq y\} = 1 - x + y$. In this case $x \Rightarrow y$ for $y = 0$ yields $1 - x$ and $1 - (1 - x) = x$, that is, involution holds.

3. Let $\varphi : [0, 1] \to [l, h]$ be a bijection from $[0, 1]$ and $[l, h]$.

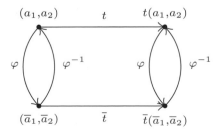

Then $\forall \bar{a}_1, \bar{a}_2 \in [l, h]$ we have $\bar{t}(\bar{a}_1, \bar{a}_2) = \varphi(t(\varphi^{-1}(\bar{a}_1), \varphi^{-1}(\bar{a}_2)))$, that is

$$\bar{t}(\cdot, \cdot) = \varphi \circ t \circ (\varphi^{-1}(\cdot), \varphi^{-1}(\cdot))$$

Now, consider the interval defined by $l = -1$ and $h = -1$. The function $\varphi(\cdot)$ in this case is defined by $\bar{a} = \varphi(a) = 2a - 1$, which clearly a bijective map which also returns $a = \varphi^{-1}(\bar{a}) = (1 + \bar{a})/2$. If we consider the p-norm, then we have

$$\bar{t}(\bar{a}_1, \bar{a}_2) = 2\frac{1 + \bar{a}_1}{2}\frac{1 + \bar{a}_2}{2} - 1 = \frac{(1 + \bar{a}_1)(1 + \bar{a}_2)}{2} - 1.$$

Likewise, for the $\neg$ operator, the function $t_\neg(a) = 1 - a$ follows the transformation

$$\bar{t}_\neg(\bar{a}) = (\varphi \circ t_\neg \circ \varphi^{-1})(\bar{a}) = 2\left(1 - \frac{1 + \bar{a}}{2}\right) - 1 = -\bar{a}.$$

Now, consider the Łukasiewicz logic and let us explore the transformation of the strong disjunction $\oplus$, that is $a_1 \oplus a_2 = \min\{1, a_1 + a_2\}$. Its remapping yields

$$\bar{a}_1 \oplus \bar{a}_2 = 2\min\{1, \frac{1 + \bar{a}_1}{2} + \frac{1 + \bar{a}_2}{2}\} - 1 = 2\min\{1, 1 + \frac{\bar{a}_1 + \bar{a}_2}{2}\} - 1.$$

4. From De Morgan's rule $a \vee b = \neg(\neg a \wedge \neg b)$, and therefore, according to the above results, $a \vee b \mapsto= 1 - (1 - a)(1 - b)/2$. Hence we can easily check that $a_1 \Leftrightarrow a_2 \mapsto a_1 a_2 - (1 - a_1^2)(1 - a_2^2)/8$.

Section 6.2.4

3. We start with the weak disjunction. Let $0 \leq \mu \leq 1$ and $x_1, x_2 \in X$. Then $\forall x = \mu x_1 + (1 - \mu)x_2$ we have

$$\begin{aligned}
(\psi_1 \vee \psi_2)(x) &= \max\{\psi_1(x), \psi_2(x)\} \\
&= \max\{\psi_1(\mu x_1 + (1 - \mu)x_2), \psi_2(\mu x_1 + (1 - \mu)x_2)\} \\
&\leq \max\{\mu\psi_1(x_1) + (1 - \mu)\psi_1(x_2), \mu\psi_2(x_1) + (1 - \mu)\psi_2(x_2)\} \\
&\leq \mu\max\{\psi_1(x_1) + \psi_2(x_1)\} + (1 - \mu)\max\{\psi_1(x_2) + \psi_2(x_2)\} \\
&= \mu(\psi_1 \vee \psi_2)(x_1) + (1 - \mu)(\psi_1 \vee \psi_2)(x_2).
\end{aligned}$$

Notice that the conclusion can also be drawn when considering that $\psi_1 \vee \psi_2 = \neg(\neg\psi_1 \wedge \neg\psi_2)$. Since ψ_1, ψ_2 are convex, $\neg\psi_1, \neg\psi_2$ are concave, and therefore we draw the conclusion by relying on the result given in Section 6.2.4 on the closure of concavity. Likewise the conclusion on the convexity of $\psi_1 \otimes \psi_2$ comes from $\psi_1 \otimes \psi_2 = \neg(\neg\psi_1 \oplus \neg\psi_2)$.

4. It suffices to notice that $(\neg\psi_1 \vee \psi_2) \to \neg\psi_3 = (\psi_1 \wedge \neg\psi_2) \oplus \neg\psi_3$, which is clearly in $(\wedge, \oplus)^*$. Interestingly, $(\neg\psi_1 \vee \psi_2) \to \neg\psi_3$ is not a Horn clause.

Section 6.3
Section 6.3.1

1. For any $n > 0$, let $b_n := [n > 0](a_{n+1} - a_n)$. This is a telescopic sequence, for which we have $\sum_{n=0}^{m} b_n = a_{m+1} - a_0$. We can calculate the generating function $B(z)$ by

$$B(z) = \sum_{n \geq 0} z^n b_n = \sum_{n \geq 0} z^n (a_{n+1} - a_n) = z^{-1} \sum_{n \geq 0} z^{n+1} a_{n+1} - A(z)$$

$$= (z^{-1} - 1) A(z) - a_0.$$

Now we have

$$\lim_{\mathbb{R} \ni z \to 1} B(z) = \sum_{n \geq 0} b_n = \lim_{n \to \infty} a_n - a_0 = \lim_{\mathbb{R} \ni z \to 1} (z^{-1} - 1) A(z) - a_0.$$

Finally, we conclude that

$$\lim_{n \to \infty} a_n = \lim_{\mathbb{R} \ni z \to 1} \frac{1 - z}{z} A(z),$$

from which the wanted result follows straightforwardly.

2. Suppose that $s(z, x) = wz + ux$, where $w, u \in \mathbb{R}$. Then the transition of the state equations for the two nodes are $z_1 = wz_2 + ux_1$ and $z_2 = wz_1 + ux_2$. We determine the solutions by exploring the convergence of $z_1(t + 1) = wz_2(t) + ux_1$, $z_2(t + 1) = wz_1(t) + ux_2$, which can be rewritten as

$$\begin{pmatrix} z_1 \\ z_2 \end{pmatrix} (t + 1) = \begin{pmatrix} 0 & w \\ w & 0 \end{pmatrix} \begin{pmatrix} z_1 \\ z_2 \end{pmatrix} (t) + u \begin{pmatrix} x_1 \\ x_2 \end{pmatrix}.$$

If we set $z := \begin{pmatrix} z_1 \\ z_2 \end{pmatrix}$, $A := \begin{pmatrix} 0 & w \\ w & 0 \end{pmatrix}$, $U := \begin{pmatrix} u & 0 \\ 0 & u \end{pmatrix}$ and $x := \begin{pmatrix} x_1 \\ x_2 \end{pmatrix}$, then we have

$$z(t + 1) = Az(t) + Ux. \tag{$*$}$$

Now from $\rho(W) < 1$ we can easily see that $\rho(A) < 1$, since $\text{eigen}(A) = \{-w, w\}$. Hence this linear system is asymptotically stable. Let $z_0 = 0$. If we compute the generating function of the sequence, we get $zZ(z) = AZ(z) + (1 - z)^{-1} Ux$. Hence we get

$$Z(z) = \frac{(zI - A)^{-1} Ux}{1 - z}.$$

If we use the final value theorem (see Exercise 1), finally, we get

$$\lim_{\mathbb{R} \ni z \to 1} (1 - z) Z(z) = \lim_{t \to \infty} z(t) = (I - A)^{-1} Ux.$$

This allows us to conclude that it is also BIBO (Bounded Input Bounded Output) stable, which makes the corresponding constraint well-defined. Clearly, if $|w| > 1$, system $(*)$ is unstable and the relational constraint is ill-posed.

Section 6.3.2

1. The satisfaction of the constraints yields:

$$z_1 = \frac{d^2 + 3d + 2}{d^2 + 2d + 2}, \quad z_2 = \frac{2d^2 + 2d + 2}{d^2 + 2d + 2}, \quad z_3 = \frac{d + 2}{d^2 + 2d + 2}.$$

We can immediately see that $z_1 + z_2 + z_3 = 3$. A detailed analysis to understand when $\sum_{v \in V} z_v = |V|$ is given in [M. Bianchini, M. Gori, F. Scarselli, *ACM Trans. Internet Technol.* **5** (2005), 92–128].

3. Any eigenvalue λ of A is such that $Av = \lambda v$. Let $\alpha \in \mathbb{R}$. Then we have $(\alpha A)v = (\alpha \lambda)v$. Now let $M > 0$ be such that $|\lambda| \leq M$. If we choose $\alpha < 1/M$ then we conclude that $\rho(\alpha A) < 1$.

Section 6.3.3

2. From Eq. (7), we get

$$A \bowtie (W_1, W_2)$$

$$\times \begin{pmatrix} 0 & 0 & 0 & 0 & 0 & 0 & w_{1,1,1} & w_{1,2,1} \\ 0 & 0 & 0 & 0 & 0 & 0 & w_{2,1,1} & w_{2,2,1} \\ w_{1,1,1} & w_{1,2,1} & 0 & 0 & 0 & 0 & 0 & 0 \\ w_{2,1,1} & w_{2,2,1} & 0 & 0 & 0 & 0 & 0 & 0 \\ 0 & 0 & w_{1,1,1} & w_{1,2,1} & 0 & 0 & 0 & 0 \\ 0 & 0 & w_{2,1,1} & w_{2,2,1} & 0 & 0 & 0 & 0 \\ 0 & 0 & w_{1,1,1} & w_{1,2,1} & w_{1,1,2} & w_{1,2,2} & 0 & 0 \\ 0 & 0 & w_{2,1,1} & w_{2,2,1} & w_{2,1,2} & w_{2,2,2} & 0 & 0 \end{pmatrix}$$

Section 6.4
Section 6.4.1

1. Let's start with a classic different expression of the Laplacian. We have

$$(\Delta q)(v) = \frac{1}{2} \sum_{e \vdash v} \frac{1}{\sqrt{d}} \left(\frac{\partial}{\partial e} \sqrt{d} \frac{\partial q}{\partial e} \right) \Big|_v = \frac{1}{2\sqrt{d}} \sum_{e \vdash v} \left(\frac{\partial}{\partial e} \sqrt{d} \frac{\partial q}{\partial e} \right) \Big|_v$$

$$= \frac{1}{2\sqrt{d}} \sum_{e \vdash v} \sqrt{\frac{w_{u,v}}{d_v}} \left(\sqrt{d} \frac{\partial q}{\partial e} \right) \Big|_v - \sqrt{\frac{w_{u,v}}{d_u}} \left(\sqrt{d} \frac{\partial q}{\partial e} \right) \Big|_u \right)$$

$$= \frac{1}{2\sqrt{d}} \sum_{e \vdash v} \sqrt{w_{u,v}} \left(\frac{\partial q}{\partial e}\Big|_v - \frac{\partial q}{\partial e}\Big|_u \right) = \frac{1}{\sqrt{d}} \sum_{e \vdash v} \sqrt{w_{u,v}} \frac{\partial q}{\partial e}\Big|_v$$

$$= \frac{1}{\sqrt{d}} \sum_{e \vdash v} \left(\frac{w_{u,v}}{\sqrt{d_v}} q_v - \frac{w_{u,v}}{\sqrt{d_u}} q_u \right) = q_v - \sum_{e \vdash v} \frac{w_{u,v}}{d_u d_v} q_u.$$

Now we have

$$R_d = \frac{1}{2} \sum_{v \in V} \|\nabla q_v\|^2 = \sum_{v \in V} \sum_{e \vdash v} \left(\frac{\partial q}{\partial e}\Big|_v \right)^2 = \frac{1}{2} \sum_{v \in V} \sum_{e \vdash v} \left(\sqrt{\frac{w_{u,v}}{d_v}} q_v - \sqrt{\frac{w_{u,v}}{d_u}} q_u \right)^2$$

$$= \frac{1}{2} \sum_{v \in V} \sum_{e \vdash v} \left(\frac{w_{u,v}}{d_v} q_v^2 + \frac{w_{u,v}}{d_u} q_u^2 - 2 \frac{w_{u,v}}{d_u d_v} q_u q_v \right)$$

$$= \frac{1}{2} \sum_{v \in V} q_v^2 + \frac{1}{2} \sum_{u \in V} \sum_{v \in V} \frac{w_{u,v}}{d_u} q_u^2 - \sum_{v \in V} \frac{w_{u,v}}{d_u d_v} q_u q_v \right)$$

$$= \sum_{v \in V} \left(q_v^2 - \sum_{e \vdash v} \frac{w_{u,v}}{d_u d_v} q_u q_v \right),$$

from which we finally conclude that

$$\langle q, \Delta q \rangle = q' \Delta q = \frac{1}{2} \sum_{v \in V} \|\nabla q_v\|^2 = R_d.$$

2. Based on results found in Exercise 1, let's define

$$Sq = (I - L)q = \sum_{e \vdash v} \frac{w_{u,v}}{d_u d_v} q_u,$$

so that Eq. (1) can be rewritten as $(I - S)q = \alpha(q)$, that is,

$$q_v = \sum_{e \vdash v} \frac{w_{u,v}}{d_u d_v} q_u + \alpha(q), \tag{$*$}$$

where we set $\alpha(q) := (1/\mu_d) \sum_{\psi \in \mathcal{C}_\psi}^{\backprime} (\partial/\partial q) V_\psi(q)$. Now, to reproduce a random walk, suppose $w_{u,v} \equiv 1$, so that $d_u = |\operatorname{pa}(u)|$. Then from Eq. $(*)$ we get

$$(Sq)_v = \frac{1}{|\operatorname{pa}(v)|} \sum_{u \in \operatorname{pa}(v)} \frac{q_u}{\operatorname{pa}(u)},$$

which yields a random walk.

Section 6.4.2

1. When using the definition, we have
(symmetry)

$$k(X_i, X_j) = \langle P\beta(\cdot, X_i), P\beta(\cdot, X_j) \rangle = \langle P\beta(\cdot, X_j), P\beta(\cdot, X_i) \rangle = k(X_j, X_i).$$

Of course, the symmetry is kept also in case of degeneration of the sets to a single point.

(nonnegativeness) Since g is Green and a kernel function, there exists a feature map ϕ such that $\forall x, z\ g(x, z) = \langle \phi(x), \phi(z) \rangle$. We distinguish three cases:

(*i*) $|X_i| > 0, |X_j| > 0$. From the definition of k, we have

$$
\begin{aligned}
k(X_i, X_j) &= \int_X \int_X g(x, z)\, 1_{X_i}(z)\, 1_{X_j}(x)dxdz \\
&= \int_X \int_X \langle \phi(x), \phi(z) \rangle\, 1_{X_i}(z)\, 1_{X_j}(x)dxdz \\
&= \int_X \int_X \langle \phi(x)\, 1_{X_j}(x), \phi(z)\, 1_{X_i}(z) \rangle dxdz \\
&= \langle \int_X \phi(x)\, 1_{X_j}(x)dx, \int_X \phi(z)\, 1_{X_i}(z)dz \rangle \\
&= \langle \Phi(X_i), \Phi(X_j) \rangle,
\end{aligned}
$$

where $\Phi(X_i) := \int_X \phi(x)\, 1_{X_i}(x)dx$ and $\Phi(X_j) := \int_X \phi(x)\, 1_{X_j}(x)dx$.
(*ii*) $|X_i| > 0, X_j = \{x_j\}$. In this case, we have

$$
\begin{aligned}
k(X_i, x_j) &= \int_X \int_X g(x, z)\delta(x - x_i)\, 1_{X_j}(z)dxdz \\
&= \int_X \int_X \langle \phi(x), \phi(z) \rangle\delta(x - x_i)\, 1_{X_j}(z)dxdz \\
&= \langle \int_X \phi(x)\delta(x - x_i)dx, \int_X \phi(z)\, 1_{X_j}(z)dz \rangle \\
&= \langle \phi(x_i), \Phi(X_j) \rangle.
\end{aligned}
$$

(*iii*) $X_i = \{x_i\}, X_j = \{x_j\}$. In this case, we have

$$
\begin{aligned}
k(x_i, x_j) &= \int_X \int_X g(x, z)\delta(x - x_i)\delta(z - x_i)dxdz \\
&= \int_X \int_X \langle \phi(x), \phi(z) \rangle\delta(x - x_i)\delta(z - x_i)dxdz \\
&= \langle \int_X \phi(x)\delta(x - x_i)dx, \int_X \phi(z)\delta(z - x_j)dx \rangle \\
&= \langle \phi(x_i), \phi(x_j) \rangle.
\end{aligned}
$$

Finally, the feature map factorization expressed by (i), (ii), and (iii) guarantees the nonnegative definiteness.

Section 6.4.3

2. This is a special case of the problem of supervised learning under the linear constraints $Af(x) = b(x)$. In this case, we simply have $A(x) = (1, \ldots, 1)$ and $b(x) = 1$. However, in order to guarantee the existence of $\langle Pf, Pf \rangle$, $b(x)$ needs to vanish as $x \to \infty$. Let us choose $b(x) = [\|x\| \leq B] + [\|x\| > B]e^{-\frac{\|x\|}{\sigma^2}}$, where $B > 0$ can be chosen arbitrarily large. When considering that $L = \sum_{\kappa=1}^{k}(-1)^\kappa \rho_\kappa \nabla^{2\kappa}$, Eq. (8) on the Lagrangian multipliers yields

$$\lambda(x) = \frac{1}{n}Lb(x) + \frac{1}{n\ell}\sum_{\kappa=1}^{\ell}\sum_{j=1}^{n}(\bar{f}_j(x) - y_\kappa)\delta(x - x_\kappa).$$

Now we have

$$f_j^*(x) = \frac{1}{n}[\|x\| \leq B]g * \rho_0 + \frac{1}{n}[\|x\| > B]g * \left((L - \rho_0)e^{-\frac{\|x\|}{\sigma^2}}\right)$$

$$+ \frac{1}{n\ell}\sum_{\kappa=1}^{\ell}\sum_{j=1}^{n}(\bar{f}_j(x) - y_\kappa)g(x - x_\kappa)$$

$$= [\|x\| \leq B]\left(c + \sum_{\kappa=1}^{\ell}\lambda_{j\kappa}g(x - x_\kappa)\right) + \frac{1}{n}[\|x\| > B]g * \left((L - \rho_0)e^{-\frac{\|x\|}{\sigma^2}}\right).$$

Hence in the ball defined by $[\|x\| \leq B]$ the solution is given by a shifted kernel expansion. Hence learning can be based by ordinary kernel machine mathematical and algorithmic apparatus.

Section 6.4.4

1. See the solution is based on Fredholm Kernels [see G. Gnecco, M. Gori, S. Melacci, M. Sanguineti, *Neural Comput.* **27** (2015), 388–480].

5. No, it isn't. This promptly comes from the solution of Exercise 4 when considering that we are dealing with a graph composed of a collection of disconnected vertexes.

8. The solution is based on noticing that we can still use the backpropagation chaining rule — which leads to space locality — and that the term $\zeta_{t,i,j,k}$ defined in RTRL degenerates to $\zeta_{t,i,k} = \frac{\partial z_{t,i}}{\partial w_{i,k}}$. Details can be found in M. Gori, Y. Bengio, R. De Mori, *Proc. of the Int. Joint Conf. on Neural Netw.* (1989), 643–644, M. Gori, *Proc. of Neuro-Nimes* (1989), 83–93 and in Y. Bengio, R. De Mori, M. Gori, Special Issue on Artificial Neural Netw., Pattern Recognit. Lett. **13** (1992), 375–386.

Constrained optimization

Support Vector Machine (SVM) is the most popular machine learning approach, where representation and learning involves Lagrangian multipliers. The theory relies on classic constrained optimization, namely on the solution of the problem

$$\min_{w \in W} p(w),$$

$$W = \{w \in \mathbb{R} : (h(w) = 0) \wedge (g(w) \geq 0)\}.$$

In the MMP, the parsimony function is $p(w) = w^2$, while g represents the ℓ unilateral constraints $g_\kappa(w) = y_\kappa \hat{w}' \widehat{\phi}(x_\kappa) - 1 \geq 0$, $\kappa = 1, \ldots, \ell$, while there is no bilateral constraint $h(w) = 0$.

There's an elegant solution to this problem, which is based on the Lagrangian approach to optimization. A first insight on the solution comes from distinguishing the case in which the minimum is in the interior or on the border, where constraints are active — at least one of them. Clearly, in case there is no active constraint, the problems can be faced by imposing $\nabla p(w) = 0$ so as to determine critical points that need to be classified by inspecting higher order derivatives. Things get more involved when the minimum is located on the border. Lagrangian multipliers arise from a clever idea to determine stationary points located on the border. Let us start by assuming that only one (bilateral) constraint is active — say $h(w) = 0$. Let us denote by w^* a critical point on the border ∂W of W. We have that $\nabla h(w^*)$ is orthogonal to the tangent of the surface $h(w) = 0$ at w^*. Let τ be a vector in the tangent plane to the surface $h(w) = 0$ at w^*. Now, any variation of p along the direction defined by τ is null if and only if $\langle \nabla p(w^*), \tau \rangle = 0$, which happens whenever $\nabla p(w^*)$ is on orthogonal line to τ. As a consequence, since $\nabla h(w^*)$ and $\nabla h(w^*)$ are both orthogonal to τ, they are collinear. This means that there exists $\mu \in \mathbb{R}$ such that

$$\nabla p(w^*) + \mu \nabla h(w^*) = 0. \tag{1}$$

What if two or more constraints are simultaneously active? Let $h_1(x) = 0$ and $h_2(x) = 0$ be active and consider the domain generated by their intersection. Now, let us denote by $H_1 = \{w \mid h_1(w) = 0\}$ and by $H_2 = \{w \mid h_2(w) = 0\}$ the domains defined by the constraints. This time $\langle \nabla p(w^*), \tau \rangle = 0$ holds true whenever τ is orthogonal to $H_1 \cap H_2$ at w^*. In the previous case, the condition of orthogonality to $H = \{w \mid h(w) = 0\}$ at w^* univocally defines the orthogonal direction, which is in fact that of $\nabla h(w^*)$ and $\nabla p(w^*)$. Clearly, the intersection $H_1 \cap H_2$ in general reduces the measure of the feasible set, so as when adding $h_2(x) = 0$ to the first constraint $h_1(x) = 0$ the measure of $H_1 \cap H_2$ becomes smaller than the measure of H_1. On the other hand, the measure of the orthogonal space increases. In particular, the orthogonal space at a given $w^* \in H_1 \cap H_2$ is defined by

$$\perp (H_1 \cap H_2) = \{ w : w = \beta_1 \nabla h_1(w^*) + \beta_2 \nabla h_2(w^*), \quad \mu_1, \mu_2 \in \mathbb{R} \}. \tag{2}$$

This can promptly be understood when considering that for w^* to be a feasible point, the tangent v must be orthogonal to $\nabla h_1(w^*)$ and $\nabla h_2(w^*)$. Finally, $\nabla p(w^*) \in \perp (H_1 \cap H_2)$, that is, there exist $\mu_1, \mu_2 \in \mathbb{R}$ such that

$$\nabla p(w^*) + \mu_1^* \nabla h_1(w^*) + \mu_2^* \nabla h_2(w^*) = 0.$$

To sum up, one more constraint increases the dimension of the orthogonal space, which corresponds to the number of constraints. Clearly, this holds for any number of jointly active constraints. In addition, this property is not restricted to the bilateral constraint, it holds for also unilateral constraints in case they become active. If we introduce the Lagrangian

$$L(w, \mu, \lambda) := p(w) + \sum_{h=1}^{n} \mu_h g_h(w) + \sum_{\kappa=1}^{\ell} \lambda_\kappa g_\kappa(w), \tag{3}$$

then the previous stationarity condition can be compactly translated into the condition $\nabla L(w^*, \mu^*, \lambda^*) = 0$. Notice that this condition only sheds light on the critical points, but it doesn't provide any information on their type. Now let us consider a unilateral constraint $g(w) \geq 0$ in the case it is active. The Lagrangian is $L(w, \lambda) = p(w) + \lambda g(w)$ and, therefore, the search for minima still needs to satisfy $\nabla L(w, \lambda) = 0$. Interestingly, the fact that we are dealing with a unilateral constraint provides additional information on the nature of the critical point w^*. Let us analyze the case in which w^* is a minimum. Since $g(w) \geq 0$, we can promptly see that in any ball $B(w^*, \epsilon)$ of radius ϵ centered at w^*, we have that $\nabla_w g(w)$ points to the interior of the feasible set $G = \{ w : g(w) \geq 0 \}$. Since w^* is a minimum, also $\nabla_w p(w^*)$ points to the interior and, therefore, $\langle \nabla_w g(w^*), \nabla_w p(w^*) \rangle \geq 0$. Now, since we also have $\nabla p(w^*) + \lambda \nabla g(w^*) = 0$, we conclude that $\lambda \leq 0$. Hence, while for bilateral constraints the Lagrangian multipliers are reals, for the unilateral constraint $g(w) \geq 0$ we gain additional information which results in the sign constraint on λ.

A substantial progress comes from solving[1] $\nabla L(w, \lambda) = 0$ in the *dual space*. The case of SVM is a nice example of an important class of problems in which the separation between the weights w and the Lagrangian multiplier λ is possible. In that case, the Lagrangian can be rewritten in terms of Lagrangian multipliers only by

---

[1] The analysis which arises from Eq. (2) when summing up the contributions from Lagrangian multipliers clearly holds also in case bilateral and unilateral constraints are jointly present. For the sake of simplicity, here, we restrict attention to the case of a single unilateral constraint.

function $\theta(\lambda) = L(w, \lambda)$. In general, a similar function θ might be hard to determine because of difficulties arising in the elimination of w in $L(w, \lambda)$. However, in general, we can define

$$\theta(\lambda) = \inf_{w \in W} L(w, \lambda). \tag{4}$$

Clearly, the case $\forall w \in W$ $\theta(\lambda) = L(w, \lambda)$ is just a special instance of the above definition. When dealing with minima, we can establish a fundamental upper bound on θ. This can be obtained when noticing that, for any minimum $\lambda \leq 0$ and for any feasible point w, we have $g(w) \geq 0$. These conditions yield $\lambda g(w) \leq 0$ and, therefore, for any λ we have

$$\theta(\lambda) = \inf_{w \in W} L(w, \lambda) = \inf_{w \in W} (p(w) + \lambda g(w)) \leq \inf_{w \in W} p(w). \tag{5}$$

This immediately yields

$$\sup_{\lambda \leq 0} \theta(\lambda) \leq \inf_{w \in W} p(w). \tag{6}$$

Interestingly, $\sup_{\lambda \leq 0} \theta(\lambda)$ turns out to be an approximation of $\inf_{w \in W} p(w)$, but, in general, there is a positive difference between the primal $\inf_{w \in W} p(w)$ and the dual $\sup_{\lambda \leq 0} \theta(\lambda)$ bound, which is referred to as *duality gap*. Whenever this gap is null, the search for $\inf_{w \in W} p(w)$ can conveniently be replaced with the problem of determining the dual bound $\sup_{\lambda \leq 0} \theta(\lambda)$. When inspecting inequality (5), we notice that if $\lambda g(w) = 0$, we end up with a null duality gap, that is, $\sup_{\lambda \leq 0} \theta(\lambda) = \inf_{w \in W} p(w)$. Hence, the sufficient conditions

$$\nabla L(w, \lambda) = 0,$$
$$\lambda g(w) = 0,$$
$$\lambda \leq 0,$$

which are referred to as the *Karush–Kuhn–Tucker* (KTT) conditions, guarantee that the duality gap is zero. As shown in Exercises 1 and 2, the KKT conditions, in general, are not necessary.

However, KKT hold true for the optimization problem associated with the MMP defined by Eq. (2), which allows us to conclude that there is no duality gap. Now, let us have a look at L in (w^*, λ^*). We can promptly see that this point is a saddle for L. It is in fact a minimum when moving along w^*, whereas it is a maximum when moving toward any λ_κ direction.

Regularization operators

B

KERNEL MACHINES can be nicely presented within a regularization framework based on differential operators. Here we give an introduction to differential and pseudo-differential operators. A natural way of imposing the development of a "smooth solution" f of a learning problem is to think of a special expression of the parsimony principle which relies on restricting the quick variations of f.

In the simplest case in which $f : X \subset \mathbb{R} \to \mathbb{R}$ and $f \in L^2(X)$, one can introduce the index

$$\mathcal{R} = \int_X [f'(x)]^2 \, dx = \int_X \frac{d}{dx} f(x) \cdot \frac{d}{dx} f(x) \, dx = \int_X Pf(x) \cdot Pf(x) \, dx = \langle f', f' \rangle$$
$$= \|f\|_P^2.$$

The index $\|f\|_P^2 \geq 0$ is a seminorm in $L^2(X)$. It has all the properties of a norm, except for the fact that $\|f\|_P = 0$ does not imply $f \equiv 0$, since this clearly holds for constant functions $f(x) \equiv c$, too. In case $X = \mathbb{R}$, we can promptly see that this way of measuring the degree of parsimony of f makes strong conditions on the asymptotic behavior of f. If $X = [a, b]$, then

$$\int_a^b [f'(x)]^2 \, dx = \int_a^b \frac{d}{dx} f(x) df(x) = [f^2(x)]_a^b - \int_a^b f'(x) f(x) \, dx.$$

If $f(a) = f(b) = 0$, then

$$\int_a^b [f'(x)]^2 \, dx = -\int_a^b f'(x) f(x) \, dx = \langle f, -\frac{d}{dx} \frac{d}{dx} f \rangle = \langle f, P^* P f \rangle, \quad (1)$$

where $P^* := -d/dx$ is the adjoint operator of $P = d/dx$. Interestingly, once we assume the boundary condition that f is null on its border, it turns out that $\|f\|_P$ is related to $L = P^* P = -d^2/dt^2$.

Now, let us consider the case of $X \subset \mathbb{R}^d$ in which we replace $P = d/dx$ with $P = \nabla$. Like for the case $d = 1$, we still assume to analyze functions in $L^2(X)$. Assume $f, u \in L^2(X)$. We have

$$\nabla \cdot (u \nabla f) = \nabla f \cdot \nabla u + u \nabla^2 f.$$

505

Now, like in the case of a single dimension, let us assume as boundary condition that u vanishes on the boundary ∂X of X. Then we get

$$\int_X \nabla f \cdot \nabla u \, dx = \int_X \left(\nabla \cdot (u \nabla f) - u \nabla^2 f \right) dx$$
$$= \int_{\partial X} u \nabla f \cdot dS - \int_X u \nabla^2 f \, dx = - \int_X u \nabla^2 f \, dx$$

This can be rewritten as

$$\langle \nabla f, \nabla u \rangle = \langle u, -\nabla \cdot \nabla f \rangle = \langle u, \nabla^* \nabla f \rangle,$$

and then $\nabla^* = -\nabla \cdot$. Now for $f = u$, we get

$$\langle \nabla f, \nabla f \rangle = \langle f, -\nabla^2 f \rangle. \tag{2}$$

Of course, like for $P = d/dx$, the above expression for $\|f\|_P$, which clearly generalizes (1), holds in case function f is identically null on its border.

Now, we consider the case $P = \Delta = \nabla^2$. Interestingly, we can analyze this case by invoking the result discovered for $P = \nabla$. Given $u, v \in L^2(X)$, if $(\nabla u = 0) \wedge (\nabla v = 0)$ on ∂X, then we have

$$\langle \nabla u, \nabla v \rangle = -\langle \nabla^2 u, v \rangle.$$

If we exchange u with v, we get $\langle \nabla v, \nabla u \rangle = -\langle \nabla^2 v, u \rangle$. Since $\langle \nabla u, \nabla v \rangle = \langle \nabla v, \nabla u \rangle$, we get $\langle \nabla^2 u, v \rangle = \langle \nabla^2 v, u \rangle$, that is, Δ is self-adjoint. As a consequence, we can determine $\|f\|_\Delta^2$ since

$$\langle \Delta f, \Delta f \rangle = \langle f, \Delta(\Delta f) \rangle = \langle f, \Delta^2 f \rangle = \langle f, \nabla^4 f \rangle.$$

Of course, this holds whenever $\nabla f = 0$ on ∂X. Now, it is interesting to see what happens when we consider higher order differential operators. A crucial remark concerns the periodic structure that emerges in P^m. Beginning from $P = \nabla$ and $P^2 = \nabla \cdot \nabla$, it becomes natural to define $P^3 = \nabla(\nabla \cdot \nabla)$ and, therefore, the sequence

$$P^0 = I, \quad P^1 = \nabla, \quad P^2 = \nabla \cdot \nabla, \quad P^3 = \nabla \nabla \cdot \nabla, \quad P^4 = \nabla \cdot \nabla \nabla \cdot \nabla, \dots$$

Now, let $a_k \in \mathbb{R}^+$, with $\kappa \in \mathbb{N}_m$, and consider

$$\odot_m^e = \sum_{h=0}^{m/2} a_{2h} \nabla^{2h} \quad \text{and} \quad \odot_m^o = \sum_{h=0}^{m/2} a_{2h+1} \nabla \nabla^{2h},$$

where $P^{2h} = \Delta^h = \nabla^{2h}$ and $P^{2h+1} = \nabla \nabla^{2h}$ for $h = 0, \ldots, m$. Now, the operator $\odot_m^e$ leads to

$$\langle \odot_m^e f, \odot_m^e f \rangle = \left\langle \sum_{h=0}^{m/2} a_{2h} \nabla^{2h} f, \sum_{\kappa=0}^{m/2} a_{2\kappa} \nabla^{2\kappa} f \right\rangle$$

$$= \sum_{h=0}^{m/2} \sum_{\kappa=0}^{m/2} a_{2h} a_{2\kappa} \langle \nabla^{2\kappa} f, \nabla^{2h} f \rangle$$

$$= \sum_{h=0}^{m/2} \sum_{\kappa=0}^{m/2} a_{2h} a_{2\kappa} \langle f, \nabla^{2\kappa} \nabla^{2h} f \rangle$$

$$= \sum_{h=0}^{m/2} \sum_{\kappa=0}^{m/2} a_{2h} a_{2\kappa} \langle f, \nabla^{2(h+\kappa)} f \rangle.$$

Likewise, for $\odot_m^o$ we have

$$\langle \odot_m^o f, \odot_m^o f \rangle = \left\langle \sum_{h=0}^{m/2} a_{2h+1} \nabla \nabla^{2h} f, \sum_{\kappa=0}^{m/2} a_{2\kappa+1} \nabla \nabla^{2\kappa} f \right\rangle$$

$$= \sum_{h=0}^{m/2} \sum_{\kappa=0}^{m/2} a_{2h+1} a_{2\kappa+1} \langle \nabla \nabla^{2h} f, \nabla \nabla^{2\kappa} f \rangle$$

$$= -\sum_{h=0}^{m/2} \sum_{\kappa=0}^{m/2} a_{2h+1} a_{2\kappa+1} \langle \nabla^{2h} f, \nabla \cdot \nabla \nabla^{2\kappa} f \rangle$$

$$= -\sum_{h=0}^{m/2} \sum_{\kappa=0}^{m/2} a_{2h+1} a_{2\kappa+1} \langle f, \nabla^{2h} \nabla^{2(\kappa+1)} f \rangle$$

$$= -\sum_{h=0}^{m/2} \sum_{\kappa=0}^{m/2} a_{2h+1} a_{2\kappa+1} \langle f, \nabla^{2(h+\kappa+1)} f \rangle.$$

By definition, these operators give rise to the norm

$$\| f \|_{\odot_m}^2 := \| f \|_{\odot_m^e}^2 + \| f \|_{\odot_m^o}^2.$$

The following proposition helps determine the adjoint of $\odot_m$.

Proposition 1. *Let* $u, v \in C^{2n}(X \subset \mathbb{R}^d, \mathbb{R})$ *be such that* $\forall n \in \mathbb{N}$ *and* $\forall x \in \partial X$, $\nabla^n u(x) = v(x) = 0$. *If* $h = 2n$ *then* $(P^h)^* = P^h$, *and if* $h = 2n + 1$ *then* $(P^h)^* = -\nabla \cdot \overline{\nabla}^{2n}$, *where* $\overline{\nabla} \doteq \nabla \nabla \cdot$.

Proof. We start noting that the proposition holds trivially for $h = 0$; in this case P^h reduces to the identity. Then we discuss even and odd terms separately. We prove that for the *even* terms P^{2n} is Hermitian. The proof is given by induction on n.

- *Basis of induction.* For $n = 1$, $P^2 = \nabla^2$ and P^2 is self-adjoint.
- *Induction step.* Since ∇^2 is self-adjoint (basis of induction), because of the induction hypothesis $\langle \nabla^{2(n-1)}u, v \rangle = \langle u, \nabla^{2(n-1)}v \rangle$, and because of the conditions on the border ∂X, we have $\langle \nabla^{2n}u, v \rangle = \langle \nabla^2(\nabla^{2(n-1)}u), v \rangle = \langle \nabla^{2(n-1)}u, \nabla^2 v \rangle = \langle u, \nabla^{2(n-1)}\nabla^2 v \rangle = \langle u, \nabla^{2n}v \rangle$.

Now, for the odd terms, we prove that $(\nabla^{2n+1})^* = -\nabla \cdot \overline{\nabla}^{2n}$.

- *Basis of induction.* For $n = 0$, we have $P^1 = \nabla$ and $\nabla^* = -\nabla\cdot$.
- *Induction step.* We get $\langle \nabla^{2n+1}u, v \rangle = \langle \nabla\nabla^{2n}u, v \rangle = \langle \nabla^{2n}u, -\nabla \cdot v \rangle = \langle u, -\nabla^{2n}\nabla \cdot v \rangle = \langle u, -\nabla \cdot \overline{\nabla}^{2n}v \rangle$. $\qquad\square$

Corollary 1. *Let $u, v \colon X \subset \mathbb{R}^d \to \mathbb{R}$ be two analytic functions such that $\forall h \in \mathbb{N}$ and $\forall x \in \partial X$, $\nabla^{2h}u(x) = v(x) = 0$. Then*

$$(\odot_m^e)^* = \odot_m = \sum_{h=0}^{m/2} a_{2h} \nabla^{2h} \quad and \quad (\odot_m^o)^* = -\sum_{h=0}^{m/2} a_{2h+1} \nabla \cdot \overline{\nabla}^{2h}.$$

Proof. For $m = 2r$, given any two functions that satisfy the hypotheses, from Proposition 1, we have

$$\langle \odot_m u, v \rangle = \left\langle \sum_{h=0}^r a_{2h} \nabla^{2h} u, v \right\rangle = \left\langle u, \sum_{h=0}^m a_{2h} \nabla^{2h} v \right\rangle = \langle u, \odot_m^* v \rangle.$$

Likewise, for $m = 2r + 1$ we have

$$\langle \odot_m u, v \rangle = \left\langle \sum_{h=0}^r a_{2h+1} \nabla^{2h+1} u, v \right\rangle = \left\langle u, -\sum_{h=0}^r a_{2h+1} \nabla \cdot \overline{\nabla}^{2h} v \right\rangle = \langle u, \odot_m^* v \rangle.$$

The distinct definitions of $\odot_m^e$ and $\odot_m^o$ for even and odd integers with the corresponding adjoint operators $(\odot_m^e)^*$ and $(\odot_m^o)^*$ makes it possible to compute

$$\langle \odot_m f, \odot_m f \rangle = \sum_{h=0}^{m/2} \left(a_{2h}^2 \langle \nabla^{2h} f, \nabla^{2h} f \rangle + a_{2h+1}^2 \langle \nabla\nabla^{2h} f, \nabla\nabla^{2h} f \rangle \right). \qquad\square$$

Proposition 2. *Let m be an even number. Then*

$$\langle \odot_m f, \odot_m f \rangle = \langle f, (\odot_m^* \odot_m) f \rangle = \langle f, \sum_{h=0}^{m+1} (-1)^h a_h^2 \nabla^{2h} f \rangle.$$

Proof. From straightforward application of the above propositions to $\odot_m$,

$$\langle \odot_m f, \odot_m f \rangle = \sum_{h=0}^{m/2} \left(a_{2h}^2 \langle \nabla^{2h} f, \nabla^{2h} f \rangle + a_{2h+1}^2 \langle \nabla \nabla^{2h} f, \nabla \nabla^{2h} f \rangle \right)$$

$$= \sum_{h=0}^{m/2} \left(a_{2h}^2 \langle f, \nabla^{4h} f \rangle + a_{2h+1}^2 \langle f, -\nabla^{4h+2} f \rangle \right)$$

$$= \langle f, \sum_{h=0}^{m+1} (-1)^h a_h^2 \nabla^{2h} f \rangle.$$

Now we discuss a more general case in which

$$\diamond_m = \sum_{h=0}^{m} a_h \left(\frac{\partial}{\partial x_1} + \frac{\partial}{\partial x_2} + \cdots + \frac{\partial}{\partial x_d} \right)^h = \sum_{h=0}^{m} a_h D^h = \sum_{h=0}^{m} a_h \sum_{|\alpha|=h} \frac{h!}{\alpha!} \left(\frac{\partial}{\partial x} \right)^\alpha \tag{3}$$

with $|\alpha| = \alpha_1 + \alpha_2 + \cdots + \alpha_d$. $\qquad\square$

Proposition 3. *Let us consider the differential operator given by (3). Then*

$$(\diamond_m)^* = \sum_{h=0}^{m} (-1)^h a_h \sum_{|\alpha|=h} \frac{h!}{\alpha!} \left(\frac{\partial}{\partial x} \right)^\alpha. \tag{4}$$

Proof. Since the adjoint of a sum of operators is the sum of the adjoints, we can restrict the proof to the operator ∂_x^α. So we just need to prove that $(\partial_x^\alpha)^* = (-1)^{|\alpha|} \partial_x^\alpha$. For $|\alpha| = 0$, the proof is trivial. For $|\alpha| > 0$, under some regularity conditions on the space on which the operator acts, we can always write $\partial_x^\alpha = \partial_{x_{i_1}} \partial_{x_{i_2}} \cdots \partial_{x_{i_{|\alpha|}}}$, where the indices $i_1, \ldots, i_{|\alpha|}$ belong to $\{1, 2, \ldots, d\}$; for example, $\partial_{x_1}^{|\alpha|} = \partial_{x_1} \partial_{x_1} \cdots \partial_{x_1}$. From here it is immediate to see that

$$\langle \partial_{x_{i_1}} \partial_{x_{i_2}} \cdots \partial_{x_{i_{|\alpha|}}} u, v \rangle = (-1)^{|\alpha|} \langle u, \partial_{x_{i_1}} \partial_{x_{i_2}} \cdots \partial_{x_{i_{|\alpha|}}} v \rangle.$$

This ends the proof. $\qquad\square$

Let M be the set of d-dimensional multiindices with length between 0 and m, $M = \{ \alpha = (\alpha_1, \ldots, \alpha_d) : 0 \leq |\alpha| \leq m \}$. Then $\diamond_m = \sum_{\alpha \in M} b_\alpha \partial_x^\alpha$, and the regularization term deriving from $\diamond_m$ is

$$\langle \diamond_m f, \diamond_m f \rangle = \left\langle \sum_{\alpha \in M} b_\alpha \partial_x^\alpha f, \sum_{\beta \in M} b_\beta \partial_x^\beta f \right\rangle = \sum_{\alpha \in M} \sum_{\beta \in M} (-1)^{|\alpha|} b_\alpha b_\beta \langle f, \partial_x^\alpha \partial_x^\beta f \rangle.$$

The above discussion on differential operators can be enriched at least two different directions. First, we can consider an infinite number of differential terms ($m \to \infty$) and, second, we can replace the a_k coefficients with functions $a_k : X \to \mathbb{R}$.

Calculus of variations

C.1 Functionals and variations

Let us start describing functionals and related classic problems by some examples.

Example 1. Let $y \in F := C^1(A \subset \mathbb{R}, B \subset \mathbb{R})$ be and consider

$$\mathcal{L}_1(y) = \int_{x_1}^{x_2} \sqrt{1 + (y'(x))^2} dx \tag{1}$$

$\mathcal{L}_1$ is referred to as a *functional* and operates on the functional space F to which y belongs. One might be interested in determining when $\mathcal{L}_1(f)$ gets the minimum in $F((x_1, y_1), (x_2, y_2)) = \{y \in F : y(x_1) = y_1, y(x_2) = y_2\}$.

Example 2. Let us consider the problem of minimizing the time to get from $P_1 = (x_1, y_1)$ down to $P_2 = (x_2, y_2)$ beginning at $y_1 > y_2$ when sliding from a curve described by $x \mapsto y(x)$. Obviously, it is

$$\mathcal{L}_2(y) = \frac{1}{\sqrt{2g}} \int_{x_1}^{x_2} \frac{\sqrt{1 + (y'(x))^2}}{\sqrt{y_1 - y(x)}} dx.$$

This is referred to as the brachistochrone problem.

Example 3. Let us consider

$$\mathcal{L}_3(f) = \int_{x_1}^{x_2} \sqrt{1 + y'(x)^2 + z'(x)^2} dx,$$
$$\phi(x, y, z) = 0, \tag{2}$$

where now we have $z = f(x, y)$, which must be determined under the constraint $\phi(x, y, z) = 0$. This corresponds to approaching the geodesic problem, which consists of determining the minimum path on the surface $\phi(x, y, z) = 0$ given any two points $P_1 \equiv (x_1, y_1, z_1)$ and $P_2 \equiv (x_2, y_2, z_2)$.

Example 4. Another nice example of a constrained problem is that of finding a closed curve $\gamma := (x(t), y(t)), t \in [t_1, t_2)$ with maximum area $\mathcal{L}_4(\gamma) = 2\mathcal{A}(\mathcal{S}_\gamma)$ and a given length L. That is,

$$\mathcal{L}_4(\gamma) = \int_{t_1}^{t_2} (x(t)\dot{y}(t) - \dot{x}(t)y(t))\, dt, \qquad L(\gamma) = \int_{t_1}^{t_2} \sqrt{\dot{x}(t)^2 + \dot{y}(t)^2}\, dt. \quad (3)$$

This is referred to as the isoperimetric problem. Now, L is the perimeter of γ with $\tau = (\dot{x}(t), \dot{y}(t))'$ being the associated tangent vector. We can also see that $\mathcal{L}_4(\gamma)$ is the area of the surface $\mathcal{S}_\gamma$ bounded by the same curve. Let $n := (0, 0, 1)'$ and let us consider the field $u = (-y, x, 0)'/2$. We have that $\nabla \times u = (0, 0, 1)'$. From Stokes theorem,

$$\begin{aligned}
\mathcal{A}(\mathcal{S}_\gamma) &= \int_{\mathcal{S}_\gamma} (0, 0, 1)' \cdot (0, 0, 1)\, ds \\
&= \int_{\mathcal{S}_\gamma} \nabla \times u \cdot n\, ds \\
&= \oint_\gamma u \cdot \tau\, dl \\
&= \frac{1}{2} \int_{t_1}^{t_2} (-y(t), x(t), 0)'(\dot{x}(t), \dot{y}(t), 0)\, dt \\
&= \frac{1}{2} \int_{t_1}^{t_2} (x(t)\dot{y}(t) - \dot{x}(t)y(t))\, dt.
\end{aligned}$$

Hence, we end up with the problem of determining the curve $\gamma := (x(t), y(t)), t \in [t_1, t_2)$ with a prescribed length $L(\gamma)$ that has the maximum area $\mathcal{A}(\mathcal{S}_\gamma)$.

These classic problems are instances of more general functionals like

- $\int_{x_1}^{x_2} F(x, y(x), y'(x))dx$,
- $\int_{x_1}^{x_2} F(x, y(x), y'(x), \ldots, y^{(n)}(x))dx$,
- $\int_{x_1}^{x_2} F(x, y_1(x), \ldots, y_n(x), y_1'(x), \ldots, y_n'(x))dx$,
- $\iint_D F(x, y, z(x, y), \frac{\partial z}{\partial x}, \frac{\partial y}{\partial z})dxdy$,

where in the last couple of examples they are paired with subsidiary conditions. In the optimization of functionals, it is very important to make clear the precise functional space on with we operate. For instance, the functional spaces involved in

$$\int_{x_1}^{x_2} F(x, y, y'), dx, \qquad \int_{x_1}^{x_2} F(x, y, y', y')\, dx$$

might be different, since in the first case one is likely to assume to operate in $C^1(\mathbb{R}, \mathbb{R})$, whereas in the second one the likely space is $C^2(\mathbb{R}, \mathbb{R})$.

C.2 Basic notion on variations

Definition 1. Let U and W be Banach spaces and $U \subset V$ be an open set of V. A function $F: V \to W$ is Frechet differentiable at $x \in U$ if there exists a linear bounded operator $A_x : U \to W$ such that

$$\lim_{h \to 0} \frac{\|f(x+h) - f(x) - A_x(h)\|_W}{\|h\|_V} = 0.$$

The limit here is meant in the usual sense of a limit of a function defined on a metric space. This is a generalization of the ordinary notion of derivative for real-valued functions also in the multivariate case. Basically, when dealing with finite dimensional spaces, the Frechet derivative corresponds with the Jacobian matrix. Another related notion of derivative explores directly the directional changes.

Definition 2. A function $f : V \to W$ is Gateaux differentiable if f admits directional derivatives along all directions at x, that is, if there exists $g : U \subset V \to W$ such that

$$g(h) = \lim_{\alpha \to 0} \frac{f(x + \alpha h) - f(x)}{h}.$$

If f is Frechet differentiable at x, it is also Gateaux differentiable there, and g is just the linear operator $A = Df(x)$. However, the converse does not hold. This can be shown by counterexamples.

Example 5. Let us consider $f: \mathbb{R} \to \mathbb{R}, x \to |x|$. This function is Gateaux differentiable at $x = 0$, since

$$\lim_{\alpha \to 0} \frac{|0 + \alpha h| - |0|}{\alpha} = |h|.$$

Now, the existence of $g(h) = |h|$ means that f is Gateaux differentiable, but since $g(h)$ is nonlinear, we conclude that $|x|$ is not Frechet differentiable. Basically, the function is not differentiable in the classic sense at the origin.

Example 6. Let us consider a function $f(x, y)$ which is null in the origin $(0, 0)$ and

$$f(x, y) = \frac{x^3}{x^2 + y^2}$$

otherwise. We can immediately see that f is not Frechet differentiable at $(0, 0)$. The linear operator is the gradient which, for $(x, y) \neq (0, 0)$, is

$$\frac{\partial f}{\partial x} = \frac{x^4 + 3x^2 y^2}{(x^2 + y^2)^2}, \quad \frac{\partial f}{\partial y} = -\frac{2yx^3}{(x^2 + y^2)^2}.$$

Now

$$\lim_{x \to (0,0)} \frac{x^4 + 3x^2 y^2}{(x^2 + y^2)^2}$$

does not exist. If we approach $(0,0)$ with $x \to 0$ we have $\partial f / \partial x \to 0$, whereas as $y \to 0$ we have $\partial f / \partial x \to 1$. Interestingly, f is Gateaux differentiable at $(0,0)$. Let $h = (a,b)$. We have

$$g(a,b) = \lim_{\alpha \to 0} \frac{1}{\alpha} \left(\frac{(x+\alpha a)^3}{(x+\alpha a)^2 + (y+\alpha b)^2} - \frac{x^3}{x^2 + y^2} \right).$$

Now, it is easy to see that $g(0,0) = 0$, while for $(a,b) \neq (0,0)$,

$$g(a,b) = \frac{a^3}{a^2 + b^2},$$

which is the directional derivative for any $(x,y) \neq (0,0)$ along the (x,y) direction, since

$$\frac{1}{(x^2+y^2)^2} \langle \left(x^4 + 3x^2 y^2, -2yx^3 \right), (x,y)' \rangle = \frac{x^3}{x^2+y^2}.$$

Notice that, again the function g is not linear, which is the reason why f is only Gateaux differentiable and not Frechet differentiable.

Example 7. Let f be a function which is null in the origin $(0,0)$ and

$$f(x,y) = \frac{2xy}{\sqrt{x^2+y^2}}$$

otherwise. Now we have that f is Gateaux differentiable at $(0,0)$; if $h = (a,b)$, we have

$$g(a,b) = 2 \frac{ab}{\sqrt{a^2+b^2}}.$$

However, f is not Frechet differentiable since g is not linear and the gradient does not exist at $(0,0)$.

To understand these extended notions of a derivative, we need to analyze the meaning of the $\| \cdot \|_V$ and $\| \cdot \|_W$. For our purposes $\| \cdot \|_W$ is typically the Euclidean norm of $\mathbb{R}^n$, while V is a functional space and the introduction of the metrics needs some clarification. Given any two functions f_1 and f_2, we can use the p-norm,

$$\|f_1 - f_2\|_p := \sqrt[p]{\int_V (f_1(x) - f_2(x))^p \, dx}.$$

As $p \to \infty$ this norm reduces to

$$\max_{x \in V} |f_1(x) - f_2(x)|.$$

In variational calculus, one might be interested in different measures of distance between functions which involve derivatives of different order, namely

$$d_Q(f_1, f_2) = \sum_{q=1}^{Q} \max_{x \in V} |f_1^q(x) - f_2^q(x)|.$$

Consequently, we talk about *weak* and *strong* derivatives depending on the kind of norm we use. Strong derivatives just involve $\max_{x \in V} |f_1(x) - f_2(x)|$, whereas weak derivatives involve $d_Q(f_1, f_2)$. Obviously, any function which is strongly differentiable is weakly differentiable, but the converse does not hold. In most common problems encountered in variational calculus, weak derivatives at least of the first order are assumed. Concerning the structure of the functionals, we can define the related concepts of strong and weak extrema depending again on the measure of closeness between the functions. It is quite common to deal with the notion of weak extrema of the first order, which means that an extreme at a given point (function) requires ensuring the property in the neighborhood only moving by variations that are in C^1.

C.3 Euler–Lagrange equations

We start with an important result that is referred to as the *fundamental lemma of variational calculus*.

Lemma 1. *Given $g \in C^1([x_1, x_2], \mathbb{R})$, let us assume that $\forall h \in C^1([x_1, x_2], \mathbb{R})$ the following condition holds:*

$$\int_{x_1}^{x_2} g(x)h(x)dx = 0. \tag{4}$$

Then $\forall x \in [x_1, x_2]\ g(x) = 0$.

Proof. The proof is given by contradiction. Let us assume that there exists $g \neq 0$ such that

$$\int_{x_1}^{x_2} g(x)h(x)dx = 0$$

holds $\forall h \in C^1([x_1, x_2], \mathbb{R})$. Let $\hat{x} \in [x_1, x_2]$ be such that $g(\hat{x}) \neq 0$ and, for $\sigma > 0$, let us choose $h(x) = 1(\sigma - |x - \hat{x}|)/\sigma = (\sigma - |x - \hat{x}|)/\sigma$. Then

$$\lim_{\sigma \to 0} \frac{1}{\sigma} \int_{x_1}^{x_2} g(x)\, 1(\sigma - |x - \hat{x}|)\, dx = 0,$$

which is impossible, unless $g \notin C^1([x_1, x_2], \mathbb{R})$, contradicting the hypothesis. $\qquad \square$

Remark 1. Notice that if we consider g, h as elements of the Hilbert space F then $\int_{x_1}^{x_2} g(x)h(x)dx$ is an inner product $\langle g, h \rangle$ and condition (4) can be rewritten $\forall h \in F$ $\langle g, h \rangle = 0$. The lemma states that $g = 0$, which is a classic property of inner products.

Theorem 1. *Let us consider*

$$\mathcal{I}[y] := \int_{x_1}^{x_2} F(x, y(x), y'(x))\, dx, \tag{5}$$

with $F \in C^1([x_1, x_2] \times \mathbb{R} \times \mathbb{R}, \mathbb{R})$. Then

$$F_y(x, y(x), y'(x)) - \frac{d}{dx} F_{y'}(x, y(x), y'(x)) = 0$$

is a necessary condition for its extrema.

Proof. Let $h \in C^1([x_1, x_2], \mathbb{R})$, $\epsilon > 0$ be such that $h(x_1) = h(x_2) = 0$ and consider the variation corresponding to $y + \epsilon h$. We have

$$\frac{1}{\epsilon}\delta\mathcal{I}[y] = \frac{1}{\epsilon}(\mathcal{I}[y + \epsilon h] - \mathcal{I}[y])$$

$$= \frac{1}{\epsilon}\int_{x_1}^{x_2} \Big(F(x, y(x) + \epsilon h(x), y'(x) + \epsilon h'(x)) - F(x, y(x), y'(x))\Big)\, dx$$

$$= \frac{1}{\epsilon}\int_{x_1}^{x_2} F_y(x, y(x), y'(x))h(x)dx + \frac{1}{\epsilon}\int_{x_1}^{x_2} F_{y'}(x, y(x), y'(x))h'(x)\, dx.$$

Now, if we integrate by parts, we obtain

$$\int_{x_1}^{x_2} F_{y'}(x, y(x), y'(x))h'(x)dx = \big|F_{y'}(x, y(x), y'(x))h(x)\big|_{x_1}^{x_2}$$

$$- \int_{x_1}^{x_2} \frac{d}{dx} F_{y'}(x, y(x), y'(x))h(x)dx$$

$$= -\int_{x_1}^{x_2} \frac{d}{dx} F_{y'}(x, y(x), y'(x))h(x)\, dx.$$

Hence, because of Lemma 1, the condition

$$\frac{1}{\epsilon}\delta\mathcal{I}[y] = \int_{x_1}^{x_2} \left(F_y(x, y(x), y'(x)) - \frac{d}{dx} F_{y'}(x, y(x), y'(x))\right)h(x)dx = 0$$

yields the claim.   □

x-independence. Whenever $F(x, y, y') = F(y, y')$, we have

$$F_y - \frac{d}{dx} F_{y'} = F_y - F_{yy'} \cdot y' - F_{y'y'} \cdot y' = 0,$$

which yields

$$y' F_y - F_{yy'} \cdot [y']^2 - F_{y'y'} \cdot y'y' = 0,$$

that is,

$$\frac{d}{dx}\left(F - y' F_{y'}\right) = 0.$$

Finally, this means that there exists $C \in \mathbb{R}$ such that

$$F - y' F_{y'} = C. \qquad (6)$$

Application to the previous examples. We start by considering Example (1). When applying the Euler–Lagrange equations, we get

$$\frac{y'}{\sqrt{1 + (y'(x))^2}} = 0,$$

which yields $y(x) = mx + q$, where m and q can be determined by imposing that the given points (x_1, y_1) and (x_2, y_2) belong to the prescribed solution, which is a line.

Now, let us consider the functional in (2) and, without loss of generality, let P_1 be placed in the center of the axes and y be placed in the reverse direction. Then we get

$$\mathcal{I}_1[y(x)] := \frac{1}{\sqrt{2g}} \int_{x_1}^{x_2} \frac{\sqrt{1 + y'(x)^2}}{\sqrt{y}} \, dx,$$

$$y(0) = 0,$$

$$y(x_2) = y_2.$$

From (6), we get

$$\frac{\sqrt{1 + [y']^2}}{\sqrt{y}} - \frac{[y']^2}{\sqrt{y(1 + [y']^2)}} = C,$$

that is, $y(1 + [y']^2) = C_1$, where $C_1 := C^{-2}$. To solve this differential equation, let $y' \cdot \tan t = 1$. Then

$$y = C_1 \sin^2 t = \frac{C_1}{2}(1 - \cos 2t)$$

and

$$\frac{dy}{y'} = 2C_1 \sin t \cos t \cdot \tan t \, dt = 2C_1 \sin^2 t \, dt = C_1(1 - \cos 2t) \, dt.$$

Hence, we get the parametric equations

$$x = \frac{C_1}{2}(2t - \sin 2t) + C_2, \qquad y = \frac{C_1}{2}(1 - \cos 2t).$$

When imposing the condition $y(0) = x(0) = 0$ and $\tau = 2t$, we end up with

$$x = \frac{C_1}{2}(\tau - \sin \tau), \qquad y = \frac{C_1}{2}(1 - \cos \tau).$$

Finally, from the conditions $y(\tau) = y_2$ and $x(\tau) = x_2$ we can determine C_1 and τ. Interestingly, the trajectory is a section of cycloid created by a circle of radius $C_1/2$.

C.4 Variational problems with subsidiary conditions

Now we consider the problem of determining the extreme of

$$\mathcal{J}[x, y_1, \ldots, y_p] = \int_{x_1}^{x_2} F(x, y_1, \ldots, y_p, y_1', \ldots, y_p')dx \qquad (7)$$

under the constraint $\varphi_i(x, y_1, \ldots, y_p) = 0$, $i = 1, \ldots, q$ and $q < p$. Following the case of finite dimensions, we construct the associated functional

$$\mathcal{J}^* := \int_{x_1}^{x_2} (F + \sum_{i=1}^{q} \lambda_j(x) \cdot \varphi_i)dx = \int_{x_1}^{x_2} F^* dx \qquad (8)$$

where $\lambda_i : [x_1, x_2] \to \mathbb{R}$. Instead of facing the original constrained problem, we determine the extremes of the associated unconstrained functional $\mathcal{J}^*$. We can directly write down the Euler–Lagrange equations for $\mathcal{J}^*$ as follows:

$$F_{y_j}^* - \frac{d}{dx} F_{y_j'}^* = 0, \quad j = 1, \ldots, p,$$
$$\varphi_i = 0, \quad i = 1, \ldots, q. \qquad (9)$$

For any x, we get $p + q$ equations in $p + q$ variables y_j and λ_i that can be uniquely determined. Notice that the need to introduce functions as multipliers can be understood in terms of existence of the solution of (9). The solutions that are determined in this way are also solutions of the original problem. When imposing $\partial F^*/\partial \lambda_j = 0$, we get $\varphi_i = 0$, $i = 1, \ldots, q$, and therefore, $\mathcal{J}^* = \mathcal{J}$. Hence the above Euler–Lagrange equations determine the extremes of $\mathcal{J}^*$ but also those of $\mathcal{J}$. However, the previous claim does not allow us to conclude that any solution of the original constrained problem is also a solution of the associated problem (9).

Theorem 2. *Let us assume that $\forall x \in (x_1, x_2)$ we can find a permutation of q functions[1] such that the Jacobian is not singular, that is,*

$$\frac{D(\varphi_1, \ldots, \varphi_q)}{D(y_1, \ldots, y_q)} \neq 0. \qquad (10)$$

---

[1] Without loss of generality, for simplicity they are numbered using the first $q < p$ functions.

Then there exist a set of functions $\lambda_i(x)$, $i = 1, \ldots, q$ *such that the extremes of functional* (7) *under the constraints* $\varphi_i(x, y_1, \ldots, y_p) = 0$, $i = 1, \ldots, q$ *and* $q < p$ *can be determined as extremes of* $\mathcal{J}^*$. *In particular, the functions* λ_i *and* y_j *can be determined by solving* (9).

Proof. Let $h_j \in C^1([x_1, x_2], \mathbb{R})$, $\epsilon > 0$ be such that $h_j(x_1) = h_j(x_2) = 0$ and consider the Gateaux derivatives associated with $y + \epsilon h$. We have already proved that any extrema of $\mathcal{J}$ must satisfy

$$\int_{x_1}^{x_2} \sum_{j=1}^{p} (F_{y_j} - \frac{d}{dx} F_{y'_j}) h_j \cdot dx = 0. \tag{11}$$

Now, unlike for unconstrained optimization problem, we cannot apply the fundamental lemma of variational calculus since, due to the presence of constraints, h_j are not independent. They must in fact be consistent with

$$\frac{\partial \varphi}{\partial h} = \lim_{\epsilon \to 0} \frac{1}{\epsilon} \big(\varphi(x, y_1 + \epsilon h_1, \ldots, y_p + \epsilon y_p) - \varphi(x, y_1, \ldots, y_p) \big) = 0,$$

that is,

$$\sum_{j=1}^{p} \frac{\partial \varphi_i}{\partial y_j} \cdot h_j = 0, \quad i = 1, \ldots, q.$$

As a consequence, only $p - q$ variations h_j are independent, while the others are constrained to satisfy the above equation. Now, let us consider a set of q functions λ_i. We get

$$\langle \lambda, \sum_{j=1}^{p} \frac{\partial \varphi}{\partial y_j} \cdot h_j \rangle = \int_{x_1}^{x_2} \lambda_i(x) \sum_{j=1}^{p} \frac{\partial \varphi_i}{\partial y_j} h_j \cdot dx = 0.$$

When summing up Eqs. (11), we get

$$\int_{x_1}^{x_2} \sum_{j=1}^{p} (F_{y_j} - \frac{d}{dx} F_{y'_j} + \lambda_i \sum_{j=1}^{p} \frac{\partial \varphi_i}{\partial y_j}) h_j \cdot dx = \int_{x_1}^{x_2} \sum_{j=1}^{p} (F^*_{y_j} - \frac{d}{dx} F^*_{y'_j}) h_j \cdot dx = 0,$$

where

$$F^* := F + \sum_{i=1}^{q} \lambda(x) \cdot \varphi_i.$$

Now, $\forall x \in (x_1, x_2)$ we choose $\lambda_i(x)$, $i = 1, \ldots, q$ such that

$$F^*_{y_j} - \frac{d}{dx} F^*_{y'_j} = 0, \quad j = 1, \ldots, q. \tag{12}$$

Interestingly, $\forall x \in (x_1, x_2)$, this is a linear system in λ_i, since it can be rewritten as

$$\sum_{i=1}^{q} \frac{\partial \varphi_i}{\partial y_j} \lambda_i(x) = \frac{d}{dx} F_{y_j'} - F_{y_j}. \tag{13}$$

From hypothesis (10), there exists a unique solution

$$\lambda_1(x), \lambda_2(x), \ldots, \lambda_q(x)$$

which reduces the search for extremes to

$$\int_{x_1}^{x_2} \sum_{j=q+1}^{p} (F_{y_j}^* - \frac{d}{dx} F_{y_j'}^*) h_j \cdot dx = 0.$$

Now, $h_{q+1}, \ldots, h_p$ are independent variations, and, therefore, we can invoke the fundamental theorem of variational calculus, which yields

$$F_{y_j}^* - \frac{d}{dx} F_{y_j'}^* = 0, \quad j = q, \ldots, p.$$

Finally, these equations, along with (12), yield the claim (9).   □

Index to notation

If not otherwise stated, letters that appear without any other kind of specification have the following meaning:

κ — nonnegative integer-valued arithmetic expression
j, k, m, n — integer-valued arithmetic expression
x, y — real-valued or boolean arithmetic expression
z — complex-valued arithmetic expression
f — integer-valued, real-valued, or complex-valued function (task)
ψ — integer-valued, real-valued, or complex-valued function (constraint)
X, Y — scts
$\mathbb{N}, \mathbb{R}$ — the set of natural and real numbers

Formal symbolism	Meaning	Where defined
a_n	the n-th element of a vector	
A_{nm}	the element of row n and column m of a matrix	
$[x, y]$	closed interval: $\{a \mid x \leq a \leq y\}$	
(x, y)	open interval: $\{a \mid x < a < y\}$	
$[x, y)$	half open interval: $\{a \mid x \leq a < y\}$	
$(x, y]$	half-closed interval: $\{a \mid x < a \leq y\}$	
$[R]$	characteristic function of relation R: denotes 1 if the relation is true, 0 if R is false	§1.1.4
δ_{jk}	Kronecker delta: $[j = k]$	§1.1.1
$\lvert X \rvert$	cardinality: The number of elements in X	
$\langle X_n \rangle$	the infinite sequence $X_0, X_1, X_2, \ldots$	
$\lVert x \rVert_p$	L^p norm: $(\sum_{i=1}^{n} \lvert x_i \rvert^p)^{1/p}$	
$\lVert x \rVert_\infty$	maximum norm: $\max_i \lvert x_i \rvert$	
$\langle \cdot, \cdot \rangle_H$	inner product in a Hilbert or pre-Hilbert space H	
$\lVert f \rVert_H$	norm in H: $\sqrt{\langle f, f \rangle_H}$	
$\lVert f \rVert_P$	norm induced by the operator P: $\lVert Pf \rVert$	
$k(x, y)$	kernel: A symmetric semi-positive definite function of x and y	§4.2.1
H_k^0	pre-Hilbert space induced by kernel k	§4.3.3
$\langle \cdot, \cdot \rangle_k^0$	Inner product in H_k^0	§4.3.3
$\lVert f \rVert_k^0$	norm in H_k^0: $\sqrt{\langle f, f \rangle_k^0}$	

Formal symbolism	Meaning	Where defined
H_k	Hilbert space induced by kernel k	§4.3.3
$\langle \cdot, \cdot \rangle_k$	Inner product in H_k	§4.3.3
$\|f\|_k$	norm in H_k: $\sqrt{\langle f, f \rangle_k}$	
$\mathcal{N}_f$	Kernel of function $f : X \to Y$: $\{x \in X : f(x) = 0\}$	
$\operatorname{rank} A$	rank of matrix A	
$\lfloor x \rfloor$	floor of x, greatest integer function	
$\lceil x \rceil$	ceiling of x, least integer function	
$\lfloor x \rceil$	round off of x, nearest integer function	§1.2.1
$x \bmod y$	mod function: $x[y = 0] + (x - y\lfloor x/y \rfloor)[y \neq 0]$	
$S \setminus T$	set difference: $\{s : s \text{ in } S \text{ but not in } T\}$	
$a \cdot b$	inner product of a with b: $a \cdot b = \sum_{i=1}^{n} a_i b_i$	
A'	transpose of matrix A: $(A')_{nm} = A_{mn}$	
$\log_b x$	logarithm, base b, of x (for $x > 0$, $b > 0$ and $b \neq 1$): y such that $b^y = x$	
$\ln x$	natural logarithm: $\log_e x$	
$\log x$	binary logarithm: $\log_2 x$	
$\exp x$	exponential of x: $e^x = \sum_{k=0}^{\infty} x^k / k!$	
$f'(x)$	derivative of f at x	
$f''(x)$	second derivative of f at x	
$f^{(n)}(x)$	n-th derivative of f at x	
$\frac{\partial f}{\partial x_i}(x)$	partial derivative along x_i of f in x	
$\partial_{x_i} f(x)$	$\frac{\partial f}{\partial x_i}(x)$	
$\partial_i f(x)$	$\partial_{x_i} f(x)$	
$\nabla f(x)$	gradient of f at x: $(\nabla f(x))_i = \partial_i f(x)$	
$\nabla^2 f(x)$	laplacian of f at x: $\nabla^2 f(x) = \sum_{i=1}^{n} \partial_i^2 f(x)$	
$\nabla \cdot f(x)$	divergence of f at x: $\nabla \cdot f(x) = \sum_{i=1}^{n} \partial_i f_i(x)$	
$\hat{f}(\xi)$	Fourier transform of f: $\frac{1}{(2\pi)^d} \int_{\mathbb{R}^d} f(x) \exp(-ix'\xi)$	
$f * g$	convolution of f and g: $(f * g)(x) = \int_X f(z)g(x - z)\,dz$	
$\bigotimes_{i=1}^{n} f_i$	n-fold convolution: $f_1 * f_2 * \cdots * f_n$	§4.3.4
$\dot{x}$	temporal derivative of $x(t)$: $x'(t)$	
$\ddot{x}$	second temporal derivative of $x(t)$: $x''(t)$	
F_n	Fibonacci number: $n[n \leq 1] + (F_{n-1} + F_{n-2})[n > 1]$	§1.2.1
$\Gamma(x)$	gamma function: $\int_0^{\infty} e^{-t} t^{x-1}\,dt$	§1.1.4
ϕ	golden ratio: $(1 + \sqrt{5})/2$	§1.2.1
$x^{\bar{k}}$	rising factorial power: $x(x + 1) \cdots (x + k - 1)$	
$x^{\underline{k}}$	falling factorial power: $x(x - 1) \cdots (x - k + 1)$	
$|\alpha|$	length of multi-index α: $\alpha_1 + \alpha_2 + \alpha_d$ if $\alpha = \alpha_1 \alpha_2 \cdots \alpha_d$	
$\alpha!$	factorial of $\alpha = \alpha_1 \alpha_2 \cdots \alpha_d$: $\alpha_1! \alpha_2! \cdots \alpha_d!$	
x^{α}	$x = (x_1, \ldots, x_d)$ to the power of $\alpha = \alpha_1 \alpha_2 \cdots \alpha_d$: $x_1^{\alpha_1} x_2^{\alpha_2} \cdots x_d^{\alpha_d}$	
$F\left(\begin{smallmatrix} a,b \\ c \end{smallmatrix} \middle| z\right)$	Gaussian hypergeometric: $\sum_{k \geq 0} (a^{\bar{k}} b^{\bar{k}} z^k)/(c^{\bar{k}} k!)$	
$O(f(n))$	big-oh of $f(n)$, as the variable $n \to \infty$	

Formal symbolism	Meaning	Where defined
$\Omega(f(n))$	big-omega of $f(n)$, as the variable $n \to \infty$	
$\Theta(f(n))$	big-theta of $f(n)$, as the variable $n \to \infty$	
$X^\sharp$	discrete sampling of the set X	§1.1.1
$H(x)$	Heaviside function: $[x \geq 0]$	§3.2
$\mathrm{smx}_i(x_1, \ldots, x_n)$	softmax function: $\exp x_i / \sum_{j=1}^n \exp x_j$	§1.3.3
$\delta(x)$	Dirac delta function: $\langle \delta, f \rangle = f(0)$	
$:=$	equal by definition	
$\neg x$ or $\bar{x}$	complement: $1 - x$	
$x \wedge y$	and: $x \cdot y$	
$x \vee y$	or	
$x \oplus y$	xor: $(x + y) \bmod 2$	
$x \Rightarrow y$	implication: $\neg x \vee y$	
$x \Leftrightarrow y$	equivalence: $(x \Rightarrow y) \wedge (y \Rightarrow x)$	
$x_1, \ldots, x_n \vdash c$	formal deduction: $(x_1 \wedge \cdots \wedge x_n) \Rightarrow c$	
$x_1, \ldots, x_n \models c$	deduction in the environment	§6.1.4
$x_1, \ldots, x_n \models^* c$	parsimonious deduction	§6.1.4
$u \odot v$	Hadamard product: $(u \odot v)_i = u_i v_i$	
$\mathcal{G} \sim (V, A)$	graph with vertices in V and arcs in A	
$u \text{---} v$	the arc that connects u and v	
$u \longrightarrow v$	oriented arc in A that connects u to v	
$\mathrm{pa}(v)$	parents of v: $\{u \in V : u \longrightarrow v \in A\}$	
$\mathrm{ch}(v)$	children of v: $\{u \in V : v \longrightarrow u \in A\}$	
$\blacksquare$	end of algorithm or agent	§3.4.1

Index

As a result, his books tend to be delayed,
but the indexes tend to be pretty good.
DONALD E. KNUTH, The TeXbook (1986)